Modern Systems Analysis and Design

SEVENTH EDITION
INTERNATIONAL EDITION

Jeffrey A. Hoffer
University of Dayton

Joey F. George
Iowa State University

Joseph S. Valacich
University of Arizona

PEARSON

Boston   Columbus   Indianapolis   New York   San Francisco   Upper Saddle River
Amsterdam   Cape Town   Dubai   London   Madrid   Milan   Munich   Paris   Montréal   Toronto
City   São Paulo   Sydney   Hong Kong   Seoul   Singapore   Taipei   Tokyo

Editor in Chief: Stephanie Wall
Executive Editor: Bob Horan
Senior Acquisitions Editor, International Edition:
 Steven Jackson
Editorial Project Manager: Kelly Loftus
Editorial Assistant: Ashlee Bradbury
Director of Marketing: Maggie Moylan
Executive Marketing Manager: Anne Fahlgren
Marketing Manager, International: Dean Erasmus
Marketing Assistant: Gianna Sandri
Senior Managing Editor: Judy Leale

Production Project Manager: Jane Bonnell
Senior Manufacturing Controller, Production,
 International: Trudy Kimber
Creative Director: Blair Brown
Art Director: Steve Frim
Interior Designer: Jill Lehan
Cover Designer: Jodi Notowitz
Cover Art: © cozyta
Media Project Manager, Editorial: Alana Coles
Media Project Manager, Production: Lisa Rinaldi

Pearson Education Limited
Edinburgh Gate
Harlow
Essex CM20 2JE
England

and Associated Companies throughout the world

Visit us on the World Wide Web at: www.pearson.com/uk

© Pearson Education Limited 2014

ISBN 10: 0-273-78709-8
ISBN 13: 978-0-273-78709-9

British Library Cataloguing-in-Publication Data
A catalogue record for this book is available from the British Library

10 9 8 7 6 5 4 3 2
15 14

Typeset in New BaskervilleStd by PreMediaGlobal
Printed and bound by Courier/Kendallville in United States of America

The publisher's policy is to use paper manufactured from sustainable forests.

To Patty, for her sacrifices, encouragement, and support.
To my students, for being receptive and critical, and challenging me to be
a better teacher.

—Jeff

In memory of Tom Clark, mentor and friend.

—Joey

To Jackie, Jordan, James, and the rest of my family. Your love and support
are my greatest inspiration.

—Joe

BRIEF CONTENTS

PREFACE   19

PART ONE   FOUNDATIONS FOR SYSTEMS DEVELOPMENT   27

1   The Systems Development Environment   29
2   The Origins of Software   55
3   Managing the Information Systems Project   73

Appendix: Object–Oriented Analysis and Design:
                  Project Management   107

PART TWO   PLANNING   115

4   Identifying and Selecting Systems Development Projects   117
5   Initiating and Planning Systems Development Projects   141

PART THREE   ANALYSIS   175

6   Determining System Requirements   177
7   Structuring System Process Requirements   212

Appendix 7A: Object–Oriented Analysis and Design: Use Cases   247
Appendix 7B: Object–Oriented Analysis and Design: Activity Diagrams   262
Appendix 7C: Object–Oriented Analysis and Design: Sequence Diagrams   266
Appendix 7D: Business Process Modeling   275

8   Structuring System Data Requirements   282

Appendix: Object–Oriented Analysis and Design: Object Modeling—
                  Class Diagrams   317

PART FOUR   DESIGN   337

9   Designing Databases   339
10   Designing Forms and Reports   380
11   Designing Interfaces and Dialogues   407
12   Designing Distributed and Internet Systems   443

PART FIVE   IMPLEMENTATION AND MAINTENANCE   479

13   System Implementation   481
14   Maintaining Information Systems   516

GLOSSARY OF TERMS   535
GLOSSARY OF ACRONYMS   544
INDEX   546

CONTENTS

PREFACE   19

PART ONE   FOUNDATIONS FOR SYSTEMS DEVELOPMENT

AN OVERVIEW OF PART ONE   28

1   **The Systems Development Environment**   29

A Modern Approach to Systems Analysis and Design   32

Developing Information Systems and the Systems Development Life Cycle   33

A Specialized Systems Development Life Cycle   39

The Heart of the Systems Development Process   40

    The Traditional Waterfall SDLC   41

Different Approaches to Improving Development   42

    CASE Tools   42

    Rapid Application Development   44

Agile Methodologies   45

    eXtreme Programming   47

Object-Oriented Analysis and Design   48

Our Approach to Systems Development   50

Summary   51

Key Terms   52

Review Questions   53

Problems and Exercises   53

Field Exercises   53

References   54

2   **The Origins of Software**   55

Systems Acquisition   56

    Outsourcing   56

    Sources of Software   58

    Choosing Off-the-Shelf Software   64

    Validating Purchased Software Information   66

Reuse   67

Summary   69

Key Terms   70

Review Questions   70

Problems and Exercises   70

Field Exercises   70

References   71

PE CASE: THE ORIGINS OF SOFTWARE   72
Case Questions   72

3   **Managing the Information Systems Project**   73

Pine Valley Furniture Company Background   74

Managing the Information Systems Project   75
   Initiating a Project   79
   Planning the Project   81
   Executing the Project   89
   Closing Down the Project   91

Representing and Scheduling Project Plans   92
   Representing Project Plans   94
   Calculating Expected Time Durations Using PERT   95
   Constructing a Gantt Chart and Network Diagram at Pine Valley Furniture   95

Using Project Management Software   99
   Establishing a Project Start Date   100
   Entering Tasks and Assigning Task Relationships   100
   Selecting a Scheduling Method to Review Project Reports   101

Summary   102

Key Terms   102

Review Questions   103

Problems and Exercises   104

Field Exercises   105

References   106

Appendix: Object-Oriented Analysis and Design: Project Management   107

Define the System as a Set of Components   107
   Complete Hard Problems First   109
   Using Iterations to Manage the Project   110
   Don't Plan Too Much Up Front   110
   How Many and How Long Are Iterations?   111
   Project Activity Focus Changes over the Life of a Project   111

Summary   113

Review Question   113

Problems and Exercises   113

PE CASE: MANAGING THE INFORMATION SYSTEMS PROJECT   114
Case Questions   114

PART TWO   PLANNING

AN OVERVIEW OF PART TWO   116

4   **Identifying and Selecting Systems Development Projects**   117

Identifying and Selecting Systems Development Projects   118

The Process of Identifying and Selecting IS Development Projects   119

Deliverables and Outcomes   123

Corporate and Information Systems Planning   124

Corporate Strategic Planning   125

Information Systems Planning   127

Electronic Commerce Applications: Identifying and Selecting Systems Development Projects   134

Internet Basics   134

Pine Valley Furniture WebStore   135

Summary   136

Key Terms   136

Review Questions   137

Problems and Exercises   137

Field Exercises   138

References   138

PE CASE: IDENTIFYING AND SELECTING SYSTEMS DEVELOPMENT PROJECTS   140

Case Questions   140

5   **Initiating and Planning Systems Development Projects**   141

Initiating and Planning Systems Development Projects   142

The Process of Initiating and Planning Is Development Projects   143

Deliverables and Outcomes   144

Assessing Project Feasibility   145

Assessing Economic Feasibility   145

Assessing Technical Feasibility   154

Assessing Other Feasibility Concerns   156

Building and Reviewing the Baseline Project Plan   157

Building the Baseline Project Plan   158

Reviewing the Baseline Project Plan   162

Electronic Commerce Applications: Initiating and Planning Systems Development Projects   167

Initiating and Planning Systems Development Projects for Pine Valley Furniture's WebStore   168

Summary   169

Key Terms   170

Review Questions   170

Problems and Exercises   171

Field Exercises   171

References   172

PE CASE: INITIATING AND PLANNING SYSTEMS DEVELOPMENT PROJECTS 173
Case Questions 173

PART THREE ANALYSIS

AN OVERVIEW OF PART THREE 176

6 Determining System Requirements 177

Performing Requirements Determination 178
  The Process of Determining Requirements 178
  Deliverables and Outcomes 179

Traditional Methods for Determining Requirements 180
  Interviewing and Listening 180
  Interviewing Groups 184
  Directly Observing Users 185
  Analyzing Procedures and Other Documents 187

Contemporary Methods for Determining System Requirements 192
  Joint Application Design 192
  Using Prototyping during Requirements Determination 195

Radical Methods for Determining System Requirements 197
  Identifying Processes to Reengineer 198
  Disruptive Technologies 198

Requirements Determination Using Agile Methodologies 199
  Continual User Involvement 199
  Agile Usage-Centered Design 200
  The Planning Game from eXtreme Programming 201

Electronic Commerce Applications: Determining System Requirements 203
  Determining System Requirements for Pine Valley Furniture's WebStore 203

Summary 206

Key Terms 206

Review Questions 207

Problems and Exercises 207

Field Exercises 208

References 209

PE CASE: DETERMINING SYSTEM REQUIREMENTS 210
Case Questions 211

7 Structuring System Process Requirements 212

Process Modeling 213
  Modeling a System's Process for Structured Analysis 213
  Deliverables and Outcomes 214

Data Flow Diagramming Mechanics 214
  Definitions and Symbols 215
  Developing DFDs: An Example 217
  Data Flow Diagramming Rules 219
  Decomposition of DFDs 220
  Balancing DFDs 223

An Example DFD   225

Using Data Flow Diagramming in the Analysis Process   228
  Guidelines for Drawing DFDs   228
  Using DFDs as Analysis Tools   230
  Using DFDs in Business Process Reengineering   231

Modeling Logic With Decision Tables   233

Electronic Commerce Application:
Process Modeling Using Data Flow Diagrams   236
  Process Modeling for Pine Valley Furniture's WebStore   237

Summary   238

Key Terms   239

Review Questions   240

Problems and Exercises   240

Field Exercises   246

References   246

Appendix 7A: Object-Oriented Analysis and Design: Use Cases   247

  Use Cases   248
    What Is a Use Case?   248
    Use Case Diagrams   249
    Definitions and Symbols   249

  Written Use Cases   252
    Level   253
    The Rest of the Template   254

  Electronic Commerce Application: Process Modeling
  Using Use Cases   256

  Writing Use Cases for Pine Valley Furniture's WebStore   257

  Summary   259

  Key Terms   260

  Review Questions   260

  Problems and Exercises   260

  Field Exercise   261

  References   261

Appendix 7B: Object-Oriented Analysis and Design: Activity Diagrams   262
  When to Use an Activity Diagram   264
  Problems and Exercises   264
  Reference   265

Appendix 7C: Object-Oriented Analysis and Design: Sequence Diagrams   266
  Dynamic Modeling: Sequence Diagrams   267
  Designing a Use Case With a Sequence Diagram   268

  A Sequence Diagram for Hoosier Burger   271
  Summary   273
  Key Terms   273

Review Questions   273

Problems and Exercises   274

Field Exercise   274

References   274

Appendix 7D: Business Process Modeling   275

Basic Notation   276

Example   278

Summary   279

Key Terms   279

Review Questions   279

Problems and Exercises   279

Field Exercises   279

References   279

PE CASE: STRUCTURING SYSTEM PROCESS REQUIREMENTS   280

Case Questions   281

8   Structuring System Data Requirements   282

Conceptual Data Modeling   284

The Conceptual Data Modeling Process   284

Deliverables and Outcomes   285

Gathering Information for Conceptual Data Modeling   286

Introduction to E-R Modeling   288

Entities   289

Attributes   291

Candidate Keys and Identifiers   292

Other Attribute Types   292

Relationships   294

Conceptual Data Modeling and the E-R Model   295

Degree of a Relationship   295

Cardinalities in Relationships   297

Naming and Defining Relationships   299

Associative Entities   299

Summary of Conceptual Data Modeling with E-R Diagrams   302

Representing Supertypes and Subtypes   302

Business Rules   303

Domains   304

Triggering Operations   305

Role of Packaged Conceptual Data Models—Database Patterns   306

Universal Data Models   306

Industry-Specific Data Models   307

Benefits of Database Patterns and Packaged Data Models   307

Electronic Commerce Application: Conceptual Data Modeling   308

Conceptual Data Modeling for Pine Valley Furniture's WebStore   308

Summary   312

Key Terms   312

Review Questions   313

Problems and Exercises   313

Field Exercises   316

References   316

Appendix: Object-Oriented Analysis and Design: Object Modeling—Class Diagrams   317

Representing Objects and Classes   317

Types of Operations   319

Representing Associations   319

Representing Associative Classes   321

Representing Stereotypes for Attributes   323

Representing Generalization   323

Representing Aggregation   326

 An Example of Conceptual Data Modeling at Hoosier Burger   326

Summary   330

Key Terms   330

Review Questions   331

Problems and Exercises   331

References   332

PE CASE: STRUCTURING SYSTEM DATA REQUIREMENTS   333
Case Questions   334

PART FOUR   DESIGN

AN OVERVIEW OF PART FOUR   338

9   Designing Databases   339

Database Design   340

The Process of Database Design   341

Deliverables and Outcomes   342

Relational Database Model   345

Well-Structured Relations   346

Normalization   347

Rules of Normalization   347

Functional Dependence and Primary Keys   347

Second Normal Form   348

Third Normal Form   348

Transforming E-R Diagrams into Relations   350

Represent Entities   350

Represent Relationships   351

Summary of Transforming E-R Diagrams to Relations   354

Merging Relations   354

An Example of Merging Relations   355

View Integration Problems   355

Logical Database Design for Hoosier Burger   356

Physical File and Database Design   359

Designing Fields   359
   Choosing Data Types   359
   Controlling Data Integrity   361

Designing Physical Tables   362
   Arranging Table Rows   364
   Designing Controls for Files   368

Physical Database Design for Hoosier Burger   370

Electronic Commerce Application: Designing Databases   371
   Designing Databases for Pine Valley Furniture's WebStore   371

Summary   373

Key Terms   374

Review Questions   375

Problems and Exercises   375

Field Exercises   377

References   377

PE CASE: DESIGNING DATABASES   378
Case Questions   379

10   **Designing Forms and Reports**   380

Designing Forms and Reports   381
   The Process of Designing Forms and Reports   382
   Deliverables and Outcomes   383

Formatting Forms and Reports   386
   General Formatting Guidelines   387
   Highlighting Information   388
   Color versus No Color   390
   Displaying Text   391
   Designing Tables and Lists   392
   Paper versus Electronic Reports   396

Assessing Usability   397
   Usability Success Factors   398
   Measures of Usability   398

Electronic Commerce Applications: Designing Forms and Reports for
Pine Valley Furniture's WebStore   399
   General Guidelines   399
   Designing Forms and Reports at Pine Valley Furniture   400
   Lightweight Graphics   401
   Forms and Data Integrity Rules   401
   Template-Based HTML   401

Summary   402

Key Terms   402

Review Questions   402

Problems and Exercises   403

Field Exercises   403

References   404

PE CASE: DESIGNING FORMS AND REPORTS   405
Case Questions   405

11   Designing Interfaces and Dialogues   407

Designing Interfaces and Dialogues   408
    The Process of Designing Interfaces and Dialogues   408
    Deliverables and Outcomes   408

Interaction Methods and Devices   409
    Methods of Interacting   409
    Hardware Options for System Interaction   417

Designing Interfaces   418
    Designing Layouts   419
    Structuring Data Entry   422
    Controlling Data Input   423
    Providing Feedback   425
    Providing Help   426

Designing Dialogues   429
    Designing the Dialogue Sequence   429
    Building Prototypes and Assessing Usability   433

Designing Interfaces and Dialogues in Graphical Environments   433
    Graphical Interface Design Issues   433
    Dialogue Design Issues in a Graphical Environment   435

Electronic Commerce Application: Designing Interfaces and Dialogues for
Pine Valley Furniture's WebStore   435
    General Guidelines   436
    Designing Interfaces and Dialogues at Pine Valley Furniture   436
    Menu-Driven Navigation with Cookie Crumbs   436

Summary   438

Key Terms   438

Review Questions   439

Problems and Exercises   439

Field Exercises   440

References   440

PE CASE: DESIGNING INTERFACES AND DIALOGUES   441
Case Questions   442

12   Designing Distributed and Internet Systems   443

Designing Distributed and Internet Systems   444
    The Process of Designing Distributed and Internet Systems   444
    Deliverables and Outcomes   445

Designing Distributed Systems   446
    Designing Systems for LANs   446
    Designing Systems for a Client/Server Architecture   448
    Alternative Designs for Distributed Systems   449

Designing Internet Systems   453
   Internet Design Fundamentals   454
   Site Consistency   456
   Design Issues Related to Site Management   458
   Managing Online Data   461

Electronic Commerce Application: Designing a Distributed Advertisement Server
for Pine Valley Furniture's WebStore   469
   Advertising on Pine Valley Furniture's WebStore   469
   Designing the Advertising Component   470
   Designing the Management Reporting Component   471

Summary   472

Key Terms   472

Review Questions   473

Problems and Exercises   474

Field Exercises   475

References   475

PE CASE: DESIGNING DISTRIBUTED AND INTERNET SYSTEMS   476
Case Questions   476

PART FIVE   IMPLEMENTATION AND MAINTENANCE

AN OVERVIEW OF PART FIVE   480

13   System Implementation   481

System Implementation   482
   Coding, Testing, and Installation Processes   483
   Deliverables and Outcomes from Coding, Testing, and Installation   483
   Deliverables and Outcomes from Documenting the System,
      Training Users, and Supporting Users   485

Software Application Testing   485
   Seven Different Types of Tests   487
   The Testing Process   489
   Combining Coding and Testing   491
   Acceptance Testing by Users   492

Installation   493
   Direct Installation   493
   Parallel Installation   493
   Single-Location Installation   494
   Phased Installation   495
   Planning Installation   495

Documenting the System   496
   User Documentation   497

Training and Supporting Users   499
   Training Information Systems Users   499
   Supporting Information Systems Users   501
   Support Issues for the Analyst to Consider   503

Organizational Issues in Systems Implementation   503
  Why Implementation Sometimes Fails   503
  Security Issues   506

Electronic Commerce Application: System Implementation and
Operation for Pine Valley Furniture's WebStore   508
  Developing Test Cases for the WebStore   508
  Alpha and Beta Testing the WebStore   509
  WebStore Installation   510

Project Closedown   510

Summary   511

Key Terms   511

Review Questions   512

Problems and Exercises   513

Field Exercises   513

References   514

PE CASE: SYSTEM IMPLEMENTATION   515
Case Questions   515

14   **Maintaining Information Systems   516**

Maintaining Information Systems   517
  The Process of Maintaining Information Systems   517
  Deliverables and Outcomes   519

Conducting Systems Maintenance   519
  Types of Maintenance   520
  The Cost of Maintenance   521
  Managing Maintenance   523
  Role of CASE and Automated Development Tools in Maintenance   527

WebSite Maintenance   528

Electronic Commerce Application: Maintaining an Information
System for Pine Valley Furniture's WebStore   529
  Maintaining Pine Valley Furniture's WebStore   529

Summary   531

Key Terms   531

Review Questions   532

Problems and Exercises   532

Field Exercises   533

References   533

GLOSSARY OF TERMS   535
GLOSSARY OF ACRONYMS   544
INDEX   546

PREFACE

DESCRIPTION

Modern Systems Analysis and Design, Seventh Edition, covers the concepts, skills, methodologies, techniques, tools, and perspectives essential for systems analysts to successfully develop information systems. The primary target audience is upper-division undergraduates in a management information systems (MIS) or computer information systems curriculum; a secondary target audience is MIS majors in MBA and MS programs. Although not explicitly written for the junior college and professional development markets, this book can also be used by these programs.

We have over 75 years of combined teaching experience in systems analysis and design and have used that experience to create this newest edition of *Modern Systems Analysis and Design*. We provide a clear presentation of the concepts, skills, and techniques that students need to become effective systems analysts who work with others to create information systems for businesses. We use the systems development life cycle (SDLC) model as an organizing tool throughout the book to provide students with a strong conceptual and systematic framework. The SDLC in this edition has five phases and a circular design.

With this text, we assume that students have taken an introductory course on computer systems and have experience designing programs in at least one programming language. We review basic system principles for those students who have not been exposed to the material on which systems development methods are based. We also assume that students have a solid background in computing literacy and a general understanding of the core elements of a business, including basic terms associated with the production, marketing, finance, and accounting functions.

NEW TO THE SEVENTH EDITION

The following features are new to the Seventh Edition:

- *New material.* The most dramatic change in this edition is the introduction of a new end-of-chapter case. Gone is the Broadway Entertainment Company, which has served us well for many years. It has been replaced with Petrie Electronics (following all chapters except 1 and 14). Although the former case lent itself to a focus on in-house development, the new case focuses on finding an existing system and adapting it to the company's needs. A second completely new addition to this edition is an appendix on business process modeling (Appendix 7D). The appendix is based on the Business Process Modeling Notation (BPMN), which is maintained by the Object Management Group (although BPMN is not part of Unified Modeling Language [UML]). The current edition also includes a new section on Microsoft's Security Development Life Cycle (Chapter 1) and a new section on choosing the right enterprise resource planning (ERP) system (Chapter 2). Throughout the book, figures, tables, and related content have been updated and refreshed.

- *Updated content.* Throughout the book, the content in each chapter has been updated where appropriate. We have expanded our coverage of cloud computing (Chapter 2). Examples of updates in other chapters include revising the information on the information services (IS)/information technology job market in Chapter 1. Another example is Chapter 13, where we have updated the examples of system implementation failure and the leading security risks companies have reported. All screenshots come from current versions of leading software products. We have also made a special effort to update our reference lists, purging out-of-date material and including current references.

- *Dropped material*. In our efforts to keep the book current and to streamline it, the coverage of some things was dropped from this edition. Chapter 1 no longer includes the section on service oriented architecture (SOA). Chapter 4 no longer mentions intranet and extranet, including instead a discussion of more current electronic-commerce terms such as business-to-business and business-to-consumer.

- *Organization*. We have retained the organization of the book first introduced in the Sixth Edition. We have 14 chapters and six appendices. The first appendix follows Chapter 1. Four appendices follow Chapter 7, including the new one on business process modeling. The sixth appendix follows Chapter 8. This streamlined organization worked well in the Sixth Edition, so we decided to continue with it.

- *Approach to presentation of object-oriented material*. We retain our approach to object-orientation (OO) from the last edition. Brief appendices related to the object-oriented approach continue to appear immediately after related chapters. The OO appendices appear as follows: Chapter 3 features a special OO section on IS project management. Chapter 7 now has three OO appendices: one on use cases; one on sequence diagrams; and one about activity diagrams. Chapter 8 has a special section on object-oriented database design. The rationale for this organization is the same as in the past: to cleanly separate out structured and object-oriented approaches, so that instructors not teaching OO can bypass it. On the other hand, instructors who want to expose their students to object-orientation can now do so with minimal effort devoted to finding the relevant OO material.

- *Updated illustrations of technology*. Screen captures have been updated throughout the text to show examples using the latest versions of programming and Internet development environments, including the latest versions of .NET, Visio, and Microsoft Office; and user interface designs. Many references to websites are provided for students to stay current with technology trends that affect the analysis and design of information systems.

Themes of *Modern Systems Analysis and Design*

1. Systems development is firmly rooted in an organizational context. The successful systems analyst requires a broad understanding of organizations, organizational culture, and organizational operations.

2. Systems development is a practical field. Coverage of current practices as well as accepted concepts and principles is essential in a textbook.

3. Systems development is a profession. Standards of practice, a sense of continuing personal development, ethics, and a respect for and collaboration with the work of others are general themes in the textbook.

4. Systems development has significantly changed with the explosive growth in databases, data-driven systems architectures, rapid development, the Internet, and agile methodologies. Systems development and database management can be and should be taught in a highly coordinated fashion. The text is compatible with the Hoffer, Ramesh, and Topi database text, *Modern Database Management,* Eleventh Edition, also published by Pearson. Linking of these two textbooks is a strategic opportunity to meet the needs of the IS academic field.

5. Success in systems analysis and design requires not only skills in methodologies and techniques, but also project management skills for managing time, resources, and risks. Thus, learning systems analysis and design requires a thorough understanding of the process as well as the techniques and deliverables of the profession.

Given these themes, this textbook emphasizes the following:

- A business, rather than a technology, perspective
- The role, responsibilities, and mind-set of the systems analyst as well as the systems project manager, rather than those of the programmer or business manager
- The methods and principles of systems development, rather than the specific tools or tool-related skills of the field

DISTINCTIVE FEATURES

The following are some of the distinctive features of *Modern Systems Analysis and Design*:

1. This book is organized in parallel to the Hoffer, Ramesh, and Topi database text, *Modern Database Management*, Eleventh Edition, which will facilitate consistency of frameworks, definitions, methods, examples, and notations to better support systems analysis and design and database courses adopting both texts. Even with the strategic compatibilities between this text and *Modern Database Management*, each of these books is designed to stand alone as a market leader.

2. The grounding of systems development in the typical architecture for systems in modern organizations, including database management and Web-based systems.

3. A clear linkage of all dimensions of systems description and modeling—process, decision, and data modeling—into a comprehensive and compatible set of systems analysis and design approaches. Such a broad coverage is necessary so that students understand the advanced capabilities of the many systems development methodologies and tools that are automatically generating a large percentage of code from design specifications.

4. Extensive coverage of oral and written communication skills, including systems documentation, project management, team management, and a variety of systems development and acquisition strategies (e.g., life cycle, prototyping, Rapid Application Development [RAD], object orientation, Joint Application Development [JAD], systems reengineering, and agile methodologies).

5. Consideration of standards for the methodologies of systems analysis and the platforms on which systems are designed.

6. Discussion of systems development and implementation within the context of change management, conversion strategies, and organizational factors in systems acceptance.

7. Careful attention to human factors in systems design that emphasize usability in both character-based and graphical user interface situations.

8. Visual development products are illustrated and the current limitations technologies are highlighted.

9. The text includes a separate chapter on systems maintenance. Given the type of job many graduates first accept and the large installed base of systems, this chapter covers an important and often neglected topic in systems analysis and design texts.

PEDAGOGICAL FEATURES

The pedagogical features of *Modern Systems Analysis and Design* reinforce and apply the key content of the book.

Three Illustrative Fictional Cases

The text features three fictional cases, described below.

Pine Valley Furniture (PVF): In addition to demonstrating an electronic business-to-consumer shopping website, several other systems development activities from PVF are used to illustrate key points. PVF is introduced in Chapter 3 and revisited throughout the book. As key systems development life cycle concepts are presented, they are applied and illustrated with this descriptive case. For example, in Chapter 5, we explore how PVF plans a development project for a customer tracking system. A margin icon identifies the location of the case segments.

Hoosier Burger (HB): This second illustrative case is introduced in Chapter 7 and revisited throughout the book. HB is a fictional fast-food restaurant in Bloomington, Indiana. We use this case to illustrate how analysts would develop and implement an automated food-ordering system. A margin icon identifies the location of the case segments.

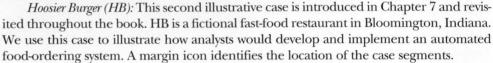

Petrie Electronics: This fictional retail electronics company is used as an extended project case at the end of 12 of the 14 chapters, beginning with Chapter 2. Designed to bring the chapter concepts to life, this case illustrates how a company initiates, plans, models, designs, and implements a customer loyalty system. Discussion questions are included to promote critical thinking and class participation. Suggested solutions to the discussion questions are provided in the Instructor's Manual.

End-of-Chapter Material

We developed an extensive selection of end-of-chapter materials that are designed to accommodate various learning and teaching styles.

- *Chapter Summary.* Reviews the major topics of the chapter and previews the connection of the current chapter with future ones.
- *Key Terms.* Designed as a self-test feature, students match each key term in the chapter with a definition.
- *Review Questions.* Test students' understanding of key concepts.
- *Problems and Exercises.* Test students' analytical skills and require them to apply key concepts.
- *Field Exercises.* Give students the opportunity to explore the practice of systems analysis and design in organizations.
- *Margin Term Definitions.* Each key term and its definition appear in the margin. Glossaries of terms and acronyms appear at the back of the book.
- *References.* References are located at the end of each chapter. The total number of references in this text amounts to over 100 books, journals, and websites that can provide students and faculty with additional coverage of topics.

USING THIS TEXT

As stated earlier, this book is intended for mainstream systems analysis and design courses. It may be used in a one-semester course on systems analysis and design or over two quarters (first in a systems analysis and then in a systems design course). Because this book text parallels *Modern Database Management*, chapters from this book and from *Modern Database Management* can be used in various sequences suitable for your curriculum. The book will be adopted typically in business schools or departments, not in computer science programs. Applied computer science or computer technology programs may also adopt the book.

The typical faculty member who will find this book most interesting is someone who

- Has a practical, rather than technical or theoretical, orientation
- Has an understanding of databases and the systems that use databases
- Uses practical projects and exercises in their courses

More specifically, academic programs that are trying to better relate their systems analysis and design and database courses as part of a comprehensive understanding of systems development will be especially attracted to this book.

The outline of the book generally follows the systems development life cycle, which allows for a logical progression of topics; however, it emphasizes that various approaches (e.g., prototyping and iterative development) are also used, so what appears to be a logical progression often is a more cyclic process. Part One provides an overview of systems development and previews the remainder of the book. Part One also introduces students to the many sources of software that they can draw on to build their systems and to manage projects. The remaining four parts provide thorough coverage of the five phases of a generic systems development life cycle, interspersing coverage of alternatives to the SDLC as appropriate. Some chapters may be skipped depending on the orientation of the instructor or the students' background. For example, Chapter 3 (Managing the Information Systems Project) can be skipped or quickly reviewed if students have completed a course on project management. Chapter 4 (Identifying and Selecting Systems Development Projects) can be skipped if the instructor wants to emphasize systems development once projects are identified or if there are fewer than 15 weeks available for the course. Chapters 8 (Structuring System Data Requirements) and 9 (Designing Databases) can be skipped or quickly scanned (as a refresher) if students have already had a thorough coverage of these topics in a previous database or data structures course. The sections on object orientation in Chapters 3, 7, and 8 can be skipped if faculty wish to avoid object-oriented topics. Finally, Chapter 14 (Maintaining Information Systems) can be skipped if these topics are beyond the scope of your course.

Because the material is presented within the flow of a systems development project, it is not recommended that you attempt to use the chapters out of sequence, with a few exceptions: Chapter 9 (Designing Databases) can be taught after Chapters 10 (Designing Forms and Reports) and 11 (Designing Inferfaces and Dialogues), but Chapters 10 and 11 should be taught in sequence.

THE SUPPLEMENT PACKAGE: WWW.PEARSONINTERNATIONALEDITIONS.COM/HOFFER

A comprehensive and flexible technology support package is available to enhance the teaching and learning experience. All instructor supplements are available on the text website: www.pearsoninternationaleditions.com/hoffer.

For Instructors

- An *Instructor's Resource Manual* provides chapter-by-chapter instructor objectives; teaching suggestions; and answers to all text review questions, problems, and exercises.
- The *Test Item File* and *TestGen* include a comprehensive set of test questions in multiple-choice, true/false, and short-answer format; questions are ranked according to level of difficulty and referenced with page numbers from the text. The Test Item File is available in Microsoft Word and as the computerized TestGen software. Pearson Education's test-generating software is available from www.pearsoninternationaleditions.com/hoffer. The software is PC/MAC compatible and preloaded with all of the Test Item File questions. You can

manually or randomly view test questions and drag and drop to create a test. You can add or modify test-bank questions as needed.

- *PowerPoint Presentation Slides* feature lecture notes that highlight key text terms and concepts. Professors can customize the presentation by adding their own slides or by editing the existing ones.

- The *Image Library* is a collection of the text art organized by chapter. This collection includes all of the figures, tables, and screenshots (as permission allows) from the book. These images can be used to enhance class lectures and PowerPoint slides.

ACKNOWLEDGMENTS

The authors have been blessed by considerable assistance from many people on all aspects of preparation of this text and its supplements. We are, of course, responsible for what eventually appears between the covers, but the insights, corrections, contributions, and proddings of others have greatly improved our manuscript. The people we recognize here all have a strong commitment to students, to the IS field, and to excellence. Their contributions have stimulated us, frequently prompting us to include new topics and innovative pedagogy.

We would like to recognize the efforts of the many faculty and practicing systems analysts who have been reviewers of this and its associated text, *Essentials of Systems Analysis and Design*. We have tried to deal with each reviewer comment, and although we did not always agree with specific points (within the approach we wanted to take with this book), all reviewers made us stop and think carefully about what and how we were writing. The reviewers were:

Eric Ackerman, *Nova Southeastern University*

Barbara Allen, *Douglas College*

Bay Arinze, *Drexel University*

Janine Aronson, *University of Georgia*

Susan Athey, *Colorado State University*

Sultan Bhimjee, *San Francisco State*

Bill Boroski, *Trident Technical College*

Nora Braun, *Augsburg College*

Penny Brunner, *University of North Carolina, Asheville*

Richard Burkhard, *San Jose State University*

Pedro Cabrejos, *Champlain College*

Donald Chand, *Bentley College*

Phyllis Chasser, *Nova Southeastern University*

Joselina Cheng, *University of Central Oklahoma*

Suzanne R. Clayton, *Drake University*

Amir Dabirian, *California State University, Fullerton*

Jason Deane, *Virginia Tech*

Thomas W. Dillon, *James Madison University*

Mark Dishaw, *University of Wisconsin at Oshkosh*

Jerry Dubyk, *Northern Alabama Institute of Technology*

Bob Foley, *DeVry Institute of Technology*

Barry Frew, *Naval Post-Graduate School*

Jim Gifford, *University of Wisconsin*

Richard Glass, *Bryant University*

Bonnie C. Glassberg, *University of Buffalo*

Mike Godfrey, *California State University, Long Beach*

Dave Groff, *Washington University*

Dale Gust, *Central Michigan University*

John Haney, *Walla Walla College*

Alexander Hars, *University of Southern California*

Kathleen Hartzel, *Duquesne University*

Ellen Hoadley, *Loyola College-Baltimore*

Monica Holmes, *Central Michigan University*

Yujong Hwang, *DePaul University*

Robert Jackson, *Brigham Young University*

Murray Jennex, *University of Phoenix*

Len Jessup, *Washington State University*

Robert Keim, *Arizona State University*

Gene Klawikowski, *Nicolet Area Technical College*

Mat Klempa, *California State University at Los Angeles*

Ned Kock, *Temple University*

Rebecca Koop, *Wright State University*

Sophie Lee, *University of Massachusetts at Boston*

Chang-Yang Lin, *Eastern Kentucky University*

Michael Martel, *Ohio University*

Nancy Martin, *USA Group, Indianapolis, Indiana*

Roger McHaney, *Kansas State University*

Nancy Melone, *University of Oregon*

David Paper, *Utah State University*

G. Premkumar, *Iowa State University*

Mary Prescott, *University of South Florida*

Stephen Priest, *Daniel Webster College*

Harry Reif, *James Madison University*

Steven Ross, *Western Washington University*

Terence Ryan, *Southern Illinois University*

Robert Saldarini, *Bergen Community College*

Arthur Santoianni, *University of Dayton*

Elaine Seeman, *Pitt Community College*

Leiser Silva, *University of Houston*

Taverekere Srikantaiah, *University of Maryland*

Eugene Stafford, *Iona College*

Gary Templeton, *Mississippi State University*

Bob Tucker, *Antares Alliance, Plano, Texas*

Merrill Warkentin, *Northeastern University*

Cheryl Welch, *Ashland University*

Connie Wells, *Nicholls State University*

Chris Westland, *University of Southern California*

Charles Winton, *University of North Florida*

Liang Yu, *San Francisco State University*

Terry Zuechow, *EDS Corporation, Piano, Texas*

We extend a special note of thanks to Jeremy Alexander, who was instrumental in conceptualizing and writing the PVF WebStore feature that appears in Chapters 4 through 14. The addition of this feature has helped make those chapters more modern and innovative. We would also like to thank Jeremy and Saonee Sarker of Washington State University for their assistance in designing installation guides and tutorials for using Oracle development tools in conjunction with prior versions of the book; Darren Nicholson for the help he provided with the Visual Basic and .NET-related materials in previous editions; and Ryan Wright, of the University of Massachusetts, Amherst, who helped with the .NET screen captures in previous editions and who also carefully reviewed each of the book's chapters before they were revised. Thanks to Jeff Jenkins, of the University of Arizona, for his help with the Visual Basic screenshots in the current edition.

We also wish to thank Atish Sinha of the University of Wisconsin–Milwaukee for writing the original version of some of the object-oriented analysis and design material. Dr. Sinha, who has been teaching this topic for several years to both undergraduates and MBA students, executed a challenging assignment with creativity and cooperation.

We are also indebted to our undergraduate and MBA students at the University of Dayton, Iowa State University, and the University of Arizona, who have given us many helpful comments as they worked with drafts of this text, and our thanks go to Fred McFadden (University of Colorado, Colorado Springs), Mary Prescott (University of South Florida), Ramesh Venkataraman (Indiana University), and Heikki Topi (Bentley University) for their assistance in coordinating this text with its companion book, *Modern Database Management,* also by Pearson Education.

Finally, we have been fortunate to work with a large number of creative and insightful people at Pearson have added much to the development, format, and production of this text. We have been thoroughly impressed with their commitment to this text and to the IS education market. These people include: Bob Horan (Executive Editor), Kelly Loftus (Editorial Project Manager), Ashlee Bradbury

(Editorial Assistant), Jane Bonnell (Production Project Manager), Anne Fahlgren (Senior Marketing Manager), and Alana Coles (Media Project Manager). We would also like to thank Haylee Schwenk and the crew at PreMediaGlobal.

The writing of this text has involved thousands of hours of time from the authors and from all of the people listed previously. Although our names will be visibly associated with this book, we know that much of the credit goes to the individuals and organizations listed here for any success it might achieve. It is important for the reader to recognize all the individuals and organizations that have been committed to the preparation and production of this book.

Jeffrey A. Hoffer, Dayton, Ohio
Joey F. George, Ames, Iowa
Joseph S. Valacich, Tucson, Arizona

PART ONE

Foundations for Systems Development

Chapter 1
The Systems Development Environment

Chapter 2
The Origins of Software

Chapter 3
Managing the Information Systems Project

PART ONE

Foundations for Systems Development

You are beginning a journey that will enable you to build on every aspect of your education and experience. Becoming a systems analyst is not a goal; it is a path to a rich and diverse career that will allow you to exercise and continue to develop a wide range of talents. We hope that this introductory part of the text helps open your mind to the opportunities of the systems analysis and design field and to the engaging nature of systems work.

Chapter 1 shows you what systems analysis and design is all about and how it has evolved over the past several decades. As businesses and systems have become more sophisticated and more complex, there has been an increasing emphasis on speed in systems analysis and design. Systems development began as an art, but most businesspeople soon realized this was not a tenable long-term solution to developing systems to support business processes. Systems development became more structured and more like engineering, and managers stressed the importance of planning, project management, and documentation. Now, we are witnessing a reaction against excesses in all three of these areas, and the focus has shifted to agile development. The evolution of systems analysis and design and the current focus on agility are explained in Chapter 1. It is also important, however, that you remember that systems analysis and design exists within a multifaceted organizational context that involves other organizational members and external parties. Understanding systems development requires an understanding not only of each technique, tool, and method, but also of how these elements cooperate, complement, and support each other within an organizational setting.

As you read this book you'll also discover that the systems analysis and design field is constantly adapting to new situations due to a strong commitment to constant improvement. Our goal in this book is to provide you with a mosaic of the skills needed to work effectively in whatever environment you find yourself, armed with the knowledge to determine the best practices for that situation and argue for them effectively.

Chapter 2 presents an introduction to the many sources from which software and software components can be obtained. Back when systems analysis and design was an art, all systems were written from scratch by in-house experts. Businesses had little choice. Now there is little excuse for in-house development, so it becomes crucial that systems analysts understand the software industry and the many different sources of software. Chapter 2 provides an initial map of the software industry landscape and explains most of the many choices available to systems analysts.

Chapter 3 addresses a fundamental characteristic of life as a systems analyst: working within the framework of projects with constrained resources. All systems-related work demands attention to deadlines, working within budgets, and coordinating the work of various people. The very nature of the systems development life cycle (SDLC) implies a systematic approach to a project, which is a group of related activities leading to a final deliverable. Projects must be planned, started, executed, and completed. The planned work of the project must be represented so that all interested parties can review and understand it. In your job as a systems analyst, you will have to work within the schedule and other project plans, and thus it is important to understand the management process controlling your work.

Finally, Part I introduces the Petrie Electronics case. The Petrie case helps demonstrate how what you learn in each chapter might fit into a practical organizational situation. The case begins after Chapter 2; the remaining book chapters through Chapter 13 each have an associated case installment. The first section introduces the company and its existing information systems. This introduction to Petrie provides insights into the company, which will help you understand the company more completely when we look at the requirements and design for new systems in later case sections.

CHAPTER ONE

The Systems Development Environment

Learning Objectives

After studying this chapter, you should be able to:

- Define *information systems analysis and design.*
- Describe the information systems development life cycle (SDLC).
- Explain Rapid Application Development (RAD) and computer-aided software engineering (CASE) tools.
- Describe the Agile Methodologies and eXtreme Programming.
- Explain object-oriented analysis and design and the Rational Unified Process (RUP).

Introduction

Information systems analysis and design is a complex, challenging, and stimulating organizational process that a team of business and systems professionals uses to develop and maintain computer-based information systems. Although advances in information technology continually give us new capabilities, the analysis and design of information systems is driven from an organizational perspective. An organization might consist of a whole enterprise, specific departments, or individual work groups. Organizations can respond to and anticipate problems and opportunities through innovative uses of information technology. Information systems analysis and design is, therefore, an organizational improvement process. Systems are built and rebuilt for organizational benefits. Benefits result from adding value during the process of creating, producing, and supporting the organization's products and services. Thus, the analysis and design of information systems is based on your understanding of the organization's objectives, structure, and processes, as well as your knowledge of how to exploit information technology for advantage.

In the current business environment, the Internet, especially the World Wide Web, has been firmly integrated into an organization's way of doing business. Although you are probably most familiar

Information systems analysis and design
The complex organizational process whereby computer-based information systems are developed and maintained.

with marketing done on the Web and Web-based retailing sites, such as eBay or Amazon.com, the overwhelming majority of business use of the Web is business-to-business applications. These applications run the gamut of everything businesses do, including transmitting orders and payments to suppliers, fulfilling orders and collecting payments from customers, maintaining business relationships, and establishing electronic marketplaces where businesses can shop online for the best deals on resources they need for assembling their products and services. Although the Internet seems to pervade business these days, it is important to remember that many of the key aspects of business—offering a product or service for sale, collecting payment, paying employees, maintaining supplier and client relationships—have not changed. Understanding the business and how it functions is still the key to successful systems development, even in the fast-paced, technology-driven environment that organizations find themselves in today.

Careers in information technology (IT) present a great opportunity for you to make a significant and visible impact on business. The demand for skilled information technology workers is growing. According to the U.S. Bureau of Labor Statistics, the professional IT workforce will grow by more than 22 percent between 2010 and 2020 (Thibodeau, 2012). The fastest growth will come for software developers (32 percent) and database administrators (31 percent). One particular aspect of the information technology industry, cloud computing, is predicted to lead to the creation of almost 14 million technology and related jobs by 2015 (McDougall, 2012). Annual revenues from cloud computing will be over $1.1 trillion (USD) starting that year. And the growth will be global, with the number of cloud computing jobs in Brazil increasing by 186 percent, the number of jobs in China and India almost doubling, and growth in cloud-related jobs increasing by 66 percent in the United States. (See more about cloud computing in Chapter 2.) With the challenges and opportunities of dealing with rapid advances in technology, it is difficult to imagine a more exciting career choice than information technology, and systems analysis and design is a big part of the IT landscape. Furthermore, analyzing and designing information systems will give you the chance to understand organizations at a depth and breadth that might take many more years to accomplish in other careers.

Application software
Computer software designed to support organizational functions or processes.

An important (but not the only) result of systems analysis and design is **application software**, software designed to support a specific organizational function or process, such as inventory management, payroll, or market analysis. In addition to application software, the total information system includes the hardware and systems software on which the application software runs, documentation and training materials, the specific job roles associated with the overall system, controls, and the people who use the software along with their work methods. Although we will address all of these various dimensions of the overall system, we will emphasize application software development—your primary responsibility as a systems analyst.

In the early years of computing, analysis and design was considered an art. Now that the need for systems and software has become so great, people in industry and academia have developed work methods that make analysis and design a disciplined process. Our goal is to help you develop the knowledge and skills needed to understand and follow such software engineering processes. Central to software engineering processes (and to this book) are various methodologies, techniques, and tools that have been developed, tested, and widely used over the years to assist people like you during systems analysis and design.

Methodologies are comprehensive, multiple-step approaches to systems development that will guide your work and influence the quality of your final product—the information system. A methodology adopted by an organization will be consistent with its general management style (e.g., an organization's orientation toward consensus management will influence its choice of systems development methodology). Most methodologies incorporate several development techniques.

Techniques are particular processes that you, as an analyst, will follow to help ensure that your work is well thought out, complete, and comprehensible to others on your project team. Techniques provide support for a wide range of tasks, including conducting thorough interviews to determine what your system should do, planning and managing the activities in a systems development project, diagramming the system's logic, and designing the reports your system will generate.

Tools are typically computer programs that make it easy to use and benefit from techniques and to faithfully follow the guidelines of the overall development methodology. To be effective, techniques and tools must both be consistent with an organization's systems development methodology. Techniques and tools must make it easy for systems developers to conduct the steps called for in the methodology. These three elements—methodologies, techniques, and tools—work together to form an organizational approach to systems analysis and design (see Figure 1-1).

Although many people in organizations are responsible for systems analysis and design, in most organizations the **systems analyst** has the primary responsibility. When you begin your career in systems development, you will most likely begin as a systems analyst or as a programmer with some systems analysis responsibilities. The primary role of a systems analyst is to study the problems and needs of an organization in order to determine how people, methods, and information technology can best be combined to bring about improvements in the organization. A systems analyst helps system users and other business managers define their requirements for new or enhanced information services. As such, a systems analyst is an agent of change and innovation.

In the rest of this chapter, we will examine the systems approach to analysis and design. You will learn how systems analysis and design has changed over the decades as computing has become more central to business. You will learn about the systems development life cycle, which provides the basic overall structure of the systems development process and of this book. This chapter ends with a discussion of some of the methodologies, techniques, and tools created to support the systems development process.

Systems analyst
The organizational role most responsible for the analysis and design of information systems.

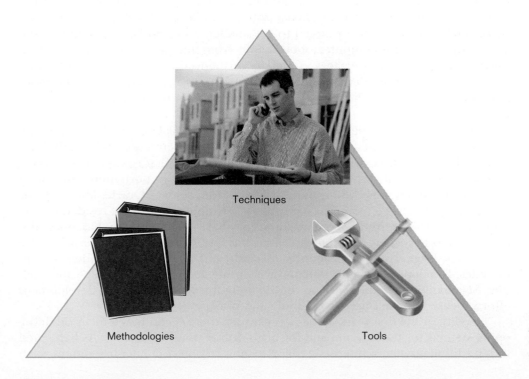

FIGURE 1-1
An organizational approach to systems analysis and design is driven by methodologies, techniques, and tools

A MODERN APPROACH TO SYSTEMS ANALYSIS AND DESIGN

The analysis and design of computer-based information systems began in the 1950s. Since then, the development environment has changed dramatically, driven by organizational needs as well as by rapid changes in the technological capabilities of computers. In the 1950s, development focused on the processes the software performed. Because computer power was a critical resource, efficiency of processing became the main goal. Computers were large, expensive, and not very reliable. Emphasis was placed on automating existing processes, such as purchasing or paying, often within single departments. All applications had to be developed in machine language or assembly language, and they had to be developed from scratch because there was no software industry. Because computers were so expensive, computer memory was also at a premium, so system developers conserved as much memory for data storage as possible.

The first procedural, or third-generation, computer programming languages did not become available until the beginning of the 1960s. Computers were still large and expensive, but the 1960s saw important breakthroughs in technology that enabled the development of smaller, faster, less expensive computers—minicomputers—and the beginnings of the software industry. Most organizations still developed their applications from scratch using their in-house development staffs. Systems development was more an art than a science. This view of systems development began to change in the 1970s, however, as organizations started to realize how expensive it was to develop customized information systems for every application. Systems development came to be more disciplined as many people worked to make it more like engineering. Early database management systems, using hierarchical and network models, helped bring discipline to the storage and retrieval of data. The development of database management systems helped shift the focus of systems development from processes first to data first.

The 1980s were marked by major breakthroughs in computing in organizations, as microcomputers became key organizational tools. The software industry expanded greatly as more and more people began to write off-the-shelf software for microcomputers. Developers began to write more and more applications in fourth-generation languages, which, unlike procedural languages, instructed a computer on what to do instead of how to do it. Computer-aided software engineering (CASE) tools were developed to make systems developers' work easier and more consistent. As computers continued to get smaller, faster, and cheaper, and as the operating systems for computers moved away from line prompt interfaces to windows- and icon-based interfaces, organizations moved to applications with more graphics. Organizations developed less software in-house and bought relatively more from software vendors. The systems developer's job went through a transition from builder to integrator.

The systems development environment of the late 1990s focused on systems integration. Developers used visual programming environments, such as PowerBuilder or Visual Basic, to design the user interfaces for systems that run on client/server platforms. The database, which may be relational or object-oriented, and which may have been developed using software from firms such as Oracle, Microsoft, or Ingres, resided on the server. In many cases, the application logic resided on the same server. Alternatively, an organization may have decided to purchase its entire enterprise-wide system from companies such as SAP AG or Oracle. Enterprise-wide systems are large, complex systems that consist of a series of independent system modules. Developers assemble systems by choosing and implementing specific modules. Starting in the middle years of the 1990s, more and more systems development efforts focused on the Internet, especially the Web.

Today, in the first years of the new century, there is continued focus on developing systems for the Internet and for firms' intranets and extranets. As happened with traditional systems, Internet developers now rely on computer-based tools, such as ColdFusion, to speed and simplify the development of Web-based systems. Many CASE tools, such as those developed by Oracle, now directly support Web application development. More and more, systems implementation involves a three-tier design, with the database on one server, the application on a second server, and client logic located on user machines. Another important development in the early years of the new century is the move to wireless system components. Wireless devices, such as cell phones and personal digital assistants (PDAs; e.g., Palm Pilots or Pocket PCs), can access Web-based applications from almost anywhere. Finally, the trend continues toward assembling systems from programs and components purchased off the shelf. In many cases, organizations do not develop the application in-house. They don't even run the application in-house, choosing instead to use the application on a per-use basis by accessing it through an application service provider (ASP).

DEVELOPING INFORMATION SYSTEMS AND THE SYSTEMS DEVELOPMENT LIFE CYCLE

Most organizations find it beneficial to use a standard set of steps, called a **systems development methodology**, to develop and support their information systems. Like many processes, the development of information systems often follows a life cycle. For example, a commercial product follows a life cycle in that it is created, tested, and introduced to the market. Its sales increase, peak, and decline. Finally, the product is removed from the market and replaced by something else. The **systems development life cycle (SDLC)** is a common methodology for systems development in many organizations; it features several phases that mark the progress of the systems analysis and design effort. Every textbook author and information systems development organization uses a slightly different life-cycle model, with anywhere from 3 to almost 20 identifiable phases.

The life cycle can be thought of as a circular process in which the end of the useful life of one system leads to the beginning of another project that will develop a new version or replace an existing system altogether (see Figure 1-2). At first glance, the life cycle appears to be a sequentially ordered set of phases, but it is not. The specific steps and their sequence are meant to be adapted as required for a project, consistent with management approaches. For example, in any given SDLC phase, the project can return to an earlier phase if necessary. Similarly, if a commercial product does not perform well just after its introduction, it may be temporarily removed from the market and improved before being reintroduced. In the SDLC, it is also possible to complete some activities in one phase in parallel with some activities of

Systems development methodology
A standard process followed in an organization to conduct all the steps necessary to analyze, design, implement, and maintain information systems.

Systems development life cycle (SDLC)
The traditional methodology used to develop, maintain, and replace information systems.

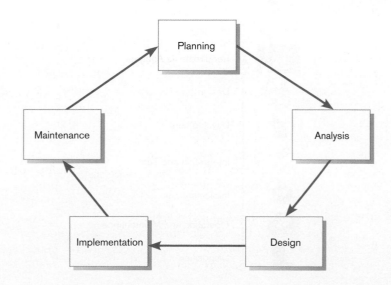

FIGURE 1-2
Systems development life cycle

FIGURE 1-3
Evolutionary model

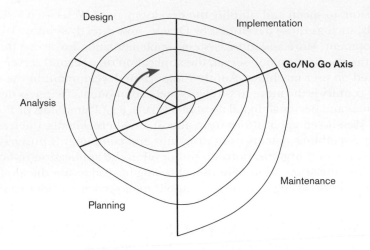

another phase. Sometimes the life cycle is iterative; that is, phases are repeated as required until an acceptable system is found. Some people consider the life cycle to be a spiral, in which we constantly cycle through the phases at different levels of detail (see Figure 1-3). However conceived, the systems development life cycle used in an organization is an orderly set of activities conducted and planned for each development project. The skills required of a systems analyst apply to all life-cycle models. Software is the most obvious end product of the life cycle; other essential outputs include documentation about the system and how it was developed, as well as training for users.

Every medium to large corporation and every custom software producer will have its own specific life cycle or systems development methodology in place (see Figure 1-4). Even if a particular methodology does not look like a cycle, and Figure 1-4 does not, you will probably discover that many of the SDLC steps are performed and SDLC techniques and tools are used. Learning about systems analysis and design from the life cycle approach will serve you well no matter which systems development methodology you use.

FIGURE 1-4
U.S. Department of Justice's systems development life cycle

(*Source:* Diagram based on *www.usdoj.gov/ jmd/irm/lifecycle/ch1.htm#para1.2.*)

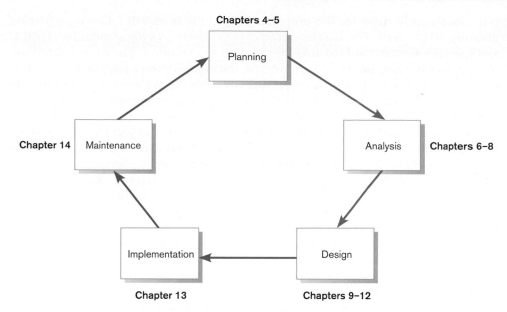

FIGURE 1-5
SDLC-based guide to this book

When you begin your first job, you will likely spend several weeks or months learning your organization's SDLC and its associated methodologies, techniques, and tools. In order to make this book as general as possible, we follow a rather generic life-cycle model, as described in more detail in Figure 1-5. Notice that our model is circular. We use this SDLC as one example of a methodology but, more important, as a way to arrange the topics of systems analysis and design. Thus, what you learn in this book, you can apply to almost any life cycle you might follow. As we describe this SDLC throughout the book, you will see that each phase has specific outcomes and deliverables that feed important information to other phases. At the end of each phase, a systems development project reaches a milestone and, as deliverables are produced, they are often reviewed by parties outside the project team. In the rest of this section, we provide a brief overview of each SDLC phase. At the end of the section, we summarize this discussion in a table that lists the main deliverables or outputs from each SDLC phase.

The first phase in the SDLC is **planning**. In this phase, someone identifies the need for a new or enhanced system. In larger organizations, this recognition may be part of a corporate and systems planning process. Information needs of the organization as a whole are examined, and projects to meet these needs are proactively identified. The organization's information system needs may result from requests to deal with problems in current procedures, from the desire to perform additional tasks, or from the realization that information technology could be used to capitalize on an existing opportunity. These needs can then be prioritized and translated into a plan for the information systems department, including a schedule for developing new major systems. In smaller organizations (as well as in large ones), determination of which systems to develop may be affected by ad hoc user requests submitted as the need for new or enhanced systems arises, as well as from a formalized information planning process. In either case, during project identification and selection, an organization determines whether resources should be devoted to the development or enhancement of each information system under consideration. The outcome of the project identification and selection process is a determination of which systems development projects should be undertaken by the organization, at least in terms of an initial study.

Two additional major activities are also performed during the planning phase: the formal, yet still preliminary, investigation of the system problem or opportunity at hand and the presentation of reasons why the system should or should not be developed by the organization. A critical step at this point is determining the scope of the proposed system. The project leader and initial team of systems analysts also

Planning
The first phase of the SDLC in which an organization's total information system needs are identified, analyzed, prioritized, and arranged.

produce a specific plan for the proposed project the team will follow using the remaining SDLC steps. This baseline project plan customizes the standardized SDLC and specifies the time and resources needed for its execution. The formal definition of a project is based on the likelihood that the organization's information systems department is able to develop a system that will solve the problem or exploit the opportunity and determine whether the costs of developing the system outweigh the benefits it could provide. The final presentation of the business case for proceeding with the subsequent project phases is usually made by the project leader and other team members to someone in management or to a special management committee with the job of deciding which projects the organization will undertake.

Analysis
The second phase of the SDLC in which system requirements are studied and structured.

The second phase in the SDLC is **analysis**. During this phase, the analyst thoroughly studies the organization's current procedures and the information systems used to perform organizational tasks. Analysis has two subphases. The first is requirements determination. In this subphase, analysts work with users to determine what the users want from a proposed system. The requirements determination process usually involves a careful study of any current systems, manual and computerized, that might be replaced or enhanced as part of the project. In the second part of analysis, analysts study the requirements and structure them according to their interrelationships, and eliminate any redundancies. The output of the analysis phase is a description of (but not a detailed design for) the alternative solution recommended by the analysis team. Once the recommendation is accepted by those with funding authority, the analysts can begin to make plans to acquire any hardware and system software necessary to build or operate the system as proposed.

Design
The third phase of the SDLC in which the description of the recommended solution is converted into logical and then physical system specifications.

The third phase in the SDLC is **design**. During design, analysts convert the description of the recommended alternative solution into logical and then physical system specifications. The analysts must design all aspects of the system, from input and output screens to reports, databases, and computer processes. The analysts must then provide the physical specifics of the system they have designed, either as a model or as detailed documentation, to guide those who will build the new system. That part of the design process that is independent of any specific hardware or software platform is referred to as **logical design**. Theoretically, the system could be implemented on any hardware and systems software. The idea is to make sure that the system functions as intended. Logical design concentrates on the business aspects of the system and tends to be oriented to a high level of specificity.

Logical design
The part of the design phase of the SDLC in which all functional features of the system chosen for development in analysis are described independently of any computer platform.

Physical design
The part of the design phase of the SDLC in which the logical specifications of the system from logical design are transformed into technology-specific details from which all programming and system construction can be accomplished.

Once the overall high-level design of the system is worked out, the analysts begin turning logical specifications into physical ones. This process is referred to as **physical design**. As part of physical design, analysts design the various parts of the system to perform the physical operations necessary to facilitate data capture, processing, and information output. This can be done in many ways, from creating a working model of the system to be implemented to writing detailed specifications describing all the different parts of the system and how they should be built. In many cases, the working model becomes the basis for the actual system to be used. During physical design, the analyst team must determine many of the physical details necessary to build the final system, from the programming language the system will be written in, to the database system that will store the data, to the hardware platform on which the system will run. Often the choices of language, database, and platform are already decided by the organization or by the client, and at this point these information technologies must be taken into account in the physical design of the system. The final product of the design phase is the physical system specifications in a form ready to be turned over to programmers and other system builders for construction. Figure 1-6 illustrates the differences between logical and physical design.

Implementation
The fourth phase of the SDLC in which the information system is coded, tested, installed, and supported in the organization.

The fourth phase in the SDLC is **implementation**. The physical system specifications, whether in the form of a detailed model or as detailed written specifications, are turned over to programmers as the first part of the implementation phase. During implementation, analysts turn system specifications into a working system that is tested and then put into use. Implementation includes coding, testing, and

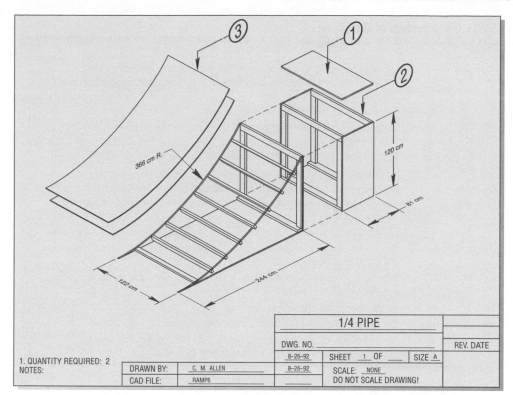

FIGURE 1-6
Difference between logical design and
physical design
(a) A skateboard ramp blueprint (logical
design)
(*Sources: www.tumyeto.com/tydu/skatebrd/
organizations/plans/14pipe.jpg; www
.tumyeto.com/tydu/skatebrd/organizations/
plans/iuscblue.html.* Accessed
September 16, 1999. Reprinted by
permission of the International
Association of Skateboard Companies.)

(b) A skateboard ramp (physical design)

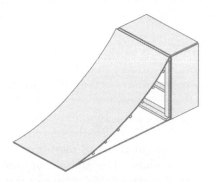

installation. During coding, programmers write the programs that make up the system. Sometimes the code is generated by the same system used to build the detailed model of the system. During testing, programmers and analysts test individual programs and the entire system in order to find and correct errors. During installation, the new system becomes part of the daily activities of the organization. Application software is installed, or loaded, on existing or new hardware, and users are introduced to the new system and trained. Testing and installation should be planned for as early as the project initiation and planning phase; both testing and installation require extensive analysis in order to develop exactly the right approach.

Implementation activities also include initial user support such as the finalization of documentation, training programs, and ongoing user assistance. Note that documentation and training programs are finalized during implementation; documentation is produced throughout the life cycle, and training (and education) occurs from the inception of a project. Implementation can continue for as long as the system exists because ongoing user support is also part of implementation. Despite the best efforts of analysts, managers, and programmers, however, installation is not always a simple process. Many well-designed systems have failed because the installation process was faulty. Even a well-designed system can fail if implementation is not well managed. Because the project team usually manages implementation, we stress implementation issues throughout this book.

TABLE 1-1   Products of SDLC Phases

Phase	Products, Outputs, or Deliverables
Planning	Priorities for systems and projects; an architecture for data, networks, and selection hardware, and information systems management are the result of associated systems
	Detailed steps, or work plan, for project
	Specification of system scope and planning and high-level system requirements or features
	Assignment of team members and other resources
	System justification or business case
Analysis	Description of current system and where problems or opportunities are with a general recommendation on how to fix, enhance, or replace current system
	Explanation of alternative systems and justification for chosen alternative
Design	Functional, detailed specifications of all system elements (data, processes, inputs, and outputs)
	Technical, detailed specifications of all system elements (programs, files, network, system software, etc.)
	Acquisition plan for new technology
Implementation	Code, documentation, training procedures, and support capabilities
Maintenance	New versions or releases of software with associated updates to documentation, training, and support

Maintenance
The final phase of the SDLC in which an information system is systematically repaired and improved.

The fifth and final phase in the SDLC is **maintenance**. When a system (including its training, documentation, and support) is operating in an organization, users sometimes find problems with how it works and often think of better ways to perform its functions. Also, the organization's needs with respect to the system change over time. In maintenance, programmers make the changes that users ask for and modify the system to reflect changing business conditions. These changes are necessary to keep the system running and useful. In a sense, maintenance is not a separate phase but a repetition of the other life cycle phases required to study and implement the needed changes. One might think of maintenance as an overlay on the life cycle rather than as a separate phase. The amount of time and effort devoted to maintenance depends a great deal on the performance of the previous phases of the life cycle. There inevitably comes a time, however, when an information system is no longer performing as desired, when maintenance costs become prohibitive, or when an organization's needs have changed substantially. Such problems indicate that it is time to begin designing the system's replacement, thereby completing the loop and starting the life cycle over again. Often the distinction between major maintenance and new development is not clear, which is another reason maintenance often resembles the life cycle itself.

The SDLC is a highly linked set of phases whose products feed the activities in subsequent phases. Table 1-1 summarizes the outputs or products of each phase based on the in-text descriptions. The chapters on the SDLC phases will elaborate on the products of each phase as well as on how the products are developed.

Throughout the SDLC, the systems development project itself must be carefully planned and managed. The larger the systems project, the greater the need for project management. Several project management techniques have been developed over the past decades, and many have been made more useful through automation. Chapter 3 contains a more detailed treatment of project planning and management techniques. Next, we will discuss some of the criticisms of the SDLC and present alternatives developed to address those criticisms. First, however, we will introduce you to a specialized SDLC that focuses on security during development.

A SPECIALIZED SYSTEMS DEVELOPMENT LIFE CYCLE

Although the basic SDLC provides an overview of the systems development process, the concept of the SDLC can also be applied to very specific aspects of the process. As mentioned previously, the maintenance phase can be described in terms of the SDLC. Another example of a specialized SDLC is Microsoft's Security Development Lifecycle (SDL) (see http://www.microsoft.com/security/sdl/default.aspx for details). The Security Development Lifecycle is depicted in Figure 1-7. First note how the five basic phases of the development life cycle (in green) are not exactly the same as the five phases of the SDLC we will use in this book. Three of the five phases are almost identical to the phases in our SDLC. The Microsoft SDL starts with requirements, which is similar to analysis; this is followed by the design phase, which is followed by implementation. Our life cycle starts with planning and ends with maintenance. Both of these phases are peculiar to systems development in an organizational context, where the information systems that are procured are used inside the organization. Which systems will be developed has to be planned carefully for an organization. Each system developed is an investment, and if the organization invests in a particular system, it cannot invest in some alternative system or in something else, such as a new store. Investment funds are limited, after all. Maintenance is also peculiar to an organizational context. Once systems go into general use, the organization needs to earn as much of a return on those investments as it can, so it is important that the systems run as long as possible. Companies such as Microsoft, which develop systems for others to use, do not need to worry about internal planning and maintenance in their product life cycles. Instead, as they have limited control over systems once they have been sold, they are concerned about selling a mature and reliable product. Therefore, they have two phases after implementation: verification and release. Verification involves quality assurance for products before they are released. Release involves all of the activities related to making the product available for general use. Next, note the two parts of the SDL that precede and follow the main development phases: training (in blue) and response (in orange). Two things make this particular SDLC specialized to security issues: the two unique phases that begin and end the life cycle (training and response), and the particular security activities associated with each phase in the development life cycle.

Training in Microsoft's SDL refers to the training each member of a development team receives about security basics and trends in security. The idea behind the training—indeed the idea behind a specialized security development life cycle—is to have security become a part of the development process from the beginning and not suddenly appear at the end of the SDLC. By training team members about security and how it can be addressed throughout the life cycle, security measures can be built into the system throughout its development. The response at the end of the SDL refers to a response plan developed during the release phase. If there is a security threat to a particular product, then the previously developed response plan is executed. The security-related activities that take place throughout the development life cycle vary by phase. Listing and explaining each activity are beyond the scope of this chapter, but we can provide some examples. One specialized activity performed during the requirements phase is a separate analysis

FIGURE 1-7
Microsoft's Security Development Lifecycle (SDL)

(*Source: http://www.microsoft.com/security/sdl/default.aspx.* Used by permission.)

of requirements related to both security and privacy. During design, developers can model threats to a system and consider how those threats differ with different design options. During implementation, project team members can conduct static analyses of source code, looking for security threats. During verification, they can conduct dynamic analyses. As part of the release phase, team members develop the incident response plan, mentioned previously, and they conduct a final security review. By adhering to a specialized life cycle devoted to security, project team members can ensure not only that security is addressed but that it is addressed in a planned, systematic manner.

THE HEART OF THE SYSTEMS DEVELOPMENT PROCESS

The SDLC provides a convenient way to think about the processes involved in systems development and the organization of this book. The different phases are clearly defined, their relationships to each other are well specified, and the sequencing of phases from one to the next, from beginning to end, has a compelling logic. In many ways, though, the SDLC is fiction. Although almost all systems development projects adhere to some type of life cycle, the exact location of activities and the specific sequencing of steps can vary greatly from one project to the next. Current practice combines the activities traditionally thought of as belonging to analysis, design, and implementation into a single process. Instead of systems requirements being produced in analysis, systems specifications being created in design, and coding and testing being done at the beginning of implementation, current practice combines all of these activities into a single analysis–design–code–test process (Figure 1-8). These activities are the heart of systems development, as we suggest in Figure 1-9. This combination of activities started with Rapid Application Development (RAD) and is seen in current practices such as Agile Methodologies. A well-known instance of one of the Agile Methodologies is eXtreme Programming, although there are other variations. We will introduce you to RAD, Agile Methodologies, and eXtreme Programming, but first it is important that you learn about the problems with the traditional SDLC. You will read about these problems next. Then you will read about CASE tools, RAD, Agile Methodologies, and eXtreme Programming.

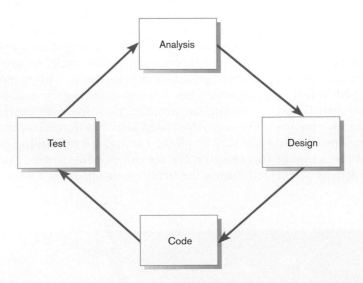

FIGURE 1-8
Analysis–design–code–test loop

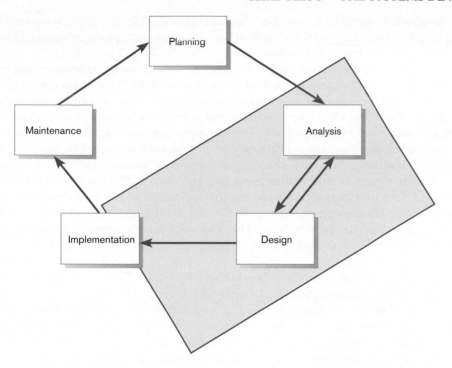

FIGURE 1-9
Heart of systems development

The Traditional Waterfall SDLC

There are several criticisms of the traditional life cycle approach to systems development; one relates to the way the life cycle is organized. To better understand these criticisms, it is best to see the form in which the life cycle has traditionally been portrayed, the so-called waterfall (Figure 1-10). Note how the flow of the project begins in the planning phase and from there runs "downhill" to each subsequent phase, just like a stream that runs off a cliff. Although the original developer of the waterfall model, W. W. Royce, called for feedback between phases in the waterfall,

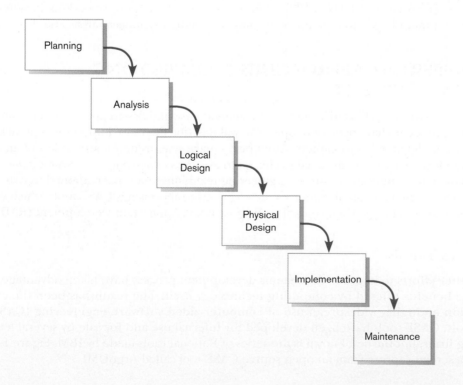

FIGURE 1-10
Traditional waterfall SDLC

this feedback came to be ignored in implementation (Martin, 1999). It became too tempting to ignore the need for feedback and to treat each phase as complete unto itself, never to be revisited once finished.

Traditionally, one phase ended and another began once a milestone had been reached. The milestone usually took the form of some deliverable or prespecified output from the phase. For example, the design deliverable is the set of detailed physical design specifications. Once the milestone had been reached and the new phase initiated, it became difficult to go back. Even though business conditions continued to change during the development process and analysts were pressured by users and others to alter the design to match changing conditions, it was necessary for the analysts to freeze the design at a particular point and go forward. The enormous amount of effort and time necessary to implement a specific design meant that it would be very expensive to make changes in a system once it was developed. The traditional waterfall life cycle, then, had the property of locking users into requirements that had been previously determined, even though those requirements might have changed.

Yet another criticism of the traditional waterfall SDLC is that the role of system users or customers was narrowly defined (Kay, 2002). User roles were often delegated to the requirements determination or analysis phases of the project, where it was assumed that all of the requirements could be specified in advance. Such an assumption, coupled with limited user involvement, reinforced the tendency of the waterfall model to lock in requirements too early, even after business conditions had changed.

In addition, under the traditional waterfall approach, nebulous and intangible processes such as analysis and design are given hard-and-fast dates for completion, and success is overwhelmingly measured by whether those dates are met. The focus on milestone deadlines, instead of on obtaining and interpreting feedback from the development process, leads to too little focus on doing good analysis and design. The focus on deadlines results in systems that do not match users' needs and that require extensive maintenance, unnecessarily increasing development costs. Finding and fixing a software problem after the delivery of the system is often 100 times more expensive than finding and fixing it during analysis and design (Griss, 2003). The result of focusing on deadlines rather than on good practice is unnecessary rework and maintenance effort, both of which are expensive. According to some estimates, maintenance costs account for 40 to 70 percent of systems development costs (Dorfman and Thayer, 1997). Given these problems, people working in systems development began to look for better ways to conduct systems analysis and design.

DIFFERENT APPROACHES TO IMPROVING DEVELOPMENT

In the continuing effort to improve the systems analysis and design process, several different approaches have been developed. We will describe the more important approaches in more detail in later chapters. Attempts to make systems development less of an art and more of a science are usually referred to as *systems engineering* or *software engineering*. As the names indicate, rigorous engineering techniques have been applied to systems development. One manifestation of the engineering approach is CASE tools, which you will read about next. Then you will read about Rapid Application Development (RAD).

CASE Tools

Other efforts to improve the systems development process have taken advantage of the benefits offered by computing technology itself. The result has been the creation and fairly widespread use of **computer-aided software engineering (CASE)** tools. CASE tools have been developed for internal use and for sale by several leading firms, but the best known is the series of Rational tools made by IBM. Figure 1-11 is a screen capture from an open source CASE tool called ArgoUML.

Computer-aided software engineering (CASE) tools
Software tools that provide automated support for some portion of the systems development process.

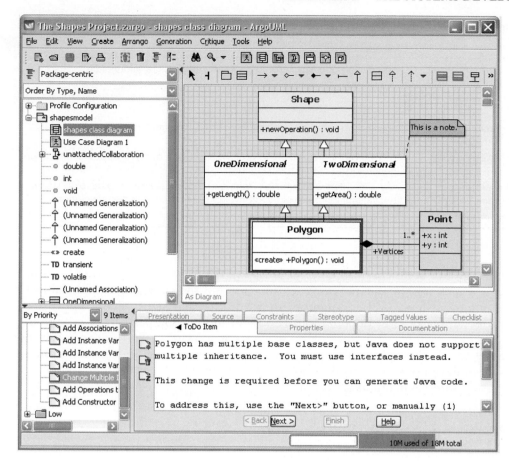

FIGURE 1-11
Screen shot of ArgoUML, an open source
CASE tool

(*Source: http://argouml.tigris.org/*)

CASE tools are used to support a wide variety of SDLC activities. CASE tools can be used to help in the project identification and selection; project initiation and planning, analysis, and design phases; and/or in the implementation and maintenance phases of the SDLC. An integrated and standard database called a *repository* is the common method for providing product and tool integration, and has been a key factor in enabling CASE to more easily manage larger, more complex projects and to seamlessly integrate data across various tools and products. The idea of a central repository of information about a project is not new—the manual form of such a repository is called a *project dictionary* or *workbook*.

The general types of CASE tools are listed below:

- Diagramming tools enable system process, data, and control structures to be represented graphically.
- Computer display and report generators help prototype how systems "look and feel." Display (or form) and report generators make it easier for the systems analyst to identify data requirements and relationships.
- Analysis tools automatically check for incomplete, inconsistent, or incorrect specifications in diagrams, forms, and reports.
- A central repository enables the integrated storage of specifications, diagrams, reports, and project management information.
- Documentation generators produce technical and user documentation in standard formats.
- Code generators enable the automatic generation of program and database definition code directly from the design documents, diagrams, forms, and reports.

CASE helps programmers and analysts do their jobs more efficiently and more effectively by automating routine tasks. However, many organizations that use CASE

TABLE 1-2   Examples of CASE Usage within the SDLC

SDLC Phase	Key Activities	CASE Tool Usage
Project identification and selection	Display and structure high-level organizational information	Diagramming and matrix tools to create and structure information
Project initiation and planning	Develop project scope and feasibility	Repository and documentation generators to develop project plans
Analysis	Determine and structure system requirements	Diagramming to create process, logic, and data models
Logical and physical design	Create new system designs	Form and report generators to prototype designs; analysis and documentation generators to define specifications
Implementation	Translate designs into an information system	Code generators and analysis, form and report generators to develop system; documentation generators to develop system and user documentation
Maintenance	Evolve information system	All tools are used (repeat life cycle)

tools do not use them to support all phases of the SDLC. Some organizations may extensively use the diagramming features but not the code generators. Table 1-2 summarizes how CASE is commonly used within each SDLC phase. There are a variety of reasons why organizations choose to adopt CASE partially or not use it at all. These reasons range from a lack of vision for applying CASE to all aspects of the SDLC, to the belief that CASE technology will fail to meet an organization's unique system development needs. In some organizations, CASE has been extremely successful, whereas in others it has not.

Rapid Application Development

Rapid Application Development (RAD) is an approach to developing information systems that promises better and cheaper systems and more rapid deployment by having systems developers and end users work together jointly in real time to develop systems. RAD grew out of the convergence of two trends: (1) the increased speed and turbulence of doing business in the late 1980s and early 1990s, and (2) the ready availability of high-powered, computer-based tools to support systems development and easy maintenance. As the conditions of doing business in a changing, competitive global environment became more turbulent, management in many organizations began to question if it made sense to wait two to three years to develop systems (in a methodical, controls-rich process) that would be obsolete upon completion.

The ready availability of increasingly powerful software tools created to support RAD also increased interest in this approach. RAD has become a legitimate way to develop information systems. Today, the focus is increasingly on the rapid development of Web-based systems. RAD tools, and software created to support rapid development, almost all provide for the speedy creation of Web-based applications.

As Figure 1-12 shows, the same phases that are followed in the traditional SDLC are also followed in RAD, but the phases are shortened and combined with each other to produce a more streamlined development technique. Planning and design phases in RAD are shortened by focusing work on system function and user interface requirements at the expense of detailed business analysis and concern for system performance issues. Also, RAD usually looks at the system being developed in isolation from other systems, thus eliminating the time-consuming activities of coordinating with existing standards and systems during design and development. The emphasis in RAD is generally less on the sequence and structure of processes in the life cycle and more on doing different tasks in parallel with each other and on using prototyping extensively. Notice also that the iteration in the RAD life cycle is limited to the design and development phases. This is where the bulk of the work in a RAD approach takes place.

Rapid Application Development (RAD)
Systems development methodology created to radically decrease the time needed to design and implement information systems. RAD relies on extensive user involvement, prototyping, integrated CASE tools, and code generators.

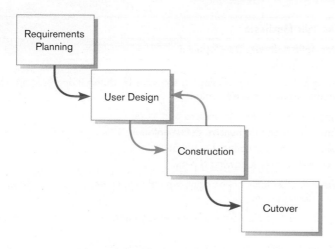

FIGURE 1-12
RAD life cycle

To succeed, RAD relies on bringing together several systems development practices. RAD depends on extensive user involvement. End users are involved from the beginning of the development process, when they participate in application planning; through requirements determination; and then into design and implementation, when they work with system developers to validate final elements of the system's design. Much end-user involvement takes place in the *prototyping* process, when users and analysts work together to design interfaces and reports for new systems. Prototyping involves the development of a working model of a system (more in Chapter 6). CASE tools can be used to build prototypes. Alternatively, RAD may employ visual development environments instead of CASE tools with code generators, but the benefits from prototyping are the same. User design ends with agreement on a computer-based design. The gap between the end of design and the handing over of the new system to users might only take 3 months instead of the usual 18.

In construction, the same information systems professionals who created the design now generate code using the CASE tools' code generator. End users also participate, validating screens and other aspects of the design as the application system is being built. Construction can be combined with user design into one phase when developing smaller systems.

Cutover is the delivery of the new system to its end users. Because the RAD approach is so fast, planning for cutover must begin early in the RAD process. Cutover involves many of the traditional activities of implementation, including testing the system, training users, dealing with organizational changes, and running the new and old systems in parallel; but all of these activities occur on an accelerated basis.

The primary advantage of RAD is obvious—information systems developed in as little as one-quarter of the usual time. An additional advantage is that shorter development cycles also mean cheaper systems because fewer organizational resources need to be devoted to develop any particular system. Martin (1991) points out that RAD also involves smaller development teams, which results in even more savings. Finally, because there is less time between the end of design and conversion, the new system is closer to current business needs and is therefore of higher quality than would be a similar system developed the traditional way. Others point out, however, that, although RAD works, it only works well for systems that have to be developed quickly (Gibson and Hughes, 1994).

AGILE METHODOLOGIES

RAD is just one reaction to the problems with the traditional waterfall methodology for systems development. As you might imagine, many other approaches to systems analysis and design have been developed over the years. In February 2001, many of

TABLE 1-3   The Agile Manifesto

The Manifesto for Agile Software Development

Seventeen anarchists agree:

We are uncovering better ways of developing software by doing it and helping others do it. Through this work we have come to value:

- *Individuals and interactions* over processes and tools.
- *Working software* over comprehensive documentation.
- *Customer collaboration* over contract negotiation.
- *Responding to change* over following a plan.

That is, while we value the items on the right, we value the items on the left more. We follow the following principles:

- Our highest priority is to satisfy the customer through early and continuous delivery of valuable software.
- Welcome changing requirements, even late in development. Agile processes harness change for the customer's competitive advantage.
- Deliver working software frequently, from a couple of weeks to a couple of months, with a preference to the shorter timescale.
- Businesspeople and developers work together daily throughout the project.
- Build projects around motivated individuals. Give them the environment and support they need, and trust them to get the job done.
- The most efficient and effective method of conveying information to and within a development team is face-to-face conversation.
- Working software is the primary measure of progress.
- Continuous attention to technical excellence and good design enhances agility.
- Agile processes promote sustainable development. The sponsors, developers, and users should be able to maintain a constant pace indefinitely.
- Simplicity—the art of maximizing the amount of work not done—is essential.
- The best architectures, requirements, and designs emerge from self-organizing teams.
- At regular intervals, the team reflects on how to become more effective, then tunes and adjusts its behavior accordingly.

—Kent Beck, Mike Beedle, Arie van Bennekum, Alistair Cockburn, Ward Cunningham, Martin Fowler, James Grenning, Jim Highsmith, Andrew Hunt, Ron Jeffries, Jon Kern, Brian Marick, Robert C. Martin, Steve Mellor, Ken Schwaber, Jeff Sutherland, Dave Thomas (*www .agileAlliance.org*)

(*Source: http://agilemanifesto.org/* © 2001, the above authors this declaration may be freely copied in any form, but only in its entirety through this notice.)

the proponents of these alternative approaches met in Utah and reached a consensus on several of the underlying principles their various approaches contained. This consensus turned into a document they called "The Agile Manifesto" (Table 1-3). According to Fowler (2003), the Agile Methodologies share three key principles: (1) a focus on adaptive rather than predictive methodologies, (2) a focus on people rather than roles, and (3) a focus on self-adaptive processes.

The Agile Methodologies group argues that software development methodologies adapted from engineering generally do not fit with real-world software development (Fowler, 2003). In engineering disciplines, such as civil engineering, requirements tend to be well understood. Once the creative and difficult work of design is completed, construction becomes very predictable. In addition, construction may account for as much as 90 percent of the total project effort. For software, on the other hand, requirements are rarely well understood, and they change continually during the lifetime of the project. Construction may account for as little as 15 percent of the total project effort, with design constituting as much as 50 percent. Applying techniques that work well for predictable, stable projects, such as bridge building, tend not to work well for fluid, design-heavy projects such as writing software, say the Agile Methodology proponents. What is needed are methodologies that embrace change and that are able to deal with a lack of predictability. One mechanism for dealing with a lack of predictability, which all Agile Methodologies share, is iterative development (Martin, 1999). Iterative development focuses on the frequent production of working versions of a system that have a subset of the total number of required features. Iterative development provides feedback to customers and developers alike.

The Agile Methodologies' focus on people is an emphasis on individuals rather than on the roles that people perform (Fowler, 2003). The roles that people fill, of systems analyst or tester or manager, are not as important as the individuals who fill those roles. Fowler argues that the application of engineering principles to systems development has resulted in a view of people as interchangeable units instead of a view of people as talented individuals, each bringing something unique to the development team.

The Agile Methodologies promote a self-adaptive software development process. As software is developed, the process used to develop it should be refined and improved. Development teams can do this through a review process, often associated with the completion of iterations. The implication is that, as processes are adapted, one would not expect to find a single monolithic methodology within a given corporation or enterprise. Instead, one would find many variations of the methodology, each of which reflects the particular talents and experience of the team using it.

Agile Methodologies are not for every project. Fowler (2003) recommends an agile or adaptive process if your project involves:

- Unpredictable or dynamic requirements
- Responsible and motivated developers
- Customers who understand the process and will get involved

A more engineering-oriented, predictable process may be called for if the development team exceeds 100 people or if the project is operating under a fixed-price or fixed scope contract. In fact, whether a systems development project is organized in terms of Agile or more traditional methodologies depends on many different considerations. If a project is considered to be high risk, highly complex, and has a development team made up of hundreds of people, then more traditional methods will apply. Less risky, smaller, and simpler development efforts lend themselves more to Agile methods. Other determining factors include organizational practice and standards, and the extent to which different parts of the system will be contracted out to others for development. Obviously, the larger the proportion of the system that will be farmed out, the more detailed the design specifications will need to be so that subcontractors can understand what is needed. Although not universally agreed upon, the key differences between these development approaches are listed in Table 1-4, which is based on work by Boehm and Turner (2004). These differences can be used to help determine which development approach would work best for a particular project.

Many different individual methodologies come under the umbrella of Agile Methodologies. Fowler (2003) lists the Crystal family of methodologies, Adaptive Software Development, Scrum, Feature Driven Development, and others as Agile Methodologies. Perhaps the best known of these methodologies, however, is eXtreme Programming, discussed next.

eXtreme Programming

eXtreme Programming is an approach to software development put together by Beck & Andres (2004). It is distinguished by its short cycles, incremental planning approach, focus on automated tests written by programmers and customers to monitor the development process, and a reliance on an evolutionary approach to development that lasts throughout the lifetime of the system. Key emphases of eXtreme Programming are its use of two-person programming teams, described later, and having a customer on-site during the development process. The relevant parts

TABLE 1-4   Five Critical Factors That Distinguish Agile and Traditional Approaches to Systems Development

Factor	Agile Methods	Traditional Methods
Size	Well matched to small products and teams. Reliance on tacit knowledge limits scalability.	Methods evolved to handle large products and teams. Hard to tailor down to small projects.
Criticality	Untested on safety-critical products. Potential difficulties with simple design and lack of documentation.	Methods evolved to handle highly critical products. Hard to tailor down to products that are not critical.
Dynamism	Simple design and continuous refactoring are excellent for highly dynamic environments but a source of potentially expensive rework for highly stable environments.	Detailed plans and Big Design Up Front, excellent for highly stable environment but a source of expensive rework for highly dynamic environments.
Personnel	Requires continuous presence of a critical mass of scarce experts. Risky to use no-agile people.	Needs a critical mass of scarce experts during project definition but can work with fewer later in the project, unless the environment is highly dynamic.
Culture	Thrives in a culture where people feel comfortable and empowered by having many degrees of freedom (thriving on chaos).	Thrives in a culture where people feel comfortable and empowered by having their roles defined by clear practices and procedures (thriving on order).

(*Source:* Boehm, Barry; Turner, Richard, *Balancing Agility and Discipline: A Guide for the Perplexed,* 1st Ed., © 2004. Reprinted and electronically reproduced by permission of Pearson Education, Inc. Upper Saddle River, New Jersey.)

of eXtreme Programming that relate to design specifications are (1) how planning, analysis, design, and construction are all fused into a single phase of activity; and (2) its unique way of capturing and presenting system requirements and design specifications. With eXtreme Programming, all phases of the life cycle converge into a series of activities based on the basic processes of coding, testing, listening, and designing.

Under this approach, coding and testing are intimately related parts of the same process. The programmers who write the code also develop the tests. The emphasis is on testing those things that can break or go wrong, not on testing everything. Code is tested very soon after it is written. The overall philosophy behind eXtreme Programming is that the code will be integrated into the system it is being developed for and tested within a few hours after it has been written. If all the tests run successfully, then development proceeds. If not, the code is reworked until the tests are successful.

Another part of eXtreme Programming that makes the code-and-test process work more smoothly is the practice of pair programming. All coding and testing is done by two people working together who write code and develop tests. Beck says that pair programming is not one person typing while the other one watches; rather, the two programmers work together on the problem they are trying to solve, exchanging information and insight and sharing skills. Compared to traditional coding practices, the advantages of pair programming include: (1) more (and better) communication among developers, (2) higher levels of productivity, (3) higher-quality code, and (4) reinforcement of the other practices in eXtreme Programming, such as the code-and-test discipline (Beck & Andres, 2004). Although the eXtreme Programming process has its advantages, just as with any other approach to systems development, it is not for everyone and is not applicable to every project.

OBJECT-ORIENTED ANALYSIS AND DESIGN

Object-oriented analysis and design (OOAD)
Systems development methodologies and techniques based on objects rather than data or processes.

There is no question that **object-oriented analysis and design (OOAD)** is becoming more and more popular (we elaborate on this approach later throughout the book). OOAD is often called the third approach to systems development, after the process-oriented and data-oriented approaches. The object-oriented approach combines data and processes (called *methods*) into single entities called **objects**. Objects usually correspond to the real things an information system deals with, such as customers, suppliers, contracts, and rental agreements. Putting data and processes together in one place recognizes the fact that there are a limited number of operations for any given data structure, and it makes sense even though typical systems development keeps data and processes independent of each other. The goal of OOAD is to make systems elements more reusable, thus improving system quality and the productivity of systems analysis and design.

Another key idea behind object orientation is **inheritance**. Objects are organized into **object classes**, which are groups of objects sharing structural and behavioral characteristics. Inheritance allows the creation of new classes that share some of the characteristics of existing classes. For example, from a class of objects called "person," you can use inheritance to define another class of objects called "customer." Objects of the class "customer" would share certain characteristics with objects of the class "person": They would both have names, addresses, phone numbers, and so on. Because "person" is the more general class and "customer" is more specific, every customer is a person but not every person is a customer.

As you might expect, a computer programming language is required that can create and manipulate objects and classes of objects in order to create object-oriented information systems. Several object-oriented programming languages have been created (e.g., C++, Eiffel, and Java). In fact, object-oriented languages were developed first, and object-oriented analysis and design techniques followed. Because OOAD is still relatively new, there is little consensus or standardization among the many OOAD techniques available. In general, the primary task of object-oriented analysis is identifying objects and defining their structure and behavior and their relationships. The primary tasks of object-oriented design are modeling the details of the objects' behavior and communication with other objects so that system requirements are met, and reexamining and redefining objects to better take advantage of inheritance and other benefits of object orientation.

The object-oriented approach to systems development shares the iterative development approach of the Agile Methodologies. Some say that the current focus on agility in systems development is nothing more than the mainstream acceptance of object-oriented approaches that have been germinating for years, so this similarity should come as no surprise (Fowler, 2003). One of the most popular realizations of the iterative approach for object-oriented development is the **Rational Unified Process (RUP)**, which is based on an iterative, incremental approach to systems development. RUP has four phases: inception, elaboration, construction, and transition (see Figure 1-13).

In the inception phase, analysts define the scope, determine the feasibility of the project, understand user requirements, and prepare a software development plan. In the elaboration phase, analysts detail user requirements and develop a baseline architecture. Analysis and design activities constitute the bulk of the elaboration phase. In the construction phase, the software is actually coded, tested, and documented. In the transition phase, the system is deployed, and the users are trained and supported. As is evident from Figure 1-13, the construction phase is generally the longest and the most resource intensive. The elaboration phase is also long, but less resource intensive. The transition phase is resource intensive but short. The inception phase is short and the least resource intensive. The areas of the rectangles in Figure 1-13 provide an estimate of the overall resources allocated to each phase.

Each phase can be further divided into iterations. The software is developed incrementally as a series of iterations. The inception phase will generally entail a single iteration. The scope and feasibility of the project is determined at this stage. The elaboration phase may have one or two iterations and is generally considered the most critical of the four phases (Kruchten, 2000). The elaboration phase is mainly about systems analysis and design, although other activities are also involved. At the

Object
A structure that encapsulates (or packages) attributes and methods that operate on those attributes. An object is an abstraction of a real-world thing in which data and processes are placed together to model the structure and behavior of the real-world object.

Inheritance
The property that occurs when entity types or object classes are arranged in a hierarchy and each entity type or object class assumes the attributes and methods of its ancestors, that is, those higher up in the hierarchy. Inheritance allows new but related classes to be derived from existing classes.

Object class
A logical grouping of objects that have the same (or similar) attributes and behaviors (methods).

Rational Unified Process (RUP)
An object-oriented systems development methodology. RUP establishes four phases of development: inception, elaboration, construction, and transition. Each phase is organized into a number of separate iterations.

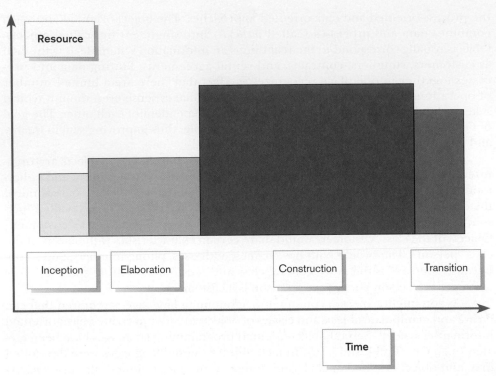

FIGURE 1-13
Phases of OOSAD-based development

end of the elaboration phase, the architecture of the project should have been developed. The architecture includes a vision of the product, an executable demonstration of the critical pieces, a detailed glossary and a preliminary user manual, a detailed construction plan, and a revised estimate of planned expenditures.

Although the construction phase mainly involves coding, which is accomplished in several iterations, revised user requirements could require analysis and design. The components are developed or purchased and used in the code. As each executable is completed, it is tested and integrated. At the end of the construction phase, a beta version of the project is released that should have operational capabilities. The transition phase entails correcting problems, beta testing, user training, and conversion of the product. The transition phase is complete when the project objectives meet the acceptance criteria. Once acceptance criteria have been met, the product can then be released for distribution.

OUR APPROACH TO SYSTEMS DEVELOPMENT

Much of the criticism of the SDLC has been based on abuses of the life cycle perspective, both real and imagined. One of the criticisms, one based in reality, is that reliance on the life cycle approach forced intangible and dynamic processes, such as analysis and design, into timed phases that were doomed to fail (Martin, 1999). Developing software is not like building a bridge, and the same types of engineering processes cannot always be applied (Fowler, 2003), even though viewing software development as a science rather than an art has no doubt resulted in vast improvements in the process and the resulting products. Another criticism with its basis in fact is that life cycle reliance has resulted in massive amounts of process and documentation, much of which seems to exist for its own sake. Too much process and documentation does slow down development, hence the streamlining that underlies RAD and the admonition from Agile Developers that source code is enough documentation. A criticism of the SDLC that is based more on fiction is that all versions of the SDLC are waterfall-like, with no feedback between steps. Another false criticism is that a life cycle approach necessarily limits the involvement of users to the earliest stages of the process. Yet Agile

Methodologies, and eXtreme Programming in particular, advocate an analysis–design–code–test sequence that is a cycle (Figure 1-8), and users can be and are involved in every step of this cycle; thus, cycles in and of themselves do not necessarily limit user involvement.

Whether the criticisms have been based on fact or not, however, it is true that the traditional SDLC waterfall approach has problems, and we applaud the changes taking place in the systems development community. These changes are allowing problems with traditional approaches to be fixed, and without a doubt the result is better software produced more expertly and more quickly.

However, despite the criticisms of a life cycle approach to systems analysis and design, the view of systems analysis and design taking place in a cycle continues to be pervasive, and we think, true as well. There are many types of cycles, from the waterfall to the analysis–design–code–test cycle, and they all capture the iterative nature of systems development. The waterfall approach may be losing its relevance, but the cycle in Figure 1-8 is gaining in popularity, and the analysis–design–code–test cycle is embedded in a larger organizational cycle. Although we typically use the terms *systems analysis and design* and *systems development* interchangeably, perhaps it is better to think about systems analysis and design as being the cycle in Figure 1-8 and systems development as being the larger cycle in Figure 1-2. The analysis–design–code–test cycle largely ignores the organizational planning that precedes it and the organizational installation and systems maintenance that follow, yet they are all important aspects of the larger systems development effort. And to us, the best, clearest way to think about both efforts is in terms of cycles.

Therefore, in this book you will see Figure 1-2 at the beginning of almost every chapter. We will use our SDLC as an organizing principle in this book, with activities and processes arranged according to whether they fit under the category of planning, analysis, design, implementation, or maintenance. To some extent, we will artificially separate activities and processes so that each one can be individually studied and understood. Once individual components are clearly understood, it is easier to see how they fit with other components, and eventually it becomes easy to see the whole. Just as we may artificially separate activities and processes, we may also construct artificial boundaries between phases of the SDLC. Our imposition of boundaries should never be interpreted as hard and fast divisions. In practice, as we have seen with the Agile Methodologies and in the introduction to OOAD, phases and parts of phases may be combined for speed, understanding, and efficiency. Our intent is to introduce the pieces in a logical manner, so that you can understand all the pieces and how to assemble them in the best way for your systems development purposes. Yet the overall structure of the cycle, of iteration, remains throughout. Think of the cycle as an organizing and guiding principle.

SUMMARY

This chapter introduced you to information systems analysis and design, the complex organizational process whereby computer-based information systems are developed and maintained. You read about how systems analysis and design in organizations has changed over the past several decades. You also learned about the basic framework that guides systems analysis and design—the systems development life cycle (SDLC), with its five major phases: planning, analysis, design, implementation, and maintenance. The SDLC life cycle has had its share of criticism, which you read about, and other frameworks have been developed to address the life cycle's problems. These approaches include Rapid Application Design (RAD), which depends on prototyping; computer-aided software engineering (CASE) tools; and the Agile Methodologies, the most famous of which is eXtreme Programming. You were also briefly introduced to object-oriented analysis and design, an approach that is becoming more and more popular. Throughout all of these approaches is the underlying idea of iteration, as manifested in the systems development life cycle and the analysis–design–code–test cycle of the Agile Methodologies.

KEY TERMS

1. Analysis
2. Application software
3. Computer-aided software engineering (CASE) tools
4. Design
5. Implementation
6. Information systems analysis and design

7. Inheritance
8. Logical design
9. Maintenance
10. Object
11. Object class
12. Object-oriented analysis and design (OOAD)
13. Physical design

14. Planning
15. Rapid Application Development (RAD)
16. Rational Unified Process (RUP)
17. Systems analyst
18. Systems development life cycle (SDLC)
19. Systems development methodology

Match each of the key terms above with the definition that best fits it.

_____ The complex organizational process whereby computer–based information systems are developed and maintained.

_____ Computer software designed to support organizational functions or processes.

_____ The organizational role most responsible for the analysis and design of information systems.

_____ A standard process followed in an organization to conduct all the steps necessary to analyze, design, implement, and maintain information systems.

_____ The traditional methodology used to develop, maintain, and replace information systems.

_____ The first phase of the SDLC in which an organization's total information system needs are identified, analyzed, prioritized, and arranged.

_____ The second phase of the SDLC in which system requirements are studied and structured.

_____ The third phase of the SDLC in which the description of the recommended solution is converted into logical and then physical system specifications.

_____ The part of the design phase of the SDLC in which all functional features of the system chosen for development are described independently of any computer platform.

_____ The part of the design phase of the SDLC in which the logical specifications of the system from logical design are transformed into technology-specific details from which all programming and system construction can be accomplished.

_____ The fourth phase of the SDLC in which the information system is coded, tested, installed, and supported in the organization.

_____ The final phase of the SDLC in which an information system is systematically repaired and improved.

_____ Software tools that provide automated support for some portion of the systems development process.

_____ Systems development methodology created to radically decrease the time needed to design and implement information systems. This methodology relies on extensive user involvement, prototyping, integrated CASE tools, and code generators.

_____ Systems development methodologies and techniques based on objects rather than data or processes.

_____ A structure that encapsulates (or packages) attributes and the methods that operate on those attributes. It is an abstraction of a real-world thing in which data and processes are placed together to model the structure and behavior of the real-world object.

_____ The property that occurs when entity types or object classes are arranged in a hierarchy and each entity type or object class assumes the attributes and methods of its ancestors; that is, those higher up in the hierarchy. The property allows new but related classes to be derived from existing classes.

_____ A logical grouping of objects that have the same (or similar) attributes and behaviors (methods).

_____ An object-oriented systems development methodology. This methodology establishes four phases of development, each of which is organized into a number of separate iterations: inception, elaboration, construction, and transition.

REVIEW QUESTIONS

1. What is information systems analysis and design?

2. How has systems analysis and design changed over the past four decades?

3. List and explain the different phases in the SDLC.

4. List and explain some of the problems with the traditional waterfall SDLC.

5. What are CASE tools?

6. Describe each major component of a comprehensive CASE system. Is any component more important than any other?

7. Describe how CASE is used to support each phase of the SDLC.

8. What is RAD?

9. Explain what is meant by Agile Methodologies.

10. What is eXtreme Programming?

11. When would you use Agile methodologies versus an engineering-based approach to development?

12. What is object-oriented analysis and design?

PROBLEMS AND EXERCISES

1. Why is it important to use systems analysis and design methodologies when building a system? Why not just build the system in whatever way appears to be "quick and easy"? What value is provided by using an "engineering" approach?

2. How might prototyping be used as part of the SDLC?

3. Compare Figures 1-2 and 1-3. What similarities and differences do you see?

4. Compare Figures 1-2 and 1-4. Can you match steps in Figure 1-4 with phases in Figure 1-2? How might you explain the differences between the two figures?

5. Compare Figures 1-2 and 1-12. How do they differ? How are they similar? Explain how Figure 1-12 conveys the idea of speed in development.

6. Compare Figures 1-2 and 1-9. How does Figure 1-9 illustrate some of the problems of the traditional waterfall approach that are not illustrated in Figure 1-2? How does converting Figure 1-9 into a circle (like Figure 1-2) fix these problems?

7. Explain how object-oriented analysis and design differs from the traditional approach. Why isn't RUP (Figure 1-13) represented as a cycle? Is that good or bad? Explain your response.

FIELD EXERCISES

1. Choose an organization that you interact with regularly and list as many different "systems" (computer-based or not) as you can that are used to process transactions, provide information to managers and executives, help managers and executives make decisions, aid group decision making, capture knowledge and provide expertise, help design products and/or facilities, and assist people in communicating with each other. Draw a diagram that shows how these systems interact (or should interact) with each other. Are these systems well integrated?

2. Imagine an information system built without using a systems analysis and design methodology and without any thinking about the SDLC. Use your imagination and describe any and all problems that might occur, even if they seem a bit extreme or absurd. (The problems you will describe have probably occurred in one setting or another.)

3. Choose a relatively small organization that is just beginning to use information systems. What types of systems are being used? For what purposes? To what extent are these systems integrated with each other? With systems outside the organization? How are these systems developed and controlled? Who is involved in systems development, use, and control?

4. Interview information systems professionals who use CASE tools and find out how they use the tools throughout the SDLC process. Ask them what advantages and disadvantages they see in using the tools that they do.

5. Go to a CASE tool vendor's website and determine the product's price, functionality, and advantages. Try to find information related to any future plans for the product. If changes are planned, what changes and/or enhancements are planned for future versions? Why are these changes being made?

6. Use the Web to find out more about the Agile Methodologies. Write a report on what the movement toward agility means for the future of systems analysis and design.

7. You may want to keep a personal journal of ideas and observations about systems analysis and design while you are studying this book. Use this journal to record comments you hear, summaries of news stories or professional articles you read, original ideas or hypotheses you create, and questions that require further analysis. Keep your eyes and ears open for anything related to systems analysis and design. Your instructor may ask you to turn in a copy of your journal from time to time in order to provide feedback and reactions. The journal is an unstructured set of personal notes that will supplement your class notes and can stimulate you to think beyond the topics covered within the time limitations of most courses.

REFERENCES

Beck, K., and C. Andres. 2004. *eXtreme Programming eXplained.* Upper Saddle River, NJ: Addison-Wesley.

Boehm, B., and R. Turner. 2004. *Balancing Agility and Discipline.* Boston: Addison-Wesley.

Dorfman, M., and R. M. Thayer (eds). 1997. *Software Engineering.* Los Alamitos, CA: IEEE Computer Society Press.

Fowler, M. 2003. "The New Methodologies." December. Available at *http://martinfowler.com/articles/newMethodology.html.* Accessed February 3, 2009.

Fowler, M., and J. Highsmith. 2001. "The Agile Manifesto." Available at *www.ddj.com/architect/184414755.* Accessed March 19, 2009.

Gibson, M. L., and C. T. Hughes. 1994. *Systems Analysis and Design: A Comprehensive Methodology with CASE.* Danvers, MA: Boyd & Fraser Publishing Company.

Griss, M. 2003. "Ranking IT Productivity Improvement Strategies." Available at *http://martin.griss.com/pub/WPGRISS01.pdf.* Accessed February 3, 2009.

Kay, R. 2002. "QuickStudy: System Development Life Cycle." *Computerworld,* May 14. Available at *www.computerworld.com.* Accessed February 3, 2009.

Kruchten, P. 2000. "From Waterfall to Iterative Lifecycle—A Tough Transition for Project Managers." Rational Software White Paper: TP-173 5/00. Available at *www.ibm.com/developerworks/rational.* Accessed February 3, 2009.

Martin, J. 1991. *Rapid Application Development.* New York: Macmillan Publishing Company.

Martin, R. C. 1999. "Iterative and Incremental Development I." Available at *http://www.objectmentor.com/resources/articles/IIDI.pdf.* Accessed October 12, 2012.

McDougall, P. 2012. "Cloud Will Create 14 Million Jobs, Study Says." InformationWeek, 3/5/12. Available at *http://www.informationweek.com/news/windows/microsoft_news/232601993.* Accessed March 13, 2012.

Mearian, L. 2002. "Merrill Lynch Unit Puts Software Development Process to the Test." *Computerworld,* October 14. Available at *www.computerworld.com.* Accessed February 3, 2009.

Thibodeau, P. 2012. "IT jobs will grow 22% through 2020, says U.S." Computerworld, 3/29/12. Available at *http://www.computerworld.com/s/article/print/9225673/IT_jobs_will_grow_22_through_2020_says_U.S.?taxonomyName=Management+and+Careers&taxonomyId=14.* Accessed March 3, 2012.

The Origins of Software

Learning Objectives

After studying this chapter, you should be able to:

- Explain outsourcing.
- Describe six different sources of software.
- Discuss how to evaluate off-the-shelf software.
- Explain reuse and its role in software development.

Introduction

As you learned in Chapter 1, there was a time, not too long ago, when there were no systems analysts and no symbolic computer programming languages. Yet people still wrote and programmed applications for computers. You read in Chapter 1 about how things have changed over the last 50-plus years. Even though today's systems analyst has dozens of programming languages and development tools to work with, you could easily argue that systems development is even more difficult now than it was 50 years ago. Then, as well as even more recently, one issue was decided for you: If you wanted to write application software, you did it in-house, and you wrote the software from scratch. Today, there are many different sources of software, and many of you reading this book will end up working for firms that produce software rather than in the information systems department of a corporation. But for those of you who do go on to work in a corporate information systems department, the focus is no longer exclusively on in-house development. Instead, the focus will be on where to obtain the many pieces and components that you will combine into the application system you have been asked to create. You and your peers will still write code, mainly to make all the different pieces work together, but more and more of your application software will be written by someone else. Even though you will not write the code, you will still use the basic structure and processes of the systems analysis and design life cycle to build the application systems your organization demands. The organizational process of systems development remains the focus for the rest of the book, but first you need to know more about where software originates in today's development environment.

In this chapter, you will learn about the various sources of software for organizations. The first source considered is outsourcing, in which all or part of an organization's information systems, their development, and their maintenance are given over to another organization. You will then read about six different sources of software: (1) information technology services firms, (2) packaged software providers, (3) vendors of enterprise-wide solution software, (4) cloud computing, (5) open-source software, and (6) the organization itself when it develops software in-house. You will learn of criteria to evaluate software from these different sources. The chapter closes with a discussion of reuse and its impact on software development.

SYSTEMS ACQUISITION

Although there will always be some debate about when and where the first administrative information system was developed, there is general agreement that the first such system in the United Kingdom was developed at J. Lyons & Sons. In the United States, the first administrative information system was General Electric's (GE) payroll system, which was developed in 1954 (Computer History Museum, 2003). At that time, and for many years afterward, obtaining an information system meant one thing only: in-house development. The software industry did not even come into existence until a decade after GE's payroll system was implemented.

Since GE's payroll system was built, in-house development has become a progressively smaller piece of all the systems development work that takes place in and for organizations. Internal corporate information systems departments now spend a smaller and smaller proportion of their time and effort on developing systems from scratch. Companies continue to spend relatively little time and money on traditional software development and maintenance. Instead, they invest in packaged software, open-source software, and outsourced services. Organizations today have many choices when seeking an information system. We will start with a discussion of outsourcing development and operation and then move on to a presentation on the sources of software.

Outsourcing

Outsourcing
The practice of turning over responsibility for some to all of an organization's information systems applications and operations to an outside firm.

If one organization develops or runs a computer application for another organization, that practice is called **outsourcing**. Outsourcing includes a spectrum of working arrangements. At one extreme is having a firm develop and run your application on its computers—all you do is supply input and take output. A common example of such an arrangement is a company that runs payroll applications for clients so that clients do not have to develop an independent in-house payroll system. Instead, they simply provide employee payroll information to the company, and, for a fee, the company returns completed paychecks, payroll accounting reports, and tax and other statements for employees. For many organizations, payroll is a very cost-effective operation when outsourced in this way. Another example of outsourcing would be if you hired a company to run your applications at your site on your computers. In some cases, an organization employing such an arrangement will dissolve some or all of its IS unit and fire all of its information systems employees. Often the company brought in to run the organization's computing will rehire many of the organization's original IS unit employees.

Outsourcing is big business. Some organizations outsource the IT development of many of their IT functions, at a cost of billions of dollars. Most organizations outsource at least some aspect of their information systems activities. One example of the extent of outsourcing is Shell Oil. In 2008, Shell signed outsourcing contracts with EDS, T-Systems, and AT&T worth $3.2 billion USD. In addition, Shell also signed application support deals with IBM, Logica, Wipro, and Accenture. In 2011, Shell outsourced all of its SAP-based human resources and payroll application management services to Accenture. More than 90,000 Shell employees in 60 countries around the world use these systems. Accenture delivers these services through outsourcing centers in India and the Philippines. Individual outsourcing vendors, such as EDS, IBM, and Accenture, typically sign large contracts for their services. These vendors have multiple outsourcing contracts in place with many different firms all over the world.

Why would an organization outsource its information systems operations? As we saw in the payroll example, outsourcing may be cost effective. If a company specializes in running payroll for other companies, it can leverage the economies of scale it achieves from running one stable computer application for many organizations into very low prices. Outsourcing also provides a way for firms to leapfrog their current position in information systems and to turn over development and operations to outside staff who possess knowledge and skills not found internally (Ketler and Willems, 1999). Other reasons for outsourcing include:

- Freeing up internal resources
- Increasing the revenue potential of the organization
- Reducing time to market
- Increasing process efficiencies
- Outsourcing noncore activities

An organization may move to outsourcing and dissolve its entire information processing unit for political reasons as well, such as overcoming operating problems the organization faces in its information systems unit. For example, the city of Grand Rapids, Michigan, hired an outside firm to run its computing center 40 years ago in order to better manage its computing center employees. Union contracts and civil service constraints then in force made it difficult to fire people, so the city brought in a facilities management organization to run its computing operations, and it was able to get rid of problem employees at the same time. As mentioned earlier, another reason for total outsourcing is that an organization's management may feel its core mission does not involve managing an information systems unit and that it might achieve more effective computing by turning over all of its operations to a more experienced, computer-oriented company. Kodak decided in the late 1980s that it was not in the computer applications business and turned over management of its mainframes to IBM and management of its personal computers to Businessland (Applegate and Montealagre, 1991).

Although you have most likely heard about outsourcing in terms of IT jobs from all over the world going to India, it turns out that the global outsourcing marketplace is much more complicated. According to a 2011 report by ATKearney (Petersen et al., 2011), India is the number-1 outsourcing nation, while China is close behind, and Malaysia is third. Despite much turmoil in the overall outsourcing market over the years, the top three rankings have not changed since ATKearney's first report on outsourcing in 2003. Not all of the 2011 top 10 outsourcing countries are located in Asia. Although seven are in Asia, two are in Latin America (Mexico and Chile) and one is in Africa (Egypt). Even the United States is an outsourcing nation, number 18 on the ATKearney list. In fact, Indian outsourcing firms, such as Wipro, Infosys, and Tata Consulting, operate offices in the United States. As Indian firms have become so successful at offshoring, and as currencies have fluctuated, it has become more expensive for firms to contract with Indian companies, so many firms have started to look elsewhere. Many U.S. firms have turned to what is called *nearshoring*, or contracting with companies in Latin American countries. Many of these countries are no more than one time zone away, and they maintain some of the

FIGURE 2-1
Sources of application software

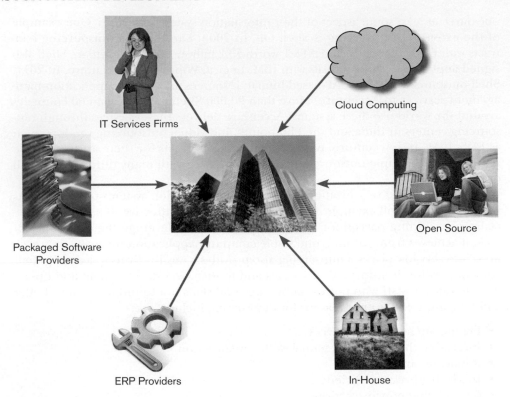

labor cost advantages that are eroding in India (King, 2008a). Mexico is increasingly seen as a complement to India and other offshore locations and is listed as number 6 in the ATKearney 2011 list. It is also becoming more common for firms to distribute their outsourcing work across vendors in several countries at the same time.

Analysts need to be aware of outsourcing as an alternative. When generating alternative system development strategies for a system, as an analyst you should consult organizations in your area that provide outsourcing services. It may well be that at least one such organization has already developed and is running an application very close to what your users are asking for. Perhaps outsourcing the replacement system should be one of your alternatives. Knowing what your system requirements are before you consider outsourcing means that you can carefully assess how well the suppliers of outsourcing services can respond to your needs. However, should you decide not to consider outsourcing, you need to determine whether some software components of your replacement system should be purchased and not built in-house.

Sources of Software

We can group the sources of software into six major categories: information technology services firms, packaged software producers, enterprise-wide solutions, cloud computing vendors, open-source software, and in-house developers (Figure 2-1). These various sources represent points along a continuum of options, with many hybrid combinations along the way.

Information Technology Services Firms   If a company needs an information system but does not have the expertise or the personnel to develop the system in-house, and a suitable off-the-shelf system is not available, the company will likely consult an information technology (IT) services firm. IT services firms help companies develop custom information systems for internal use, or they develop, host, and run applications for customers, or they provide other services. Note in Table 2-1, that many of

TABLE 2-1   Leading Software Firms and Their Development Specializations

Specialization	Example Firms or Websites
IT Services	Accenture
	Computer Sciences Corporation (CSC)
	IBM
	HP
Packaged Software Providers	Intuit
	Microsoft
	Oracle
	SAP AG
	Symantec
Enterprise Software Solutions	Oracle
	SAP AG
Cloud Computing	Amazon.com
	Google
	IBM
	Microsoft
	Salesforce.com
Open Source	SourgeForge.net

the leading software companies in the world specialize in services, which include custom systems development. These firms employ people with expertise in the development of information systems. Their consultants may also have expertise in a given business area. For example, consultants who work with banks understand financial institutions as well as information systems. Consultants use many of the same methodologies, techniques, and tools that companies use to develop systems in-house.

It may surprise you to see IBM listed as a top global software producer; some people still think of it as primarily a hardware company. Yet IBM has been moving away from a reliance on hardware development for many years. The purchase of the IT consulting arm of PricewaterhouseCoopers by IBM in 2002 solidified its move into services and consulting. IBM is also well known for its development of Web server and middleware software. Other leading IT services firms include traditional consulting firms, such as Computer Sciences Corp., Accenture, and HP (Hewlett-Packard). HP, another company formerly focused on hardware, has made the transition to an IT services firm. In 2008, HP bought EDS, continuing its transition to a services-oriented company.

Packaged Software Producers   The growth of the software industry has been phenomenal since its beginnings in the mid-1960s. Some of the largest computer companies in the world are companies that produce software exclusively. A good example is Microsoft, probably the best known software company in the world. Almost 87 percent of Microsoft's revenue comes from its software sales, mostly for its Windows operating systems and its personal productivity software, the Microsoft Office Suite. Also listed in Table 2-1, Oracle is exclusively a software company known primarily for its database software, but Oracle also makes enterprise systems. Another company on the list, SAP, is also a software-focused company that develops enterprise-wide system solutions. You will read more about Oracle and SAP shortly, in the section on enterprise systems.

Software companies develop what are sometimes called *prepackaged* or *off-the-shelf systems*. Microsoft's Project (Figure 2-2) and Intuit's Quicken, QuickPay, and QuickBooks are popular examples of such software. The packaged software development industry serves many market segments. Their software offerings range from general, broad-based packages, such as general ledgers, to very narrow, niche packages, such as software to help manage a day care center. Software companies develop software to run on many different computer platforms, from microcomputers to large mainframes. The companies range in size from just a few people to thousands of employees.

FIGURE 2-2
Microsoft Project
(*Source:* Microsoft Corporation.)

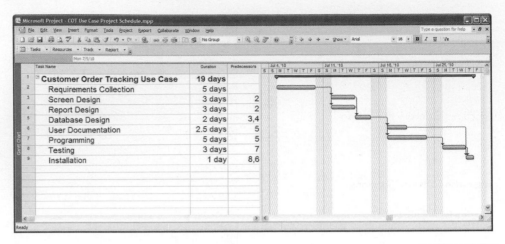

Software companies consult with system users after the initial software design has been completed and an early version of the system has been built. The systems are then tested in actual organizations to determine whether there are any problems or if any improvements can be made. Until testing is completed, the system is not offered for sale to the public.

Some off-the-shelf software systems cannot be modified to meet the specific, individual needs of a particular organization. Such application systems are sometimes called *turnkey systems.* The producer of a turnkey system will make changes to the software only when a substantial number of users ask for a specific change. However, other off-the-shelf application software can be modified or extended, by the producer or by the user, to more closely fit the needs of the organization. Even though many organizations perform similar functions, no two organizations do the same thing in quite the same way. A turnkey system may be good enough for a certain level of performance, but it will never perfectly match the way a given organization does business. A reasonable estimate is that off-the-shelf software can at best meet 70 percent of an organization's needs. Thus, even in the best case, 30 percent of the software system does not match the organization's specifications.

Enterprise Solutions Software   As mentioned in Chapter 1, many firms have chosen complete software solutions, called *enterprise solutions* or **enterprise resource planning (ERP) systems**, to support their operations and business processes. These ERP software solutions consist of a series of integrated modules. Each module supports an individual, traditional business function, such as accounting, distribution, manufacturing, or human resources. The difference between the modules and traditional approaches is that the modules are integrated to focus on business processes rather than on business functional areas. For example, a series of modules will support the entire order entry process, from receiving an order to adjusting inventory to shipping to billing to after-the-sale service. The traditional approach would use different systems in different functional areas of the business, such as a billing system in accounting and an inventory system in the warehouse. Using enterprise software solutions, a firm can integrate all parts of a business process in a unified information system. All aspects of a single transaction occur seamlessly within a single information system, rather than as a series of disjointed, separate systems focused on business functional areas.

Enterprise resource planning (ERP) systems
A system that integrates individual traditional business functions into a series of modules so that a single transaction occurs seamlessly within a single information system rather than several separate systems.

The benefits of the enterprise solutions approach include a single repository of data for all aspects of a business process and the flexibility of the modules. A single repository ensures more consistent and accurate data, as well as less maintenance. The modules are flexible because additional modules can be added as needed once the basic system is in place. Added modules are immediately integrated into the existing system. However, there are disadvantages to

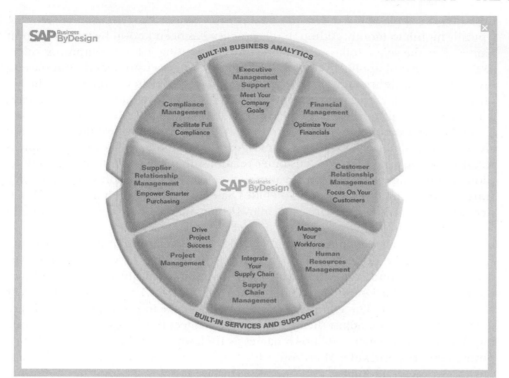

enterprise solutions software. The systems are very complex, so implementation can take a long time to complete. Organizations typically do not have the necessary expertise in house to implement the systems, so they must rely on consultants or employees of the software vendor, which can be very expensive. In some cases, organizations must change how they do business in order to benefit from a migration to enterprise solutions.

Several major vendors provide enterprise solution software. The best known is probably SAP AG, the German firm mentioned earlier, known for its flagship product R/3. SAP stands for Systems, Applications, and Products in Data Processing. SAP AG was founded in 1972, but most of its growth has occurred since 1992. In 2010, SAP was one of the largest suppliers of software in the world. The other major vendor of enterprise solutions is Oracle Corp., a U.S.-based firm, perhaps better known for its database software. Oracle captured a large share of the ERP market through its own financial systems and through the acquisition of other ERP vendors. At the end of 2004, Oracle acquired PeopleSoft, Inc., a U.S. firm founded in 1987. PeopleSoft began with enterprise solutions that focused on human resources management and expanded to cover financials, materials management, distribution, and manufacturing before Oracle acquired them. Just before being purchased by Oracle, PeopleSoft had boosted its corporate strength in 2003 through acquiring another ERP vendor, J.D. Edwards. Together, SAP and Oracle control about 42 percent of the ERP market (Panorama Consulting Group, 2011). The world market for ERP software is projected to reach $67.8 billion USD by 2015 (PRWeb, 2010). Because the higher end of the market has become saturated with ERP systems, most ERP vendors are looking to medium and small businesses for growth. For example, SAP's offering for medium and small businesses is called SAP Business By Design (Figure 2-3).

Cloud Computing Another method for organizations to obtain applications is to rent them or license them from third-party providers who run the applications at remote sites. Users have access to the applications through the Internet or through virtual private networks. The application provider buys, installs, maintains, and

Cloud computing
The provision of computing resources, including applications, over the Internet, so customers do not have to invest in the computing infrastructure needed to run and maintain the resources.

upgrades the applications. Users pay on a per-use basis or they license the software, typically month to month. Although this practice has been known by many different names over the years, today it is called **cloud computing**. Cloud computing refers to the provision of applications over the Internet, where customers do not have to invest in the hardware and software resources needed to run and maintain the applications. You may have seen the Internet referred to as a cloud in other contexts, which comes from how the Internet is depicted on computer network diagrams. A well-known example of cloud computing is Google Drive, where users can store anything they want on Google's servers (Figure 2-4). Another well-known example is Salesforce.com, which provides customer relationship management software online. Cloud computing includes many areas of technology, including software as a service (often referred to as *SaaS*), which includes Salesforce.com, and hardware as a service, which includes Google Drive and allows companies to order server capacity and storage on demand.

Merrill Lynch has predicted that by 2013, 12 percent of the world's corporate computing will be done by cloud computing (King, 2008b). Microsoft and IDC predicted that cloud computing will create 14 million new jobs by 2015 and that the total global market for cloud computing will reach $1.1 trillion USD that year (McDougall, 2012). The companies that are most likely to profit immediately are those that can quickly adjust their product lines to meet the needs of cloud computing. These include such well-known names as IBM, which has built nine cloud computing centers worldwide; Microsoft, which in 2008 announced its Azure platform to support the development and operation of business applications and consumer services on its own servers; and Amazon.com, which provides storage and capacity from its own servers to customers.

As these growth forecasts indicate, taking the cloud computing route has its advantages. The top three reasons for choosing to go with cloud computing, all of which result in benefits for the company, are (1) freeing internal IT staff, (2) gaining access to applications faster than via internal development, and (3) achieving lower-cost access to corporate-quality applications. Especially appealing is the ability to gain access to large and complex systems without having to go through the expensive and time-consuming process of implementing the systems themselves in-house. Getting your computing through a cloud also makes it easier to walk away from an

FIGURE 2-4
An introduction to Google Docs, part of Google Apps
(*Source:* Google.)

unsatisfactory systems solution. Other reasons include cost effectiveness, speed to market, and better performance (Moyle & Kelley, 2011).

IT managers do have some concerns about cloud computing, however. The primary concern is over security. Concerns over security are based on storing company data on machines one does not own and that others can access. In fact, the top two reasons for not using cloud services are concerns about unauthorized access to proprietary information and unauthorized access to customer information (Moyle & Kelley, 2011). Another concern is reliability. Some warn that the cloud is actually a network of networks, and as such, it is vulnerable to unexpected risks due to its complexity (kfc, 2012). Still another concern is compliance with government regulations, such as Sarbanes-Oxley. Experts recommend a three-step process for secure migration to the cloud (Moyle & Kelley, 2011). First, have the company's security experts involved early in the migration process, so that a vendor who understands the company's security and regulatory requirements can be selected. Second, set realistic security requirements. Make sure the requirements are clearly spelled out as part of the bidding process. Third, do an honest risk assessment. Determine which data will be migrated and pay attention to how it will be managed by the cloud vendor. Once migration has occurred, it is important for companies to continue to monitor their data and systems and actively work with the vendor to maintain security.

Open-Source Software   Open-source software is unlike the other types of software you have read about so far. Open-source software is different because it is freely available, not just the final product but the source code itself. It is also different because it is developed by a community of interested people instead of by employees of a particular company. Open-source software performs the same functions as commercial software, such as operating systems, e-mail, database systems, Web browsers, and so on. Some of the most well-known and popular open-source software names are Linux, the operating system; mySQL, a database system; and Firefox, a Web browser. Open source also applies to software components and objects. Open source is developed and maintained by communities of people, and sometimes these communities can be very large. Developers often use common Web resources, such as SourceForge.net, to organize their activities. As of April 2012, SourceForge.net hosted 324,000 projects and had over 3.4 million registered users. There is no question that the open-source movement would not be having the success it enjoys without the availability of the Internet for providing access and organizing development activities.

If the software is free, you might wonder how anybody makes any money by developing open-source software. There are two primary ways companies and individuals can make money with open source: (1) by providing maintenance and other services or (2) by providing one version of the software free and selling a more fully featured version. Some open-source solutions have more of an impact on the software industry than others. Linux, for example, has been very successful in the server software market, where it is estimated to have 18 percent of the market share (Vaughn-Nichols, 2012). In the desktop operating systems market, Linux has about 1 percent market share. Other open-source software products, such as mySQL, have also been successful, and open source's share of the software industry seems destined to continue to grow.

In-House Development   We have talked about several different types of external organizations that serve as sources of software, but in-house development remains an option. In-house development has become a progressively smaller piece of all systems development work that takes place in and for organizations. As you read earlier in this chapter, internal corporate information systems departments now spend a smaller and smaller proportion of their time and effort on developing systems from scratch. In-house development can lead to a larger maintenance burden than other development methods, such as packaged applications. A study by Banker, Davis, and Slaughter found that using a code generator as the basis for in-house development

TABLE 2-2   Comparison of Six Different Sources of Software Components

Producers	When to Go to This Type of Organization for Software	Internal Staffing Requirements
IT services firms	When task requires custom support and system can't be built internally or system needs to be sourced	Internal staff may be needed, depending on application
Packaged software producers	When supported task is generic	Some IS and user staff to define requirements and evaluate packages
Enterprise-wide solutions vendors	For complete systems that cross functional boundaries	Some internal staff necessary but mostly need consultants
Cloud computing	For instant access to an application; when supported task is generic	Few; frees up staff for other IT work
Open-source software	When supported task is generic but cost is an issue	Some IS and user staff to define requirements and evaluate packages
In-house developers	When resources and staff are available and system must be built from scratch	Internal staff necessary though staff size may vary

was related to an increase in maintenance hours, whereas using packaged applications was associated with a decrease in maintenance effort.

Of course, in-house development need not entail development of all of the software that will comprise the total system. Hybrid solutions involving some purchased and some in-house software components are common. If you choose to acquire software from outside sources, this choice is made at the end of the analysis phase. The choice between a package and an external supplier will be determined by your needs, not by what the supplier has to sell. As we will discuss, the results of your analysis study will define the type of product you want to buy and will make working with an external supplier much easier, more productive, and worthwhile. Table 2-2 compares the six different software sources discussed in this section.

Choosing Off-the-Shelf Software

Once you have decided to purchase off-the-shelf software rather than write some or all of the software for your new system, how do you decide what to buy? There are several criteria to consider, and special criteria may arise with each potential software purchase. For each criterion, an explicit comparison should be made between the software package and the process of developing the same application in-house. The most common criteria include the following:

- Cost
- Functionality
- Vendor support
- Viability of vendor
- Flexibility
- Documentation
- Response time
- Ease of installation

These criteria are presented in no particular order. The relative importance of the criteria will vary from project to project and from organization to organization. If you had to choose two criteria that would always be among the most important, those two would probably be vendor viability and vendor support. You do not want to get involved with a vendor that might not be in business tomorrow. Similarly, you do not want to license software from a vendor with a reputation for poor support. How you rank the importance of the remaining criteria will very much depend on the specific situation in which you find yourself.

Cost involves comparing the cost of developing the same system in-house with the cost of purchasing or licensing the software package. You should include a comparison of the cost of purchasing vendor upgrades or annual license fees with the costs you would incur to maintain your own software. Costs for purchasing and developing in-house can be compared based on economic feasibility measures (e.g., a present value can be calculated for the cash flow associated with each alternative).

Functionality refers to the tasks the software can perform and the mandatory, essential, and desired system features. Can the software package perform all or just some of the tasks your users need? If only some, can it perform the necessary core tasks? Note that meeting user requirements occurs at the end of the analysis phase because you cannot evaluate packaged software until user requirements have been gathered and structured. Purchasing application software is not a substitute for conducting the systems analysis phase; rather, purchasing software is part of one design strategy for acquiring the system identified during analysis.

As we said earlier, vendor support refers to whether and how much support the vendor can provide. Support occurs in the form of assistance to install the software, to train user and systems staff on the software, and to provide help as problems arise after installation. Recently, many software companies have significantly reduced the amount of free support they will provide customers, so the cost to use telephone, on-site, fax, or computer bulletin board support facilities should be considered. Related to support is the vendor's viability. You do not want to get stuck with software developed by a vendor that might go out of business soon. This latter point should not be minimized. The software industry is quite dynamic, and innovative application software is created by entrepreneurs working from home offices—the classic cottage industry. Such organizations, even with outstanding software, often do not have the resources or business management ability to stay in business very long. Further, competitive moves by major software firms can render the products of smaller firms outdated or incompatible with operating systems. One software firm we talked to while developing this book was struggling to survive just trying to make its software work on any supposedly IBM-compatible PC (given the infinite combination of video boards, monitors, BIOS chips, and other components). Keeping up with hardware and system software changes may be more than a small firm can handle, and good off-the-shelf application software can be lost.

Flexibility refers to how easy it is for you, or the vendor, to customize the software. If the software is not very flexible, your users may have to adapt the way they work to fit the software. Are they likely to adapt in this manner? Purchased software can be modified in several ways. Sometimes the vendor will be willing to make custom changes for you, if you are willing to pay for the redesign and programming. Some vendors design the software for customization. For example, the software may include several different ways of processing data and, at installation time, the customer chooses which to initiate. Also, displays and reports may be easily redesigned if these modules are written in a fourth-generation language. Reports, forms, and displays may be easily customized using a process whereby your company name and chosen titles for reports, displays, forms, column headings, and so forth, are selected from a table of parameters you provide. You may want to employ some of these same customization techniques for systems developed in-house so that the software can be easily adapted for different business units, product lines, or departments.

Documentation includes the user's manual as well as technical documentation. How understandable and up to date is the documentation? What is the cost for multiple copies, if required? Response time refers to how long it takes the software package to respond to the user's requests in an interactive session. Another measure of time would be how long it takes the software to complete running a job. Finally, ease of installation is a measure of the difficulty of loading the software and making it operational.

Of course, the criteria for software acquisition will vary with the type of system you are acquiring. For example, if you are thinking about licensing an ERP system, you will certainly take all of the prior criteria into account, but you will also want to investigate criteria that are specific to ERP systems. Verville and colleagues (2005) studied organizations that had acquired ERP systems to discover what the critical factors were for success. They found 10 success factors, with 5 related to the acquisition process, and 5 related to the people in the process. They found the acquisition process had to be highly planned and structured, and it had to be rigorous. For the process to be successful, nothing could be overlooked during planning. It was important that two of the five success factors related to the process be completed before ERP vendors were contacted. These two factors were determining all of the system requirements, and establishing the selection and evaluation criteria. These two factors helped the organizations compose clear descriptions of their needs and evaluate bids from vendors. The fifth process-related criterion was obtaining accurate information. Information sources needed to be verified and cross-validated.

The other five success factors dealt with the people involved in the acquisition process. The first factor was clear and unambiguous authority. The person in charge of the process needed to be objective and a strong leader. Second, the composition of the acquisition team was important. The team needed to be diverse, with each member having a particular skill set that was complementary to those of the other team members. Third, it was considered important to approach the relationship with the vendor as a partnership, as opposed to an adversarial or neutral relationship. Given the complexity and cost of ERP systems, members of the acquiring organization would be working with the vendors for several years, so a comfortable working relationship was essential. Fourth, future users of the ERP system were active participants in the acquisition process. Lastly, the fifth success factor related to people in the process was user buy-in. In the companies studied, user buy-in often translated into user acceptance of the system and even enthusiasm and excitement about it.

Validating Purchased Software Information

One way to get all of the information you want about a software package is to collect it from the vendor. Some of this information may be contained in the software documentation and technical marketing literature. Other information can be provided upon request. For example, you can send prospective vendors a questionnaire, asking specific questions about their packages. This may be part of a **request for proposal (RFP)** or a request for quote (RFQ) process your organization requires when major purchases are made. Space does not permit us to discuss the topic of RFPs and RFQs here; you may wish to refer to purchasing and marketing texts if you are unfamiliar with such processes (additional references about RFPs and RFQs are found at the end of this chapter).

There is, of course, no replacement for actually using the software yourself and running it through a series of tests based on your software selection criteria. Remember to test not only the software, but also the documentation, training materials, and even the technical support facilities. One requirement you can place on prospective software vendors as part of the bidding process is that they install (free or at an agreed-upon cost) their software for a limited amount of time on your computers. This way you can determine how their software works in your environment, not in some optimized environment they have for demonstration purposes.

One of the most reliable and insightful sources is other users of the software. Vendors will usually provide a list of customers (remember, they will naturally tell you about satisfied customers, so you may have to probe for a cross section of customers) and people who are willing to be contacted by prospective customers. And here is where your personal network of contacts, developed through professional groups, college friends, trade associations, or local business clubs, can be a resource;

Request for proposal (RFP)
A document provided to vendors that asks them to propose hardware and system software that will meet the requirements of a new system.

do not hesitate to find some contacts on your own. Such current or former customers can provide a depth of insight on the use of a package at their organizations.

To gain a range of opinions about possible packages, you can use independent software testing and abstracting services that periodically evaluate software and collect user opinions. Such surveys are available for a fee either as subscription services or on demand (two popular services are Auerbach Publishers and DataPro); occasionally, unbiased surveys appear in trade publications. Often, however, articles in trade publications, even software reviews, are actually seeded by the software manufacturer and are not unbiased.

If you are comparing several software packages, you can assign scores for each package on each criterion and compare the scores using the quantitative method we demonstrate in Chapter 4 for comparing alternative system design strategies.

REUSE

Reuse is the use of previously written software resources in new applications. Because so many bits and pieces of applications are relatively generic across applications, it seems intuitive that great savings can be achieved in many areas if those generic bits and pieces do not have to be written anew each time they are needed. Reuse should increase programmer productivity because being able to use existing software for some functions means they can perform more work in the same amount of time. Reuse should also decrease development time, minimizing schedule overruns. Because existing pieces of software have already been tested, reusing them should also result in higher-quality software with lower defect rates, decreasing maintenance costs.

Reuse
The use of previously written software resources, especially objects and components, in new applications.

Although reuse can conceivably apply to many different aspects of software, typically it is most commonly applied to two different development technologies: object-oriented and component-based development. You were briefly introduced to object-oriented development in Chapter 1. For example, consider an object class created to model an employee. The object class Employee would contain both the data about employees and the instructions necessary for calculating payroll for a variety of job types. The object class could be used in any application that dealt with employees, but if changes had to be made in calculating payroll for different types of employees, the changes would have to be made only to the object class and not to the various applications that used it. By definition, using the Employee object class in more than one application constitutes reuse.

Component-based development is similar to object-oriented development in that the focus is on creating general-purpose pieces of software that can be used interchangeably in many different programs. Components can be as small as objects or as large as pieces of software that handle single business functions, such as currency conversion. The idea behind component-based development is the assembly of an application from many different components at many different levels of complexity and size. Many vendors are working on developing libraries of components that can be retrieved and assembled, as needed, into desired applications.

Some evidence suggests that reuse can be effective, especially for object classes. For example, one laboratory study found that reuse of object class libraries resulted in increased productivity, reduced defect density, and reduced rework (Basili et al., 1996). For HP, a reuse program resulted in cutting time to market for certain products by a factor of three or more, from 18 months to less than 5 months (Griss, 2003). However, for reuse to work in an organizational setting, many different issues must be addressed. Technical issues include the current lack of a methodology for creating and clearly defining and labeling reusable components for placement in a library and the small number of reusable and reliable software resources currently available. Key organizational issues include the lack of commitment to reuse, as well as the lack of proper training and rewards needed to promote it, the lack of organizational support for institutionalizing reuse, and the difficulty in measuring the economic gains from reuse. Royce (1998) argues that, due to the considerable costs

FIGURE 2-5
Investments necessary to achieve
reusable components

(*Source:* Royce, Walker, *Software Project
Management: A Unified Framework,* 1st ed.,
©1998. Reprinted and Electronically
reproduced by permission of Pearson
Education, Inc. Upper Saddle River, New
Jersey.)

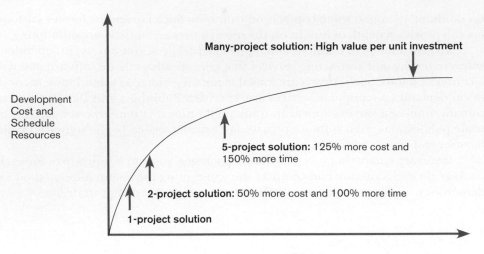

of developing a reusable component, most organizations cannot compete economically with established commercial organizations that focus on selling components as their main line of business. Success depends on being able to leverage the cost of components across a large user and project base (Figure 2-5). There are also key legal and contractual issues concerning the reuse of object classes and components originally used in other applications (Kim and Stohr, 1998).

When an organization's management decides to pursue reuse as a strategy, it is important for the organization to match its approach to reuse with its strategic business goals (Griss, 2003). The benefits of reuse grow as more corporate experience is gained from it, but so do the costs and the amount of resources necessary for reuse to work well. Software reuse has three basic steps: abstraction, storage, and recontextualization (Grinter, 2001). Abstraction involves the design of a reusable piece of software, starting from existing software assets or from scratch. Storage involves making software assets available for others to use. Although it sounds like a simple problem, storage can actually be very challenging. The problem is not simply putting software assets on a shelf; the problem is correctly labeling and cataloging assets so that others can find the ones they want to use. Once an asset has been found, recontextualization becomes important. This involves making the reusable asset understandable to developers who want to use it in their systems. Software is complex, and a software asset developed for a particular system under system-specific circumstances may not at all be the asset it appears to be. What appears to be a generic asset called "customer" may actually be something quite different, depending on the context in which it was developed. It may often appear to be easier to simply build your own assets rather than invest the time and energy it takes to establish a good understanding of software someone else has developed. A key part of a reuse strategy, as mentioned previously, is establishing rewards, incentives, and organizational support for reuse to help make it more worthwhile than developing your own assets.

According to Griss (2003), an organization can take one of four approaches to reuse (Table 2-3). The ad hoc approach to reuse is not really an approach at all, at least from an official organizational perspective. With this approach, individuals are free to find or develop reusable assets on their own, and there are few, if any, organizational rewards for reusing assets. Storage is not an issue because individuals keep track of and distribute their own software assets. For such an ad hoc, individually driven approach, it is difficult to measure any potential benefits to the company.

Another approach to reuse is *facilitated* reuse. With this approach, developers are not required to practice reuse, but they are encouraged to do so. The organization makes available some tools and techniques that enable the development

TABLE 2-3 **Four Approaches to Reuse**

Approach	Reuse Level	Cost	Policies & Procedures
Ad hoc	None to low	Low	None
Facilitated	Low	Low	Developers are encouraged to reuse but are not required to do so.
Managed	Moderate	Moderate	Development, sharing, and adoption of reusable assets are mandated; organizational policies are established for documentation, packaging, and certification.
Designed	High	High	Reuse is mandated; policies are put in place so that reuse effectiveness can be measured; code must be designed for reuse during initial development, regardless of the application it is originally designed for; there may be a corporate office for reuse.

(*Source:* Based on Flashline, Inc. and Griss, 2003.)

and sharing of reusable assets, and one or more employees may be assigned the role of evangelist to publicize and promote the program. Very little is done to track the quality and use of reusable assets, however, and the overall corporate investment is small.

Managed reuse is a more structured, and more expensive, mode of managing software reuse. With managed reuse, the development, sharing, and adoption of reusable assets is mandated. The organization establishes processes and policies for ensuring that reuse is practiced and that the results are measured. The organization also establishes policies and procedures for ensuring the quality of its reusable assets. The focus is on identifying existing assets that can be potentially reused from various sources, including from utility asset libraries that come with operating systems, from companies that sell assets, from the open source community, from internal repositories, from scouring existing legacy code, and so on.

The most expensive and extensive approach to reuse is *designed reuse*. In addition to mandating reuse and measuring its effectiveness, the designed reuse approach takes the extra step of mandating that assets be designed for reuse as they are being designed for specific applications. The focus is more on developing reusable assets than on finding existing assets that might be candidates for reuse. A corporate reuse office may be established to monitor and manage the overall methodology. Under such an approach, as much as 90 percent of software assets may be reused across different applications.

Each approach to reuse has its advantages and disadvantages. No single approach is a silver bullet that will solve the reuse puzzle for all organizations and for all situations. Successful reuse requires an understanding of how reuse fits within larger organizational goals and strategies as well as an understanding of the social and technical world into which the reusable assets must fit.

SUMMARY

As a systems analyst, you must be aware of where you can obtain software that meets some or all of an organization's needs. You can obtain application (and system) software from information technology services firms, packaged software providers, vendors of enterprise-wide solution software, cloud computing vendors, and open-source software providers, as well as from internal systems development resources, including the reuse of existing software components. You can even hire an organization to handle all of your systems development needs, which is called *outsourcing*. You must also know the criteria to use when choosing among off-the-shelf software products. These criteria include cost, functionality, vendor support, vendor viability, flexibility, documentation, response time, and ease of installation. Requests for proposals are one way you can collect more information about system software, its performance, and its costs.

KEY TERMS

1. Cloud computing
2. Enterprise resource planning (ERP) systems
3. Outsourcing
4. Request for proposal (RFP)
5. Reuse

Match each of the key terms above with the definition that best fits it.

_____ The practice of turning over responsibility of some to all of an organization's information systems applications and operations to an outside firm.

_____ A system that integrates individual traditional business functions into a series of modules so that a single transaction occurs seamlessly within a single information system, rather than several separate systems.

_____ A document that is provided to vendors to ask them to propose hardware and system software that will meet the requirements of your new system.

_____ The use of previously written software resources, especially objects and components, in new applications.

_____ The provision of computing resources, including applications, over the Internet so customers do not have to invest in the computing infrastructure needed to run and maintain computing resources.

REVIEW QUESTIONS

1. Describe and compare the various sources of software.
2. How can you decide among various off-the-shelf software options? What criteria should you use?
3. What is an RFP and how do analysts use one to gather information on hardware and system software?
4. What methods can a systems analyst employ to verify vendor claims about a software package?
5. What are ERP systems? What are the benefits and disadvantages of such systems as a design strategy?
6. Explain reuse and its advantages and disadvantages.
7. Compare and contrast the four approaches to reuse.

PROBLEMS AND EXERCISES

1. Research how to prepare an RFP.
2. Review the criteria for selecting off-the-shelf software presented in this chapter. Use your experience and imagination and describe other criteria that are or might be used to select off-the-shelf software in the real world. For each new criterion, explain how use of this criterion might be functional (i.e., it is useful to use this criterion), dysfunctional, or both.
3. In the section on choosing off-the-shelf software, eight criteria are proposed for evaluating alternative packages.

Suppose the choice was between alternative custom software developers rather than prewritten packages. What criteria would be appropriate to select and compare among competing bidders for custom development of an application? Define each of these criteria.

4. How might the project team recommending an ERP design strategy justify its recommendation as compared with other types of design strategies?

FIELD EXERCISES

1. Interview businesspeople who participate in the purchase of off-the-shelf software in their organizations. Review with them the criteria for selecting off-the-shelf software presented in this chapter. Have them prioritize the list of criteria as they are used in their organization and provide an explanation of the rationale for the ranking of each criterion. Ask them to list and describe any other criteria that are used in their organization.
2. Obtain copies of actual RFPs used for information systems developments and/or purchases. If possible, obtain RFPs from

public and private organizations. Find out how they are used. What are the major components of these proposals? Do these proposals seem to be useful? Why or why not? How and why do RFPs from public and private organizations differ?

3. Contact an organization that has or is implementing an integrated ERP application. Why did it choose this design strategy? How has it managed this development project differently from prior large projects? What organizational changes have occurred due to this design strategy? How long did the implementation last and why?

REFERENCES

Applegate, L. M., and R. Montealegre. 1991. "Eastman Kodak Company: Managing Information Systems Through Strategic Alliances." Harvard Business School case 9-192-030. Cambridge, MA: President and Fellows of Harvard College.

Banker, R. D., G. B. Davis, and S. A. Slaughter. 1998. "Software Development Practices, Software Complexity, and Software Maintenance Performance: A Field Study." *Management Science* 44(4): 433–50.

Basili, V. R., L. C. Briand, and W. L. Melo. 1996. "How Reuse Influences Productivity in Object-Oriented Systems." *Communications of the ACM* 39(10): 104–16.

Computer History Museum. 2003. Timeline of Computer History. Available at *www.computerhistory.org*. Accessed February 14, 2009.

Grinter, R. E. 2001. "From Local to Global Coordination: Lessons from Software Reuse." In *Proceedings of Group '01*, 144–153. Boulder, CO: Association for Computing Machinery SIGGROUP.

Griss, M. 2003. "Reuse Comes in Several Flavors." Flashline white paper. Available at *www.flashline.com*. Accessed February 10, 2009.

Ketler, K., and J. R. Willems. 1999. "A Study of the Outsourcing Decision: Preliminary Results." *Proceedings of SIGCPR '99*, New Orleans, LA: 182–89.

kfc. 2012. "The Hidden Risk of a Meltdown in the Cloud." *Technology Review*. Available at *http://www.technologyreview.com/printer_friendly_blog.aspx?id=27642*. Accessed April 17, 2012.

Kim, Y., and E. A. Stohr. 1998. "Software Reuse: Survey and Research Directions." *Journal of MIS* 14(4): 113–47.

King, L. 2010. "Shell standardising operations in $3bn saving drive." *Computerworld UK*, 2/4/10. Available at *http://www.computerworlduk.com/news/applications/18669/*. Accessed January 11, 2012.

King, R. 2008a. "The New Economics of Outsourcing." *BusinessWeek* online, April 7, 2008. Available at *http://www.businessweek.com/stories/2008-04-07/the-new-economics-of-outsourcingbusinessweek-business-news-stock-market-and-financial-advice* Accessed August 25, 2012.

King, R. 2008b. "How Cloud Computing is Changing the World." *BusinessWeek* online, August 4, 2008. Available at *www.businessweek.com/print/technology/content/aug2008/tc2008082_445669.htm*. Accessed January 29, 2009.

McDougall, P. 2012. "Cloud Will Create 14 Million Jobs, Study Says." InformationWeek, 3/5/12. Available at *http://www.informationweek.com/news/windows/microsoft_news/232601993*. Accessed March 13, 2012.

Moyle, E., and D. Kelley. 2011. "Cloud Security: Understand the Risks Before You Make the Move." InformationWeek Analytics. Available at *http://analytics.informationweek.com*. Accessed April 14, 2012.

Nguyen, A. 2012. "Shell completes transition of HR application services to Accenture." *Computerworld UK*, 1/5/12. From *http://www.computerworld.com/s/article/print/9223159/Shell_completes_transition_of_HR_application_services_to_Accenture*. Accessed January 11, 2012.

Panorama Consulting Group. 2011. "2011 Guide to ERP Systems and Vendors." Available at *http://panorama-consulting.com/*. Accessed May 11, 2012.

Petersen, E., J. Gott, and S. King. 2011. "Offshoring Opportunities Amid Economic Turbulence." Chicago, IL: ATKearney Inc. Available at *http://www.atkearney.com/index.php/Publications/global-services-location-index-gsli.html*. Accessed 4/17/11.

PRWeb. 2010. "Global ERP Software Market to Reach US$67.8 Billion by 2015, According to New Report by Global Industry Analysts, Inc." Available at *http://www.prweb.com/printer/3772994.htm*. Accessed April 17, 2012.

Royce, W. 1998. *Software Project Management: A Unified Framework*. Boston: Addison-Wesley.

Vaughn-Nichols, S. J. 2012. "Linux servers keep growing, Windows & Unix keep shrinking." 3/15/12. Available at *http://www.zdnet.com/blog/open-source/linux-servers-keep-growing-windows-unix-keep-shrinking/10616*. Accessed May 11, 2012.

Verville, J., C. Bernadas, and A. Halingten. 2005. "So You're Thinking of Buying an ERP? Ten Critical Factors for Successful Acquisition." *Journal of Enterprise Information Management* 18(6), 665–77.

PETRIE ELECTRONICS

Chapter 2: The Origins of Software

Jim Watanabe looked around his new office. He couldn't believe that he was the assistant director of information technology at Petrie Electronics, his favorite consumer electronics retail store. He always bought his new DVDs and video games for his Xbox 360 at Petrie. In fact, he had bought his BluRay player and his Xbox 360 at Petrie, along with his surround sound system and his 40-inch flat-screen HD LED TV. And now he worked there, too. The employee discount was a nice perk[1] of his new job, but he was also glad that his technical and people skills were finally recognized by the people at Petrie. He had worked for five years at Broadway Entertainment Company as a senior systems analyst, and it was clear that he was not going to be promoted there. He was really glad he had put his resume up on Monster.com and that now he had a bigger salary and a great job with more responsibility at Petrie.

Petrie Electronics had started as a single electronics store in 1984 in San Diego, California. The store was started by Jacob Rosenstein in a strip mall. It was named after Rob Petrie, the TV writer played by Dick Van Dyke in the TV show of the same name. Rosenstein always liked that show. When he had grown the store to a chain of 13 stores in the Southern California area, it was too much for Rosenstein to handle. He sold out in 1992, for a handsome profit, to the Matsutoya Corporation, a huge Japanese conglomerate that saw the chain of stores as a place to sell its many consumer electronics goods in the United States.

Matsutoya aggressively expanded the chain to 218 stores nationwide by the time they sold the chain in 2002, for a handsome profit, to Sam and Harry's, a maker and seller of ice cream. Sam and Harry's was looking for a way to diversify and invest the considerable cash they had make creating and selling ice cream, with flavors named after actors and actresses, like their best-selling Lime Neeson and Jim Carrey-mel. Sam and Harry's brought in professional management to run the chain, and since they bought it, they had added 15 more stores, including one in Mexico and three in Canada. Even though they originally wanted to move the headquarters to their home base state of Delaware, Sam and Harry decided to keep Petrie headquartered in San Diego.

The company had made some smart moves and had done well, Jim knew, but he also knew that competition was fierce. Petrie competitors included big electronics retail chains like BestBuy. In California, Fry's was a ferocious competitor. Other major players in the arena included the electronics departments of huge chains like Walmart and Target and online vendors like Amazon.com. Jim knew that part of his job in IT was to help the company grow and prosper and beat the competition—or at least survive.

Just then, as Jim was trying to decide if he needed a bigger TV, Ella Whinston, the CEO at Petrie, walked into his office. "How's it going, Jim? Joe keeping you busy?" Joe was Joe Swanson, Jim's boss, the director of IT. Joe was away for the week, at a meeting in Tucson, Arizona. Jim quickly pulled his feet off his desk.

"Hi, Ella. Oh, yeah, Joe keeps me busy. I've got to get through the entire corporate strategic IT plan before he gets back—he's going to quiz me—and then there's the new help desk training we are going to start next week."

"I didn't know we had a strategic IT plan," Ella teased. "Anyway, what I came in here for is to give you some good news. I want you to be the project manager for a project that is crucial to our corporate survival."

"Me?" Jim said, "But I just got here."

"Who better than you? You have a different perspective, new ideas. You aren't chained down by the past and by the Petrie way of doing things, like the rest of us. Not that it matters, since you don't have a choice. Joe and I both agree that you are the best person for the job."

"So," Jim asked, "What's the project about?"

"Well," Ella began, "the executive team has decided that the number-1 priority we have right now is to not only survive but to thrive and to prosper, and the way to do that is to develop closer relationships with our customers. The other person on the executive team who is even more excited about this than me is John (John Smith, the head of marketing). We want to attract new customers, like all of our competitors. But also like our competitors, we want to keep our customers for life, kind of like a frequent flier program, but better. Better for us and for our loyal customers. And we want to reward most the customers who spend the most. We are calling the project 'No Customer Escapes.'"

"I hope that's only an internal name," Jim joked. "Seriously, I can see how something like this would be good for Petrie, and I can see how IT would play an important, no, crucial role in making something like this happen. So, what's the next step in getting the project approved?"

Case Questions

1. How do information systems projects get started in organizations?
2. How are organizational information systems related to company strategy? How does strategy affect the information systems a company develops and uses?
3. Research customer loyalty programs in retail firms. How common are they? What are their primary features?
4. What do you think Jim's next step would be? Why?
5. Why would a systems analyst new to a company be a good choice to lead an important systems development effort?

[1]perquisite

Managing the Information Systems Project

Learning Objectives

After studying this chapter, you should be able to:

- Explain the process of managing an information systems project.

- Describe the skills required to be an effective project manager.

- List and describe the skills and activities of a project manager during project initiation, project planning, project execution, and project closedown.

- Explain what is meant by critical path scheduling and describe the process of creating Gantt charts and network diagrams.

- Explain how commercial project management software packages can be used to assist in representing and managing project schedules.

Introduction

In Chapters 1 and 2, we introduced the five phases of the systems development life cycle (SDLC) and explained how an information systems project moves through those five phases, in some cases repeatedly. In this chapter, we focus on the systems analyst's role as project manager of an information systems project. Throughout the SDLC, the project manager is responsible for initiating, planning, executing, and closing down the systems development project. Project management is arguably the most important aspect of an information systems development project. Effective project management helps to ensure that systems development projects meet customer expectations and are delivered within budget and time constraints.

Today, there is a shift in the types of projects most firms are undertaking, which makes project management much more difficult and even more critical to project success (Fuller et al., 2008; King, 2003; Kirsch, 2000). For example, in the past, organizations focused much of their development on very large, custom-designed, stand-alone applications. Today, much of the systems development effort in organizations focuses on implementing packaged software such as enterprise resource planning (ERP) and data warehousing systems. Existing legacy applications are also being modified so that business-to-business transactions can occur seamlessly over the Internet. New Web-based interfaces are being added to existing legacy systems so that a broader range of users, often distributed globally, can access corporate information and systems. Additionally, software developed by global outsourcing partners that must be integrated into an organization's existing portfolio of applications is now common practice (King, 2003). Working with vendors to supply applications, with customers or suppliers to integrate systems, or with a broader and more diverse user community requires that project managers be highly skilled. Consequently, it is important that you gain an understanding of the project management process; this will become a critical skill for your future success.

In this chapter, we focus on the systems analyst's role in managing information systems projects and will refer to this role as the *project manager*. The first section will provide the background for Pine Valley Furniture (PVF), a manufacturing company that we will visit throughout the remainder of the book. We will then provide you with an understanding of the project manager's role and the project management process. The discussion then turns to techniques for reporting project plans using Gantt charts and network diagrams. The chapter will conclude with a discussion of the use of commercially available project management software that can be used to assist with a wide variety of project management activities.

PINE VALLEY FURNITURE COMPANY BACKGROUND

PVF manufactures high-quality wood furniture and distributes it to retail stores throughout the United States. Its product lines include dinette sets, stereo cabinets, wall units, living room furniture, and bedroom furniture. In the early 1980s, PVF's founder, Alex Schuster, started to make and sell custom furniture in his garage. Alex managed invoices and kept track of customers by using file folders and a filing cabinet. By 1984, business expanded and Alex had to rent a warehouse and hire a part-time bookkeeper. PVF's product line had multiplied, sales volume had doubled, and staff had increased to 50 employees. By 1990, PVF moved into its third and present location. Due to the added complexity of the company's operations, Alex reorganized the company into the following functional areas:

- Manufacturing, which was further subdivided into three separate functions— Fabrication, Assembling, and Finishing
- Sales
- Orders
- Accounting
- Purchasing

Alex and the heads of the functional areas established manual information systems, such as accounting ledgers and file folders, which worked well for a time. Eventually, however, PVF selected and installed a network server to automate invoicing, accounts receivable, and inventory control applications.

When the applications were first computerized, each separate application had its own individual data files tailored to the needs of each functional area. As is typical in such situations, the applications closely resembled the manual systems on which they were based. Three computer applications at PVF are depicted in Figure 3-1: order filling, invoicing, and payroll. In the late 1990s, PVF formed a task force to study the possibility of moving to a database approach. After a preliminary study, management decided to convert its information systems to such an approach. The company upgraded its network server and implemented a centralized database management system. Today, PVF has successfully deployed an integrated, company-wide database and has converted its applications to work with the database. However, PVF is continuing to grow at a rapid rate, putting pressure on its current application systems.

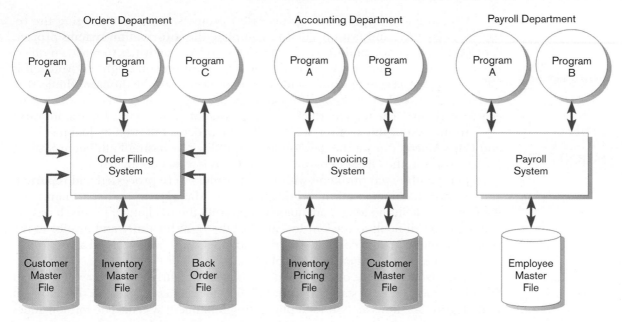

FIGURE 3-1
Three computer applications at PVF:
order filling, invoicing, and payroll
(*Source:* Hoffer, Ramesh, and Topi, 2011.)

The computer-based applications at PVF support its business processes. When customers order furniture, their orders must be processed appropriately: Furniture must be built and shipped to the right customer and the right invoice mailed to the right address. Employees have to be paid for their work. Given these tasks, most of PVF's computer-based applications are located in the accounting and financial areas. The applications include order filling, invoicing, accounts receivable, inventory control, accounts payable, payroll, and general ledger. At one time, each application had its own data files. For example, there was a customer master file, an inventory master file, a back-order file, an inventory pricing file, and an employee master file. The order filling system used data from three files: customer master, inventory master, and back order. Today, however, all systems are designed and integrated through a company-wide database in which data are organized around entities, or subjects, such as customers, invoices, and orders.

PVF, like many firms, decided to develop its application software in-house; that is, it hired staff and bought the computer hardware and software necessary to build application software suited to its own needs. (Other methods used to obtain application software were discussed in Chapter 2.) Although PVF continues to grow at a rapid rate, market conditions are becoming extremely competitive, especially with the advent of the Internet and the Web. Let's see how a project manager plays a key role in developing a new information system for PVF.

MANAGING THE INFORMATION SYSTEMS PROJECT

Project management is an important aspect of the development of information systems and a critical skill for a systems analyst. The focus of project management is to ensure that systems development projects meet customer expectations and are delivered within budget and time constraints.

The **project manager** is a systems analyst with a diverse set of skills—management, leadership, technical, conflict management, and customer relationship—who is responsible for initiating, planning, executing, and closing down a project. As a project manager, your environment is one of continual change and problem solving. In some organizations, the project manager is a very experienced systems analyst, whereas in others, both junior and senior analysts are expected to take on this role, managing parts of a project or actively supporting a more senior colleague who assumes the project manager role. Understanding the project management process is a critical skill for your future success.

Project manager
A systems analyst with a diverse set of skills—management, leadership, technical, conflict management, and customer relationship—who is responsible for initiating, planning, executing, and closing down a project.

Project
A planned undertaking of related activities to reach an objective that has a beginning and an end.

Deliverable
An end product of an SDLC phase.

Creating and implementing successful projects requires managing the resources, activities, and tasks needed to complete the information systems project. A **project** is a planned undertaking of a series of related activities to reach an objective that has a beginning and an end. The first question you might ask yourself is "Where do projects come from?" and, after considering all the different things that you could be asked to work on within an organization, "How do I know which projects to work on?" The ways in which each organization answers these questions vary.

In the rest of this section, we describe the process followed by Juanita Lopez and Chris Martin during the development of PVF's Purchasing Fulfillment System. Juanita works in the Order department, and Chris is a systems analyst.

Juanita observed problems with the way orders were processed and reported: Sales growth had increased the workload for the Manufacturing department, and the current systems no longer adequately supported the tracking of orders. It was becoming more difficult to track orders and get the right furniture and invoice to the right customers. Juanita contacted Chris, and together they developed a system that corrected these Order department problems.

The first **deliverable**, or end product, produced by Chris and Juanita was a System Service Request (SSR), a standard form PVF uses for requesting systems development work. Figure 3-2 shows an SSR for a purchasing fulfillment system. The form includes the name and contact information of the person requesting the system, a statement of the problem, and the name and contact information of the liaison and sponsor.

Pine Valley Furniture
System Service Request

REQUESTED BY ___Juanita Lopez___   DATE ___October 1, 2014___

DEPARTMENT ___Purchasing, Manufacturing Support___

LOCATION ___Headquarters, 1-322___

CONTACT ___Tel: 4-3267   FAX: 4-3270   e-mail: jlopez___

TYPE OF REQUEST   URGENCY

[X]   New System   [   ]   Immediate – Operations are impaired or
                                                            opportunity lost
[   ]   System Enhancement   [   ]   Problems exist, but can be worked around
[   ]   System Error Correction   [X]   Business losses can be tolerated until new
                                                            system installed

PROBLEM STATEMENT

Sales growth at PVF has caused greater volume of work for the manufacturing support unit within Purchasing. Further, more concentration on customer service has reduced manufacturing lead times, which puts more pressure on purchasing activities. In addition, cost-cutting measures force Purchasing to be more aggressive in negotiating terms with vendors, improving delivery times, and lowering our investments in inventory. The current modest systems support for Manufacturing/Purchasing is not responsive to these new business conditions. Data are not available, information cannot be summarized, supplier orders cannot be adequately tracked, and commodity buying is not well supported. PVF is spending too much on raw materials and not being responsive to manufacturing needs.

SERVICE REQUEST

I request a thorough analysis of our current operations with the intent to design and build a completely new information system. This system should handle all purchasing transactions, support display and reporting of critical purchasing data, and assist purchasing agents in commodity buying.

IS LIAISON ___Chris Martin (Tel: 4-6204   FAX: 4-6200   e-mail: cmartin)___

SPONSOR ___Sal Divario, Director, Purchasing___

- TO BE COMPLETED BY SYSTEMS PRIORITY BOARD -
[   ]   Request approved   Assigned to _____
                                Start date _____
[   ]   Recommend revision
[   ]   Suggest user development
[   ]   Reject for reason _____
        _____
        _____

FIGURE 3-2
System Service Request for Purchasing Fulfillment System with name and contact information of the person requesting the system, a statement of the problem, and the name and contact information of the liaison and sponsor

FIGURE 3-3
A graphical view of the five steps
followed during the project initiation of
the Purchasing Fulfillment System

1. Juanita observed problems with the existing purchasing system.

2. Juanita contacted Chris within the IS development group to initiate a System Service Request.

3. SSR was reviewed and approved by Systems Priority Board.

4. Steering committee was assigned to oversee project.

5. Detailed project plan was developed and executed.

This request was then evaluated by the Systems Priority Board of PVF. Because all organizations have limited time and resources, not all requests can be approved. The board evaluates development requests in relation to the business problems or opportunities the system will solve or create; it also considers how the proposed project fits within the organization's information systems architecture and long-range development plans. The review board selects those projects that best meet overall organizational objectives (we learn more about organizational objectives in Chapter 4). In the case of the Purchasing Fulfillment System request, the board found merit in the request and approved a more detailed feasibility study. A **feasibility study**, which is conducted by the project manager, involves determining if the information system makes sense for the organization from an economic and operational standpoint. The study takes place before the system is constructed. Figure 3-3 is a graphical view of the steps followed during the project initiation of the Purchasing Fulfillment System.

Feasibility study
A study that determines if the proposed information system makes sense for the organization from an economic and operational standpoint.

In summary, systems development projects are undertaken for two primary reasons: to take advantage of business opportunities and to solve business problems. Taking advantage of an opportunity might mean providing an innovative service to customers through the creation of a new system. For example, PVF may want to create a website so that customers can easily access its catalog and place orders at any time. Solving a business problem could involve modifying the way an existing system processes data so that more accurate or timely information is provided to users. For example, a company such as PVF may create a password-protected intranet site that contains important announcements and budget information. Of course, projects are not always initiated for the aforementioned rational reasons (taking advantage of business opportunities or solving business problems). For example, in some instances, organizations and government undertake projects to spend resources, to attain or pad budgets, to keep people busy, or to help train people and develop their skills. Our focus in this chapter is not on how and why organizations identify projects but on the management of projects once they have been identified.

Once a potential project has been identified, an organization must determine the resources required for its completion. This is done by analyzing the scope of the project and determining the probability of successful completion. After getting

FIGURE 3-4
A project manager juggles numerous
activities

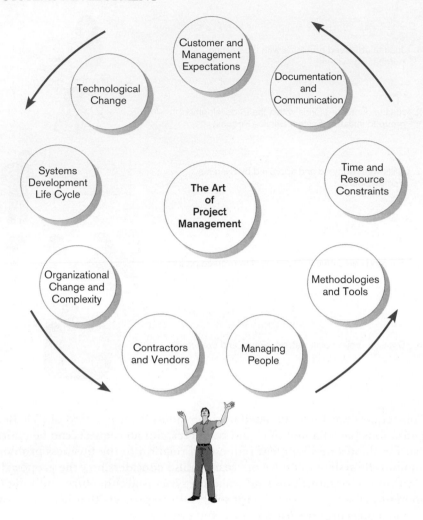

this information, the organization can then determine whether taking advantage of an opportunity or solving a particular problem is feasible within time and resource constraints. If deemed feasible, a more detailed project analysis is then conducted. As you will see, the ability to determine the size, scope, and resource requirements of a project is just one of the many skills that a project manager must possess. A project manager is often thought of as a juggler keeping aloft many balls, which reflect the various aspects of a project's development, as depicted in Figure 3-4.

To successfully orchestrate the construction of a complex information system, a project manager must have interpersonal, leadership, and technical skills. Table 3-1 lists the project manager's common skills and activities. Note that many of the skills are related to personnel or general management, not simply technical skills. Table 3-1 shows that not only does an effective project manager have varied skills, but he or she is also the most instrumental person to the successful completion of any project.

The remainder of this chapter will focus on the **project management** process, which involves four phases:

1. Initiating the project
2. Planning the project
3. Executing the project
4. Closing down the project

Several activities must be performed during each of these four phases. Following this formal project management process greatly increases the likelihood of project success.

Project management
A controlled process of initiating, planning, executing, and closing down a project.

TABLE 3-1 **Common Activities and Skills of a Project Manager**

Activity	Description	Skill
Leadership	Influencing the activities of others toward the attainment of a common goal through the use of intelligence, personality, and abilities	Communication; liaison between management, users, and developers; assigning activities; monitoring progress
Management	Getting projects completed through the effective utilization of resources	Defining and sequencing activities; communicating expectations; assigning resources to activities; monitoring outcomes
Customer relations	Working closely with customers to ensure that project deliverables meet expectations	Interpreting system requests and specifications; site preparation and user training; contact point for customers
Technical problem solving	Designing and sequencing activities to attain project goals	Interpreting system requests and specifications; defining activities and their sequence; making trade-offs between alternative solutions; designing solutions to problems
Conflict management	Managing conflict within a project team to assure that conflict is not too high or too low	Problem solving; smoothing out personality differences; compromising; goal setting
Team management	Managing the project team for effective team performance	Communication within and between teams; peer evaluations; conflict resolution; team building; self-management
Risk and change management	Identifying, assessing, and managing the risks and day-to-day changes that occur during a project	Environmental scanning; risk and opportunity identification and assessment; forecasting; resource redeployment

Initiating a Project

During **project initiation**, the project manager performs several activities to assess the size, scope, and complexity of the project and to establish procedures to support subsequent activities. Depending on the project, some initiation activities may be unnecessary and some may be very involved. The types of activities you will perform when initiating a project are summarized in Figure 3-5 and described next.

1. *Establishing the project initiation team.* This activity involves organizing an initial core of project team members to assist in accomplishing the project initiation activities (Verma, 1996; 1997). For example, during the Purchasing Fulfillment System project at PVF, Chris Martin was assigned to support the Purchasing department. It is a PVF policy that all initiation teams consist of at least one user representative, in this case Juanita Lopez, and one member of the information systems (IS) development group. Therefore, the project initiation team consisted of Chris and Juanita; Chris was the project manager.

2. *Establishing a relationship with the customer.* A thorough understanding of your customer builds stronger partnerships and higher levels of trust. At PVF, management has tried to foster strong working relationships between business units (like

PINE VALLEY FURNITURE

Project initiation
The first phase of the project management process in which activities are performed to assess the size, scope, and complexity of the project and to establish procedures to support later project activities.

Project Initiation

1. Establishing the Project Initiation Team
2. Establishing a Relationship with the Customer
3. Establishing the Project Initiation Plan
4. Establishing Management Procedures
5. Establishing the Project Management Environment and Project Workbook
6. Developing the Project Charter

FIGURE 3-5
Six project initiation activities

Purchasing) and the IS development group by assigning a specific individual to work as a liaison between both groups. Because Chris had been assigned to the Purchasing unit for some time, he was already aware of some of the problems with the existing purchasing systems. PVF's policy of assigning specific individuals to each business unit helped to ensure that both Chris and Juanita were comfortable working together prior to the initiation of the project. Many organizations use a similar mechanism for establishing relationships with customers.

3. *Establishing the project initiation plan.* This step defines the activities required to organize the initiation team while it is working to define the goals and scope of the project (Abdel-Hamid et al., 1999). Chris's role was to help Juanita translate her business requirements into a written request for an improved information system. This required the collection, analysis, organization, and transformation of a lot of information. Because Chris and Juanita were already familiar with each other and their roles within a development project, they next needed to define when and how they would communicate, define deliverables and project steps, and set deadlines. Their initiation plan included agendas for several meetings. These steps eventually led to the creation of their SSR form.

4. *Establishing management procedures.* Successful projects require the development of effective management procedures. Within PVF, many of these management procedures had been established as standard operating procedures by the Systems Priority Board and the IS development group. For example, all project development work is charged back to the functional unit requesting the work. In other organizations, each project may have unique procedures tailored to its needs. Yet, in general, when establishing procedures, you are concerned with developing team communication and reporting procedures, job assignments and roles, project change procedures, and determining how project funding and billing will be handled. It was fortunate for Chris and Juanita that most of these procedures were already established at PVF, allowing them to move on to other project activities.

5. *Establishing the project management environment and project workbook.* The focus of this activity is to collect and organize the tools that you will use while managing the project and to construct the project workbook. Diagrams, charts, and system descriptions provide much of the project workbook contents. Thus, the project workbook serves as a repository for all project correspondence, inputs, outputs, deliverables, procedures, and standards established by the project team (Rettig, 1990; Dinsmore and Cabanis-Brewin, 2006). The **project workbook** can be stored as an online electronic document or in a large three-ring binder. The project workbook is used by all team members and is useful for project audits, orientation of new team members, communication with management and customers, identifying future projects, and performing post-project reviews. The establishment and diligent recording of all project information in the workbook are two of the most important activities you will perform as project manager.

Project workbook
An online or hard-copy repository for all project correspondence, inputs, outputs, deliverables, procedures, and standards that is used for performing project audits, orienting new team members, communicating with management and customers, identifying future projects, and performing post-project reviews.

Figure 3-6 shows the project workbook for the Purchasing Fulfillment System. It consists of both a large hard-copy binder and electronic information where the system data dictionary, a catalog of data stored in the database, and diagrams are stored. For this system, all project documents can fit into a single binder. It is not unusual, though, for project documentation to be spread over several binders. As more information is captured and recorded electronically, however, fewer hard-copy binders may be needed. Many project teams keep their project workbooks on the Web. A website can be created so that all project members can easily access all project documents. This website can be a simple repository of documents or an elaborate site with password protection and security levels. The best feature of using the Web as your repository is that it enables project members and customers to review a project's status and all related information continually.

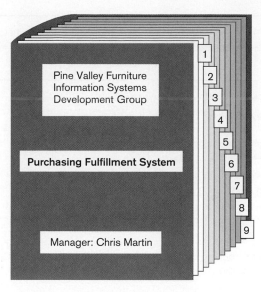

1. Project overview
2. Initiation plan and SSR
3. Project scope and risks
4. Management procedures
5. Data descriptions
6. Process descriptions
7. Team correspondence
8. Project Charter
9. Project schedule

Online copies of data dictionary, diagrams, schedules, reports, etc.

FIGURE 3-6
The project workbook for the Purchase Fulfillment System project contains nine key documents in both hard copy and electronic form

6. *Developing the project charter.* The **project charter** is a short (typically one page), high-level document prepared for the customer that describes what the project will deliver and outlines many of the key elements of the project. A project charter can vary in the amount of detail it contains, but often includes the following elements:

- Project title and date of authorization
- Project manager name and contact information
- Customer name and contact information
- Projected start and completion dates
- Key stakeholders, project role, and responsibilities
- Project objectives and description
- Key assumptions or approach
- Signature section for key stakeholders

> **Project charter**
> A short document prepared for the customer during project initiation that describes what the project will deliver and outlines generally at a high level all work required to complete the project.

The project charter ensures that both you and your customer gain a common understanding of the project. It is also a very useful communication tool; it helps to announce to the organization that a particular project has been chosen for development. A sample project charter is shown in Figure 3-7.

Project initiation is complete once these six activities have been performed. Before moving on to the next phase of the project, the work performed during project initiation is reviewed at a meeting attended by management, customers, and project team members. An outcome of this meeting is a decision to continue, modify, or abandon the project. In the case of the Purchasing Fulfillment System project at PVF, the board accepted the SSR and selected a project steering committee to monitor project progress and to provide guidance to the team members during subsequent activities. If the scope of the project is modified, it may be necessary to return to project initiation activities and collect additional information. Once a decision is made to continue the project, a much more detailed project plan is developed during the project planning phase.

Planning the Project

The next step in the project management process is **project planning**. Research has found a positive relationship between effective project planning and positive project outcomes (Guinan et al., 1998; Kirsch, 2000). Project planning involves defining

> **Project planning**
> The second phase of the project management process that focuses on defining clear, discrete activities and the work needed to complete each activity within a single project.

FIGURE 3-7
A project charter for a proposed
information systems project

Pine Valley Furniture
Project Charter

Prepared: November 2, 2014

Project Name:	Customer Tracking System
Project Manager:	Jim Woo (jwoo@pvf.com)

Customer:	Marketing
Project Sponsor:	Jackie Judson (jjudson@pvf.com)
Project Start/End (projected):	10/2/14–2/1/15

Project Overview:

This project will implement a customer tracking system for the marketing department. The purpose of this system is to automate the . . . to save employee time, reduce errors, have more timely information . . . .

Objectives:

- Minimize data entry errors
- Provide more timely information
- . . .

Key Assumptions:

- System will be built in house
- Interface will be a Web browser
- System will access customer database
- . . .

Stakeholders and Responsibilities:

Stakeholder	Role	Responsibility	Signatures
Jackie Judson	VP Marketing	Project Vision, Resources	*Jackie Judson*
Alex Datta	CIO	Monitoring, Resources	*Alex Datta*
Jim Woo	Project Manager	Planning, Monitoring, Executing Project	*Jim Woo*
James Jordan	Director of Sales	System Functionality	*James Jordan*
Mary Shide	VP Human Resources	Staff Assignments	*Mary Shide*

---

clear, discrete activities and the work needed to complete each activity within a single project. It often requires you to make numerous assumptions about the availability of resources such as hardware, software, and personnel. It is much easier to plan nearer-term activities than those occurring in the future. In actual fact, you often have to construct longer-term plans that are more general in scope and nearer-term plans that are more detailed. The repetitive nature of the project management process requires that plans be constantly monitored throughout the project and periodically updated (usually after each phase), based upon the most recent information.

Figure 3-8 illustrates the principle that nearer-term plans are typically more specific and firmer than longer-term plans. For example, it is virtually impossible to rigorously plan activities late in the project without first completing the earlier activities. Also, the outcome of activities performed earlier in the project is likely to affect later activities. This means that it is very difficult, and very likely inefficient, to try to plan detailed solutions for activities that will occur far into the future.

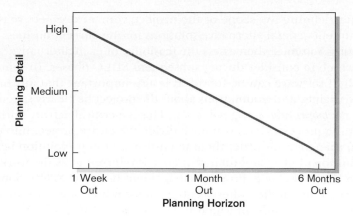

FIGURE 3-8
Level of project planning detail should be high in the short term, with less detail as time goes on

As with the project initiation process, varied and numerous activities must be performed during project planning. For example, during the Purchasing Fulfillment System project, Chris and Juanita developed a 10-page plan. However, project plans for very large systems may be several hundred pages in length. The types of activities that you can perform during project planning are summarized in Figure 3-9 and are described in the following list:

1. *Describing project scope, alternatives, and feasibility.* The purpose of this activity is to understand the content and complexity of the project. Within PVF's systems development methodology, one of the first meetings must focus on defining a project's scope. Although project scope information was not included in the SSR developed by Chris and Juanita, it was important that both shared the same vision for the project before moving too far along. During this activity, you should reach agreement on the following questions:

   - What problem or opportunity does the project address?
   - What are the quantifiable results to be achieved?
   - What needs to be done?
   - How will success be measured?
   - How will we know when we are finished?

Project Planning

1. Describing Project Scope, Alternatives, and Feasibility
2. Dividing the Project into Manageable Tasks
3. Estimating Resources and Creating a Resource Plan
4. Developing a Preliminary Schedule
5. Developing a Communication Plan
6. Determining Project Standards and Procedures
7. Identifying and Assessing Risk
8. Creating a Preliminary Budget
9. Developing a Project Scope Statement
10. Setting a Baseline Project Plan

FIGURE 3-9
Ten project planning activities

After defining the scope of the project, your next objective is to identify and document general alternative solutions for the current business problem or opportunity. You must then assess the feasibility of each alternative solution and choose which to consider during subsequent SDLC phases. In some instances, off-the-shelf software can be found. It is also important that any unique problems, constraints, and assumptions about the project be clearly stated.

2. *Dividing the project into manageable tasks.* This is a critical activity during the project planning process. Here, you must divide the entire project into manageable tasks and then logically order them to ensure a smooth evolution between tasks. The definition of tasks and their sequence is referred to as the **work breakdown structure** (PMBOK, 2008; Project Management Institute, 2002). Some tasks may be performed in parallel, whereas others must follow one another sequentially. Task sequence depends on which tasks produce deliverables needed in other tasks, when critical resources are available, the constraints placed on the project by the client, and the process outlined in the SDLC.

For example, suppose that you are working on a new development project and need to collect system requirements by interviewing users of the new system and reviewing reports they currently use to do their job. A work breakdown for these activities is represented in a Gantt chart in Figure 3-10. A **Gantt chart** is a graphical representation of a project that shows each task as a horizontal bar whose length is proportional to its time for completion. Different colors, shades, or shapes can be used to highlight each kind of task. For example, those activities on the critical path (defined later) may be in red and a summary task could have a special bar. Note that the black horizontal bars—rows 1, 2, and 6 in Figure 3-10—represent summary tasks. Planned versus actual times or progress for an activity can be compared by parallel bars of different colors, shades, or shapes. Gantt charts do not (typically) show how tasks must be ordered (precedence), but simply show when an activity should begin and end. In Figure 3-10, the task duration is shown in the second column by days, "d," and necessary prior tasks are noted in the third column as predecessors. Most project management software tools support a broad range of task durations, including minutes, hours, days, weeks, and months. As you will learn in later chapters, the SDLC consists of several phases that you will need to break down into activities. Creating a work breakdown structure requires that you decompose phases into activities—summary tasks—and activities into specific tasks. For example, Figure 3-10 shows that the activity Interviewing consists of three tasks: design interview form, schedule appointments, and conduct interviews.

Defining tasks in too much detail will make the management of the project unnecessarily complex. You will develop the skill of discovering the optimal level of detail for representing tasks through experience. For example, it may be very difficult

Work breakdown structure
The process of dividing the project into manageable tasks and logically ordering them to ensure a smooth evolution between tasks.

Gantt chart
A graphical representation of a project that shows each task as a horizontal bar whose length is proportional to its time for completion.

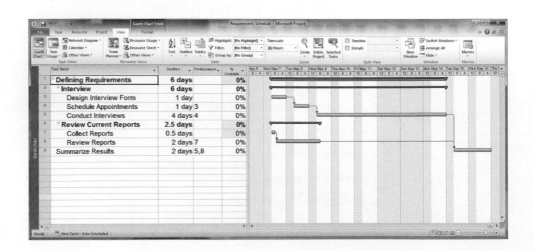

FIGURE 3-10
Gantt chart showing project tasks, duration times for those tasks, and predecessors

to list tasks that require less than one hour of time to complete in a final work break-down structure. Alternatively, choosing tasks that are too large in scope (e.g., several weeks long) will not provide you with a clear sense of the status of the project or of the interdependencies between tasks. What are the characteristics of a "task"? A task

- Can be done by one person or a well-defined group
- Has a single and identifiable deliverable (The task is, however, the process of creating the deliverable.)
- Has a known method or technique
- Has well-accepted predecessor and successor steps
- Is measurable so that the percentage completed can be determined

3. *Estimating resources and creating a resource plan.* The goal of this activity is to estimate resource requirements for each project activity and to use this information to create a project resource plan. The resource plan helps assemble and deploy resources in the most effective manner. For example, you would not want to bring additional programmers onto the project at a rate faster than you could prepare work for them. Project managers use a variety of tools to assist in making estimates of project size and costs. The most widely used method is called CO-COMO (**CO**nstructive **CO**st **MO**del), which uses parameters that were derived from prior projects of differing complexity (Boehm et al., 2000). COCOMO uses these different parameters to predict human resource requirements for basic, intermediate, and very complex systems (see Figure 3-11).

People are the most important, and expensive, part of project resource planning. Project time estimates for task completion and overall system quality are significantly influenced by the assignment of people to tasks. It is important to give people tasks that allow them to learn new skills. It is equally important to make sure that project members are not "in over their heads" or working on a task that is not well suited to their skills. Resource estimates may need to be revised based upon the skills of the actual person (or people) assigned to a particular activity. Figure 3-12 indicates the relative programming speed versus the

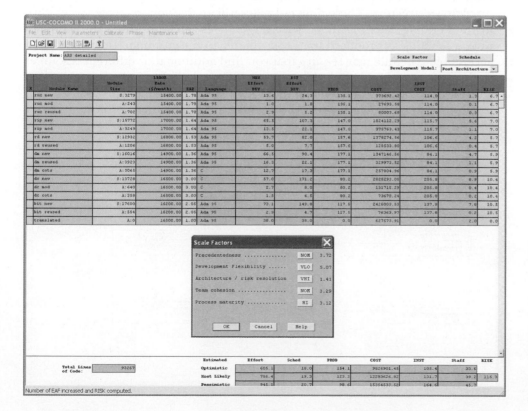

FIGURE 3-11

COCOMO is used by many project managers to estimate project resources

(*Source:* Boehm, Barry; W.; Abts, Chris; Brown, A. Winson; Chulani, Sunitya; Clark, Bradford K.; Horowitz Ellis; Madachy, Ray; Reifer, Donald J.; Steece, Bert, *Software Cost Estimation with Cocomo II,* 1st Ed., © 2000. Reprinted and electronically reproduced by permission of Pearson Education, Inc. Upper Saddle River, New Jersey.)

FIGURE 3-12
Trade-offs between the quality of the program code versus the speed of programming

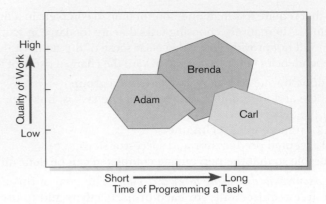

relative programming quality of three programmers. The figure suggests that Carl should not be assigned tasks in which completion time is critical and that Brenda should be assigned tasks in which high quality is most vital.

One approach to assigning tasks is to assign a single task type (or only a few task types) to each worker for the duration of the project. For example, you could assign one worker to create all computer displays and another to create all system reports. Such specialization ensures that both workers become efficient at their own particular tasks. A worker may become bored if the task is too specialized or is long in duration, so you could assign workers to a wider variety of tasks. However, this approach may lead to lowered task efficiency. A middle ground would be to make assignments with a balance of both specialization and task variety. Assignments depend on the size of the development project and the skills of the project team. Regardless of the manner in which you assign tasks, make sure that each team member works only on one task at a time. Exceptions to this rule can occur when a task occupies only a small portion of a team member's time (e.g., testing the programs developed by another team member) or during an emergency.

4. *Developing a preliminary schedule.* During this activity, you use the information on tasks and resource availability to assign time estimates to each activity in the work breakdown structure. These time estimates will enable you to create target starting and ending dates for the project. Target dates can be revisited and modified until a schedule is produced that is acceptable to the customer. Determining an acceptable schedule may require that you find additional or different resources or that the scope of the project be changed. The schedule may be represented as a Gantt chart, as illustrated in Figure 3-10, or as a network diagram, as illustrated in Figure 3-13. A **network diagram** is a graphical depiction of project tasks and their interrelationships. As with a Gantt chart, each type of task can be highlighted by different features on the network diagram. The distinguishing feature

Network diagram

A diagram that depicts project tasks and their interrelationships.

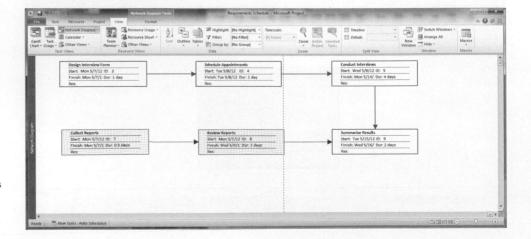

FIGURE 3-13

A network diagram illustrates tasks with rectangles (or ovals) and the relationships and sequences of those activities with arrows

(*Source:* Microsoft Corporation.)

of a network diagram is that the ordering of tasks is shown by connecting tasks—depicted as rectangles or ovals—with their predecessor and successor tasks. However, the relative size of a node (representing a task) or a gap between nodes does not imply the task's duration. Only the individual task items are drawn on a network diagram, which is why the summary tasks 1, 2, and 6—the black bars—from Figure 3-10 are not shown in Figure 3-13. We describe both of these charts later in this chapter.

5. *Developing a communication plan.* The goal of this activity is to outline the communication procedures among management, project team members, and the customer. The communication plan includes when and how written and oral reports will be provided by the team, how team members will coordinate work, what messages will be sent to announce the project to interested parties, and what kinds of information will be shared with vendors and external contractors involved with the project. It is important that free and open communication occur among all parties with respect to proprietary information and confidentiality with the customer (Fuller et al., 2008; Kettelhut, 1991; Kirsch, 2000; Verma, 1996). When developing a communication plan, numerous questions must be answered in order to assure that the plan is comprehensive and complete, including:

- Who are the stakeholders for this project?
- What information does each stakeholder need?
- When, and at what interval, does this information need to be produced?
- What sources will be used to gather and generate this information?
- Who will collect, store, and verify the accuracy of this information?
- Who will organize and package this information into a document?
- Who will be the contact person for each stakeholder should any questions arise?
- What format will be used to package this information?
- What communication medium will be most effective for delivering this information to the stakeholder?

Once these questions are answered for each stakeholder, a comprehensive communication plan can be developed. In this plan, a summary of communication documents, work assignments, schedules, and distribution methods will be outlined. Additionally, a project communication matrix can be developed that provides a summary of the overall communication plan (see Figure 3-14). This matrix can be easily shared among team members, and verified by stakeholders outside the project team, so that the right people are getting the right information at the right time, and in the right format.

FIGURE 3-14
The project communication matrix provides a high-level summary of the communication plan

Stakeholder	Document	Format	Team Contact	Date Due
Team Members	Project Status Report	Project Intranet	Juan Kim	First Monday of Month
Management Supervisor	Project Status Report	Hard Copy	Juan Kim	First Monday of Month
User Group	Project Status Report	Hard Copy	James Kim	First Monday of Month
Internal IT Staff	Project Status Report	E-Mail	Jackie James	First Monday of Month
IT Manager	Project Status Report	Hard Copy	Juan Jeremy	First Monday of Month
Contract Programmers	Software Specifications	E-Mail/Project Intranet	Jordan Kim	October 1, 2014
Training Subcontractor	Implementation and Training Plan	Hard Copy	Jordan James	January 7, 2015

6. *Determining project standards and procedures.* During this activity, you will specify how various deliverables are produced and tested by you and your project team. For example, the team must decide on which tools to use, how the standard SDLC might be modified, which SDLC methods will be used, documentation styles (e.g., type fonts and margins for user manuals), how team members will report the status of their assigned activities, and terminology. Setting project standards and procedures for work acceptance is a way to ensure the development of a high-quality system. Also, it is much easier to train new team members when clear standards are in place. Organizational standards for project management and conduct make the determination of individual project standards easier and the interchange or sharing of personnel among different projects feasible.

7. *Identifying and assessing risk.* The goal of this activity is to identify sources of project risk and to estimate the consequences of those risks (Wideman, 1992). Risks might arise from the use of new technology, prospective users' resistance to change, availability of critical resources, competitive reactions or changes in regulatory actions due to the construction of a system, or team member inexperience with technology or the business area. You should continually try to identify and assess project risk.

   The identification of project risks is required to develop PVF's new Purchasing Fulfillment System. Chris and Juanita met to identify and describe possible negative outcomes of the project and their probabilities of occurrence. Although we list the identification of risks and the outline of project scope as two discrete activities, they are highly related and often concurrently discussed.

8. *Creating a preliminary budget.* During this phase, you need to create a preliminary budget that outlines the planned expenses and revenues associated with your project. The project justification will demonstrate that the benefits are worth these costs. Figure 3-15 shows a cost-benefit analysis for a new development

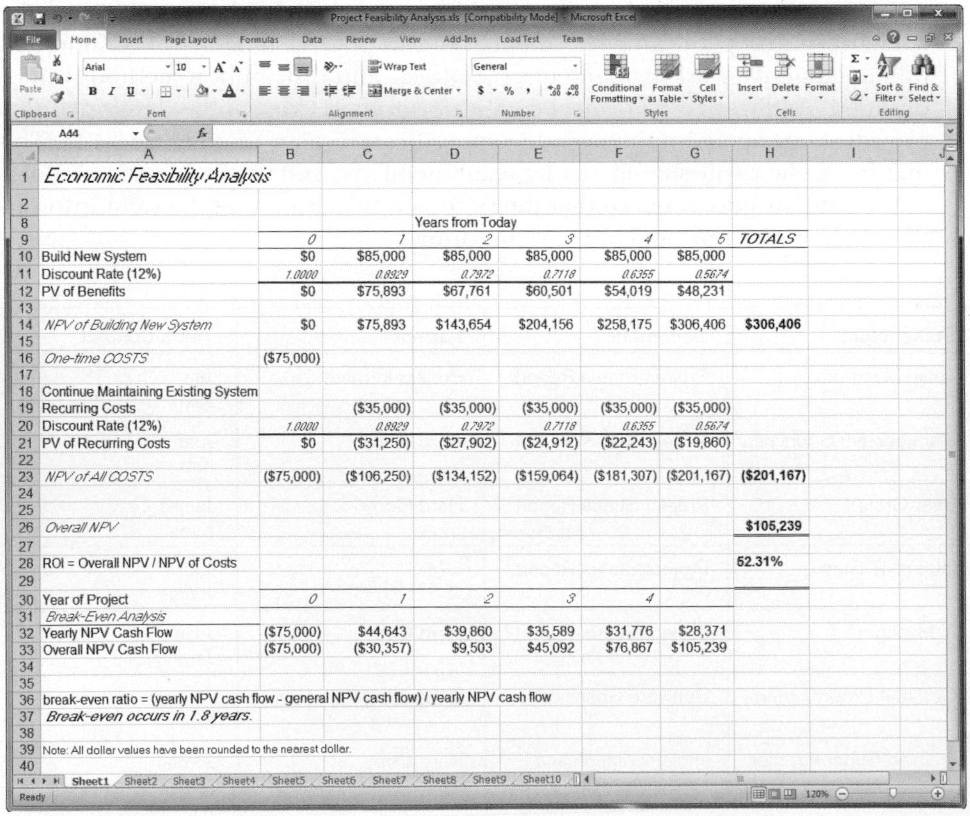

FIGURE 3-15
A financial cost and benefit analysis for a systems development project
(*Source:* Microsoft Corporation.)

	A	B	C	D	E	F	G	H
1	Economic Feasibility Analysis							
2								
8				Years from Today				
9			0	1	2	3	4	5	TOTALS
10	Build New System	$0	$85,000	$85,000	$85,000	$85,000	$85,000	
11	Discount Rate (12%)	1.0000	0.8929	0.7972	0.7118	0.6355	0.5674	
12	PV of Benefits	$0	$75,893	$67,761	$60,501	$54,019	$48,231	
13								
14	NPV of Building New System	$0	$75,893	$143,654	$204,156	$258,175	$306,406	$306,406
15								
16	One-time COSTS	($75,000)						
17								
18	Continue Maintaining Existing System							
19	Recurring Costs		($35,000)	($35,000)	($35,000)	($35,000)	($35,000)	
20	Discount Rate (12%)	1.0000	0.8929	0.7972	0.7118	0.6355	0.5674	
21	PV of Recurring Costs	$0	($31,250)	($27,902)	($24,912)	($22,243)	($19,860)	
22								
23	NPV of All COSTS	($75,000)	($106,250)	($134,152)	($159,064)	($181,307)	($201,167)	($201,167)
24								
25								
26	Overall NPV							$105,239
27								
28	ROI = Overall NPV / NPV of Costs							52.31%
29								
30	Year of Project		0	1	2	3	4	
31	Break-Even Analysis							
32	Yearly NPV Cash Flow	($75,000)	$44,643	$39,860	$35,589	$31,776	$28,371	
33	Overall NPV Cash Flow	($75,000)	($30,357)	$9,503	$45,092	$76,867	$105,239	
34								
35								
36	break-even ratio = (yearly NPV cash flow - general NPV cash flow) / yearly NPV cash flow							
37	Break-even occurs in 1.8 years.							
38								
39	Note: All dollar values have been rounded to the nearest dollar.							
40								

project. This analysis shows net present value calculations of the project's benefits and costs as well as a return on investment and cash flow analysis. We discuss project budgets fully in Chapter 5.

9. *Developing a Project Scope Statement.* An important activity that occurs near the end of the project planning phase is the development of the Project Scope Statement. Developed primarily for the customer, this document outlines work that will be done and clearly describes what the project will deliver. The Project Scope Statement is useful to make sure that you, the customer, and other project team members have a clear understanding of the intended project size, duration, and outcomes.

10. *Setting a Baseline Project Plan.* Once all of the prior project planning activities have been completed, you will be able to develop a Baseline Project Plan. This baseline plan provides an estimate of the project's tasks and resource requirements and is used to guide the next project phase—execution. As new information is acquired during project execution, the baseline plan will continue to be updated.

At the end of the project planning phase, a review of the Baseline Project Plan is conducted to double-check all information in the plan. As with the project initiation phase, it may be necessary to modify the plan, which means returning to prior project planning activities before proceeding. As with the Purchasing Fulfillment System project, you may submit the plan and make a brief presentation to the project steering committee at this time. The committee can endorse the plan, ask for modifications, or determine that it is not wise to continue the project as currently outlined.

Executing the Project

Project execution puts the Baseline Project Plan into action. Within the context of the SDLC, project execution occurs primarily during the analysis, design, and implementation phases. During the development of the Purchasing Fulfillment System, Chris Martin was responsible for five key activities during project execution. These activities are summarized in Figure 3-16 and described in the remainder of this section:

1. *Executing the Baseline Project Plan.* As project manager, you oversee the execution of the baseline plan. This means that you initiate the execution of project activities, acquire and assign resources, orient and train new team members, keep the project on schedule, and ensure the quality of project deliverables. This is a formidable task, but a task made much easier through the use of sound project management techniques. For example, as tasks are completed during a project, they can be "marked" as completed on the project schedule. In Figure 3-17, tasks 3 and 7 are marked as completed by showing 100 percent in the "% Complete" column; task 8 is marked as being partially completed. Members of the project team will come and go. You are responsible for initiating new team members by providing them with the resources they need and helping them assimilate into the team. You may want to plan social events, regular team project status meetings, team-level reviews of project deliverables, and other group events to mold the group into an effective team.

Project execution
The third phase of the project management process in which the plans created in the prior phases (project initiation and planning) are put into action.

PINE
VALLEY
FURNITURE

Project Execution

1. Executing the Baseline Project Plan

2. Monitoring Project Progress against the Baseline Project Plan

3. Managing Changes to the Baseline Project Plan

4. Maintaining the Project Workbook

5. Communicating the Project Status

FIGURE 3-16
Five project execution activities

FIGURE 3-17
Gantt chart with tasks 3 and
7 completed, and task 8 partially
completed

(*Source:* Microsoft Corporation.)

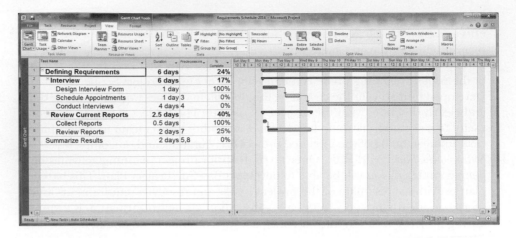

2. *Monitoring project progress against the Baseline Project Plan.* While you execute the Baseline Project Plan, you should monitor your progress. If the project gets ahead of (or behind) schedule, you may have to adjust resources, activities, and budgets. Monitoring project activities can result in modifications to the current plan. Measuring the time and effort expended on each activity will help you improve the accuracy of estimations for future projects. It is possible, with project schedule charts such as Gantt charts, to show progress against a plan, and it is easy with network diagrams to understand the ramifications of delays in an activity. Monitoring progress also means that the team leader must evaluate and appraise each team member, occasionally change work assignments or request changes in personnel, and provide feedback to the employee's supervisor.

3. *Managing changes to the Baseline Project Plan.* You will encounter pressure to make changes to the baseline plan. At PVF, policies dictate that only approved changes to the project specification can be made and all changes must be reflected in the baseline plan and project workbook, including all charts. For example, if Juanita suggests a significant change to the existing design of the Purchasing Fulfillment System, a formal change request must be approved by the steering committee. The request should explain why changes are desired and describe all possible effects on prior and subsequent activities, project resources, and the overall project schedule. Chris would have to help Juanita develop such a request. This information allows the project steering committee to more easily evaluate the costs and benefits of a significant midcourse change.

In addition to changes occurring through formal request, changes may also occur from events outside your control. In fact, numerous events may initiate a change to the Baseline Project Plan, including the following possibilities:

- A slipped completion date for an activity
- A bungled activity that must be redone
- The identification of a new activity that becomes evident later in the project
- An unforeseen change in personnel due to sickness, resignation, or termination

When an event occurs that delays the completion of an activity, you typically have two choices: devise a way to get back on schedule or revise the plan. Devising a way to get back on schedule is the preferred approach because no changes to the plan will have to be made. The ability to head off and smoothly work around problems is a critical skill that you need to master.

As you will see later in this chapter, project schedule charts are very helpful in assessing the impact of change. Using such charts, you can quickly see if the completion time of other activities will be affected by changes in the duration of a given activity or if the whole project completion date will change. Often you will have to find a way to rearrange the activities because the ultimate project completion date may be rather fixed. There may be a penalty to the organization (even legal action) if the expected completion date is not met.

4. *Maintaining the project workbook.* As in all project phases, maintaining complete records of all project events is necessary. The workbook provides the documentation new team members require to assimilate project tasks quickly. It explains why design decisions were made and is a primary source of information for producing all project reports.

5. *Communicating the project status.* The project manager is responsible for keeping all stakeholders—system developers, managers, and customers—abreast of the project status. In other words, communicating the project status focuses on the *execution* of the project communication plan and the response to any ad hoc information requests by stakeholders. A broad variety of methods can be used to distribute information, each with strengths and weakness. Some methods are easier for the information sender, but more difficult or less convenient for the receiver. With the maturing digital networks and the Internet, more and more digital communication is being exchanged. Procedures for communicating project activities vary from formal meetings to informal hallway discussions. Some procedures are useful for informing others of the project's status, others are better for resolving issues, and still others are better for keeping permanent records of information and events. Two types of information are routinely exchanged throughout the project: *work results*—the outcomes of the various tasks and activities that are performed to complete the project—and the *project plan*—the formal comprehensive document that is used to execute the project; it contains numerous items including the project charter, project schedule, budgets, risk plan, and so on. Table 3-2 lists numerous communication procedures, their level of formality, and their most likely use. Whichever procedure you use, frequent communication helps to ensure project success (Kettelhut, 1991; Kirsch, 2000; Verma, 1996).

This section outlined your role as the project manager during the execution of the Baseline Project Plan. The ease with which the project can be managed is significantly influenced by the quality of prior project phases. If you develop a high-quality project plan, it is much more likely that the project will be successfully executed. The next section describes your role during project closedown, the final phase of the project management process.

Closing Down the Project

The focus of **project closedown** is to bring the project to an end. Projects can conclude with a natural or unnatural termination. A natural termination occurs when the requirements of the project have been met—the project has been completed and is a

Project closedown
The final phase of the project management process that focuses on bringing a project to an end.

TABLE 3-2 Project Team Communication Methods

Procedure	Formality	Use
Project workbook	High	Inform
		Permanent record
Meetings	Medium to high	Resolve issues
Seminars and workshops	Low to medium	Inform
Project newsletters	Medium to high	Inform
Status reports	High	Inform
Specification documents	High	Inform
		Permanent record
Minutes of meetings	High	Inform
		Permanent record
Bulletin boards	Low	Inform
Memos	Medium to high	Inform
Brown bag lunches	Low	Inform
Hallway discussions	Low	Inform
		Resolve issues

success. An unnatural termination occurs when the project is stopped before completion (Keil et al., 2000). Several events can cause an unnatural termination of a project. For example, it may be learned that the assumption used to guide the project proved to be false, that the performance of the systems or development group was somehow inadequate, or that the requirements are no longer relevant or valid in the customer's business environment. The most likely reasons for the unnatural termination of a project relate to running out of time or money, or both. Regardless of the project termination outcome, several activities must be performed: closing down the project, conducting postproject reviews, and closing the customer contract. Within the context of the SDLC, project closedown occurs after the implementation phase. The system maintenance phase typically represents an ongoing series of projects, each of which must be individually managed. Figure 3-18 summarizes the project closedown activities that are described more fully in the remainder of this section:

1. *Closing down the project.* During closedown, you perform several diverse activities. For example, if you have several team members working with you, project completion may signify job and assignment changes for some members. You will likely be required to assess each team member and provide an appraisal for personnel files and salary determination. You may also want to provide career advice to team members, write letters to superiors praising special accomplishments of team members, and send thank-you letters to those who helped but were not team members. As project manager, you must be prepared to handle possible negative personnel issues such as job termination, especially if the project was not successful. When closing down the project, it is also important to notify all interested parties that the project has been completed and to finalize all project documentation and financial records so that a final review of the project can be conducted. You should also celebrate the accomplishments of the team. Some teams will hold a party, and each team member may receive memorabilia (e.g., a T-shirt with "I survived the X project"). The goal is to celebrate the team's effort to bring a difficult task to a successful conclusion.

2. *Conducting postproject reviews.* Once you have closed down the project, final reviews of the project should be conducted with management and customers. The objective of these reviews is to determine the strengths and weaknesses of project deliverables, the processes used to create them, and the project management process. It is important that everyone understands what went right and what went wrong in order to improve the process for the next project. Remember, the systems development methodology adopted by an organization is a living guideline that must undergo continual improvement.

3. *Closing the customer contract.* The focus of this final activity is to ensure that all contractual terms of the project have been met. A project governed by a contractual agreement is typically not completed until agreed to by both parties, often in writing. Thus, it is imperative that you gain agreement from your customer that all contractual obligations have been met and that further work is either their responsibility or covered under another SSR or contract.

Closedown is a very important activity. A project is not complete until it is closed, and it is at closedown that projects are deemed a success or failure. Completion also signifies the chance to begin a new project and to apply what you have learned. Now that you have an understanding of the project management process, the next section describes specific techniques used in systems development for representing and scheduling activities and resources.

REPRESENTING AND SCHEDULING PROJECT PLANS

A project manager has a wide variety of techniques available for depicting and documenting project plans. These planning documents can take the form of graphical or textual reports, although graphical reports have become most popular for depicting project

Project Closedown

1. Closing Down the Project

2. Conducting Postproject Reviews

3. Closing the Customer Contract

FIGURE 3-18
Three project closedown activities

plans. The most commonly used methods are Gantt charts and network diagrams. Because Gantt charts do not (typically) show how tasks must be ordered (precedence) but simply show when a task should begin and when it should end, they are often more useful for depicting relatively simple projects or subparts of a larger project, showing the activities of a single worker, or monitoring the progress of activities compared to scheduled completion dates (Figure 3-19). Recall that a network diagram shows the ordering of activities by connecting a task to its predecessor and successor tasks. Sometimes a network diagram is preferable; other times a Gantt chart more easily shows certain aspects of a project. Here are the key differences between these two charts:

- Gantt charts visually show the duration of tasks, whereas a network diagram visually shows the sequence dependencies between tasks.
- Gantt charts visually show the time overlap of tasks, whereas a network diagram does not show time overlap but does show which tasks could be done in parallel.

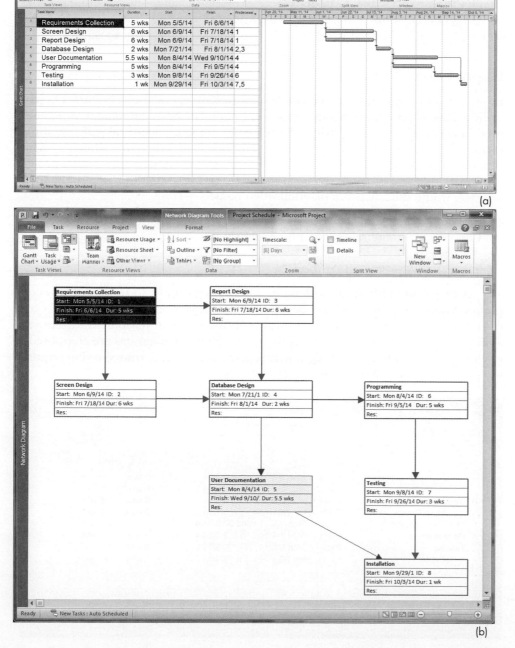

FIGURE 3-19
Graphical diagrams that depict project plans
(a) A Gantt chart
(b) A network diagram
(*Source:* Microsoft Corporation.)

- Some forms of Gantt charts can visually show slack time available within an earliest start and latest finish duration. A network diagram shows this by data within activity rectangles.

Project managers also use textual reports that depict resource utilization by task, complexity of the project, and cost distributions to control activities. For example, Figure 3-20 shows a screen from Microsoft Project for Windows that summarizes all project activities, their durations in weeks, and their scheduled starting and ending dates. Most project managers use computer-based systems to help develop their graphical and textual reports. Later in this chapter, we discuss these automated systems in more detail.

A project manager will periodically review the status of all ongoing project task activities to assess whether the activities will be completed early, on time, or late. If early or late, the duration of the activity, depicted in column 2 of Figure 3-20, can be updated. Once changed, the scheduled start and finish times of all subsequent tasks will also change. Making such a change will also alter a Gantt chart or network diagram used to represent the project tasks. The ability to easily make changes to a project is a very powerful feature of most project management environments. It enables the project manager to determine easily how changes in task duration affect the project completion date. It is also useful for examining the impact of "what if" scenarios of adding or reducing resources, such as personnel, for an activity.

Representing Project Plans

Resources
Any person, group of people, piece of equipment, or material used in accomplishing an activity.

Critical path scheduling
A scheduling technique whose order and duration of a sequence of task activities directly affect the completion date of a project.

Project scheduling and management require that time, costs, and resources be controlled. **Resources** are any person, group of people, piece of equipment, or material used in accomplishing an activity. Network diagramming is a **critical path scheduling** technique used for controlling resources. A critical path refers to a sequence of task activities whose order and durations directly affect the completion date of a project. A network diagram is one of the most widely used and best-known scheduling methods. You would use a network diagram when tasks:

- Are well defined and have a clear beginning and end point
- Can be worked on independently of other tasks
- Are ordered
- Serve the purpose of the project

A major strength of network diagramming is its ability to represent how completion times vary for activities. Because of this, it is more often used than Gantt charts to manage projects such as information systems development, where variability in the duration of activities is the norm. Network diagrams are composed of circles or rectangles representing activities and connecting arrows showing required work flows, as illustrated in Figure 3-21.

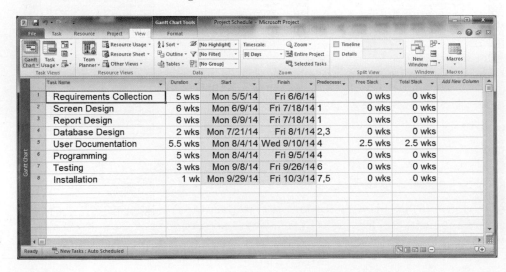

FIGURE 3-20
A screen from Microsoft Project for Windows summarizes all project activities, their durations in weeks, and their scheduled starting and ending dates

(*Source:* Microsoft Corporation.)

	Task Name	Duration	Start	Finish	Predecessor	Free Slack	Total Slack	Add New Column
1	Requirements Collection	5 wks	Mon 5/5/14	Fri 6/6/14		0 wks	0 wks	
2	Screen Design	6 wks	Mon 6/9/14	Fri 7/18/14	1	0 wks	0 wks	
3	Report Design	6 wks	Mon 6/9/14	Fri 7/18/14	1	0 wks	0 wks	
4	Database Design	2 wks	Mon 7/21/14	Fri 8/1/14	2,3	0 wks	0 wks	
5	User Documentation	5.5 wks	Mon 8/4/14	Wed 9/10/14	4	2.5 wks	2.5 wks	
6	Programming	5 wks	Mon 8/4/14	Fri 9/5/14	4	0 wks	0 wks	
7	Testing	3 wks	Mon 9/8/14	Fri 9/26/14	6	0 wks	0 wks	
8	Installation	1 wk	Mon 9/29/14	Fri 10/3/14	7,5	0 wks	0 wks	

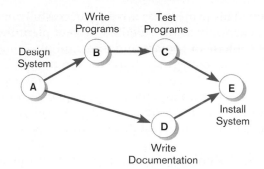

FIGURE 3-21
A network diagram showing activities (represented by circles) and sequence of those activities (represented by arrows)

Calculating Expected Time Durations Using PERT

One of the most difficult and most error-prone activities when constructing a project schedule is the determination of the time duration for each task within a work breakdown structure. It is particularly problematic to make these estimates when there is a high degree of complexity and uncertainty about a task. **PERT (Program Evaluation Review Technique)** is a technique that uses optimistic, pessimistic, and realistic time estimates to calculate the expected time for a particular task. This technique can help you to obtain a better time estimate when there is some uncertainty as to how much time a task will require to be completed.

PERT (Program Evaluation Review Technique)
A technique that uses optimistic, pessimistic, and realistic time estimates to calculate the expected time for a particular task.

The optimistic *(o)* and pessimistic *(p)* times reflect the minimum and maximum possible periods of time for an activity to be completed. The realistic *(r)* time, or most likely time, reflects the project manager's "best guess" of the amount of time the activity actually will require for completion. Once each of these estimates is made for an activity, an expected time *(ET)* can be calculated. Because the expected completion time should be closest to the realistic *(r)* time, it is typically weighted four times more than the optimistic *(o)* and pessimistic *(p)* times. Once you add these values together, it must be divided by six to determine the *ET*. This equation is shown in the following formula:

$$ET = \frac{o + 4r + p}{6}$$

where

ET = expected time for the completion for an activity
o = optimistic completion time for an activity
r = realistic completion time for an activity
p = pessimistic completion time for an activity

For example, suppose that your instructor asked you to calculate an expected time for the completion of an upcoming programming assignment. For this assignment, you estimate an optimistic time of 2 hours, a pessimistic time of 8 hours, and a most likely time of 6 hours. Using PERT, the expected time for completing this assignment is 5.67 hours. Commercial project management software such as Microsoft Project assists you in using PERT to make expected time calculations. Additionally, many commercial tools allow you to customize the weighting of optimistic, pessimistic, and realistic completion times.

Constructing a Gantt Chart and Network Diagram at Pine Valley Furniture

Although PVF has historically been a manufacturing company, it has recently entered the direct sales market for selected target markets. One of the fastest growing of these markets is economically priced furniture suitable for college students. Management has requested that a new Sales Promotion Tracking System

(SPTS) be developed. This project has already successfully moved through project initiation and is currently in the detailed project planning stage, which corresponds to the SDLC phase of project initiation and planning. The SPTS will be used to track purchases by college students for the next fall semester. Students typically purchase low-priced beds, bookcases, desks, tables, chairs, and dressers. Because PVF does not normally stock a large quantity of lower-priced items, management feels that a tracking system will help provide information about the college-student market that can be used for follow-up sales promotions (e.g., a midterm futon sale).

The project is to design, develop, and implement this information system before the start of the fall term in order to collect sales data at the next major buying period. This deadline gives the project team 24 weeks to develop and implement the system. The Systems Priority Board at PVF wants to make a decision this week based on the feasibility of completing the project within the 24-week deadline. Using PVF's project planning methodology, the project manager, Jim Woo, knows that the next step is to construct a Gantt chart and network diagram of the project to represent the Baseline Project Plan so that he can use these charts to estimate the likelihood of completing the project within 24 weeks. A major activity of project planning focuses on dividing the project into manageable activities, estimating times for each, and sequencing their order. Here are the steps Jim followed to do this:

1. *Identify each activity to be completed in the project.* After discussing the new SPTS with PVF's management, sales, and development staff, Jim identified the following major activities for the project:
   - Requirements collection
   - Screen design
   - Report design
   - Database construction
   - User documentation creation
   - Software programming
   - System testing
   - System installation

2. *Determine time estimates and calculate the expected completion time for each activity.* After identifying the major project activities, Jim established optimistic, realistic, and pessimistic time estimates for each activity. These numbers were then used to calculate the expected completion times for all project activities, as described previously using PERT. Figure 3-22 shows the estimated time calculations for each activity of the SPTS project.

ACTIVITY	TIME ESTIMATE (in weeks)			EXPECTED TIME (ET) $\frac{o + 4r + p}{6}$
	o	r	p	
1. Requirements Collection	1	5	9	5
2. Screen Design	5	6	7	6
3. Report Design	3	6	9	6
4. Database Design	1	2	3	2
5. User Documentation	2	6	7	5.5
6. Programming	4	5	6	5
7. Testing	1	3	5	3
8. Installation	1	1	1	1

FIGURE 3-22
Estimated time calculations for the SPTS project

ACTIVITY	PRECEDING ACTIVITY
1. Requirements Collection	–
2. Screen Design	1
3. Report Design	1
4. Database Design	2,3
5. User Documentation	4
6. Programming	4
7. Testing	6
8. Installation	5,7

FIGURE 3-23
Sequence of activities within the SPTS project

3. *Determine the sequence of the activities and precedence relationships among all activities by constructing a Gantt chart and network diagram.* This step helps you to understand how various activities are related. Jim starts by determining the order in which activities should take place. The results of this analysis for the SPTS project are shown in Figure 3-23. The first row of this figure shows that no activities precede requirements collection. Row 2 shows that screen design must be preceded by requirements collection. Row 4 shows that both screen and report design must precede database construction. Thus, activities may be preceded by zero, one, or more activities.

Using the estimated time and activity sequencing information from Figures 3-22 and 3-23, Jim can now construct a Gantt chart and network diagram of the project's activities. To construct the Gantt chart, a horizontal bar is drawn for each activity that reflects its sequence and duration, as shown in Figure 3-24. The Gantt chart may not, however, show direct interrelationships between activities. For example, the fact that the database design activity begins right after the screen design and report design bars finish does not imply that these two activities must finish before database design can begin. To show such precedence relationships, a network diagram must be used. The Gantt chart in Figure 3-24 does, however, show precedence relationships.

Network diagrams have two major components: arrows and nodes. Arrows reflect the sequence of activities, whereas nodes reflect activities that consume time and resources. A network diagram for the SPTS project is shown in Figure 3-25. This diagram has eight nodes labeled 1 through 8.

4. *Determine the critical path.* The critical path of a network diagram is represented by the sequence of connected activities that produce the longest overall time period. All nodes and activities within this sequence are referred to as being "on"

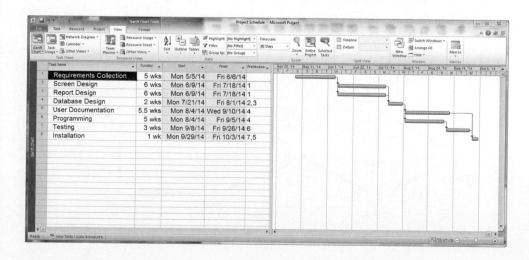

FIGURE 3-24
Gantt chart that illustrates the sequence and duration of each activity of the SPTS project

(*Source:* Microsoft Corporation.)

FIGURE 3-25
A network diagram that illustrates the activities (circles) and the sequence (arrows) of those activities

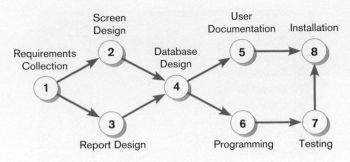

Critical path
The shortest time in which a project can be completed.

Slack time
The amount of time that an activity can be delayed without delaying the project.

the **critical path**. The critical path represents the shortest time in which a project can be completed. In other words, any activity on the critical path that is delayed in completion delays the entire project. Nodes not on the critical path, however, can be delayed (for some amount of time) without delaying the final completion of the project. Nodes not on the critical path contain **slack time** and allow the project manager some flexibility in scheduling.

Figure 3-26 shows the network diagram that Jim constructed to determine the critical path and expected completion time for the SPTS project. To determine the critical path, Jim calculated the earliest and latest expected completion time for each activity. He found each activity's earliest expected completion time (T_E) by summing the estimated time (ET) for each activity from left to right (i.e., in precedence order), starting at activity 1 and working toward activity 8. In this case, T_E for activity 8 is equal to 22 weeks. If two or more activities precede an activity, the largest expected completion time of these activities is used in calculating the new activity's expected completion time. For example, because activity 8 is preceded by both activities 5 and 7, the largest expected completion time between 5 and 7 is 21, so T_E for activity 8 is 21 + 1, or 22. The earliest expected completion time for the last activity of the project represents the amount of time the project should take to complete. Because the time of each activity can vary, however, the projected completion time represents only an estimate. The project may in fact require more or less time for completion.

The latest expected completion time (T_L) refers to the time in which an activity can be completed without delaying the project. To find the values for each activity's T_L, Jim started at activity 8 and set T_L equal to the final T_E (22 weeks). Next, he worked right to left toward activity 1 and subtracted the expected time for each activity. The slack time for each activity is equal to the difference between its latest and earliest expected completion times ($T_L - T_E$). Figure 3-27 shows the slack time calculations for all activities of the SPTS project. All activities with a slack time equal to zero are on the critical path. Thus, all activities

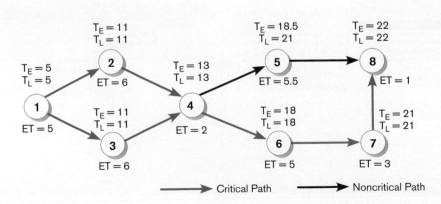

FIGURE 3-26
A network diagram for the SPTS project showing estimated times for each activity and the earliest and latest expected completion time for each activity

ACTIVITY	T_E	T_L	SLACK $T_L - T_E$	ON CRITICAL PATH
1	5	5	0	✓
2	11	11	0	✓
3	11	11	0	✓
4	13	13	0	✓
5	18.5	21	2.5	
6	18	18	0	✓
7	21	21	0	✓
8	22	22	0	✓

FIGURE 3-27
Activity slack time calculations for the SPTS project; all activities except number 5 are on the critical path

except activity 5 are on the critical path. Part of the diagram in Figure 3-26 shows two critical paths, between activities 1-2-4 and 1-3-4, because both of these parallel activities have zero slack.

In addition to the possibility of having multiple critical paths, there are actually two possible types of slack. *Free slack* refers to the amount of time a task can be delayed without delaying the early start of any immediately following tasks. *Total slack* refers to the amount of time a task can be delayed without delaying the completion of the project. Understanding free and total slack allows the project manager to better identify where trade-offs can be made if changes to the project schedule need to be made. For more information on understanding slack and how it can be used to manage tasks see *Information Systems Project Management,* by (Fuller et al., 2008).

USING PROJECT MANAGEMENT SOFTWARE

A wide variety of automated project management tools is available to help you manage a development project. New versions of these tools are continuously being developed and released by software vendors. Most of the available tools have a set of common features that include the ability to define and order tasks, assign resources to tasks, and easily modify tasks and resources. Project management tools are available to run on IBM-compatible personal computers, the Macintosh, and larger mainframe and workstation-based systems. These systems vary in the number of task activities supported, the complexity of relationships, system processing and storage requirements, and, of course, cost. Prices for these systems can range from a few hundred dollars for personal computer–based systems to more than $100,000 for large-scale, multiproject systems. Yet a lot can be done with systems such as Microsoft Project as well as public domain and shareware systems. For example, numerous shareware project management programs (e.g., OpenProj, Bugzilla, and eGroup-Ware) can be downloaded from the Web (e.g., at *www.download.com*). Because these systems are continuously changing, you should comparison shop before choosing a particular package.

We now illustrate the types of activities you would perform when using project management software. Microsoft Project for Windows is a project management system that has received consistently high marks in computer publication reviews (see *www.microsoft.com* and search for "project"—also, if you search the Web, there are many very useful tutorials for improving your Microsoft Project skills). When using this system to manage a project, you need to perform at least the following activities:

- Establish a project starting or ending date.
- Enter tasks and assign task relationships.
- Select a scheduling method to review project reports.

FIGURE 3-28
Establishing a project starting date in Microsoft Project for Windows

(*Source:* Microsoft Corporation.)

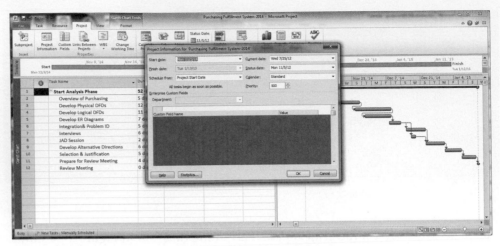

Establishing a Project Start Date

Defining the general project information includes obtaining the name of the project and the project manager, and the starting or ending date of the project. Starting and ending dates are used to schedule future activities or backdate others (see the following) based on their duration and relationships with other activities. An example from Microsoft Project for Windows of the data entry screen for establishing a project starting or ending date is shown in Figure 3-28. This screen shows PVF's Purchasing Fulfillment System project. Here, the starting date for the project is Monday, November 3, 2014.

Entering Tasks and Assigning Task Relationships

The next step in defining a project is to define project tasks and their relationships. For the Purchasing Fulfillment System project, Chris defined 11 tasks to be completed when he performed the initial system analysis activities for the project (task 1—Start Analysis Phase—is a summary task that is used to group-related tasks). The task entry screen, shown in Figure 3-29, is similar to a financial spreadsheet program. The user moves the cursor to a cell with arrow keys or the mouse and then simply enters a textual Name and a numeric Duration for each activity. Scheduled Start and Scheduled Finish are automatically entered based on the project start date and duration. To set an activity relationship, the ID number (or numbers) of the activity that must be completed before the start of the current activity is entered into the Predecessors column. Additional codes under this column make the precedence

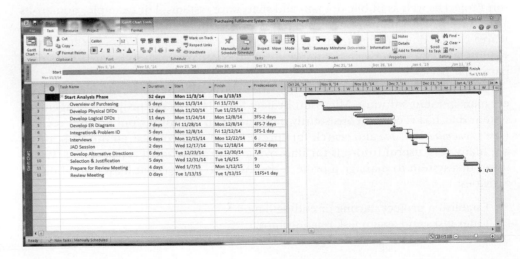

FIGURE 3-29
Entering tasks and assigning task relationships in Microsoft project for Windows

(*Source:* Microsoft Corporation.)

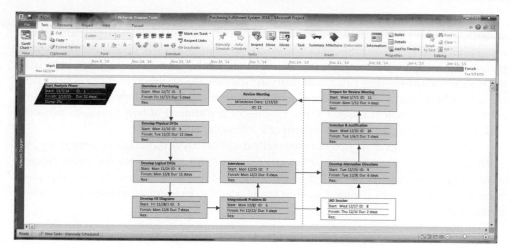

FIGURE 3-30
Viewing project information as a network diagram in Microsoft Project for Windows

(*Source:* Microsoft Corporation.)

relationships more precise. For example, consider the Predecessor column for ID 6. The entry in this cell says that activity 6 cannot start until one day before the finish of activity 5. (Microsoft Project provides many different options for precedence and delays, as demonstrated in this example, but discussion of these is beyond the scope of our coverage.) The project management software uses this information to construct Gantt charts, network diagrams, and other project-related reports.

Selecting a Scheduling Method to Review Project Reports

Once information about all the activities for a project has been entered, it is very easy to review the information in a variety of graphical and textual formats using displays or printed reports. For example, Figure 3-29 shows the project information in a Gantt chart screen, whereas Figure 3-30 shows the project information in a network diagram. You can easily change how you view the information by making a selection from the View menu shown in Figure 3-30.

As mentioned earlier, interim project reports to management will often compare actual progress with plans. Figure 3-31 illustrates how Microsoft Project shows progress with a solid line within the activity bar. In this figure, task 2 has been completed and task 3 is almost completed, but there remains a small percentage of work, as shown by the incomplete solid lines within the bar for this task. Assuming that this screen represents the status of the project on Wednesday, November 5, 2014, the third activity is approximately on schedule, but the second activity is behind its expected completion date. Tabular reports can summarize the same information.

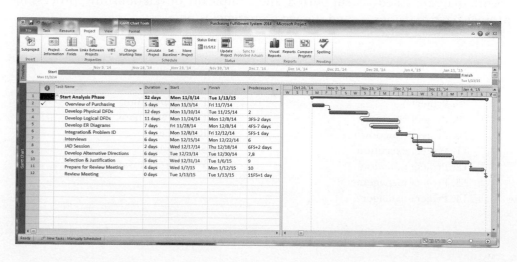

FIGURE 3-31
Gantt chart showing progress of activities (right frame) versus planned activities (left frame)

This brief introduction to project management software has only scratched the surface to show you the power and the features of these systems. Other features that are widely available and especially useful for multiperson projects relate to resource usage and utilization. Resource-related features enable you to define characteristics such as standard costing rates and daily availability via a calendar that records holidays, working hours, and vacations. These features are particularly useful for billing and estimating project costs. Often, resources are shared across multiple projects, which could significantly affect a project's schedule.

Depending on how projects are billed within an organization, assigning and billing resources to tasks can be a very time-consuming activity for most project managers. The features provided in these powerful tools can greatly ease the planning and managing of projects so that project and management resources are effectively utilized.

SUMMARY

The focus of this chapter was on managing information systems projects and the role of the project manager in this process. A project manager has both technical and managerial skills and is ultimately responsible for determining the size, scope, and resource requirements for a project. Once a project is deemed feasible by an organization, the project manager ensures that the project meets the customer's needs and is delivered within budget and time constraints. To manage the project, the project manager must execute four primary activities: project initiation, project planning, project execution, and project closedown. The focus of project initiation is on assessing the size, scope, and complexity of a project and on establishing procedures to support later project activities. The focus of project planning is on defining clear, discrete activities and the work needed to complete each activity. The focus of project execution is on putting the plans developed in project initiation and planning into action. Project closedown focuses on bringing the project to an end.

Gantt charts and network diagrams are powerful graphical techniques used in planning and controlling projects. Both Gantt charts and network diagram scheduling techniques require that a project have activities that can be defined as having a clear beginning and end, can be worked on independently of other activities, are ordered, and are such that their completion signifies the end of the project. Gantt charts use horizontal bars to represent the beginning, duration, and ending of an activity. Network diagramming is a critical path scheduling method that shows the interrelationships among activities. Critical path scheduling refers to planning methods whereby the order and duration of the project's activities directly affect the completion date of the project. These charts show when activities can begin and end, which activities cannot be delayed without delaying the whole project, how much slack time each activity has, and progress against planned activities. A network diagram's ability to use probability estimates in determining critical paths and deadlines makes it a widely used technique for very complex projects.

A wide variety of automated tools for assisting the project manager is available. Most tools have a set of common features, including the ability to define and order tasks, assign resources to tasks, and modify tasks and resources. Systems vary regarding the number of activities supported, the complexity of relationships, processing and storage requirements, and cost.

KEY TERMS

1. Critical path
2. Critical path scheduling
3. Deliverable
4. Feasibility study
5. Gantt chart
6. Network diagram
7. PERT (Program Evaluation Review Technique)
8. Project
9. Project charter
10. Project closedown
11. Project execution
12. Project initiation
13. Project management
14. Project manager
15. Project planning
16. Project workbook
17. Resources
18. Slack time
19. Work breakdown structure

Match each of the key terms above with the definition that best fits it.

_____ A systems analyst with a diverse set of skills—management, leadership, technical, conflict management, and customer relationship—who is responsible for initiating, planning, executing, and closing down a project.

_____ A planned undertaking of related activities to reach an objective that has a beginning and an end.

_____ An end product of an SDLC phase.

_____ A study that determines if the proposed information system makes sense for the organization from an economic and operational standpoint.

_____ A controlled process of initiating, planning, executing, and closing down a project.

_____ The first phase of the project management process in which activities are performed to assess the size, scope, and complexity of the project and to establish procedures to support later project activities.

_____ An online or hard-copy repository for all project correspondence, inputs, outputs, deliverables, procedures, and standards.

_____ The second phase of the project management process that focuses on defining clear, discrete activities and the work needed to complete each activity within a single project.

_____ The process of dividing the project into manageable tasks and logically ordering them to ensure a smooth evolution between tasks.

_____ A graphical representation of a project that shows each task as a horizontal bar whose length is proportional to its time for completion.

_____ A diagram that depicts project tasks and their interrelationships.

_____ The third phase of the project management process in which the plans created in the prior phases are put into action.

_____ The final phase of the project management process that focuses on bringing a project to an end.

_____ Any person, group of people, piece of equipment, or material used in accomplishing an activity.

_____ A scheduling technique whose order and duration of a sequence of task activities directly affect the completion date of a project.

_____ The shortest time in which a project can be completed.

_____ The amount of time that an activity can be delayed without delaying the entire project.

_____ A technique that uses optimistic, pessimistic, and realistic time estimates to calculate the expected completion time for a particular task.

_____ A short document prepared for the customer during project initiation that describes what the project will deliver and outlines, generally at a high level, all work required to complete the project.

REVIEW QUESTIONS

1. Contrast the following terms:
   a. Critical path scheduling, Gantt, network diagramming, slack time
   b. Project, project management, project manager
   c. Project initiation, project planning, project execution, project closedown
   d. Project workbook, resources, work breakdown structure

2. Discuss the reasons why organizations undertake information systems projects.

3. List and describe the common skills and activities of a project manager. Which skill do you think is most important? Why?

4. Describe the activities performed by the project manager during project initiation.

5. Describe the activities performed by the project manager during project planning.

6. Describe the activities performed by the project manager during project execution.

7. List various project team communication methods and describe an example of the type of information that might be shared among team members using each method.

8. Describe the activities performed by the project manager during project closedown.

9. What characteristics must a project have in order for critical path scheduling to be applicable?

10. Describe the steps involved in making a Gantt chart.

11. Describe the steps involved in making a network diagram.

12. In which phase of the SDLC does project planning typically occur? In which phase does project management occur?

13. What are some reasons why one activity may have to precede another activity before the second activity can begin? In other words, what causes precedence relationships between project activities?

PROBLEMS AND EXERCISES

1. Which of the four phases of the project management process do you feel is most challenging? Why?

2. What are some sources of risk in a systems analysis and design project and how does a project manager cope with risk during the stages of project management?

3. Search computer magazines or the Web for recent reviews of project management software. Which packages seem to be most popular? What are the relative strengths and weaknesses of each software package? What advice would you give to someone intending to buy project management software for his or her PC? Why?

4. Suppose that you have been contracted by a jewelry store to manage a project to create a new inventory tracking system. Describe your initial approach to the project. What should your first activity be? What information would you need? To whom might you need to speak?

5. Can a project have two critical paths? Why or why not? Give a brief example to illustrate your point.

6. Calculate the expected time for the following activities.

Activity	Optimistic Time	Most Likely Time	Pessimistic Time	Expected Time
A	3	7	11	
B	5	9	13	
C	1	2	9	
D	2	3	16	
E	2	4	18	
F	3	4	11	
G	1	4	7	
H	3	4	5	
I	2	4	12	
J	4	7	9	

7. A project has been defined to contain the following list of activities along with their required times for completion.

Activity No.	Immediate Activity	Time (weeks)	Predecessors
1	Collect requirements	3	
2	Analyze processes	2	1
3	Analyze data	2	2
4	Design processes	6	2
5	Design data	3	3
6	Design screens	2	3,4
7	Design reports	4	4,5
8	Program	5	6,7
9	Test and document	7	7
10	Install	2	8,9

   a. Draw a network diagram for the activities.
   b. Calculate the earliest expected completion time.
   c. Show the critical path.
   d. What would happen if activity 6 were revised to take 6 weeks instead of 2 weeks?

8. Construct a Gantt chart for the project defined in Problem and Exercise 7.

9. Look again at the activities outlined in Problem and Exercise 7. Assume that your team is in its first week of the project and has discovered that each of the activity duration estimates is wrong. Activity 2 will take only two weeks to complete. Activities 4 and 7 will each take three times longer than anticipated. All other activities will take twice as long to complete as previously estimated. In addition, a new activity, number 11, has been added. It will take one week to complete, and its immediate predecessors are activities 10 and 9. Adjust the network diagram and recalculate the earliest expected completion times.

10. Construct a Gantt chart and network diagram for a project you are or will be involved in. Choose a project of sufficient depth from work, home, or school. Identify the activities to be completed, determine the sequence of the activities, and construct a diagram reflecting the starting times, ending times, durations, and precedence (network diagram only) relationships among all activities. For your network diagram, use the procedure in this chapter to determine time estimates for each activity and calculate the expected time for each activity. Now determine the critical path and the early and late starting and finishing times for each activity. Which activities have slack time?

11. For the project you described in Problem and Exercise 10, assume that the worst has happened. A key team member has dropped out of the project and has been assigned to another project in another part of the country. The remaining team members are having personality clashes. Key deliverables for the project are now due much earlier than expected. In addition, you have just determined that a key phase in the early life of the project will now take much longer than you had originally expected. To make matters worse, your boss absolutely will not accept that this project cannot be completed by this new deadline. What will you do to account for these project changes and problems? Begin by reconstructing your Gantt chart and network diagram and determining a strategy for dealing with the specific changes and problems just described. If new resources are needed to meet the new deadline, outline the rationale that you will use to convince your boss that these additional resources are critical to the success of the project.

12. Assume that you have a project with seven activities labeled A–G (below). Derive the earliest completion time (or early finish—EF), latest completion time (or late finish—LF), and slack for each of the following tasks (begin at time = 0). Which tasks are on the critical path? Draw a Gantt chart for these tasks.

Activity	Preceding Event	Expected Duration	EF	LF	Slack	Critical Path?
A	—	5				
B	A	3				
C	A	4				
D	C	6				
E	B, C	4				
F	D	1				
G	D, E, F	5				

13. Draw a network diagram for the tasks shown in Problem and Exercise 12. Highlight the critical path.

14. Assume you have a project with ten activities labeled A–J, as shown. Derive the earliest completion time (or early finish—EF), latest completion time (or late finish—LF), and slack for each of the following tasks (begin at time = 0). Which tasks are on the critical path? Highlight the critical path on your network diagram.

Activity	Preceding Event	Expected Duration	EF	LF	Slack	Critical Path?
A	—	4				
B	A	5				
C	A	6				
D	A	7				
E	A, D	6				
F	C, E	5				
G	D, E	4				
H	E	3				
I	F, G	4				
J	H, I	5				

15. Draw a Gantt chart for the tasks shown in Problem and Exercise 14.

16. Assume you have a project with 10 activities labeled A–J. Derive the earliest completion time (or early finish—EF), latest completion time (or late finish—LF), and slack for each of the following tasks (begin at time = 0). Which tasks are on the critical path? Draw both a Gantt chart and a network diagram for these tasks, and make sure you highlight the critical path on your network diagram.

Activity	Preceding Event	Expected Duration	EF	LF	Slack	Critical Path?
A	—	3				
B	A	1				
C	A	2				
D	B, C	5				
E	C	3				
F	D	2				
G	E, F	3				
H	F, G	5				
I	G, H	5				
J	I	2				

17. Make a list of the tasks that you performed when designing your schedule of classes for this term. Develop a table showing each task, its duration, preceding event(s), and expected duration. Develop a network diagram for these tasks. Highlight the critical path on your network diagram.

18. Fully decompose a project you've done in another course (e.g., a semester project or term paper). Discuss the level of detail where you stopped decomposing and explain why.

19. Create a work breakdown structure based on the decomposition you carried out for Problem and Exercise 18.

20. Working in a small group, pick a project (it could be anything, such as planning a party, writing a group term paper, developing a database application, etc.) and then write the various tasks that need to be done to accomplish the project on Post-it Notes (one task per Post-it Note). Then use the Post-it Notes to create a work breakdown structure (WBS) for the project. Was it complete? Add missing tasks if necessary. Were some tasks at a lower level in the WBS than others? What was the most difficult part of doing this?

FIELD EXERCISES

1. Identify someone who manages an information systems project in an organization. Describe to him or her each of the skills and activities listed in Table 3-1. Determine which items he or she is responsible for on the project. Of those he or she is responsible for, determine which are the more challenging and why. Of those he or she is not responsible for, determine why not and identify who is responsible for these activities. What other skills and activities, not listed in Table 3-1, is this person responsible for in managing this project?

2. Identify someone who manages an information systems project in an organization. Describe to him or her each of the project planning elements in Figure 3-9. Determine the extent to which each of these elements is part of that person's project planning process. If that person is not able to perform some of these planning activities, or if he or she cannot spend as much time on any of these activities as he or she would like, determine what barriers are prohibitive for proper project planning.

3. Identify someone who manages an information systems project (or other team-based project) in an organization. Describe to him or her each of the project team communication methods listed in Table 3-2. Determine which types of communication methods are used for team communication and describe which he or she feels are best for communicating various types of information.

4. Identify someone who manages an information systems project in an organization. Describe to him or her each of the project execution elements in Figure 3-14. Determine the extent to which each of these elements is part of that person's project execution process. If that person does not perform some of these activities, or if he or she cannot spend much time on any of these activities, determine what barriers or reasons prevent performing all project execution activities.

5. Interview a sample of project managers. Divide your sample into two small subsamples: one for managers of information systems projects and one for managers of other types of projects. Ask each respondent to identify personal leadership attributes that contribute to successful project management and explain why these are important. Summarize your results. What seem to be the attributes most often cited as leading to successful project management, regardless of the type of project? Are there any consistent

differences between the responses in the two subsamples? If so, what are these differences? Do they make sense to you? If there are no apparent differences between the responses of the two subsamples, why not? Are there no differences in the skill sets necessary for managing information systems projects versus managing other types of projects?

6. Observe a real information systems project team in action for an extended period of time. Keep a notebook as you watch individual members performing their individual tasks, as you review the project management techniques used by the team's leader, and as you sit in on some of their meetings. What seem to be the team's strengths and weaknesses? What are some areas in which the team can improve?

REFERENCES

Abdel-Hamid, T. K., K. Sengupta, and C. Swett. 1999. "The Impact of Goals on Software Project Management: An Experimental Investigation." *MIS Quarterly* 23(4): 531–55.

Boehm, B.W., and R. Turner. 2000. *Software Cost Estimation with COCOMO II.* Upper Saddle River, NJ: Prentice Hall.

Dinsmore, P. C., and J. Cabanis-Brewin. 2006. *The AMA Handbook of Project Management: Vol. 1.* New York: AMACOM. American Management Association.

Fuller, M. A., J. S. Valacich, and J. F. George. 2008. *Information Systems Project Management.* Upper Saddle River, NJ: Prentice Hall.

George, J. F., D. Batra, J. S. Valacich, and J. A. Hoffer. 2007. *Object-Oriented Analysis and Design,* 2nd ed. Upper Saddle River, NJ: Prentice Hall.

Guinan, P. J., J. G. Cooprider, and S. Faraj. 1998. "Enabling Software Development Team Performance During Requirements Definition: A Behavioral Versus Technical Approach." *Information Systems Research* 9(2): 101–25.

Hoffer, J. A., V. Ramesh, and H. Topi. 2011. *Modern Database Management,* 10th ed. Upper Saddle River, NJ: Prentice Hall.

Keil, M., B. C. Y. Tan, K. K. Wei, T. Saarinen, V. Tuunainen, and A. Wassenaar. 2000. "A Cross-Cultural Study on Escalation of Commitment Behavior in Software Projects." *MIS Quarterly* 24(2): 631–64.

Kettelhut, M. C. 1991. "Avoiding Group-Induced Errors in Systems Development." *Journal of Systems Management* 42(12): 13–17.

King, J. 2003. "IT's Global Itinerary: Offshore Outsourcing Is Inevitable," September 15, 2003. Available at *www.cio.com.* Accessed February 21, 2006.

Kirsch, L. J. 2000. "Software Project Management: An Integrated Perspective for an Emerging Paradigm." In R. W. Zmud (ed.),*Framing the Domains of IT Management: Projecting the Future from the Past,* 285–304. Cincinnati, OH: Pinnaflex Educational Resources.

PMBOK. 2008. *A Guide to the Project Management Body of Knowledge,* 4th. ed. Newtown Squre, PA: Project Management Institute.

Project Management Institute. 2002. *Work Breakdown Structures.* Newton Square, PA: Project Management Institute.

Rettig, M. 1990. "Software Teams." *Communications of the ACM* 33(10): 23–27.

Royce, W. 1998. *Software Project Management.* Boston: Addison-Wesley.

Verma, V. K. 1996. *Human Resource Skills for the Project Manager.* Newton Square, PA: Project Management Institute.

Verma, V. K. 1997. *Managing the Project Team.* Newton Square, PA: Project Management Institute.

Wideman, R. M. 1992. *Project and Program Risk Management.* Newton Square, PA: Project Management Institute.

Object-Oriented Analysis and Design

Project Management

Learning Objectives

After studying this section, you should be able to:

- Describe the unique characteristics of an OOSAD project.

Unique Characteristics of an OOSAD Project

In this chapter, we have described how projects are managed when using a structured development approach. These concepts and techniques are very robust to a broad range of projects and development approaches. However, when developing a system using a more iterative design approach—such as prototyping or object-oriented analysis and design—there are some additional issues to consider. In this section, we will discuss some unique characteristics of these types of projects (see Fuller et al., 2008; George et al., 2007).

When a system is developed using an iterative approach, it means that, over the duration of the project, a portion of the final system is constructed during each iteration phase. In this way, the system evolves incrementally so that by the last iteration of the project, the entire system is completed (see Figure 3-32). In order for the system to evolve in this manner, the project manager must understand several unique characteristics of an OOSAD project.

DEFINE THE SYSTEM AS A SET OF COMPONENTS

In order to manage the project as a series of iterations, the project manager must subdivide the overall system into a set of components; when combined, this set will yield the entire system (see Figure 3-33). Each

FIGURE 3-32
During the OOSAD process, the system evolves incrementally over the life of the project

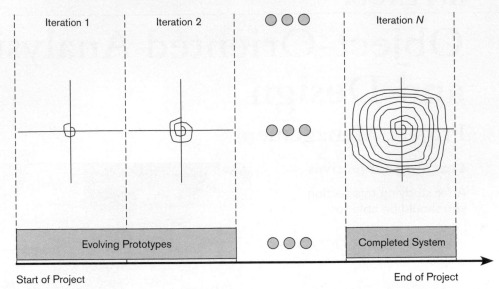

of these separate system components is often referred to as a "vertical slice" of the overall system; this is a key feature of the system that can be demonstrated to users. Alternatively, each slice should not be a subsystem that spans "horizontally" throughout the entire system because these horizontal slices typically do not focus on a specific system feature, nor are they typically good for demonstration to users. Basically, each vertical slice represents a use case of the system (see Chapter 7 for more information on use case diagrams). Also, note in Figure 3-33 that project management and planning is an activity that continues throughout the life of the project.

One outcome of defining the overall system as a collection of components is the likelihood that the components constructed earlier in the project will require greater rework than those developed later in the project. For example, during the early stages of the project, missing components or a lack of understanding of key architectural features will require that components developed early in the project be modified substantially as the project moves forward in order to integrate these

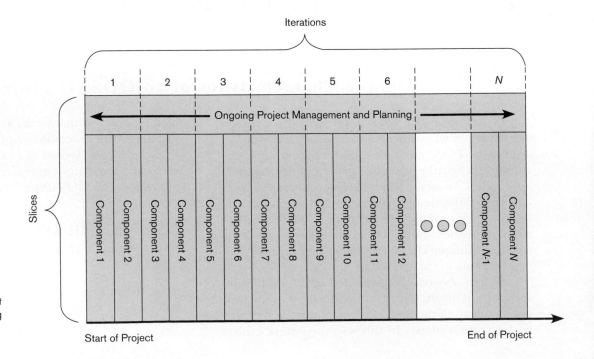

FIGURE 3-33
Object-oriented development projects are developed using ongoing management and evolving system functionality

components into a single comprehensive system successfully. This means that rework is a natural part of an OOSAD project and that one should not be overly concerned when this occurs. It is simply a characteristic of the iterative and incremental development process of OOSAD.

Complete Hard Problems First

Another characteristic of the OOSAD approach is that it tackles the hard problems first. In classic structured systems development, a hard problem, such as choosing the physical implementation environment, is addressed late in the development process. As a result, following a classic systems development approach tends to result in putting off making some of the key systems architectural decisions until late in the project. This approach is sometimes problematic because such decisions often determine whether a project is a success or a failure. On the other hand, addressing hard problems as early as possible allows the difficult problems to be examined before substantial resources have been expended. This mitigates project risk.

In addition, completing the harder problems associated with the systems architecture as early as possible helps in completing all subsequent components because most will build upon these basic architectural capabilities. (With some projects, the hardest components depend upon simpler components. In these cases, one must complete the simpler slices first before moving to the harder ones. Nonetheless, focus should be placed on the hard problems as soon as possible.) From a project planning perspective, this means that there is a natural progression and ordering of components over the life of the project. The initial iteration or two must focus on the system architecture such as the database or networking infrastructure. Once the architecture is completed, core system capabilities, such as creating and deleting records, are implemented. After the core system components are completed, detailed system features are implemented that help to fine-tune key system capabilities. During the final iteration phases, the primary focus is on activities that bring the project to a close (e.g., interface refinement, user manuals and training; see Figure 3-34).

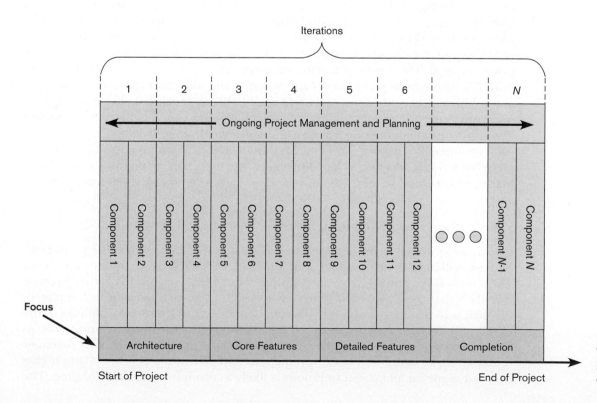

FIGURE 3-34
The focus and ordering of system components change over the life of the project

FIGURE 3-35
The workflow of an iteration
(*Source:* Based on Royce, 1998; George
et al., 2007.)

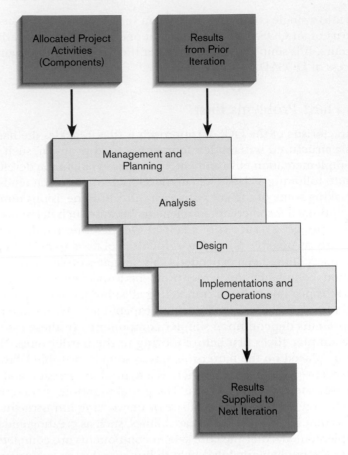

Using Iterations to Manage the Project

During each project iteration, all systems development life cycle activities are performed (see Figure 3-35). This means that each project iteration has management and planning, analysis, design, and implementation and operation activities. For each iteration, the inputs to the process are the allocated project components—vertical slices or use cases—to perform during this iteration and the results from the prior iteration. The results of this iteration are then used as inputs to the next iteration. For example, as components are designed and implemented, much is learned about how subsequent components will need to be implemented. The learning that occurs during each iteration helps the project manager gain a better understanding about how subsequent components will be designed, what problems might occur, what resources are needed, and how long and complex a component will be to complete. As a result, most experienced project managers believe that it is a mistake to make project plans too detailed early in the project when much is still unknown.

Don't Plan Too Much Up Front

During each iteration, more and more will be learned about how subsequent components will need to be designed, how long each might take to complete, and so on. Therefore, it is a mistake to make highly detailed plans far into the future because it is likely that these plans will be wrong. In OOSAD, as each iteration is completed, the goal is to learn more about the system being constructed, the capabilities of the development team, the complexity of the development environment, and so on. As this understanding is gained over the course of the project, the project manager is able to make better and better predictions and plans. As a result, making highly detailed plans for all project iterations is likely to result in a big waste of time. The

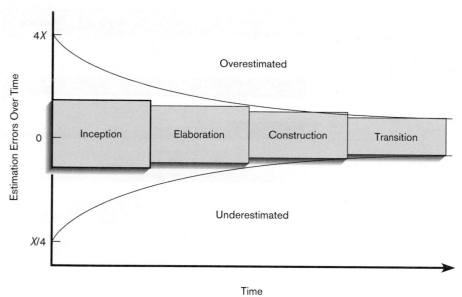

FIGURE 3-36
Planning estimation improves over time
(*Source:* Based on Royce, 1998; George et al., 2007.)

project manager should be concerned only with making highly detailed plans for the next iteration or two. As the project manager learns over the course of the project, he or she will be able to continually refine schedules, time estimates, and resource requirements with better and better estimates (see Figure 3-36).

How Many and How Long Are Iterations?

One question that many people have when first experiencing OOSAD has to do with the number and duration of iterations. Iterations are designed to be a fixed length of time, typically from two to eight weeks, but they can be as short as one week (especially for smaller projects). During a single iteration, multiple components (use cases) can be completed. However, it is important not to try to pack the development of too many components into a single iteration. Experience has shown that having more iterations with fewer components to be completed is better than having only a few iterations with many components needing to be completed. It is only by iterating—completing a full systems development cycle—that significant learning can occur to help the project manager better plan subsequent iterations.

The inception phase generally will entail one iteration, but it is not uncommon for this to require two or more iterations in large, complex projects. Likewise, elaboration often is completed in one or two iterations, but again system complexity and size can influence this. Construction can range from two to several iterations, and transition typically occurs over one or two iterations. Thus, experienced OOSAD project managers typically use from six to nine iterations when designing and constructing a system (see Figure 3-37). Note that all completed components are integrated into a comprehensive system at the conclusion of each iteration. During the first iteration, it is likely that simple component prototypes, such as file opening, closing, and saving, will be created. However, as the project progresses, the prototypes become increasingly sophisticated until the entire system is completed (see Figure 3-38).

Project Activity Focus Changes Over the Life of a Project

Over the life of a project, the project manager moves from iteration to iteration, beginning with inception and ending with the transition phase. Additionally, during all project iterations the manager engages in all phases of the systems development cycle. However, the level of activity in each phase changes over the life of the project

FIGURE 3-37
An OOSAD project typically has six to nine iterations

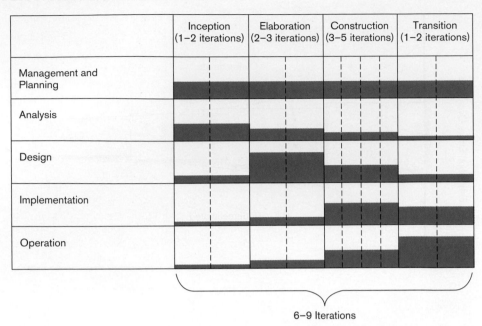

(see Figure 3-39). For example, throughout the life of the project, management and planning are an ongoing and important part of the project. Additionally, during inception the primary focus is analysis, during elaboration the primary focus is design, during construction the primary focus is implementation, and during transition the primary focus is making the system operational. In sum, although all project life cycle activities are employed during every project iteration, the mix and focus of these activities change over time.

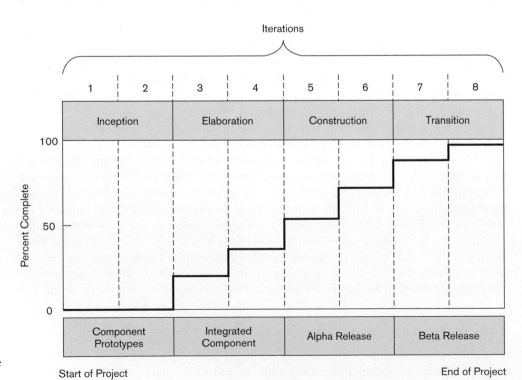

FIGURE 3-38
As the project evolves, system functionality evolves

(*Source:* Based on Royce, 1998; George et al., 2007.)

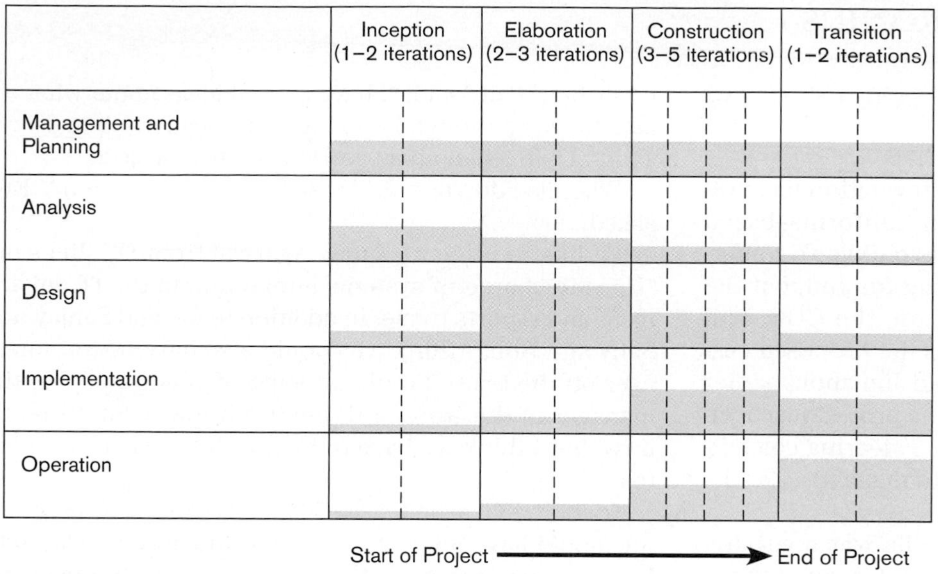

	Inception (1–2 iterations)	Elaboration (2–3 iterations)	Construction (3–5 iterations)	Transition (1–2 iterations)
Management and Planning				
Analysis				
Design				
Implementation				
Operation				

Start of Project ⟶ End of Project

FIGURE 3-39
The level and focus of activity across the systems development process change from the start to the end of the project

SUMMARY

When managing an OOSAD project, the project manager must define the project as a set of components. Once defined, these components can be analyzed and ordered so that the most difficult components are implemented first. An OOSAD project is managed by a series of iterations, and each iteration contains all phases of the systems development cycle. Over each iteration, more and more of the system is created (component by component), and more and more is learned about the system being constructed, the capabilities of the development team, and the complexity of the development environment. As this learning increases over time, the project manager is better able to plan project activities more accurately. Therefore, it is not good practice to plan long-range activities in great detail; detailed planning should occur only for the current and subsequent iteration. Most projects have six to nine iterations, but large projects could have several more. An iteration is a fixed time period, usually about two weeks, but it can be shorter or longer depending upon the characteristics of the project.

REVIEW QUESTION

1. Describe the unique characteristics of OOSAD projects that have ramifications for how these projects are managed.

PROBLEMS AND EXERCISES

1. Why should project managers complete hard problems first in an OOSAD project?

2. Why is planning too much up front a mistake in an OOSAD project?

PETRIE **PETRIE ELECTRONICS**
ELECTRONICS

Chapter 3: Managing the Information Systems Project

Jim Watanabe, assistant director of information technology at Petrie Electronics, a Southern California–based electronics retail store, walked into his building's conference room. It was early in the morning for Jim, but the meeting was important. Ella Whinston, the COO, had called the meeting. On the agenda was the proposed customer relationship project Ella had told Jim about earlier in the week. She had asked Jim to be the project manager. If the project was approved by Petrie IS steering committee, it would be Jim's first big project to manage at Petrie. He was excited about getting started.

"Hi Jim," said Ella Whinston. With Ella was a guy Jim did not know. "Jim, this is Bob Petroski. I know that the customer loyalty project has not been officially approved yet, but I am certain it will be. I'd like for Bob to be on your team, to represent me."

Jim and Bob shook hands. "Nice to meet you, Jim. I'm looking forward to working with you."

"And Bob knows how important this project is to me," Ella said, "so I expect him to keep me informed about your progress." Ella smiled.

Great, Jim thought, more pressure. That's all I need.

Just then, John Smith, the head of marketing walked into the conference room. With him was a young woman Jim recognized, but he wasn't sure from where.

"Jim," John said, "Let me introduce you to Sally Fukuyama. She is the assistant director of marketing. She will be representing marketing, and me, on your 'No Employee Escapes' project. Assuming it gets official approval, of course."

"Hi Jim," Sally said, "I have a lot of ideas about what we can do. Even though I still have my regular job to worry about, I'm excited about working on this project."

"Who else do you think should be on your team?" Ella asked.

"I'd like to bring in Sanjay Agarwal from IT," Jim said. "He is in charge of systems integration in the IT department and reports to me. In addition to me and Sanjay and Sally and Bob, I think we should also have a store manager on the team. I'd like to suggest Juanita Lopez, the manager of the store in Irvine (California). She is really busy, but I think we have to have a store manager on the team."

"Irvine?" Ella asked. "That's one of our top stores. Juanita should have a lot of insight into the issues related to keeping customers, if she is managing the Irvine store. And you are right, she is going to be very busy."

Case Questions

1. What qualities might Jim possess that would make him a successful project manager?
2. How do you think Jim should respond to Ella's implied pressure about the importance of the project to her?
3. What strategies might Jim employ to deal with a very busy team member such as Juanita Lopez?
4. What should Jim do next to complete the project initiation?
5. List five team communication methods that Jim might use throughout this project. What are some pros and cons of each?

PART TWO

Planning

Chapter 4
Identifying and Selecting Systems Development Projects

Chapter 5
Initiating and Planning Systems Development Projects

PART TWO

Planning

The demand for new or replacement systems exceeds the ability and resources of most organizations to conduct systems development projects either by themselves or with consultants. This means that organizations must set priorities and a direction for systems development that will yield development projects with the greatest net benefits. As a systems analyst, you must analyze user information requirements, and you must also help make the business case—or justify why the system should be built and the development project conducted.

The reason for any new or improved information system (IS) is to add value to the organization. As systems analysts, we must choose to use systems development resources to build the mix of systems that add the greatest value to the organization. How can we determine the business value of systems and identify those applications that provide the most critical gains? Part Two addresses this topic, the first phase of the systems development life cycle (SDLC), which we call planning. Business value comes from supporting the most critical business goals and helping the organization deliver on its business strategy. All systems, whether supporting operational or strategic functions, must be linked to business goals. The two chapters in this part of the book show how to make this linkage.

The source of systems projects is either initiatives from IS planning (proactive identification of systems) or requests from users or IS professionals (reactions to problems or opportunities) for new or enhanced systems. In Chapter 4, we outline the linkages among corporate planning, IS planning, and the identification and selection of projects. We do not include IS planning as part of the SDLC, but the results of IS planning greatly influence the birth and conduct of systems projects. Chapter 4 makes a strong argument that IS planning provides not only insights into choosing which systems an organization needs, but also describes the strategies necessary for evaluating the viability of any potential systems project.

A more frequent source of project identification originates from system service requests (SSRs) from business managers and IS professionals, usually for very focused systems or incremental improvements in existing systems. Business managers request a new or replacement system when they believe that improved information services will help them do their jobs. IS professionals may request system updates when technological changes make current system implementations obsolete or when the performance of an existing system needs improvement. In either case, the request for service must be understood by management, and a justification for the system and associated project must be developed.

We continue with the Petrie Electronics case following Chapter 4. In this case, we show how an idea for a new IS project was stimulated by a synergy between corporate strategic planning and the creativity of an individual business manager.

Chapter 5 focuses on what happens after a project has been identified and selected: the next step in making the business case, initiating and planning the proposed system request. This plan develops a better understanding of the scope of the potential system change and the nature of the needed system features. From this preliminary understanding of system requirements, a project plan is developed that shows both the detailed steps and resources needed in order to conduct the analysis phase of the life cycle and the more general steps for subsequent phases. The feasibility and potential risks of the requested system are also outlined, and an economic cost-benefit analysis is conducted to show the potential impact of the system change. In addition to the economic feasibility or justification of the system, technical, organizational, political, legal, schedule, and other feasibilities are assessed. Potential risks—unwanted outcomes—are identified, and plans for dealing with these possibilities are identified. Project initiation and planning end when a formal proposal for the systems development project is completed and submitted for approval to the person who must commit the resources to systems development. If approved, the project moves into the analysis phase of the SDLC.

We illustrate a typical project initiation and planning phase in a Petrie Electronics case following Chapter 5. In this case, we show how the company developed its project scope statement and addressed various aspects of the project's initiation and planning stage.

Identifying and Selecting Systems Development Projects

Learning Objectives

After studying this chapter, you should be able to:

- Describe the project identification and selection process.
- Describe the corporate strategic planning and information systems planning process.
- Explain the relationship between corporate strategic planning and information systems planning.
- Describe how information systems planning can be used to assist in identifying and selecting systems development projects.
- Analyze information systems planning matrices to determine affinity between information systems and IS projects and to forecast the impact of IS projects on business objectives.
- Describe the three classes of Internet electronic commerce applications: business-to-consumer, business-to-employee, and business-to-business.

Introduction

The scope of information systems today is the whole enterprise. Managers, knowledge workers, and all other organizational members expect to easily access and retrieve information, regardless of its location. Nonintegrated systems used in the past—often referred to as "islands of information"—are being replaced with cooperative, integrated enterprise systems that can easily support information sharing. Although the goal of building bridges between these "islands" will take some time to achieve, it represents a clear direction for information systems development. The use of enterprise resource planning (ERP) systems from companies such as SAP (*www.sap.com*) and Oracle (*www.oracle.com*), has enabled the linking of these "islands" in many organizations. Additionally, as the use of the Internet continues to evolve to support business activities, systems integration has become a paramount concern of organizations (Hasselbring, 2000; King, 2003; Luftman, 2004; Overby, 2006).

Obtaining integrated, enterprise-wide computing presents significant challenges for both corporate and information systems management. For example, given the proliferation of personal and departmental computing wherein disparate systems and databases have been created, how can the organization possibly control and maintain all of these systems and data? In many cases they simply cannot; it is nearly impossible to track who has which systems and what data, where there are overlaps or inconsistencies, and the accuracy of the information. The reason that personal and departmental systems and databases abound is that users are either unaware of the information that exists in corporate databases or they cannot easily get at it, so they create and maintain their own information and systems. Intelligent identification and selection of system projects, for both new and replacement systems, is a critical step in gaining control of systems and data. It is the hope of many chief information officers (CIOs) that with the advent of ERP systems, improved system integration, and the rapid deployment of corporate Internet solutions, these islands will be reduced or eliminated (Koch, 2005; Luftman, 2004; Newbold and Azuna, 2007; Ross and Feeny, 2000).

The acquisition, development, and maintenance of information systems consume substantial resources for most organizations. This suggests that organizations can benefit from following a formal process for identifying and selecting projects. The first phase of the systems development life cycle—project identification and selection—deals with this issue. In the next section, you will learn about a general method for identifying and selecting projects and the deliverables and outcomes from this process. This is followed by brief descriptions of corporate strategic planning and information systems planning, two activities that can greatly improve the project identification and selection process.

IDENTIFYING AND SELECTING SYSTEMS DEVELOPMENT PROJECTS

The first phase of the SDLC is planning, consisting of project identification and selection, and project initiation and planning (see Figure 4-1). During project identification and selection, a senior manager, a business group, an IS manager, or a steering committee identifies and assesses all possible systems development projects that an organization unit could undertake. Next, those projects deemed most likely to yield significant organizational benefits, given available resources, are selected for subsequent

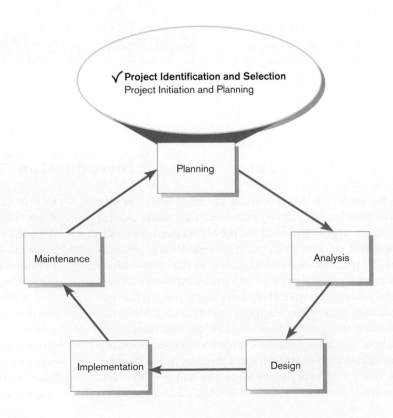

FIGURE 4-1
Systems development life cycle with project identification and selection highlighted

development activities. Organizations vary in their approach to identifying and selecting projects. In some organizations, project identification and selection is a very formal process in which projects are outcomes of a larger overall planning process. For example, a large organization may follow a formal project identification process whereby a proposed project is rigorously compared with all competing projects. Alternatively, a small organization may use informal project selection processes that allow the highest-ranking IS manager to independently select projects or allow individual business units to decide on projects after agreeing to provide project funding.

Information systems development requests come from a variety of sources. One source is requests by managers and business units for replacing or extending an existing system to gain needed information or to provide a new service to customers. Another source for requests is IS managers who want to make a system more efficient and less costly to operate, or want to move it to a new operating environment. A final source of projects is a formal planning group that identifies projects for improvement to help the organization meet its corporate objectives (e.g., a new system to provide better customer service). Regardless of how a given organization actually executes the project identification and selection process, a common sequence of activities occurs. In the following sections, we describe a general process for identifying and selecting projects and producing the deliverables and outcomes of this process.

The Process of Identifying and Selecting IS Development Projects

Project identification and selection consists of three primary activities:

1. Identifying potential development projects
2. Classifying and ranking IS development projects
3. Selecting IS development projects

Each of these steps is described below:

1. *Identifying potential development projects.* Organizations vary as to how they identify projects. This process can be performed by

   - A key member of top management, either the CEO of a small- or medium-sized organization or a senior executive in a larger organization
   - A steering committee, composed of a cross section of managers with an interest in systems
   - User departments, in which either the head of the requesting unit or a committee from the requesting department decides which projects to submit (often you, as a systems analyst, will help users prepare such requests)
   - The development group or a senior IS manager

   All methods of identification have been found to have strengths and weaknesses. Research has found, for example, that projects identified by top management more often have a strategic organizational focus. Alternatively, projects identified by steering committees more often reflect the diversity of the committee and therefore have a cross-functional focus. Projects identified by individual departments or business units most often have a narrow, tactical focus. Finally, a dominant characteristic of projects identified by the development group is the ease with which existing hardware and systems will integrate with the proposed project. Other factors, such as project cost, duration, complexity, and risk, are also influenced by the source of a given project. Characteristics of each selection method are briefly summarized in Table 4-1. In addition to who makes the decision, characteristics specific to the organization—such as the level of firm diversification, level of vertical integration, or extent of growth opportunities—can also influence any investment or project selection decision (Dewan et al., 1998; Luftman, 2004; Yoo et al., 2006; Thomas and Fernandez, 2008).

TABLE 4-1   Characteristics of Alternative Methods for Making Information Systems Identification and Selection Decisions

Selection Method	Characteristics
Top Management	Greater strategic focus
	Largest project size
	Longest project duration
	Enterprise-wide consideration
Steering Committee	Cross-functional focus
	Greater organizational change
	Formal cost-benefit analysis
	Larger and riskier projects
Functional Area	Narrow, nonstrategic focus
	Faster development
	Fewer users, management layers, and business functions involved
Development Group	Integration with existing systems focus
	Fewer development delays
	Less concern with cost-benefit analysis

(*Source:* Based on McKeen, Guimaraes, and Wetherbe, 1994; GAO, 2000.)

Of all the possible project sources, those identified by top management and steering committees most often reflect the broader needs of the organization. This occurs because top management and steering committees are likely to have a broader understanding of overall business objectives and constraints. Projects identified by top management or by a diverse steering committee are therefore referred to as coming from a *top-down source*.

Projects identified by a functional manager, business unit, or by the information systems development group are often designed for a particular business need within a given business unit. In other words, these projects may not reflect the overall objectives of the organization. This does not mean that projects identified by individual managers, business units, or the IS development group are deficient, only that they may not consider broader organizational issues. Project initiatives stemming from managers, business units, or the development group are generally referred to as coming from a *bottom-up source*. These are the types of projects in which you, as a systems analyst, will have the earliest role in the life cycle as part of your ongoing support of users. You will help user managers provide the description of information needs and the reasons for doing the project that will be evaluated in selecting, among all submitted projects, which ones will be approved to move into the project initiation and planning phase of the SDLC.

In sum, projects are identified by both top-down and bottom-up initiatives. The formality of the process of identifying and selecting projects can vary substantially across organizations. Also, because limited resources preclude the development of all proposed systems, most organizations have a process of classifying and ranking the merit of each project. Those projects deemed inconsistent with overall organizational objectives, redundant in functionality to some existing system, or unnecessary will thus be removed from consideration. This topic is discussed next.

2. *Classifying and ranking IS development projects.* The second major activity in the project identification and selection process focuses on assessing the relative merit of potential projects. As with the project identification process, classifying and ranking projects can be performed by top managers, a steering committee, business units, or the IS development group. Additionally, the criteria used when assigning the relative merit of a given project can vary. Commonly used criteria for assessing projects are summarized in Table 4-2. In any given organization, one or several criteria might be used during the classifying and ranking process.

TABLE 4-2   Possible Evaluation Criteria When Classifying and Ranking Projects

Evaluation Criteria	Description
Value Chain Analysis	Extent to which activities add value and costs when developing products and/or services
Strategic Alignment	Extent to which the project is viewed as helping the organization achieve its strategic objectives and long-term goals
Potential Benefits	Extent to which the project is viewed as improving profits, customer service, and so forth, and the duration of these benefits
Resource Availability	Amount and type of resources the project requires and their availability
Project Size/Duration	Number of individuals and the length of time needed to complete the project
Technical Difficulty/Risks	Level of technical difficulty to successfully complete the project within given time and resource constraints

As with the project identification and selection process, the actual criteria used to assess projects will vary by organization. If, for example, an organization uses a steering committee, it may choose to meet monthly or quarterly to review projects and use a wide variety of evaluation criteria. At these meetings, new project requests will be reviewed relative to projects already identified, and ongoing projects are monitored. The relative ratings of projects are used to guide the final activity of this identification process—project selection.

An important project evaluation method that is widely used for assessing information systems development projects is called **value chain analysis** (Porter, 1985; Shank and Govindarajan, 1993). Value chain analysis is the process of analyzing an organization's activities for making products and/or services to determine where value is added and costs are incurred. Once an organization gains a clear understanding of its value chain, improvements in the organization's operations and performance can be achieved. Information systems projects providing the greatest benefit to the value chain will be given priority over those with fewer benefits.

As you might have guessed, information systems have become one of the primary ways for organizations to make changes and improvements in their value chains. Many organizations, for example, are using the Internet to exchange important business information with suppliers and customers, such as orders, invoices, and receipts. To conduct a value chain analysis for an organization, think about an organization as a big input/output process (see Figure 4-2). At one end are the inputs to the organization, for example, supplies that are purchased. Within the organizations, those supplies and resources are integrated in some way to produce products and services. At the other end are the outputs, which represent the products and services that are marketed, sold, and then distributed to customers. In value chain analysis, you must first understand each activity, function, and process where value is or should be added. Next, determine the costs (and the factors that drive costs or cause them to fluctuate) within each of the areas. After understanding your value chain and costs, you can benchmark (compare) your value chain and associated costs with those of other organizations, preferably your competitors. By making these comparisons, you can identify priorities for applying information systems projects.

Value chain analysis
Analyzing an organization's activities to determine where value is added to products and/or services and the costs incurred for doing so; usually also includes a comparison with the activities, added value, and costs of other organizations for the purpose of making improvements in the organization's operations and performance.

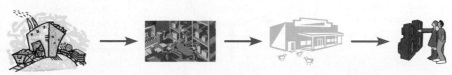

Transform Raw
Materials into
Products

Storage and
Distribution
of Products

Marketing,
Sales, and
Customer Support

FIGURE 4-2
Organizations can be thought of as a value chain, transforming raw materials into products for customers

3. *Selecting IS development projects.* The final activity in the project identification and selection process is the actual selection of projects for further development. Project selection is a process of considering both short- and long-term projects and selecting those most likely to achieve business objectives. Additionally, as business conditions change over time, the relative importance of any single project may substantially change. Thus, the identification and selection of projects is a very important and ongoing activity.

Numerous factors must be considered when making project selection decisions. Figure 4-3 shows that a selection decision requires that the perceived needs of the organization, existing systems and ongoing projects, resource availability, evaluation criteria, current business conditions, and the perspectives of the decision makers will all play a role in project selection decisions. Numerous outcomes can occur from this decision process. Of course, projects can be accepted or rejected. Acceptance of a project usually means that funding to conduct the next phase of the SDLC has been approved. Rejection means that the project will no longer be considered for development. However, projects may also be conditionally accepted; they may be accepted pending the approval or availability of needed resources or the demonstration that a particularly difficult aspect of the system can be developed. Projects may also be returned to the original requesters, who are told to develop or purchase the requested system. Finally, the requesters of a project may be asked to modify and resubmit their request after making suggested changes or clarifications.

One method for deciding among different projects, or when considering alternative designs for a given system, is illustrated in Figure 4-4. For example, suppose that, for a given system that has been identified and selected, there are three alternative designs that could be pursued—A, B, or C. Let's also suppose that early planning meetings identified three key system requirements and four key constraints that could be used to help make a decision on which alternative to pursue. In the left column of Figure 4-4, three system requirements and four constraints are listed. Because not all requirements and constraints are of equal importance, they are weighted based on their relative importance. In other words, you do not have to weight requirements and constraints equally; it is certainly possible to make requirements more or less important than constraints. Weights are arrived at in discussions among the analysis team, users, and sometimes managers. Weights tend to be fairly subjective and, for that reason, should be determined through a process of open discussion to reveal underlying assumptions, followed by an attempt to reach consensus among stakeholders. Notice that the total of the weights for both the requirements and constraints is 100 percent.

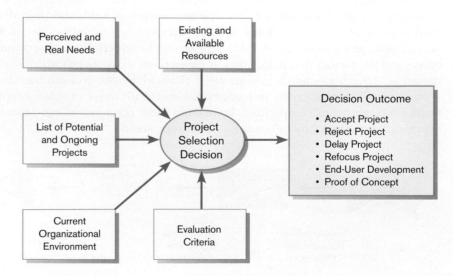

FIGURE 4-3
Project selection decisions must consider numerous factors and can have numerous outcomes

Criteria	Weight	Alternative A		Alternative B		Alternative C	
		Rating	Score	Rating	Score	Rating	Score
Requirements							
Real-time data entry	18	5	90	5	90	5	90
Automatic reorder	18	1	18	5	90	5	90
Real-time data query	14	1	14	5	70	5	70
	50		122		250		250
Constraints							
Developer costs	15	4	60	5	75	3	45
Hardware costs	15	4	60	4	60	3	45
Operating costs	15	5	75	1	15	5	75
Ease of training	5	5	25	3	15	3	15
	50		220		165		180
Total	100		342		415		430

FIGURE 4-4
Alternative projects and system design decisions can be assisted using weighted multicriteria analysis

Next, each requirement and constraint is rated on a scale of 1 to 5. A rating of 1 indicates that the alternative does not meet the requirement very well or that the alternative violates the constraint. A rating of 5 indicates that the alternative meets or exceeds the requirement or clearly abides by the constraint. Ratings are even more subjective than weights and should also be determined through open discussion among users, analysts, and managers. For each requirement and constraint, a score is calculated by multiplying the rating for each requirement and each constraint by its weight. The final step is to add the weighted scores for each alternative. Notice that we have included three sets of totals: for requirements, for constraints, and overall totals. If you look at the totals for requirements, alternative B or C is the best choice because both meet or exceed all requirements. However, if you look only at constraints, alternative A is the best choice because it does not violate any constraints. When we combine the totals for requirements and constraints, we see that the best choice is alternative C. Whether alternative C is actually chosen for development, however, is another issue. The decision makers may choose alternative A, knowing that it does not meet two key requirements, because it has the lowest cost. In short, what may appear to be the best choice for a systems development project may not always be the one that ends up being developed. By conducting a thorough analysis, organizations can greatly improve their decision-making performance.

Deliverables and Outcomes

The primary deliverable from the first part of the planning phase is a schedule of specific IS development projects, coming from both top-down and bottom-up sources, to move into the next part of the planning phase—project initiation and planning (see Figure 4-5). An outcome of this phase is the assurance that careful consideration was given to project selection, with a clear understanding of how each project can help the organization reach its objectives. Due to the principle of **incremental commitment**, a selected project does not necessarily result in a working system. After each subsequent SDLC phase, you, other members of the project team, and organizational officials will reassess your project to determine whether the business conditions have changed or whether a more detailed understanding of a system's costs, benefits, and risks would suggest that the project is not as worthy as previously thought.

Incremental commitment
A strategy in systems analysis and design in which the project is reviewed after each phase and continuation of the project is rejustified.

FIGURE 4-5
Information systems development projects come from both top-down and bottom-up initiatives

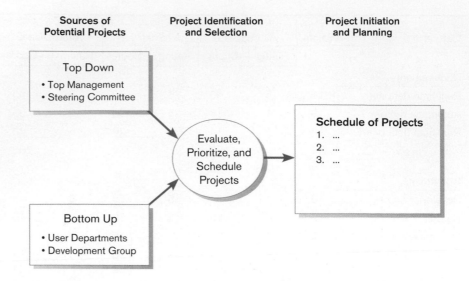

Many organizations have found that in order to make good project selection decisions, a clear understanding of overall organizational business strategy and objectives is required. This means that a clear understanding of the business and the desired role of information systems in achieving organizational goals is a precondition to improving the identification and selection process. In the next section, we provide a brief overview of the process many organizations follow, involving corporate strategic planning and information systems planning, when setting their business strategy and objectives and when defining the role of information systems in their plans.

CORPORATE AND INFORMATION SYSTEMS PLANNING

Although there are numerous motivations for carefully planning the identification and selection of projects (see Atkinson, 1990; Kelly, 2006; Luftman, 2004; Ross and Feeny, 2000), organizations have not traditionally used a systematic planning process when determining how to allocate IS resources. Instead, projects have often resulted from attempts to solve isolated organizational problems. In effect, organizations have asked the question: "What procedure (application program) is required to solve this particular problem as it exists today?" The difficulty with this approach is that the required organizational procedures are likely to change over time as the environment changes. For example, a company may decide to change its method of billing customers or a university may change its procedure for registering students. When such changes occur, it is usually necessary to again modify existing information systems.

In contrast, planning-based approaches essentially ask the question: "What information (or data) requirements will satisfy the decision-making needs or business processes of the enterprise today and well into the future?" A major advantage of this approach is that an organization's informational needs are less likely to change (or will change more slowly) than its business processes. For example, unless an organization fundamentally changes its business, its underlying data structures may remain reasonably stable for more than 10 years. However, the procedures used to access and process the data may change many times during that period. Thus, the challenge of most organizations is to design comprehensive information models containing data that are relatively independent from the languages and programs used to access, create, and update them.

To benefit from a planning-based approach for identifying and selecting projects, an organization must analyze its information needs and plan its projects carefully. Without careful planning, organizations may construct databases and systems that support individual processes but do not provide a resource that can be easily shared throughout the organization. Further, as business processes change, lack of data and systems integration will hamper the speed at which the organization can effectively make business strategy or process changes.

The need for improved information systems project identification and selection is readily apparent when we consider factors such as the following:

1. The cost of information systems has risen steadily and approaches 40 percent of total expenses in some organizations.
2. Many systems cannot handle applications that cross organizational boundaries.
3. Many systems often do not address the critical problems of the business as a whole or support strategic applications.
4. Data redundancy is often out of control, and users may have little confidence in the quality of data.
5. Systems maintenance costs are out of control as old, poorly planned systems must constantly be revised.
6. Application backlogs often extend three years or more, and frustrated end users are forced to create (or purchase) their own systems, often creating redundant databases and incompatible systems in the process.

Careful planning and selection of projects alone will certainly not solve all of these problems. We believe, however, that a disciplined approach, driven by top management commitment, is a prerequisite for most effectively applying information systems in order to reach organizational objectives. The focus of this section is to provide you with a clear understanding of how specific development projects with a broader organizational focus can be identified and selected. Specifically, we describe corporate strategic planning and information systems planning, two processes that can significantly improve the quality of project identification and selection decisions. This section also outlines the types of information about business direction and general systems requirements that can influence selection decisions and guide the direction of approved projects.

Corporate Strategic Planning

A prerequisite for making effective project selection decisions is to gain a clear idea of where an organization is, its vision of where it wants to be in the future, and how to make the transition to its desired future state. Figure 4-6 represents this as a three-step process. The first step focuses on gaining an understanding of the current enterprise. In other words, if you don't know where you are, it is impossible to tell where you are going. Next, top management must determine where it wants the enterprise to be in the future. Finally, after gaining an understanding of the current and future enterprise, a strategic plan can be developed to guide this transition. The process of developing and refining models of the current and future enterprise as well as a transition strategy is often referred to as **corporate strategic planning**. During corporate strategic planning, executives typically develop a mission statement, statements of future corporate objectives, and strategies designed to help the organization reach its objectives.

All successful organizations have a mission. The **mission statement** of a company typically states in very simple terms what business the company is in. For example, the mission statement for Pine Valley Furniture (PVF) is shown in Figure 4-7. After reviewing PVF's mission statement, it becomes clear that it is in the business of constructing and selling high-quality wood furniture to the general public, businesses, and institutions such as universities and hospitals. It is also clear that PVF is not in the business of fabricating steel file cabinets or selling its products through wholesale distributors. Based on this mission statement, you could conclude that PVF does not need a retail sales information system; instead, a high-quality human resource information system would be consistent with its goal.

Corporate strategic planning
An ongoing process that defines the mission, objectives, and strategies of an organization.

Mission statement
A statement that makes it clear what business a company is in.

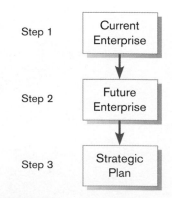

FIGURE 4-6
Corporate strategic planning is a three-step process

FIGURE 4-7
Mission statement (Pine Valley Furniture)

Pine Valley Furniture
Corporate Mission Statement

We are in the business of designing, fabricating, and selling to retail stores high-quality wood furniture for household, office, and institutional use. We value quality in our products and in our relationships with customers and suppliers. We consider our employees our most critical resource.

Objective statements

A series of statements that express an organization's qualitative and quantitative goals for reaching a desired future position.

After defining its mission, an organization can then define its objectives. **Objective statements** refer to "broad and timeless" goals for the organization. These goals can be expressed as a series of statements that are either qualitative or quantitative but that typically do not contain details likely to change substantially over time. Objectives are often referred to as *critical success factors*. Here, we will simply use the term *objectives*. The objectives for PVF are shown in Figure 4-8, with most relating to some aspect of the organizational mission. For example, the second objective relates to how PVF views its relationships with customers. This goal would suggest that PVF might want to invest in a Web-based order tracking system that would contribute to high-quality customer service. Once a company has defined its mission and objectives, a competitive strategy can be formulated.

Competitive strategy

The method by which an organization attempts to achieve its mission and objectives.

A **competitive strategy** is the method by which an organization attempts to achieve its mission and objectives. In essence, the strategy is an organization's game plan for playing in the competitive business world. In his classic book on competitive strategy, Michael Porter (1980) defined three generic strategies—low-cost producer, product differentiation, and product focus or niche—for reaching corporate objectives (see Table 4-3). These generic strategies allow you to more easily compare two companies in the same industry that may not employ the same competitive strategy. In addition, organizations employing different competitive strategies often have different informational needs to aid decision making. For example, Rolls-Royce and

Pine Valley Furniture
Statement of Objectives

1. PVF will strive to increase market share and profitability (prime objective).

2. PVF will be considered a market leader in customer service.

3. PVF will be innovative in the use of technology to help bring new products to market faster than our competition.

4. PVF will employ the fewest number of the highest-quality people necessary to accomplish our prime objective.

5. PVF will create an environment that values diversity in gender, race, values, and culture among employees, suppliers, and customers.

FIGURE 4-8
Statement of corporate objectives (Pine Valley Furniture)

TABLE 4-3 Generic Competitive Strategies

Strategy	Description
Low-Cost Producer	This strategy reflects competing in an industry on the basis of product or service cost to the consumer. For example, in the automobile industry, the South Korean–produced Hyundai is a product line that competes on the basis of low cost.
Product Differentiation	This competitive strategy reflects capitalizing on a key product criterion requested by the market (for example, high quality, style, performance, roominess). In the automobile industry, many manufacturers are trying to differentiate their products on the basis of quality (e.g., "At Ford, quality is job one.").
Product Focus or Niche	This strategy is similar to both the low-cost and differentiation strategies but with a much narrower market focus. For example, a niche market in the automobile industry is the convertible sports car market. Within this market, some manufacturers may employ a low-cost strategy and others may employ a differentiation strategy based on performance or style.

(*Source:* Based on The Free Press, a Division of Simon & Schuster Adult Publishing Group, from Porter, 1980. Copyright © 1980, 1998 by The Free Press. All rights reserved.)

Kia Motors are two car lines with different strategies: One is a high-prestige line in the ultra-luxury *niche*, whereas the other is a relatively *low-priced* line for the general automobile market. Rolls-Royce may build information systems to collect and analyze information on customer satisfaction to help manage a key company objective. Alternatively, Kia may build systems to track plant and material utilization in order to manage activities related to its low-cost strategy.

To effectively deploy resources such as the creation of a marketing and sales organization or to build the most effective information systems, an organization must clearly understand its mission, objectives, and strategy. A lack of understanding will make it impossible to know which activities are essential to achieving business objectives. From an information systems development perspective, by understanding which activities are most critical for achieving business objectives, an organization has a much greater chance to identify those activities that need to be supported by information systems. In other words, *it is only through the clear understanding of the organizational mission, objectives, and strategies that IS development projects should be identified and selected.* The process of planning how information systems can be employed to assist organizations to reach their objectives is the focus of the next section.

Information Systems Planning

The second planning process that can play a significant role in the quality of project identification and selection decisions is called **information systems planning (ISP)**. ISP is an orderly means of assessing the information needs of an organization and defining the information systems, databases, and technologies that will best satisfy those needs (Carlson et al., 1989; Cassidy, 2005; Luftman, 2004; Parker and Benson, 1989; Segars and Grover, 1999). This means that during ISP you (or, more likely, senior IS managers responsible for the IS plan) must model current and future organization informational needs and develop strategies and project plans to migrate the current information systems and technologies to their desired future state. ISP is a top-down process that takes into account the outside forces—industry, economic, relative size, geographic region, and so on—that are critical to the success of the firm. This means that ISP must look at information systems and technologies in terms of how they help the business achieve its objectives defined during corporate strategic planning.

The three key activities of this modeling process are represented in Figure 4-9. Like corporate strategic planning, ISP is a three-step process in which the first step is to assess current IS-related assets—human resources, data, processes, and technologies. Next, target blueprints of these resources are developed. These blueprints reflect the desired future state of resources needed by the organization to reach its objectives as defined during strategic planning. Finally, a series of scheduled projects is defined to help move the organization from its current to its future desired state. (Of course, scheduled projects from the ISP process are just one source for projects. Others include bottom-up requests from managers and business units, such as the SSR in Figure 3-2.)

Information systems planning (ISP)
An orderly means of assessing the information needs of an organization and defining the systems, databases, and technologies that will best satisfy those needs.

FIGURE 4-9
Information systems planning is a three-step process

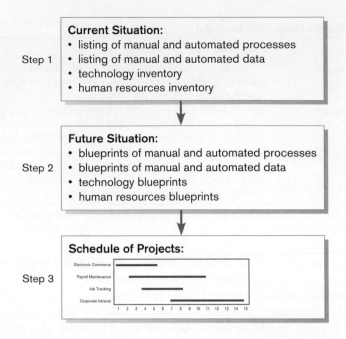

For example, a project may focus on reconfiguration of a telecommunications network to speed data communications or it may restructure work and data flows between business areas. Projects can include not only the development of new information systems or the modification of existing ones, but also the acquisition and management of new systems, technologies, and platforms. These three activities parallel those of corporate strategic planning, and this relationship is shown in Figure 4-10. Numerous methodologies such as Business Systems Planning (BSP) and Information Engineering (IE) have been developed to support the ISP process (see Segars and Grover, 1999); most contain the following three key activities:

1. *Describe the current situation.* The most widely used approach for describing the current organizational situation is generically referred to as top-down planning.

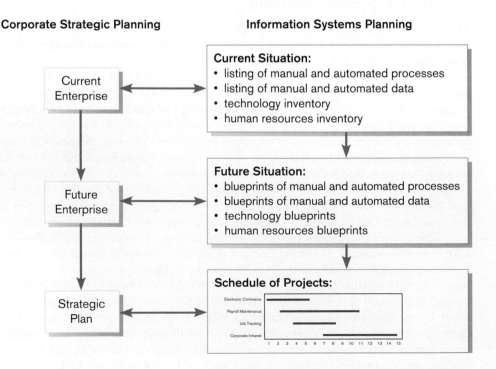

FIGURE 4-10
Parallel activities of corporate strategic planning and information systems planning

TABLE 4-4 Advantages to the Top-Down Planning Approach Over Other Planning Approaches

Advantage	Description
Broader Perspective	If not viewed from the top, information systems may be implemented without first understanding the business from general management's viewpoint.
Improved Integration	If not viewed from the top, totally new management information systems may be implemented rather than planning how to evolve existing systems.
Improved Management Support	If not viewed from the top, planners may lack sufficient management acceptance of the role of information systems in helping them achieve business objectives.
Better Understanding	If not viewed from the top, planners may lack the understanding necessary to implement information systems across the entire business rather than simply to individual operating units.

(*Source:* Based on IBM, 1982; Slater, 2002; Overby, 2008).

Top-down planning attempts to gain a broad understanding of the informational needs of the entire organization. The approach begins by conducting an extensive analysis of the organization's mission, objectives, and strategy and determining the information requirements needed to meet each objective. This approach to ISP implies by its name a high-level organizational perspective with active involvement of top-level management. The top-down approach to ISP has several advantages over other planning approaches, which are summarized in Table 4-4.

In contrast to the top-down planning approach, a **bottom-up planning** approach requires the identification of business problems and opportunities that are used to define projects. Using the bottom-up approach for creating IS plans can be faster and less costly than using the top-down approach and also has the advantage of identifying pressing organizational problems. Yet, the bottom-up approach often fails to view the informational needs of the *entire* organization. This can result in the creation of disparate information systems and databases that are redundant or not easily integrated without substantial rework.

The process of describing the current situation begins by selecting a planning team that includes executives chartered to model the existing situation. To gain this understanding, the team will need to review corporate documents; interview managers, executives, and customers; and conduct detailed reviews of competitors, markets, products, and finances. The type of information that must be collected to represent the current situation includes the identification of all organizational locations, units, functions, processes, data (or data entities), and information systems.

Within PVF, for example, organizational locations would consist of a list of all geographic areas in which the organization operates (e.g., the locations of the home and branch offices). Organizational units represent a list of people or business units that operate within the organization. Thus, organizational units would include vice president of manufacturing, sales manager, salesperson, and clerk. Functions are cross-organizational collections of activities used to perform day-to-day business operations. Examples of business functions might include research and development, employee development, purchasing, and sales. Processes represent a list of manual or automated procedures designed to support business functions. Examples of business processes might include payroll processing, customer billing, and product shipping. Data entities represent a list of the information items generated, updated, deleted, or used within business processes. Information systems represent automated and nonautomated systems used to transform data into information to support business processes. For example, Figure 4-11 shows portions of the business functions, data entities, and information systems of PVF. Once high-level

Top-down planning
A generic ISP methodology that attempts to gain a broad understanding of the information systems needs of the entire organization.

Bottom-up planning
A generic information systems planning methodology that identifies and defines IS development projects based upon solving operational business problems or taking advantage of some business opportunities.

PINE VALLEY FURNITURE

FIGURE 4-11
Information systems planning information
(Pine Valley Furniture)

FUNCTIONS:
• business planning
• product development
• marketing and sales
• production operations
• finance and accounting
• human resources
...

DATA ENTITIES:
• customer
• product
• vendor
• raw material
• order
• invoice
• equipment
...

INFORMATION SYSTEMS:
• payroll processing
• accounts payable
• accounts receivable
• time card processing
• inventory management
...

information is collected, each item can typically be decomposed into smaller units as more detailed planning is performed. Figure 4-12 shows the decomposition of several of PVF's high-level business functions into more detailed supporting functions.

After creating these lists, a series of matrices can be developed to cross reference various elements of the organization. The types of matrices typically developed include the following:

• Location-to-Function: This matrix identifies which business functions are being performed at various organizational locations.
• Location-to-Unit: This matrix identifies which organizational units are located in or interact with a specific business location.
• Unit-to-Function: This matrix identifies the relationships between organizational entities and each business function.
• Function-to-Objective: This matrix identifies which functions are essential or desirable in achieving each organizational objective.
• Function-to-Process: This matrix identifies which processes are used to support each business function.

FIGURE 4-12
Functional decomposition of information systems planning information (Pine Valley Furniture)

(*Source:* Microsoft Corporation.)

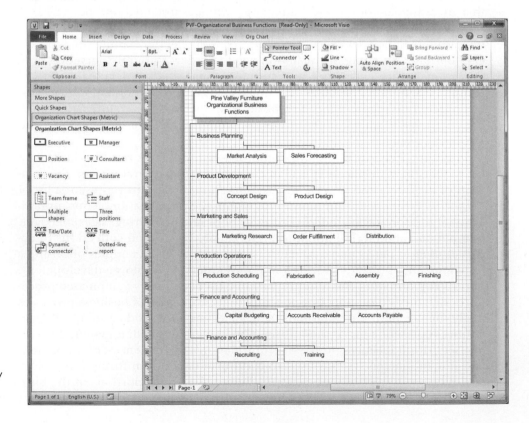

- Function-to-Data Entity: This matrix identifies which business functions utilize which data entities.
- Process-to-Data Entity: This matrix identifies which data are captured, used, updated, or deleted within each process.
- Process-to-Information System: This matrix identifies which information systems are used to support each process.
- Data Entity-to-Information System: This matrix identifies which data are created, updated, accessed, or deleted in each system.
- Information System-to-Objective: This matrix identifies which information systems support each business objective as identified during organizational planning.

Different matrices will have different relationships depending on what is being represented. For example, Figure 4-13 shows a portion of the Data Entity-to-Function matrix for PVF. The "X" in various cells of the matrix represents which business functions utilize which data entities. A more detailed picture of data utilization would be shown in the Process-to-Data Entity matrix (not shown here), in which the cells would be coded as "C" for the associated process that creates or captures data for the associated data entity, "R" for retrieve (or used), "U" for update, and "D" for delete. This means that different matrices can have different relationships depending on what is being represented. Because of this flexibility and ease of representing information, analysts use a broad range of matrices to gain a clear understanding of an organization's current situation and to plan for its future (Kerr, 1990). A primer on using matrices for ISP is provided in Figure 4-14.

2. *Describing the target situation, trends, and constraints.* After describing the current situation, the next step in the ISP process is to define the target situation that reflects the desired future state of the organization. This means that the target situation consists of the desired state of the locations, units, functions, processes, data, and IS (see Figure 4-9). For example, if a desired future state of the organization is to have several new branch offices or a new product line that requires several new employee positions, functions, processes, and data, then most lists and matrices will need to be updated to reflect this vision. The target situation must be developed in light of technology and business trends, in addition

FIGURE 4-13
Data Entity-to-Function matrix (Pine Valley Furniture)

	Customer	Product	Vendor	Raw Material	Order	Work Center	Equipment	Employees	Invoice	Work Order	...
Marketing and Sales											
Marketing Research	X	X									
Order Fulfillment	X	X			X				X		
Distribution	X	X									
Production Operation											
Production Scheduling						X	X	X		X	
Fabrication						X	X	X		X	
Assembly						X	X	X		X	
Finishing						X	X	X		X	
Finance and Accounting											
Capital Budgeting					X	X	X				
Accounts Receivable	X	X	X	X	X				X		
Accounts Payable											
...											

FIGURE 4-14
Making sense out of planning matrices

During the information systems planning process, before individual projects are identified and selected, a great deal of "behind the scenes" analysis takes place. During this planning period, which can span from six months to a year, IS planning team members develop and analyze numerous matrices like those described in the associated text. Matrices are developed to represent the current and the future views of the organization. Matrices of the "current" situation are called "as is" matrices. In other words, they describe the world "as" it currently "is." Matrices of the target or "future" situation are called "to be" matrices. Contrasting the current and future views provides insights into the relationships existing in important business information, and most important, forms the basis for the identification and selection of specific development projects. Many CASE tools provide features that will help you make sense out of these matrices in at least three ways:

1. **Management of Information.** A big part of working with complex matrices is managing the information. Using the dictionary features of the CASE tool repository, terms (such as *business functions and process* and *data entities*) can be defined or modified in a single location. All planners will therefore have the most recent information.
2. **Matrix Construction.** The reporting system within the CASE repository allows matrix reports to be easily produced. Because planning information can be changed at any time by many team members, an easy method to record changes and produce the most up-to-date reports is invaluable to the planning process.
3. **Matrix Analysis.** Possibly the most important feature CASE tools provide to planners is the ability to perform complex analyses within and across matrices. This analysis is often referred to as **affinity clustering**. Affinity refers to the extent to which information holds things in common. Thus, affinity clustering is the process of arranging matrix information so that clusters of information with some predetermined level or type of affinity are placed next to each other on a matrix report. For example, an affinity clustering of a Process-to-Data Entity matrix would create roughly a block diagonal matrix with processes that use similar data entities appearing in adjacent rows and data entities used in common by the same processes grouped into adjacent columns. This general form of analysis can be used by planners to identify items that often appear together (or should!). Such information can be used by planners to most effectively group and relate information (e.g., data to processes, functions to locations, and so on). For example, those data entities used by a common set of processes are candidates for a specific database. And those business processes that relate to a strategically important objective will likely receive more attention when managers from those areas request system changes.

Affinity clustering
The process of arranging planning matrix information so that clusters of information with a predetermined level or type of affinity are placed next to each other on a matrix report.

to organizational constraints. This means that lists of business trends and constraints should also be constructed in order to help ensure that the target situation reflects these issues.

In summary, to create the target situation, planners must first edit their initial lists and record the desired locations, units, functions, processes, data, and information systems within the constraints and trends of the organization environment (e.g., time, resources, technological evolution, competition, and so on). Next, matrices are updated to relate information in a manner consistent with the desired future state. Planners then focus on the differences between the current and future lists and matrices to identify projects and transition strategies.

3. *Developing a transition strategy and plans.* Once the creation of the current and target situations is complete, a detailed transition strategy and plan are developed by the IS planning team. This plan should be very comprehensive, reflecting broad, long-range issues in addition to providing sufficient detail to guide all levels of management concerning what needs doing, how, when, and by whom in the organization. The components of a typical information systems plan are outlined in Figure 4-15.

The IS plan is typically a very comprehensive document that looks at both short- and long-term organizational development needs. The short- and

I. **Organizational Mission, Objectives, and Strategy**
Briefly describes the mission, objectives, and strategy of the organization. The current and future views of the company are also briefly presented (i.e., where we are, where we want to be).

II. **Informational Inventory**
This section provides a summary of the various business processes, functions, data entities, and information needs of the enterprise. This inventory will view both current and future needs.

III. **Mission and Objectives of IS**
Description of the primary role IS will play in the organization to transform the enterprise from its current to future state. While it may later be revised, it represents the current best estimate of the overall role for IS within the organization. This role may be as a necessary cost, an investment, or a strategic advantage, for example.

IV. **Constraints on IS Development**
Briefly describes limitations imposed by technology and current level of resources within the company—financial, technological, and personnel.

V. **Overall Systems Needs and Long-Range IS Strategies**
Presents a summary of the overall systems needed within the company and the set of long-range (2–5 years) strategies chosen by the IS department to fill the needs.

VI. **The Short-Term Plan**
Shows a detailed inventory of present projects and systems and a detailed plan of projects to be developed or advanced during the current year. These projects may be the result of the long-range IS strategies or of requests from managers that have already been approved and are in some stage of the life cycle.

VII. **Conclusions**
Contains likely but not-yet-certain events that may affect the plan, an inventory of business change elements as presently known, and a description of their estimated impact on the plan.

FIGURE 4-15
Outline of an information systems plan

long-term developmental needs identified in the plan are typically expressed as a series of projects (see Figure 4-16). Projects from the long-term plan tend to build a foundation for later projects (such as transforming databases from old technology into newer technology). Projects from the short-term plan consist of specific steps to fill the gap between current and desired systems or respond to dynamic business conditions. The top-down (or plan-driven) projects join a set of bottom-up or needs-driven projects submitted as system service requests from managers to form the short-term systems development plan. Collectively, the short- and long-term projects set clear directions for the project selection process. The short-term plan includes not only those projects identified from the planning process but also those selected from among bottom-up requests. The overall IS plan may also influence all development projects. For example, the IS mission and IS constraints may cause projects to choose certain technologies or emphasize certain application features as systems are designed.

In this section, we outlined a general process for developing an IS plan. ISP is a detailed process and an integral part of deciding how to best deploy information systems and technologies to help reach organizational goals. It is beyond the scope

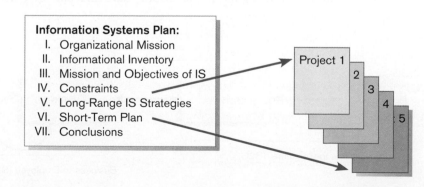

FIGURE 4-16
Systems development projects flow from the information systems plan

of this chapter, however, to extensively discuss ISP, yet it should be clear from our discussion that planning-based project identification and selection will yield substantial benefits to an organization. It is probably also clear to you that, as a systems analyst, you are not usually involved in IS planning because this process requires senior IS and corporate management participation. On the other hand, the results of IS planning, such as planning matrices like that in Figure 4-13, can be a source of very valuable information as you identify and justify projects.

ELECTRONIC COMMERCE APPLICATIONS: IDENTIFYING AND SELECTING SYSTEMS DEVELOPMENT PROJECTS

Identifying and selecting systems development projects for an Internet-based electronic commerce application is no different from the process followed for more traditional applications. Nonetheless, there are some special considerations when developing an Internet-based application. In this section, we highlight some of those issues that relate directly to the process of identifying and selecting Internet-related systems development projects.

Internet Basics

Internet
A large, worldwide network of networks that use a common protocol to communicate with each other.

Electronic commerce (EC)
Internet-based communication to support day-to-day business activities.

Business-to-consumer (B2C)
Electronic commerce between businesses and consumers.

Business-to-business (B2B)
Electronic commerce between business partners, such as suppliers and intermediaries.

Business-to-employee (B2E)
Electronic commerce between businesses and their employees.

Electronic data interchange (EDI)
The use of telecommunications technologies to directly transfer business documents between organizations.

The name **Internet** is derived from the concept of "internetworking"; that is, connecting host computers and their networks to form an even larger, global network. And that is essentially what the Internet is—a large, worldwide network of networks that use a common protocol to communicate with each other. The interconnected networks include computers running Windows, Linux, IOS, and many other network and computer types. The Internet stands as the most prominent representation of global networking. Using the Internet to support day-to-day business activities is broadly referred to as **electronic commerce (EC)**. However, not all Internet EC applications are the same. For example, there are three general classes of Internet EC applications: **business-to-consumer (B2C)**, **business-to-business (B2B)**, and **business-to-employee (B2E)**. Figure 4-17 shows three possible modes of EC using the Internet. B2C refers to business transactions between individual consumers and businesses. B2B refers to business transactions between business partners, such as suppliers and intermediaries. B2E refers to the use of the Internet within the same business, to support employee development and internal business processes. B2E is sometimes referred to as an Intranet.

B2E and B2B electronic commerce are examples of two ways organizations have been communicating via technology for years. For example, B2E is a lot like having a "global" local area network (LAN). Organizations utilizing B2E capabilities will select various applications or resources that are located on the Intranet—such as a customer contact database or an inventory-control system—that only members of the organization can access. Likewise, B2Bs use the Internet to provide similar capabilities to an established computing model, **electronic data interchange (EDI)**. EDI

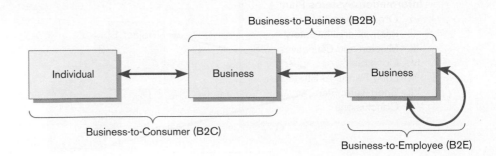

FIGURE 4-17
Three possible modes of electronic commerce

refers to the use of telecommunications technologies to directly transfer business documents between organizations. Using EDI, trading partners (suppliers, manufacturers, customers, etc.) establish computer-to-computer links that allow them to exchange data electronically. For example, a company using EDI may send an electronic purchase order instead of a paper request to a supplier. The paper order may take several days to arrive at the supplier, whereas an EDI purchase order will only take a few seconds. EDI-type data transfers over the Internet, generally referred to as B2B transactions, have become the standard by which organizations communicate with each other in the world of electronic commerce.

When developing either a B2E or B2B application, developers know who the users are, what applications will be used, the speed of the network connection, and the type of communication devices supported (e.g., Web browsers such as Firefox or Web-enabled smart phone such as the iPhone). On the other hand, when developing an Internet EC application (hereafter, simply EC), there are countless unknowns that developers have to discern in order to build a useful system. Table 4-5 lists a sample of the numerous unknowns to be dealt with when designing and building an EC application. These unknowns may result in making trade-offs based on a careful analysis of who the users are likely to be, where they are likely to be located, and how they are likely to be connected to the Internet. Even with all these difficulties to contend with, there is no shortage of Internet EC applications springing up all across the world. One company that has decided to get onto the Web with its own EC site is PVF.

Pine Valley Furniture WebStore

The board of directors of PVF has requested that a project team be created to explore the opportunity to develop an EC system. Specifically, market research has found that there is a good opportunity for online furniture purchases, especially in the areas of:

PINE
VALLEY
FURNITURE

- Corporate furniture
- Home office furniture
- Student furniture

The board wants to incorporate all three target markets into its long-term EC plan, but wants to initially focus on the corporate furniture buying system. Board members feel that this segment has the greatest potential to provide an adequate return on investment and would be a good building block for moving into the customer-based markets. Because the corporate furniture buying system will be specifically targeted to the business furniture market, it will be easier to define the system's operational requirements. Additionally, this EC system should integrate nicely with two currently existing systems: Purchasing Fulfillment and Customer Tracking. Together, these attributes make it an ideal candidate for initiating PVF's Web strategy. Throughout the remainder of the book, we will follow the evolution of the WebStore project until it becomes operational for PVF.

TABLE 4-5 Unknowns That Must Be Dealt with When Designing and Building Internet Applications

User	• Concern: Who is the user?
	• Example: Where is the user located? What is the user's expertise, education, or expectations?
Connection Speed	• Concern: What is the speed of the connection and what information can be effectively displayed?
	• Example: Modem, Cable Modem, DSL, Satellite, Broadband, Cellular
Access Method	• Concern: What is the method of accessing the net?
	• Example: Web browser, Personal Digital Assistant (PDA), Web-enabled Cellular Phone, Tablet, Web-enabled Television

SUMMARY

In this chapter, we described the first major activity of the planning phase of the SDLC—project identification and selection. Project identification and selection consists of three primary activities: identifying potential development projects, classifying and ranking projects, and selecting projects for development. A variety of organizational members or units can be assigned to perform this process, including top management, a diverse steering committee, business units and functional managers, the development group, or the most senior IS executive. Potential projects can be evaluated and selected using a broad range of criteria such as value chain analysis, alignment with business strategy, potential benefits, resource availability and requirements, and risks.

The quality of the project identification and selection process can be improved if decisions are guided by corporate strategic planning and ISP. Corporate strategic planning is the process of identifying the mission, objectives, and strategies of an organization. Crucial in this process is selecting a competitive strategy that states how the organization plans to achieve its objectives.

ISP is an orderly means for assessing the information needs of an organization and defining the systems and databases that will best satisfy those needs. ISP is a top-down process that takes into account outside forces that drive the business and the factors critical to the success of the firm. ISP evaluates the current inventory of systems and the desired future state of the organization and its system and determines which projects are needed

to transform systems to meet the future needs of the organization.

Corporate and IS planning are highly interrelated. Conceptually, these relationships can be viewed via various matrices that show how organizational objectives, locations, units, functions, processes, data entities, and systems relate to one another. Selected projects will be those viewed to be most important in supporting the organizational strategy.

The Internet is a global network consisting of thousands of interconnected individual networks that communicate with each other using a common protocol. Electronic commerce (EC) refers to the use of the Internet to support day-to-day business activities. Business-to-consumer EC refers to transactions between individual consumers and businesses. Business-to-employee EC refers to the use of the Internet within the same organization. Business-to-business EC refers to the use of the Internet between firms.

The focus of this chapter was to provide you with a clearer understanding of how organizations identify and select projects. Improved project identification and selection is needed for the following reasons: The cost of information systems is rising rapidly, systems cannot handle applications that cross organizational boundaries, systems often do not address critical organizational objectives, data redundancy is often out of control, and system maintenance costs continue to rise. Thus, effective project identification and selection is essential if organizations are to realize the greatest benefits from information systems.

KEY TERMS

1. Affinity clustering
2. Bottom-up planning
3. Business-to-business (B2B)
4. Business-to-consumer (B2C)
5. Business-to-employee (B2E)
6. Competitive strategy

7. Corporate strategic planning
8. Electronic commerce (EC)
9. Electronic data interchange (EDI)
10. Incremental commitment
11. Information systems planning (ISP)
12. Internet

13. Mission statement
14. Objective statements
15. Top-down planning
16. Value chain analysis

Match each of the key terms above with the definition that best fits it.

_____ Analyzing an organization's activities to determine where value is added to products and/or services and the costs incurred for doing so.

_____ A strategy in systems analysis and design in which the project is reviewed after each phase and continuation of the project is rejustified.

_____ An ongoing process that defines the mission, objectives, and strategies of an organization.

_____ A statement that makes it clear what business a company is in.

_____ A series of statements that express an organization's qualitative and quantitative goals for reaching a desired future position.

_____ The method by which an organization attempts to achieve its mission and objectives.

_____ An orderly means of assessing the information needs of an organization and defining the systems, databases, and technologies that will best satisfy those needs.

_____ A generic ISP methodology that attempts to gain a broad understanding of the information system needs of the entire organization.

_____ A generic ISP methodology that identifies and defines IS development projects based upon solving operational business problems or taking advantage of some business opportunities.

_____ The process of arranging planning matrix information so the clusters of information with a predetermined level or type of affinity are placed next to each other on a matrix report.

_____ A large, worldwide network of networks that use a common protocol to communicate with each other.

_____ Internet-based communication to support day-to-day business activities.

_____ Electronic commerce between businesses and consumers.

_____ Electronic commerce between business partners, such as suppliers and intermediaries.

_____ Electronic commerce between businesses and their employees.

_____ The use of telecommunications technologies to directly transfer business documents between organizations.

REVIEW QUESTIONS

1. Contrast the following terms:
   a. Mission; objective statements; competitive strategy
   b. Corporate strategic planning; ISP
   c. Top-down planning; bottom-up planning
   d. Low-cost producer; product differentiation; product focus or niche

2. Describe the project identification and selection process.

3. Describe several project evaluation criteria.

4. Describe value chain analysis and how organizations use this technique to evaluate and compare projects.

5. Discuss several factors that provide evidence for the need for improved ISP today.

6. Describe the steps involved in corporate strategic planning.

7. What are three generic competitive strategies?

8. Describe what is meant by ISP and the steps involved in the process.

9. List and describe the advantages of top-down planning over other planning approaches.

10. Briefly describe nine planning matrices that are used in ISP and project identification and selection.

11. Discuss some of the factors that must be considered when designing and building Internet applications.

PROBLEMS AND EXERCISES

1. Write a mission statement for a business that you would like to start. The mission statement should state the area of business you will be in and what aspect of the business you value highly.

2. When you are happy with the mission statement you have developed in response to the prior question, describe the objectives and competitive strategy for achieving that mission.

3. Consider an organization that you believe does not conduct adequate strategic IS planning. List at least six reasons why this type of planning is not done appropriately (or is not done at all). Are these reasons justifiable? What are the implications of this inadequate strategic IS planning? What limits, problems, weaknesses, and barriers might this present?

4. IS planning, as depicted in this chapter, is highly related to corporate strategic planning. What might those responsible for IS planning have to do if they operate in an organization without a formal corporate planning process?

5. The economic analysis carried out during the project identification and selection phase of the systems development life cycle is rather cursory. Why is this? Consequently, what factors do you think tend to be most important for a potential project to survive this first phase of the life cycle?

6. In those organizations that do an excellent job of IS planning, why might projects identified from a bottom-up process still find their way into the project initiation and planning phase of the life cycle?

7. Figure 4-14 introduces the concept of affinity clustering. Suppose that through affinity clustering it was found that three business functions provided the bulk of the use of five data entities. What implications might this have for project identification and subsequent steps in the systems development life cycle?

8. Timberline Technology manufactures membrane circuits in its Northern California plant. In addition, all circuit design and research and development work occur at this site. All finance, accounting, and human resource functions are headquartered at the parent company in the upper Midwest. Sales take place through six sales representatives located in various cities across the country. Information systems for payroll processing, accounts payable, and accounts receivable are located at the parent office while systems for inventory management and computer-integrated manufacturing are at the California plant. As best you can, list the locations, units, functions, processes, data entities, and information systems for this company.

9. For each of the following categories, create the most plausible planning matrices for Timberline Technology, described in Problem and Exercise 8: function-to-data entity, process-to-data entity, process-to-information system, data entity-to-information system. What other information systems not listed is Timberline likely to need?

10. The owners of Timberline Technology (described in Problem and Exercise 8) are considering adding a plant in Idaho and one in Arizona and six more sales representatives at various sites across the country. Update the matrices from Problem and Exercise 9 so that the matrices account for these changes.

FIELD EXERCISES

1. Obtain a copy of an organization's mission statement. (One can typically be found in an organization's annual report. Such reports are often available in university libraries or in corporate marketing brochures. If you are finding it difficult to locate this material, write or call the organization directly and ask for a copy of the mission statement.) What is this organization's area of business? What does the organization value highly (e.g., high-quality products and services, low cost to consumers, employee growth and development, etc.)? If the mission statement is well written, these concepts should be clear. Do you know anything about the information systems in this company that would demonstrate that the types of systems in place reflect the organization's mission? Explain.

2. Interview the managers of the information systems department of an organization to determine the level and nature of their strategic ISP. Does it appear to be adequate? Why or why not? Obtain a copy of that organization's mission statement. To what degree do the strategic IS plan and the organizational strategic plan fit together? What are the areas where the two plans fit and do not fit? If there is not a good fit, what are the implications for the success of the organization? For the usefulness of their information systems?

3. Choose an organization that you have contact with, perhaps your employer or university. Follow the "Outline of an information systems plan" shown in Figure 4-15 and complete a short information systems plan for the organization you chose. Write at least a brief paragraph for each of the seven categories in the outline. If IS personnel and managers are available, interview them to obtain information you need. Present your mock plan to the organization's IS manager and ask for feedback on whether or not your plan fits the IS reality for that organization.

4. Choose an organization that you have contact with, perhaps your employer or university. List significant examples for each of the items used to create planning matrices. Next, list possible relationships among various items and display these relationships in a series of planning matrices.

5. Write separate mission statements that you believe would describe Microsoft, IBM, and AT&T. Compare your mission statements with the real mission statements of these companies. Their mission statements can typically be found in their annual reports. Were your mission statements comparable to the real mission statements? Why or why not? What differences and similarities are there among these three mission statements? What information systems are necessary to help these companies deliver on their mission statements?

6. Choose an organization that you have contact with, perhaps your employer or university. Determine how information systems projects are identified. Are projects identified adequately? Are they identified as part of the ISP or the corporate strategic planning process? Why or why not?

REFERENCES

Atkinson, R. A. 1990. "The Motivations for Strategic Planning." *Journal of Information Systems Management* 7(4): 53–56.

Carlson, C. K., E. P. Gardner, and S. R. Ruth. 1989. "Technology-Driven Long-Range Planning." *Journal of Information Systems Management* 6(3): 24–29.

Cassidy, A. 2005. *A Practical Guide to Information Systems Strategic Planning.* London: CRC Press.

Dewan, S., S. C. Michael, and C-K. Min. 1998. "Firm Characteristics and Investments in Information Technology: Scale and Scope Effects." *Information Systems Research* 9(3): 219–232.

GAO. 2000. Information Technology Investment Management: A Framework for Assessing and Improving Process Maturity. U.S. Government Accountability Office. Available at *www.gao.gov/special.pubs/ai10123.pdf.* Accessed January 28, 2009.

Hasselbring, W. 2000. "Information System Integration." *Communications of the ACM* 43(6): 33–38.

IBM. 1982. "Business Systems Planning." In J. D. Couger, M. A. Colter, and R. W. Knapp (eds.), *Advanced System Development/Feasibility Techniques,* 236–314. New York: Wiley.

Kelly, R. T. 2006. "Adaptive and Aware: Strategy, Architecture, and IT Leadership in an Age of Commoditization." In P. A. Laplante and T. Costello (eds.), *CIO Wisdom II,* 249–69. Upper Saddle River, NJ: Prentice Hall.

Kerr, J. 1990. "The Power of Information Systems Planning." *Database Programming & Design* 3(12): 60–66.

King, J. 2003. "IT's Global Itinerary: Offshore Outsourcing Is Inevitable." Computerworld.com, September 15. Available at *www.computerworld.com.* Accessed February 6, 2009.

Koch, C. 2005. "Integration's New Strategy." CIO.com, September 15. Available at *www.cio.com.* Accessed February 6, 2006.

Laplante, P. A. 2006. "Software Return on Investment (ROI)." In P. A. Laplante and T. Costello (eds.), *CIO Wisdom II,* 163–76. Upper Saddle River, NJ: Prentice Hall.

Luftman, J. N. 2004. *Managing the Information Technology Resource.* With C. V. Bullen, D. Liao, E. Nash, and C. Neumann. Upper Saddle River, NJ: Prentice Hall.

McKeen, J. D., T. Guimaraes, and J. C. Wetherbe. 1994. "A Comparative Analysis of MIS Project Selection Mechanisms." *Data Base* 25(2): 43–59.

Newbold, D. L., and M. C. Azua. 2007. A Model for CIO-Led Innovation. *IBM Systems Journal* 46(4), 629–37.

Overby, S. 2006. "Big Deals, Big Savings, Big Problems." CIO. com, February 1. Available at *www.cio.com.* Accessed February 6, 2009.

Overby, S. 2008. "Tales from the Darkside: 8 IT Strategic Planning Mistakes to Avoid." January 22. Available at *www.cio .com.* Accessed February 10, 2009.

Parker, M. M., and R. J. Benson. 1989. "Enterprisewide Information Management: State-of-the-Art Strategic Planning." *Journal of Information Systems Management* 6 (Summer): 14–23.

Porter, M. 1980. *Competitive Strategy: Techniques for Analyzing Industries and Competitors.* New York: Free Press.

Porter, M. 1985. *Competitive Advantage.* New York: Free Press.

Ross, J., and D. Feeny. 2000. "The Evolving Role of the CIO." In R. W. Zmud (ed.), *Framing the Domains of IT Management: Projecting the Future from the Past,* 385–402. Cincinnati, OH: Pinnaflex Educational Resources.

Segars, A. H., and V. Grover. 1999. "Profiles of Strategic Information Systems Planning." *Information Systems Planning* 10(3): 199–232.

Shank, J. K., and V. Govindarajan. 1993. *Strategic Cost Management.* New York: Free Press.

Slater, D. 2002. Mistakes: Strategic Planning Don'ts (and Dos). June 1. Available at *www.cio.com.* Accessed February 10, 2009.

Thomas, G., and W. Fernandez. 2008. Success in IT Projects: A Matter of Definition? *International Journal of Project Management.* October: 733–42.

Yoo, M. J., R. S. Sangwan, and R. G. Qiu. 2006. "Enterprise Integration: Methods and Technologies." In P. A. Laplante and T. Costello (eds.), *CIO Wisdom II,* 107–26. Upper Saddle River, NJ: Prentice Hall.

PETRIE PETRIE ELECTRONICS

Chapter 4: Identifying and Selecting Systems Development Projects

J. K. Choi, chief financial officer for Petrie Electronics, came early to the quarterly IS Steering Committee meeting. Choi, who was the chair of the committee, took his seat at the head of the big table in the corporate conference room. He opened the cover on his tablet PC and looked at the agenda for the day's meeting. There were only a few proposed systems projects to consider today. He was familiar with the details of most of them. He briefly looked over the paperwork for each request. He didn't really think there was anything too controversial to be considered today. Most of the requests were pretty routine and involved upgrades to existing systems. The one totally new system being proposed for development was a customer loyalty system, referred to internally as "No Customer Escapes."

Choi chuckled at the name as he read through the proposal documents. "This is something we have needed for some time," he thought.

After about 15 minutes, his administrative assistant, Julie, came in. "Am I late or are you early?" she asked.

"No, you're not late," Choi said. "I wanted to come in a little early and look over the proposals. I wasn't able to spend as much time on these yesterday as I wanted."

As Julie was about to respond, the other members of the committee started to arrive. First was Ella Whinston, the chief operating officer. Choi knew that Ella was the champion for the customer loyalty project. She had talked about it for years now, it seemed to Choi. One of her people would make the presentation in support of the system. Choi knew she had buy in on the project from most of the other members of the c-suite. He also knew that Joe Swanson, Petrie director of IT, supported the project. Joe was away, but his assistant director, Jim Watanabe, would attend the meeting in his place. Ella had already let it be known that she expected Jim to be the project manager for the customer loyalty system project. Jim had just joined the company, but he had had five years of experience at Broadway Entertainment Company before its spectacular collapse. "Good thing I unloaded all that BEC stock I owned before the company went under," Choi thought. That reminded him of the meeting he had later today to plan the annual stockholders' meeting. "Better

not let the steering committee meeting run too long," he thought. "I've got more important things to do today."

Next to arrive was John Smith, the head of marketing. John, who was also a member of the steering committee, had been with Petrie for most of his career. He had been with the company longer than anyone else on the steering committee.

Just then, Jim Watanabe came speeding into the conference room. He almost ran into John Smith as he sailed into the room. It looked like he was about to drop his tablet and spill his coffee on Smith. Choi chuckled again.

"Welcome, everyone," Choi said. "I think we are all here. You all have copies of the agenda for this morning's meeting. Let's get started."

"Sorry to interrupt, JK," Ella said. "Bob Petroski is not here yet. He will be presenting the proposal on the customer loyalty system project. I don't know where he is. Maybe he got held up in traffic."

"The customer loyalty system discussion is the last item we will discuss today, so we can go ahead with the rest of the agenda. Bob does not need to be here for anything except that discussion," Choi explained.

Choi looked around the table once more. "OK, then, let's get started. Let's try to keep to the agenda as much as possible. And let's watch the clock. I know we are all busy, but I have a very important meeting this afternoon. Julie, see if you can locate Bob."

Case Questions

1. What is an IS steering committee? What are its major functions? Typically, who serves on such a committee? Why do these committees exist?
2. Where do ideas for new information systems originate in organizations?
3. What criteria are typically used to determine which new information systems projects to develop? What arguments might Bob Petroski make for developing the proposed customer loyalty system?
4. Look at Figure 4-4. What kind of information would you need to have to be able to put together a table like Figure 4-4 to present to the steering committee? How much of that information is objective? Subjective? Justify your answer.

Initiating and Planning Systems Development Projects

Learning Objectives

After studying this chapter, you should be able to:

- Describe the steps involved in the project initiation and planning process.

- Explain the need for and the contents of a Project Scope Statement and Baseline Project Plan.

- List and describe various methods for assessing project feasibility.

- Describe the differences between tangible and intangible benefits and costs and between one-time and recurring benefits and costs.

- Perform cost-benefit analysis and describe what is meant by the time value of money, present value, discount rate, net present value, return on investment, and break-even analysis.

- Describe the general rules for evaluating the technical risks associated with a systems development project.

- Describe the activities and participant roles within a structured walkthrough.

Introduction

During the first phase of the systems development life cycle (SDLC) planning, two primary activities are performed. The first, project identification and selection, focuses on the activities during which the need for a new or enhanced system is recognized. This activity does not deal with a specific project but rather identifies the portfolio of projects to be undertaken by the organization. Thus, project identification and selection is often thought of as a "preproject" step in the life cycle. This recognition of potential projects may come as part of a larger planning process, information systems planning, or from requests from managers and business units. Regardless of how a project is identified and selected, the next step is to conduct a more detailed assessment during project initiating and planning. This assessment does not focus on how the proposed system will operate but rather on understanding the scope of a proposed project and its feasibility of completion given the available resources. It is crucial that organizations understand whether resources should be devoted to a project; otherwise very expensive mistakes can be made (DeGiglio, 2002; Laplante, 2006). Thus, the focus of this chapter is on this process. Project initiation and planning is where projects are accepted for development, rejected, or redirected. This is also where you, as a systems analyst, begin to play a major role in the systems development process.

In the next section, the project initiation and planning process is briefly reviewed. Numerous techniques for assessing project feasibility are then described. We then discuss the process of building the Baseline Project Plan, which organizes the information uncovered during feasibility analysis. Once this plan is developed, a formal review of the project can be conducted. Yet, before the project can evolve to the next phase of the systems development life cycle—analysis—the project plan must be reviewed and accepted. In the final major section of the chapter, we provide an overview of the project review process.

INITIATING AND PLANNING SYSTEMS DEVELOPMENT PROJECTS

A key consideration when conducting project initiation and planning (PIP) is deciding when PIP ends and when analysis, the next phase of the SDLC, begins. This is a concern because many activities performed during PIP could also be completed during analysis. Pressman (2005) speaks of three important questions that must be considered when making this decision on the division between PIP and analysis:

1. How much effort should be expended on the project initiation and planning process?
2. Who is responsible for performing the project initiation and planning process?
3. Why is project initiation and planning such a challenging activity?

Finding an answer to the first question, how much effort should be expended on the PIP process, is often difficult. Practical experience has found, however, that the time and effort spent on initiation and planning activities easily pay for themselves later in the project. Proper and insightful project planning, including determining project scope as well as identifying project activities, can easily reduce time in later project phases. For example, a careful feasibility analysis that leads to deciding that a project is not worth pursuing can save a considerable expenditure of resources. The actual amount of time expended will be affected by the size and complexity of the project as well as by the experience of your organization in building similar systems. A rule of thumb is that between 10 and 20 percent of the entire development effort should be expended on the PIP study. Thus, you should not be reluctant to spend considerable time in PIP in order to fully understand the motivation for the requested system.

For the second question, who is responsible for performing PIP, most organizations assign an experienced systems analyst, or a team of analysts for large projects, to perform PIP. The analyst will work with the proposed customers (managers and users) of the system and other technical development staff in preparing the final plan. Experienced analysts working with customers who well understand their information services needs should be able to perform PIP without the detailed analysis typical of the analysis phase of the life cycle. Less-experienced analysts with customers who only vaguely understand their needs will likely expend more effort during PIP in order to be certain that the project scope and work plan are feasible.

As to the third question, PIP is viewed as a challenging activity because the objective of the PIP study is to transform a vague system request document into a tangible project description. This is an open-ended process. The analyst must clearly understand the motivation for and objectives of the proposed system. Therefore, effective communication among the systems analyst, users, and management is crucial to the creation of a meaningful project plan. Getting all parties to agree on the direction of a project may be difficult for cross-department projects where different parties have different business objectives. Thus, more complex organizational settings for projects will result in more time required for analysis of the current and proposed systems during PIP.

In the remainder of this chapter, we will describe the necessary activities used to answer these questions. In the next section, we will revisit the project initiation and planning activities originally outlined in Chapter 3 in the section on "Managing the Information Systems Project." This is followed by a brief description of the deliverables and outcomes from this process.

THE PROCESS OF INITIATING AND PLANNING IS DEVELOPMENT PROJECTS

As its name implies, two major activities occur during project initiation and planning (Figure 5-1). Because the steps of the project initiation and planning process were explained in Chapter 3, our primary focus in this chapter is to describe several techniques that are used when performing this process. Therefore, we will only briefly review the PIP process.

Project initiation focuses on activities designed to assist in organizing a team to conduct project planning. During initiation, one or more analysts are assigned to work with a customer—that is, a member of the business group that requested or will be affected by the project—to establish work standards and communication procedures. Examples of the types of activities performed are shown in Table 5-1. Depending upon the size, scope, and complexity of the project, some project initiation activities may be unnecessary or may be very involved. Also, many organizations have established procedures for assisting with common initiation activities. One key activity of project initiation is the development of the project charter (defined in Chapter 3).

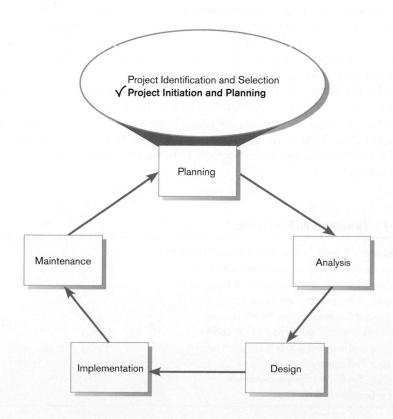

FIGURE 5-1
Systems development life cycle with project initiation and planning highlighted

TABLE 5-1   Elements of Project Initiation

- Establishing the Project Initiation Team
- Establishing a Relationship with the Customer
- Establishing the Project Initiation Plan
- Establishing Management Procedures
- Establishing the Project Management Environment and Project Workbook
- Developing the Project Charter

Project planning, the second activity within PIP, is distinct from general information systems planning, which focuses on assessing the information systems needs of the entire organization (discussed in Chapter 4). Project planning is the process of defining clear, discrete activities and the work needed to complete each activity within a single project. The objective of the project planning process is the development of a *Baseline Project Plan (BPP)* and the *Project Scope Statement (PSS)* (Morris and Sember, 2008). The BPP becomes the foundation for the remainder of the development project. The PSS produced by the team clearly outlines the objectives and constraints of the project for the customer. As with the project initiation process, the size, scope, and complexity of a project will dictate the comprehensiveness of the project planning process and resulting documents. Further, numerous assumptions about resource availability and potential problems will have to be made. Analysis of these assumptions and system costs and benefits forms a **business case**. The range of activities performed during project planning is listed in Table 5-2.

Business case

The justification for an information system, presented in terms of the tangible and intangible economic benefits and costs and the technical and organizational feasibility of the proposed system.

Baseline Project Plan (BPP)

A major outcome and deliverable from the project initiation and planning phase that contains the best estimate of a project's scope, benefits, costs, risks, and resource requirements.

Deliverables and Outcomes

The major outcomes and deliverables from the project initiation and planning phase are the Baseline Project Plan and the Project Scope Statement. The **Baseline Project Plan (BPP)** contains all information collected and analyzed during project initiation and planning. The plan reflects the best estimate of the project's scope, benefits, costs, risks, and resource requirements given the current understanding of the project. The BPP specifies detailed project activities for the next life cycle phase—analysis—and less detail for subsequent project phases (because these depend on the results of the analysis phase). Similarly, benefits, costs, risks, and resource requirements will become more specific and quantifiable as the project progresses. The BPP is used by the project selection committee to help decide whether the project should be accepted, redirected, or canceled. If selected, the BPP becomes the foundation document for all subsequent SDLC activities; however, it is also expected to evolve as the project evolves. That is, as new information is learned during subsequent SDLC phases, the baseline plan will be updated. Later in this chapter we describe how to construct the BPP.

TABLE 5-2   Elements of Project Planning

- Describing the Project Scope, Alternatives, and Feasibility
- Dividing the Project into Manageable Tasks
- Estimating Resources and Creating a Resource Plan
- Developing a Preliminary Schedule
- Developing a Communication Plan
- Determining Project Standards and Procedures
- Identifying and Assessing Risk
- Creating a Preliminary Budget
- Developing the Project Scope Statement
- Setting a Baseline Project Plan

The **Project Scope Statement (PSS)** is a short document prepared for the customer that describes what the project will deliver and outlines all work required to complete the project. The PSS ensures that both you and your customer gain a common understanding of the project. It is also a very useful communication tool. The PSS is a very easy document to create because it typically consists of a high-level summary of the BPP information (described later). Depending upon your relationship with your customer, the role of the PSS may vary. At one extreme, the PSS can be used as the basis of a formal contractual agreement outlining firm deadlines, costs, and specifications. At the other extreme, the PSS can simply be used as a communication vehicle to outline the current best estimates of what the project will deliver, when it will be completed, and the resources it may consume. A contract programming or consulting firm, for example, may establish a very formal relationship with a customer and use a PSS that is extensive and formal. Alternatively, an internal development group may develop a PSS that is only one to two pages in length and is intended to inform customers rather than to set contractual obligations and deadlines.

> **Project Scope Statement (PSS)**
> A document prepared for the customer that describes what the project will deliver and outlines generally at a high level all work required to complete the project.

ASSESSING PROJECT FEASIBILITY

All projects are feasible given unlimited resources and infinite time (Pressman, 2005). Unfortunately, most projects must be developed within tight budgetary and time constraints. This means that assessing project feasibility is a required activity for all information systems projects and is a potentially large undertaking. It requires that you, as a systems analyst, evaluate a wide range of factors. Typically, some of these factors will be more important than others for some projects and relatively unimportant for other projects. Although the specifics of a given project will dictate which factors are most important, most feasibility factors are represented by the following categories:

- Economic
- Technical
- Operational
- Scheduling
- Legal and contractual
- Political

Together, the culmination of these feasibility analyses forms the business case that justifies the expenditure of resources on the project. In the remainder of this section, we will examine various feasibility issues. We begin by looking at issues related to economic feasibility and then demonstrate techniques for conducting this analysis. This is followed by a discussion of techniques for assessing technical project risk. Finally, issues not directly associated with economic and technical feasibility, but no less important to ensuring project success, are discussed.

To help you better understand the feasibility assessment process, we will examine a project at Pine Valley Furniture (PVF). For this project, a System Service Request (SSR) was submitted by PVF's Vice President of Marketing Jackie Judson, to develop a Customer Tracking System (CTS) (Figure 5-2). Jackie feels that this system would allow PVF's marketing group to better track customer purchase activity and sales trends. She also feels that, if constructed, the CTS would provide many tangible and intangible benefits to PVF. This project was selected by PVF's Systems Priority Board for a project initiation and planning study. During project initiation, Senior Systems Analyst Jim Woo was assigned to work with Jackie to initiate and plan the project. At this point in the project, all project initiation activities have been completed. Jackie and Jim are now focusing on project planning activities in order to complete the BPP.

 PINE VALLEY FURNITURE

Assessing Economic Feasibility

The purpose of assessing **economic feasibility** is to identify the financial benefits and costs associated with the development project (Laplante, 2006). Economic feasibility

> **Economic feasibility**
> A process of identifying the financial benefits and costs associated with a development project.

Pine Valley Furniture
System Service Request

REQUESTED BY ___Jackie Judson___ DATE: ___August 20, 2014___

DEPARTMENT ___Marketing___

LOCATION ___Headquarters, 570c___

CONTACT ___Tel: 4-3290   FAX: 4-3270   E-Mail: jjudson___

TYPE OF REQUEST   URGENCY
[X] New System   [   ] Immediate: Operations are impaired or opportunity lost
[   ] System Enhancement   [   ] Problems exist, but can be worked around
[   ] System Error Correction [X] Business losses can be tolerated until new system installed

PROBLEM STATEMENT

Sales growth at PVF has caused a greater volume of work for the marketing department. This volume of work has greatly increased the volume and complexity of the data we need to deal with and understand. We are currently using manual methods and a complex PC-based electronic spreadsheet to track and forecast customer buying patterns. This method of analysis has many problems: (1) we are slow to catch buying trends as there is often a week or more delay before data can be taken from the point-of-sales system and manually enter it into our spreadsheet; (2) the process of manual data entry is prone to errors (which makes the results of our subsequent analysis suspect); and (3) the volume of data and the complexity of analyses conducted in the system seem to be overwhelming our current system—sometimes the program starts recalculating and never returns, while for others it returns information that we know cannot be correct.

SERVICE REQUEST

I request a thorough analysis of our current method of tracking and analysis of customer purchasing activity with the intent to design and build a completely new information system. This system should handle all customer purchasing activity, support display and reporting of critical sales information, and assist marketing personnel in understanding the increasingly complex and competitive business environment. I feel that such a system will improve the competitiveness of PVF, particularly in our ability to better serve our customers.

IS LIAISON   Jim Woo, 4-6207 FAX: 4-6200 E-Mail: jwoo ___

SPONSOR   Jackie Judson, Vice President, Marketing ___

-------------------- TO BE COMPLETED BY SYSTEMS PRIORITY BOARD ----------------------

[   ] Request approved   Assigned to ___
                             Start date ___
[   ] Recommend revision
[   ] Suggest user development
[   ] Reject for reason ___

FIGURE 5-2
System Service Request for Customer Tracking System (Pine Valley Furniture)

is often referred to as *cost-benefit analysis*. During project initiation and planning, it will be impossible for you to precisely define all benefits and costs related to a particular project. Yet it is important that you spend adequate time identifying and quantifying these items or it will be impossible for you to conduct an adequate economic analysis and make meaningful comparisons between rival projects. Here we will describe typical benefits and costs resulting from the development of an information system and provide several useful worksheets for recording costs and benefits. Additionally, several common techniques for making cost-benefit calculations are presented. These worksheets and techniques are used after each SDLC phase as the project is reviewed in order to decide whether to continue, redirect, or kill a project.

Determining Project Benefits   An information system can provide many benefits to an organization. For example, a new or renovated information system can automate monotonous jobs and reduce errors; provide innovative services to customers and suppliers; and improve organizational efficiency, speed, flexibility, and morale. In general, the benefits can be viewed as being both tangible and intangible. **Tangible benefits** refer to items that can be measured in dollars and with certainty. Examples of tangible benefits might include reduced personnel expenses, lower transaction costs, or higher profit margins. It is important to note that not all tangible benefits can be easily quantified. For example, a tangible benefit that allows a company to perform a task in 50 percent of the time may be difficult to quantify in terms of hard dollar savings. Most tangible benefits will fit within the following categories:

Tangible benefit
A benefit derived from the creation of an information system that can be measured in dollars and with certainty.

- Cost reduction and avoidance
- Error reduction
- Increased flexibility
- Increased speed of activity
- Improvement of management planning and control
- Opening new markets and increasing sales opportunities

Within the CTS at PVF, Jim and Jackie identified several tangible benefits, which are summarized on the tangible benefits worksheet shown in Figure 5-3. Jackie and Jim had to establish the values in Figure 5-3 after collecting information from users of the current customer tracking system. They first interviewed the person responsible for collecting, entering, and analyzing the correctness of the current customer tracking data. This person estimated that 10 percent of her time was spent correcting data entry errors. Given that this person's salary is $25,000, Jackie and Jim estimated an error-reduction benefit of $2,500. Jackie and Jim also interviewed managers who used the current customer tracking reports. Using this information they were able to estimate

TANGIBLE BENEFITS WORKSHEET Customer Tracking System Project	
	Year 1 through 5
A. Cost reduction or avoidance	$ 4,500
B. Error reduction	2,500
C. Increased flexibility	7,500
D. Increased speed of activity	10,500
E. Improvement in management planning or control	25,000
F. Other _____	0
TOTAL tangible benefits	**$50,000**

FIGURE 5-3
Tangible benefits for Customer Tracking System (Pine Valley Furniture)

TABLE 5-3 Intangible Benefits from the Development of an Information System

- Competitive necessity
- More timely information
- Improved organizational planning
- Increased organizational flexibility
- Promotion of organizational learning and understanding
- Availability of new, better, or more information
- Ability to investigate more alternatives
- Faster decision making
- More confidence in decision quality
- Improved processing efficiency
- Improved asset utilization
- Improved resource control
- Increased accuracy in clerical operations
- Improved work process that can improve employee morale or customer satisfaction
- Positive impacts on society
- Improved social responsibility
- Better usage of resources ("greener")

(*Source:* Based on Parker and Benson, 1988; Brynjolfsson and Yang, 1997; Keen, 2003; Cresswell, 2004.)

---

other tangible benefits. They learned that cost-reduction or avoidance benefits could be gained due to better inventory management. Also, increased flexibility would likely occur from a reduction in the time normally taken to manually reorganize data for different purposes. Further, improvements in management planning or control should result from a broader range of analyses in the new system. Overall, this analysis forecasts that benefits from the system would be approximately $50,000 per year.

Jim and Jackie also identified several intangible benefits of the system. Although these benefits could not be quantified, they will still be described in the final BPP. **Intangible benefits** refer to items that cannot be easily measured in dollars or with certainty. Intangible benefits may have direct organizational benefits, such as the improvement of employee morale, or they may have broader societal implications, such as the reduction of waste creation or resource consumption. Potential tangible benefits may have to be considered intangible during project initiation and planning because you may not be able to quantify them in dollars or with certainty at this stage in the life cycle. During later stages, such intangibles can become tangible benefits as you better understand the ramifications of the system you are designing. In this case, the BPP is updated and the business case revised to justify continuation of the project to the next phase. Table 5-3 lists numerous intangible benefits often associated with the development of an information system. Actual benefits will vary from system to system. After determining project benefits, project costs must be identified.

Determining Project Costs   Similar to benefits, an information system can have both tangible and intangible costs. **Tangible costs** refer to items that you can easily measure in dollars and with certainty. From an IS development perspective, tangible costs include items such as hardware costs, labor costs, and operational costs including employee training and building renovations. Alternatively, **intangible costs** are items that you cannot easily measure in terms of dollars or with certainty. Intangible costs can include loss of customer goodwill, employee morale, or operational inefficiency. Table 5-4 provides a summary of common costs associated with the development and operation of an information system. Predicting the costs associated with the development of an information system is an inexact science. IS researchers, however, have identified several guidelines for improving the cost-estimating process (see Table 5-5). Both underestimating and overestimating costs are problems you must avoid (Laplante, 2006; Lederer and Prasad, 1992; White and Lui, 2005). Underestimation results in cost overruns, whereas overestimation results in unnecessary allocation of resources that might be better utilized.

Besides tangible and intangible costs, you can distinguish IS-related development costs as either one-time or recurring (the same is true for benefits, although we do not discuss this difference for benefits). **One-time costs** refer to those associated with project initiation and development and the start-up of the system. These costs typically encompass activities such as systems development, new hardware and

---

Intangible benefit
A benefit derived from the creation of an information system that cannot be easily measured in dollars or with certainty.

Tangible cost
A cost associated with an information system that can be measured in dollars and with certainty.

Intangible cost
A cost associated with an information system that cannot be easily measured in terms of dollars or with certainty.

One-time cost
A cost associated with project start-up and development or system start-up.

TABLE 5-4   Possible Information Systems Costs

Type of Cost	Examples	Type of Cost	Examples
Procurement	Hardware, software, facilities infrastructure Management and staff Consulting and services	Project	Infrastructure replacement/improvements Project personnel Training Development activities Services and procurement Organizational disruptions Management and staff
Start-Up	Initial operating costs Management and staff Personnel recruiting	Operating	Infrastructure replacement/improvements System maintenance Management and staff User training and support

(*Source:* Based on King and Schrems, 1978; Sonje, 2008.)

TABLE 5-5   Guidelines for Better Cost Estimating

1. Have clear guidelines for creating estimates.
2. Use experienced developers and/or project managers for making estimates.
3. Develop a culture where all project participants are responsible for defining accurate estimates.
4. Use historical data to help in establishing better estimates of costs, risks, schedules, and resources.
5. Update estimates as the project progresses.
6. Monitor progress and record discrepancies to improve future estimates.

(*Source:* Based on Lederer and Prasad, 1992; Hubbard, 2007; Sonje, 2008.)

software purchases, user training, site preparation, and data or system conversion. When conducting an economic cost-benefit analysis, a worksheet should be created for capturing these expenses. For very large projects, one-time costs may be staged over one or more years. In these cases, a separate one-time cost worksheet should be created for each year. This separation will make it easier to perform present value calculations (described later). **Recurring costs** refer to those costs resulting from the ongoing evolution and use of the system. Examples of these costs typically include:

Recurring cost
A cost resulting from the ongoing evolution and use of a system.

- Application software maintenance
- Incremental data storage expenses
- Incremental communications
- New software and hardware leases
- Supplies and other expenses (e.g., paper, forms, data center personnel)

Both one-time and recurring costs can consist of items that are fixed or variable in nature. Fixed costs are costs that are billed or incurred at a regular interval and usually at a fixed rate (a facility lease payment). Variable costs are items that vary in relation to usage (long-distance phone charges).

During the process of determining project costs, Jim and Jackie identified both one-time and recurring costs for the project. These costs are summarized in Figures 5-4 and 5-5. These figures show that this project will incur a one-time cost of $42,500 and a recurring cost of $28,500 per year. One-time costs were established by discussing the system with Jim's boss, who felt that the system would require approximately four months to develop (at $5000 per month). To effectively run the new system, the marketing department would need to upgrade at least five of its current workstations (at $3000 each). Additionally, software licenses for each workstation (at $1000 each) and modest user training fees (ten users at $250 each) would be necessary.

FIGURE 5-4
One-time costs for Customer Tracking
System (Pine Valley Furniture)

ONE-TIME COSTS WORKSHEET Customer Tracking System Project	Year 0
A. Development costs	$20,000
B. New hardware	15,000
C. New (purchased) software, if any	
1. Packaged applications software	5,000
2. Other _____	0
D. User training	2,500
E. Site preparation	0
F. Other _____	0
TOTAL one-time costs	**$42,500**

As you can see from Figure 5-5, Jim and Jackie believe the proposed system will be highly dynamic and will require, on average, five months of annual maintenance, primarily for enhancements as users expect more from the system. Other ongoing expenses such as increased data storage, communications equipment, and supplies should also be expected. You should now have an understanding of the types of benefit and cost categories associated with an information systems project. It should be clear that there are many potential benefits and costs associated with a given project. Additionally, because the development and useful life of a system may span several years, these benefits and costs must be normalized into present-day values in order to perform meaningful cost-benefit comparisons. In the next section, we address the relationship between time and money.

The Time Value of Money   Most techniques used to determine economic feasibility encompass the concept of the **time value of money (TVM)**, which reflects the notion that money available today is worth more than the same amount tomorrow. As previously discussed, the development of an information system has both one-time and recurring costs. Furthermore, benefits from systems development will likely occur sometime in the future. Because many projects may be competing for the same investment dollars and may have different useful life expectancies, all costs and benefits must be viewed in relation to their present value when comparing investment options.

Time value of money (TVM)
The concept that money available today is worth more than the same amount tomorrow.

RECURRING COSTS WORKSHEET Customer Tracking System Project	Year 1 through 5
A. Application software maintenance	$25,000
B. Incremental data storage required: 20 GB × $50 (estimated cost/MB = $50)	1000
C. Incremental communications (lines, messages, . . .)	2000
D. New software or hardware leases	0
E. Supplies	500
F. Other _____	0
TOTAL recurring costs	**$28,500**

FIGURE 5-5
Recurring costs for Customer Tracking
System (Pine Valley Furniture)

A simple example will help in understanding the TVM. Suppose you want to buy a used car from an acquaintance and she asks that you make three payments of $1500 for three years, beginning next year, for a total of $4500. If she would agree to a single lump-sum payment at the time of sale (and if you had the money!), what amount do you think she would agree to? Should the single payment be $4500? Should it be more or less? To answer this question, we must consider the time value of money. Most of us would gladly accept $4500 today rather than three payments of $1500, because a dollar today (or $4500 for that matter) is worth more than a dollar tomorrow or next year, given that money can be invested. The rate at which money can be borrowed or invested is referred to as the *cost of capital*, and is called the **discount rate** for TVM calculations. Let's suppose that the seller could put the money received for the sale of the car in the bank and receive a 10 percent return on her investment. A simple formula can be used when figuring out the **present value** of the three $1500 payments:

$$PV_n = Y \times \frac{1}{(1 + i)^n}$$

where PV_n is the present value of Y dollars n years from now when i is the discount rate.

From our example, the present value of the three payments of $1500 can be calculated as

$$PV_1 = 1500 \times \frac{1}{(1 + .10)^1} = 1500 \times .9091 = 1363.65$$

$$PV_2 = 1500 \times \frac{1}{(1 + .10)^2} = 1500 \times .8264 = 1239.60$$

$$PV_3 = 1500 \times \frac{1}{(1 + .10)^3} = 1500 \times .7513 = 1126.95$$

where PV_1, PV_2, and PV_3 reflect the present value of each $1500 payment in years 1, 2, and 3, respectively.

To calculate the *net present value (NPV)* of the three $1500 payments, simply add the present values calculated previously ($NPV = PV_1 + PV_2 + PV_3 = 1363.65 + 1239.60 + 1126.95 = \$3730.20$). In other words, the seller could accept a lump-sum payment of $3730.20 as equivalent to the three payments of $1500, given a discount rate of 10 percent.

Given that we now know the relationship between time and money, the next step in performing the economic analysis is to create a summary worksheet reflecting the present values of all benefits and costs as well as all pertinent analyses. Due to the fast pace of the business world, PVF's System Priority Board feels that the useful life of many information systems may not exceed five years. Therefore, all cost-benefit analysis calculations will be made using a five-year time horizon as the upper boundary on all time-related analyses. In addition, the management of PVF has set its cost of capital to be 12 percent (i.e., PVF's discount rate). The worksheet constructed by Jim is shown in Figure 5-6.

Cell H11 of the worksheet displayed in Figure 5-6 summarizes the NPV of the total tangible benefits from the project. Cell H19 summarizes the NPV of the total costs from the project. The NPV for the project ($35,003) shows that, overall, benefits from the project exceed costs (see cell H22).

The overall return on investment (ROI) for the project is also shown on the worksheet in cell H25. Because alternative projects will likely have different benefit and cost values and, possibly, different life expectancies, the overall ROI value is very useful for making project comparisons on an economic basis. Of course, this example shows ROI for the overall project; an ROI analysis could be calculated for each year of the project.

Discount rate
The rate of return used to compute the present value of future cash flows.

Present value
The current value of a future cash flow.

FIGURE 5-6

Summary spreadsheet reflecting the present value calculations of all benefits and costs for the Customer Tracking System (Pine Valley Furniture)

(*Source:* Microsoft Corporation.)

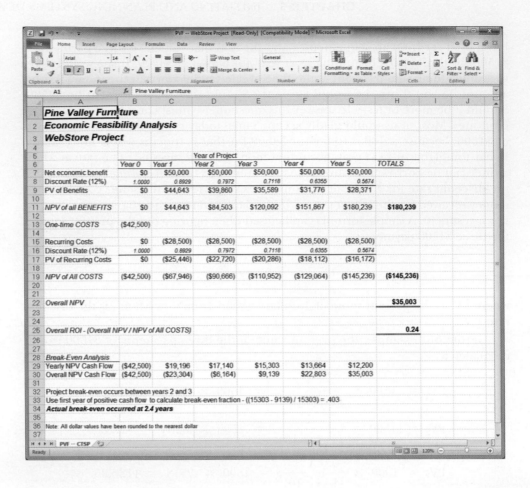

Break-even analysis

A type of cost-benefit analysis to identify at what point (if ever) benefits equal costs.

The last analysis shown in Figure 5-6 is a **break-even analysis**. The objective of the break-even analysis is to discover at what point (if ever) benefits equal costs (i.e., when breakeven occurs). To conduct this analysis, the NPV of the yearly cash flows are determined. Here, the yearly cash flows are calculated by subtracting both the one-time cost and the present values of the recurring costs from the present value of the yearly benefits. The overall NPV of the cash flow reflects the total cash flows for all preceding years. Examination of line 30 of the worksheet shows that break-even occurs between years 2 and 3. Because year 3 is the first in which the overall NPV cash flow figure is nonnegative, the identification of what point during the year break even occurs can be derived as follows:

$$\text{Break-Even Ratio} = \frac{\text{Yearly NPV Cash Flow} - \text{Overall NPV Cash Flow}}{\text{Yearly NPV Cash Flow}}$$

Using data from Figure 5-6,

$$\text{Break-Even Ratio} = \frac{15{,}303 - 9139}{15{,}303} = .403$$

Therefore, project breakeven occurs at approximately 2.4 years. A graphical representation of this analysis is shown in Figure 5-7. Using the information from the economic analysis, PVF's Systems Priority Board will be in a much better position to understand the potential economic impact of the CTS. It should be clear from this analysis that, without such information, it would be virtually impossible to know the cost-benefits of a proposed system and impossible to make an informed decision regarding approval or rejection of the service request.

TABLE 5-6   Commonly Used Economic Cost-Benefit Analysis Techniques

Analysis Technique	Description
Net Present Value (NPV)	NPV uses a discount rate determined from the company's cost of capital to establish the present value of a project. The discount rate is used to determine the present value of both cash receipts and outlays.
Return on Investment (ROI)	ROI is the ratio of the net cash receipts of the project divided by the cash outlays of the project. Trade-off analysis can be made among projects competing for investment by comparing their representative ROI ratios.
Break-Even Analysis (BEA)	BEA finds the amount of time required for the cumulative ⬚ flow ⬚ ⬚ project to equal its initial and ongoing

⬚ject's economic feasibility.
⬚ more than one year and will
⬚ year, most techniques for
⬚ TVM. Some of these cost-
⬚ers are more sophisticated.
⬚ for conducting economic
⬚ TVM or cost-benefit analysis
⬚ to review an introductory

⬚on, may not have to achieve
⬚ threshold as estimated dur-
⬚ot be able to quantify many
⬚al hurdles for a project may
⬚ an economic analysis as pos-
⬚ay be sufficient for the proj-
⬚economic analysis shown in
⬚nefit and cost estimates dur-
⬚le outcomes, along with the
⬚ting business unit, will often
⬚is phase. You must, however,
⬚ecially when investment capi-
⬚t some typical analysis phase
⬚der to clearly identify ineffi-
⬚to explain how a new system

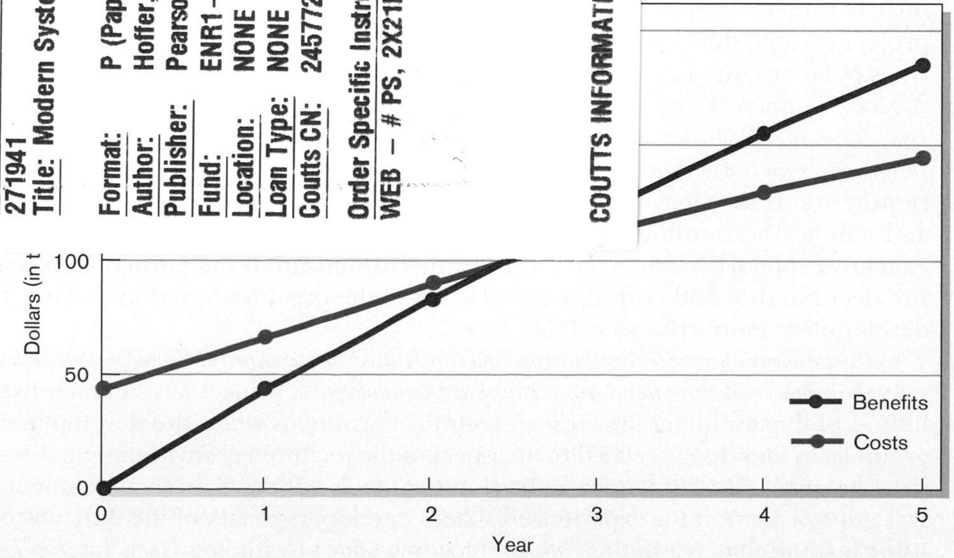

FIGURE 5-7
Break-even analysis for Customer Tracking System (Pine Valley Furniture)

will overcome these problems. Thus, building the economic case for a systems project is an open-ended activity; how much analysis is needed depends on the particular project, stakeholders, and business conditions. Also, conducting economic feasibility analyses for new types of information systems is often very difficult.

Assessing Technical Feasibility

The purpose of assessing **technical feasibility** is to gain an understanding of the organization's ability to construct the proposed system. This analysis should include an assessment of the development group's understanding of the possible target hardware, software, and operating environments to be used, as well as system size, complexity, and the group's experience with similar systems. In this section, we will discuss a framework you can use for assessing the technical feasibility of a project in which a level of project risk can be determined after answering a few fundamental questions.

It is important to note that all projects have risk and that risk is not necessarily something to avoid. Yet it is also true that, because organizations typically expect a greater return on their investment for riskier projects, understanding the sources and types of technical risks proves to be a valuable tool when you assess a project. Also, risks need to be managed in order to be minimized; you should, therefore, identify potential risks as early as possible in a project. The potential consequences of not assessing and managing risks can include the following:

- Failure to attain expected benefits from the project
- Inaccurate project cost estimates
- Inaccurate project duration estimates
- Failure to achieve adequate system performance levels
- Failure to adequately integrate the new system with existing hardware, software, or organizational procedures

You can manage risk on a project by changing the project plan to avoid risky factors, assigning project team members to carefully manage the risky aspects, and setting up monitoring methods to determine whether or not potential risk is, in fact, materializing.

The amount of technical risk associated with a given project is contingent on four primary factors: project size, project structure, the development group's experience with the application and technology area, and the user group's experience with systems development projects and the application area (see also Kirsch, 2000). Aspects of each of these risk areas are summarized in Table 5-7. Using these factors for conducting a technical risk assessment, four general rules emerge:

1. *Large projects are riskier than small projects.* Project size, of course, relates to the relative project size with which the development group typically works. A "small" project for one development group may be relatively "large" for another. The types of factors that influence project size are listed in Table 5-7.
2. *A system in which the requirements are easily obtained and highly structured will be less risky than one in which requirements are messy, ill structured, ill defined, or subject to the judgment of an individual.* For example, the development of a payroll system has requirements that may be easy to obtain due to legal reporting requirements and standard accounting procedures. On the other hand, the development of an executive support system would need to be customized to the particular executive decision style and critical success factors of the organization, thus making its development more risky (see Table 5-7).
3. *The development of a system employing commonly used or standard technology will be less risky than one employing novel or nonstandard technology.* A project has a greater likelihood of experiencing unforeseen technical problems when the development group lacks knowledge related to an aspect of the technology environment. A less risky approach is to use standard development tools and hardware environments. It is not uncommon for experienced system developers to talk of the difficulty of using leading-edge (or in their words, bleeding-edge) technology (see Table 5-7).

TABLE 5-7 **Project Risk Assessment Factors**

Risk Factor	Examples
Project Size	Number of members on the project team
	Project duration time
	Number of organizational departments involved in project
	Size of programming effort (e.g., hours, function points)
	Number of outsourcing partners
Project Structure	New system or renovation of existing system(s)
	Organizational, procedural, structural, or personnel changes resulting from system
	User perceptions and willingness to participate in effort
	Management commitment to system
	Amount of user information in system development effort
Development Group	Familiarity with target hardware, software development environment, tools, and operating system
	Familiarity with proposed application area
	Familiarity with building similar systems of similar size
User Group	Familiarity with information systems development process
	Familiarity with proposed application area
	Familiarity with using similar systems

(*Source:* Based on Applegate, Austin, and Soule, 2009; Tech Republic, 2005.)

4. *A project is less risky when the user group is familiar with the systems development process and application area than if the user group is unfamiliar with them.* Successful IS projects require active involvement and cooperation between the user and development groups. Users familiar with the application area and the systems development process are more likely to understand the need for their involvement and how this involvement can influence the success of the project (see Table 5-7).

A project with high risk may still be conducted. Many organizations look at risk as a portfolio issue: Considering all projects, it is okay to have a reasonable percentage of high-, medium-, and low-risk projects. Given that some high-risk projects will get into trouble, an organization cannot afford to have too many of these. Having too many low-risk projects may not be aggressive enough to make major breakthroughs in innovative uses of systems. Each organization must decide on its acceptable mix of projects of varying risk.

A matrix for assessing the relative risks related to the general rules just described is shown in Figure 5-8. Using the risk factor rules to assess the technical risk level of the CTS, Jim and Jackie concluded the following about their project:

1. The project is a relatively small project for PVF's development organization. The basic data for the system are readily available, so the creation of the system will not be a large undertaking.
2. The requirements for the project are highly structured and easily obtainable. In fact, an existing spreadsheet-based system is available for analysts to examine and study.
3. The development group is familiar with the technology that will likely be used to construct the system because the system will simply extend current system capabilities.
4. The user group is familiar with the application area because they are already using the PC-based spreadsheet system described in Figure 5-3.

Given this risk assessment, Jim and Jackie mapped their information into the risk framework of Figure 5-8. They concluded that this project should be viewed as having "very low" technical risk (cell 4 of the figure). Although this method is useful for gaining an understanding of technical feasibility, numerous other issues can influence the success of the project. These nonfinancial and nontechnical issues are described in the following section.

FIGURE 5-8
Effects of degree of project structure, project size, and familiarity with application area on project implementation risk
(*Source:* Based on Applegate, Austin, and Soule, 2009; Tech Republic, 2005.)

		Low Structure	High Structure
High Familiarity with Technology or Application Area	Large Project	(1) Low risk (very susceptible to mismanagement)	(2) Low risk
High Familiarity with Technology or Application Area	Small Project	(3) Very low risk (very susceptible to mismanagement)	(4) Very low risk
Low Familiarity with Technology or Application Area	Large Project	(5) Very high risk	(6) Medium risk
Low Familiarity with Technology or Application Area	Small Project	(7) High risk	(8) Medium-low risk

Assessing Other Feasibility Concerns

In this section, we will briefly conclude our discussion of project feasibility issues by reviewing other forms of feasibility that you may need to consider when formulating the business case for a system during project planning.

Operational feasibility
The process of assessing the degree to which a proposed system solves business problems or takes advantage of business opportunities.

Assessing Operational Feasiblity   The first relates to examining the likelihood that the project will attain its desired objectives, called **operational feasibility**. Its purpose is to gain an understanding of the degree to which the proposed system will likely solve the business problems or take advantage of the opportunities outlined in the System Service Request or project identification study. For a project motivated from information systems planning, operational feasibility includes justifying the project on the basis of being consistent with or necessary for accomplishing the information systems plan. In fact, the business case for any project can be enhanced by showing a link to the business or information systems plan. Your assessment of operational feasibility should also include an analysis of how the proposed system will affect organizational structures and procedures. Systems that have substantial and widespread impact on an organization's structure or procedures are typically riskier projects to undertake. Thus, it is important for you to have a clear understanding of how an information system will fit into the current day-to-day operations of the organization.

Schedule feasibility
The process of assessing the degree to which the potential time frame and completion dates for all major activities within a project meet organizational deadlines and constraints for affecting change.

Assessing Schedule Feasibility   Another feasibility concern relates to project duration and is referred to as assessing **schedule feasibility**. The purpose of assessing schedule feasibility is for you, as a systems analyst, to gain an understanding of the likelihood that all potential time frames and completion date schedules can be met and that meeting these dates will be sufficient for dealing with the needs of the organization. For example, a system may have to be operational by a government-imposed deadline, by a particular point in the business cycle (such as the beginning of the season when new products are introduced), or at least by the time a competitor is expected to introduce a similar system. Further, detailed activities may only be feasible if resources are available when called for in the schedule. For example, the schedule should not call for system testing during rushed business periods or for key project meetings during annual vacation or holiday periods. The schedule of activities produced during project initiation and planning will be very precise and detailed for the analysis phase. The estimated activities and associated times for activities after the analysis phase are typically not as detailed (e.g., it will take two weeks to program the payroll report module) but are rather at the life-cycle-phase level (e.g., it will take six weeks for physical design, four months for programming, and so on). This means that assessing schedule feasibility during project initiation and planning is more of a "rough-cut" analysis of whether the system can be completed within the constraints of the business opportunity or the desires of the users. While assessing schedule feasibility you should also evaluate scheduling trade-offs. For example, factors such as project team size, availability of key personnel, subcontracting or outsourcing activities, and changes in development environments may all be considered as having a possible impact on the eventual schedule. As with all forms of feasibility, schedule feasibility will be reassessed after each phase when you can specify with greater certainty the details of each step for the next phase.

Assessing Legal and Contractual Feasibility   A third concern relates to assessing **legal and contractual feasibility** issues. In this area, you need to gain an understanding of any potential legal ramifications due to the construction of the system. Possible considerations might include copyright or nondisclosure infringements, labor laws, antitrust legislation (which might limit the creation of systems to share data with other organizations), foreign trade regulations (e.g., some countries limit access to employee data by foreign corporations), and financial reporting standards, as well as current or pending contractual obligations. Contractual obligations may involve ownership of software used in joint ventures, license agreements for use of hardware or software, nondisclosure agreements with partners, or elements of a labor agreement (e.g., a union agreement may preclude certain compensation or work-monitoring capabilities a user may want in a system). A common situation is that development of a new application system for use on new computers may require new or expanded, and more costly, system software licenses. Typically, legal and contractual feasibility is a greater consideration if your organization has historically used an outside organization for specific systems or services that you now are considering handling yourself. In this case, ownership of program source code by another party may make it difficult to extend an existing system or link a new system with an existing purchased system.

Assessing Political Feasibility   A final feasibility concern focuses on assessing **political feasibility** in which you attempt to gain an understanding of how key stakeholders within the organization view the proposed system. Because an information system may affect the distribution of information within the organization, and thus the distribution of power, the construction of an information system can have political ramifications. Those stakeholders not supporting the project may take steps to block, disrupt, or change the intended focus of the project.

In summary, depending upon the given situation, numerous feasibility issues must be considered when planning a project. This analysis should consider economic, technical, operational, schedule, legal, contractual, and political issues related to the project. In addition to these considerations, project selection by an organization may be influenced by issues beyond those discussed here. For example, projects may be selected for construction despite high project costs and high technical risk if the system is viewed as a strategic necessity; that is, the organization views the project as being critical to the organization's survival. Alternatively, projects may be selected because they are deemed to require few resources and have little risk. Projects may also be selected due to the power or persuasiveness of the manager proposing the system. This means that project selection may be influenced by factors beyond those discussed here and beyond items that can be analyzed. Understanding the reality that projects may be selected based on factors beyond analysis, your role as a systems analyst is to provide a thorough examination of the items that can be assessed. Your analysis will ensure that a project review committee has as much information as possible when making project approval decisions. In the next section, we discuss how project plans are typically reviewed.

Legal and contractual feasibility
The process of assessing potential legal and contractual ramifications due to the construction of a system.

Political feasibility
The process of evaluating how key stakeholders within the organization view the proposed system.

BUILDING AND REVIEWING THE BASELINE PROJECT PLAN

All the information collected during project initiation and planning is collected and organized into a document called the Baseline Project Plan. Once the BPP is completed, a formal review of the project can be conducted with project clients and

BASELINE PROJECT PLAN REPORT

1.0 Introduction
  A. Project Overview—Provides an executive summary that specifies the project's scope, feasibility, justification, resource requirements, and schedules. Additionally, a brief statement of the problem, the environment in which the system is to be implemented, and constraints that affect the project are provided.
  B. Recommendation—Provides a summary of important findings from the planning process and recommendations for subsequent activities.

2.0 System Description
  A. Alternatives—Provides a brief presentation of alternative system configurations.
  B. System Description—Provides a description of the selected configuration and a narrative of input information, tasks performed, and resultant information.

3.0 Feasibility Assessment
  A. Economic Analysis—Provides an economic justification for the system using cost-benefit analysis.
  B. Technical Analysis—Provides a discussion of relevant technical risk factors and an overall risk rating of the project.
  C. Operational Analysis—Provides an analysis of how the proposed system solves business problems or takes advantage of business opportunities in addition to an assessment of how current day-to-day activities will be changed by the system.
  D. Legal and Contractual Analysis—Provides a description of any legal or contractual risks related to the project (e.g., copyright or nondisclosure issues, data capture or transferring, and so on).
  E. Political Analysis—Provides a description of how key stakeholders within the organization view the proposed system.
  F. Schedules, Time Line, and Resource Analysis—Provides a description of potential time frame and completion date scenarios using various resource allocation schemes.

4.0 Management Issues
  A. Team Configuration and Management—Provides a description of the team member roles and reporting relationships.
  B. Communication Plan—Provides a description of the communication procedures to be followed by management, team members, and the customer.
  C. Project Standards and Procedures—Provides a description of how deliverables will be evaluated and accepted by the customer.
  D. Other Project-Specific Topics—Provides a description of any other relevant issues related to the project uncovered during planning.

FIGURE 5-9
Outline of a Baseline Project Plan

other interested parties. This presentation, a walkthrough, is discussed later in this chapter. The focus of this review is to verify all information and assumptions in the baseline plan before moving ahead with the project.

Building the Baseline Project Plan

As mentioned previously, the project size and organizational standards will dictate the comprehensiveness of the project initiation and planning process as well as the

BPP. Yet most experienced systems builders have found project planning and a clear project plan to be invaluable to project success. An outline of a BPP is provided in Figure 5-9, which shows that it contains four major sections:

1. Introduction
2. System Description
3. Feasibility Assessment
4. Management Issues

The Introduction Section of the Baseline Project Plan The purpose of the *Introduction* is to provide a brief overview of the entire document and outline a recommended course of action for the project. The entire Introduction section is often limited to only a few pages. Although the Introduction section is sequenced as the first section of the BPP, it is often the final section to be written. It is only after performing most of the project planning activities that a clear overview and recommendation can be created. One activity that should be performed initially is the definition of project scope.

When defining scope for the CTS within PVF, Jim Woo first needed to gain a clear understanding of the project's objectives. To do this, Jim briefly interviewed Jackie Judson and several of her colleagues to gain a clear idea of their needs. He also spent a few hours reviewing the existing system's functionality, processes, and data use requirements for performing customer tracking activities. These activities provided him with the information needed to define the project scope and to identify possible alternative solutions. Alternative system solutions can relate to different system scopes, platforms for deployment, or approaches to acquiring the system. We elaborate on the idea of alternative solutions, called *design strategies*, when we discuss the analysis phase of the life cycle. During project initiation and planning, the most crucial element of the design strategy is the system's scope. In sum, a determination of scope will depend on the following factors:

- Which organizational units (business functions and divisions) might be affected by or use the proposed system or system change?
- With which current systems might the proposed system need to interact or be consistent, or which current systems might be changed due to a replacement system?
- Who inside and outside the requesting organization (or the organization as a whole) might care about the proposed system?
- What range of potential system capabilities will be considered?

The Project Scope Statement for the CTS project is shown in Figure 5-10.

For the CTS, project scope was defined using only textual information. It is not uncommon, however, to define project scope using diagrams such as data flow diagrams and entity-relationship models. For example, Figure 5-11 shows a context-level data flow diagram used to define system scope for PVF's Purchasing Fulfillment System. The other items in the Introduction section of the BPP are simply executive summaries of the other sections of the document.

The System Description Section of the Baseline Project Plan The second section of the BPP is the *System Description* where you outline possible alternative solutions in addition to the one deemed most appropriate for the given situation. Note that this description is at a very high level, mostly narrative in form. For example, alternatives may be stated as simply as this:

1. Web-based online system
2. Mainframe with central database
3. Local area network with decentralized databases
4. Batch data input with online retrieval
5. Purchasing of a prewritten package

Pine Valley Furniture **Project Scope Statement**	Prepared by: Jim Woo Date: September 10, 2014

General Project Information
    Project Name: Customer Tracking System
    Sponsor: Jackie Judson, VP Marketing
    Project Manager: Jim Woo

Problem/Opportunity Statement:
    Sales growth has outpaced the Marketing department's ability to accurately track and forecast customer buying trends. An improved method for performing this process must be found in order to reach company objectives.

Project Objectives:
    To enable the Marketing department to accurately track and forecast customer buying patterns in order to better serve customers with the best mix of products. This will also enable PVF to identify the proper application of production and material resources.

Project Description:
    A new information system will be constructed that will collect all customer purchasing activity, support display and reporting of sales information, aggregate data, and show trends in order to assist marketing personnel in understanding dynamic market conditions. The project will follow PVF's systems development life cycle.

Business Benefits:
    Improved understanding of customer buying patterns
    Improved utilization of marketing and sales personnel
    Improved utilization of production and materials

Project Deliverables:
    Customer tracking system analysis and design
    Customer tracking system programs
    Customer tracking documentation
    Training procedures

Estimated Project Duration:
    5 months

FIGURE 5-10
Project Scope Statement for the Customer Tracking Systems (Pine Valley Furniture)

If the project is approved for construction or purchase, you will need to collect and structure information in a more detailed and rigorous manner during the analysis phase and evaluate in greater depth these and other alternative directions for the system. At this point in the project, your objective is only to identify the most obvious alternative solutions.

When Jim and Jackie were considering system alternatives for the CTS, they focused on two primary issues. First, they discussed how the system would be acquired and considered three options: *purchase* the system if one could be found that met PVF's needs, *outsource* the development of the system to an outside organization, or *build* the system within PVF. The second issue focused on defining the comprehensiveness of the system's functionality. To complete this task, Jim asked Jackie to write

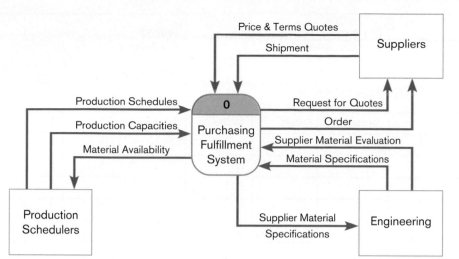

FIGURE 5-11
Context-level data flow diagram showing project scope for Purchasing Fulfillment System (Pine Valley Furniture)

a series of statements listing the types of tasks that she envisioned marketing personnel would be able to accomplish when using the CTS. This list of statements became the basis of the system description and was instrumental in helping them make their acquisition decision. After considering the unique needs of the marketing group, both decided that the best decision was to build the system within PVF.

The Feasibility Assessment Section of the Baseline Project Plan   In the third section, *Feasibility Assessment,* issues related to project costs and benefits, technical difficulties, and other such concerns are outlined. This is also the section where high-level project schedules are specified using network diagrams and Gantt charts. Recall from Chapter 3 that this process is referred to as a *work-breakdown structure.* During project initiation and planning, task and activity estimates are not generally detailed. An accurate work breakdown can be done only for the next one or two life cycle activities. After defining the primary tasks for the project, an estimate of the resource requirements can be made. As with defining tasks and activities, this activity is primarily concerned with gaining rough estimates of the human resource requirements because people are the most expensive resource element. Once you define the major tasks and resource requirements, a preliminary schedule can be developed. Defining an acceptable schedule may require that you find additional or different resources or that the scope of the project be changed. The greatest amount of project planning effort is typically expended on these Feasibility Assessment activities.

The Management Issues Section of the Baseline Project Plan   The final section, *Management Issues,* outlines a number of managerial concerns related to the project. This will be a very short section if the proposed project is going to be conducted exactly as prescribed by the organization's standard systems development methodology. Most projects, however, have some unique characteristics that require minor to major deviation from the standard methodology. In the Team Configuration and Management portion, you identify the types of people to work on the project, who will be responsible for which tasks, and how work will be supervised and reviewed (Figure 5-12). In the Communications Plan portion, you explain how the user will be kept informed about project progress (such as periodic review meetings or even a newsletter) and what mechanisms will be used to foster sharing of ideas among team members, such as some form of computer-based conference facility (Figure 5-13). An example of the type of information contained in the Project Standards and Procedures portion would be procedures for submitting and approving project change requests and any other issues deemed important for the project's success.

Project: WebStore		Prepared by: Juan Gonzales			Legend: P = Primary		
Manager: Juan Gonzales		Page: 1 of 1			S = Support		
		Responsibility Matrix					
Task ID	Task	Jordan	James	Jackie	Jeremy	Kim	Juan
A	Collect Requirements	P	S				S
B	Develop Data Model			P		S	S
C	Develop Program Interface			P		S	S
D	Build Database			S		P	S
E	Design Test Scenarios	S	S	S	P	S	S
F	Run Test Scenarios	S	S	S	S	S	P
G	Create User Documentation	P	S				S
H	Install System	S	P			S	S
I	Develop Customer Support	S	P			S	S

FIGURE 5-12
Task responsibility matrix

You should now have a feel for how a BPP is constructed and the types of information it contains. Its creation is not meant to be a project in and of itself, but rather a step in the overall systems development process. Developing the BPP has two primary objectives. First, it helps to ensure that the customer and development group share a common understanding of the project. Second, it helps to provide the sponsoring organization with a clear idea of the scope, benefits, and duration of the project.

Reviewing the Baseline Project Plan

Before the next phase of the SDLC can begin, the users, management, and development group must review the BPP in order to verify that it makes sense. This review takes place before the BPP is submitted or presented to a project approval body, such as an IS steering committee or the person who must fund the project. The objective of this review is to ensure that the proposed system conforms to organizational standards and to make sure that all relevant parties understand and agree with the information contained in the BPP. A common method for performing this review (as well as reviews during subsequent life cycle phases) is called a *structured walkthrough*. **Walkthroughs** are

Walkthrough

A peer group review of any product created during the systems development process; also called a *structured walkthrough*.

Stakeholder	Document	Format	Team Contact	Date Due
Team Members	Project Status Report	Project Intranet	Juan and Kim	First Monday of Month
Management Supervisor	Project Status Report	Hard Copy	Juan and Kim	First Monday of Month
User Group	Project Status Report	Hard Copy	James and Kim	First Monday of Month
Internal IT Staff	Project Status Report	E-Mail	Jackie and James	First Monday of Month
IT Manager	Project Status Report	Hard Copy	Juan and Jeremy	First Monday of Month
Contract Programmers	Software Specifications	E-Mail/Project Intranet	Jordan and Kim	October 4, 2014
Training Subcontractor	Implementation and Training Plan	Hard Copy	Jordan and James	January 10, 2015

FIGURE 5-13
The Project Communication Matrix provides a high-level summary of the communication plan

peer group reviews of any product created during the systems development process and are widely used by professional development organizations. Experience has shown that walkthroughs are a very effective way to ensure the quality of an information system and have become a common day-to-day activity for many systems analysts.

Most walkthroughs are not rigidly formal or exceedingly long in duration. It is important, however, that a specific agenda be established for the walkthrough so that all attendees understand what is to be covered and the expected completion time. At walkthrough meetings, there is a need to have individuals play specific roles. These roles are as follows (Yourdon, 1989):

- *Coordinator.* This person plans the meeting and facilitates a smooth meeting process. This person may be the project leader or a lead analyst responsible for the current life cycle step.
- *Presenter.* This person describes the work product to the group. The presenter is usually an analyst who has done all or some of the work being presented.
- *User.* This person (or group) makes sure that the work product meets the needs of the project's customers. This user would usually be someone not on the project team.
- *Secretary.* This person takes notes and records decisions or recommendations made by the group. This may be a clerk assigned to the project team or it may be one of the analysts on the team.
- *Standards bearer.* The role of this person is to ensure that the work product adheres to organizational technical standards. Many larger organizations have staff groups within the unit responsible for establishing standard procedures, methods, and documentation formats. These standards bearers validate the work so that it can be used by others in the development organization.
- *Maintenance oracle.* This person reviews the work product in terms of future maintenance activities. The goal is to make the system and its documentation easy to maintain.

After Jim and Jackie completed their BPP for the CTS, Jim approached his boss and requested that a walkthrough meeting be scheduled and that a walkthrough coordinator be assigned to the project. PVF assists the coordinator by providing a Walkthrough Review Form, shown in Figure 5-14. Using this form, the coordinator can more easily make sure that a qualified individual is assigned to each walkthrough role; that each member has been given a copy of the review materials; and that each member knows the agenda, date, time, and location of the meeting. At the meeting, Jim presented the BPP and Jackie added comments from a user's perspective. Once the walkthrough presentation was completed, the coordinator polled each representative for his or her recommendation concerning the work product. The results of this voting may result in validation of the work product, validation pending changes suggested during the meeting, or a suggestion that the work product requires major revision before being presented for approval. In this latter case, substantial changes to the work product are usually requested, after which another walkthrough must be scheduled before the project can be proposed to the Systems Priority Board (steering committee). In the case of the CTS, the BPP was supported by the walkthrough panel, pending some minor changes to the duration estimates in the schedule. These suggested changes were recorded by the secretary on a Walkthrough Action List (see Figure 5-15) and given to Jim to incorporate into a final version of the baseline plan to be presented to the steering committee.

As suggested by the previous discussion, walkthrough meetings are a common occurrence in most systems development groups and can be used for more activities than reviewing the BPP, including the following:

- System specifications
- Logical and physical designs
- Code or program segments
- Test procedures and results
- Manuals and documentation

Pine Valley Furniture
Walkthrough Review Form

Session Coordinator:

Project/Segment:

Coordinator's Checklist:

1. Confirmation with producer(s) that material is ready and stable: _____
2. Issue invitations, assign responsibilities, distribute materials: [ ] Y   [ ] N
3. Set date, time, and location for meeting:

Date: ____ / ____ / ____   Time: _____ A.M. / P.M. (circle one)

Location: _____

Responsibilities	Participants	Can Attend		Received Materials	
Coordinator	_____	[ ] Y	[ ] N	[ ] Y	[ ] N
Presenter	_____	[ ] Y	[ ] N	[ ] Y	[ ] N
User	_____	[ ] Y	[ ] N	[ ] Y	[ ] N
Secretary	_____	[ ] Y	[ ] N	[ ] Y	[ ] N
Standards	_____	[ ] Y	[ ] N	[ ] Y	[ ] N
Maintenance	_____	[ ] Y	[ ] N	[ ] Y	[ ] N

Agenda:
_____ 1. All participants agree to follow PVF's Rules of a Walkthrough
_____ 2. New material: walkthrough of all material
_____ 3. Old material: item-by-item checkoff of previous action list
_____ 4. Creation of new action list (contribution by each participant)
_____ 5. Group decision (see below)
_____ 6. Deliver copy of this form to the project control manager

Group Decision:
_____ Accept product as-is
_____ Revise (no further walkthrough)
_____ Review and schedule another walkthrough

Signatures		

FIGURE 5-14
Walkthrough Review Form (Pine Valley Furniture)

One of the key advantages in using a structured review process is that it ensures that formal review points occur during the project. At each subsequent phase of the project, a formal review should be conducted (and shown on the project schedule) to make sure that all aspects of the project are satisfactorily accomplished

Pine Valley Furniture Walkthrough Action List	
Session Coordinator:	
Project/Segment:	
Date and Time of Walkthrough: Date: ____ /_____ / ____ Time: _____ A.M. / P.M. (circle one)	
Fixed (✓)	**Issues raised in review:**

FIGURE 5-15
Walkthrough Action List (Pine Valley Furniture)

before assigning additional resources to the project. This conservative approach
of reviewing each major project activity with continuation contingent on successful
completion of the prior phase is called *incremental commitment*. It is much easier to
stop or redirect a project at any point when using this approach.

TABLE 5-8   Guidelines for Making an Effective Presentation

Presentation Planning	
Who is the audience?	To design the most effective presentation, you need to consider the audience (e.g., What do they know about your topic? What is their education level?).
What is the message?	Your presentation should be designed with a particular objective in mind.
What is the presentation environment?	Knowledge of the room size, shape, and lighting is valuable information for designing an optimal presentation.

Presentation Design	
Organize the sequence	Organize your presentation so that like elements or topics are found in one place, instead of scattered throughout the material in random fashion.
Keep it simple	Make sure that you don't pack too much information onto a slide so that it is difficult to read. Also, work to have as few slides as possible; in other words, only include information that you absolutely need.
Be consistent	Make sure that you are consistent in the types of fonts, font sizes, colors, design approach, and backgrounds.
Use variety	Use both textual and graphical slides to convey information in the most meaningful format.
Don't rely on the spell checker alone	Make sure you carefully review your presentation for typographical and wording errors.
Use bells and whistles sparingly	Make sure that you use familiar graphical icons to guide and enhance slides; don't lose sight of your message as you add bells and whistles. Also, take great care when making transitions between slides and elements so that "special effects" don't take away from your message.
Supplemental materials	Take care when using supplemental materials so that they don't distract the audience. For example, don't provide handouts until you want the audience to actually read this material.
Have a clear beginning and end	At the beginning, introduce yourself and your teammates (if any), thank your audience for being there, and provide a clear outline of what will be covered during the presentation. At the conclusion, have a concluding slide so that the audience clearly sees that the presentation is over.

Presentation Delivery	
Practice	Make sure that you thoroughly test your completed work on yourself and others to be sure it covers your points and presents them in an effective manner within the timeframe required.
Arrive early and cue up your presentation	It is good practice, when feasible, to have your presentation ready to go prior to the arrival of the audience.
Learn to use the "special" software keys	Using special keys to navigate the presentation will allow you to focus on your message and not on the software.
Have a backup plan	Have a backup plan in case technology fails or your presentation is lost when traveling.
Delivery	To make an effective presentation, you must become an effective public speaker through practice.
Personal appearance	Your appearance and demeanor can go a long way toward enhancing how the audience receives your presentation.

Walkthroughs are used throughout the duration of the project for briefing team members and external stakeholders. These presentations can provide many benefits to the team, but, unfortunately, are often not well done. With the proliferation of computer technology and the availability of powerful software to

TABLE 5-9   Web-Based System Costs

Cost Category	Examples
Platform Costs	• Web-hosting service
	• Web server
	• Server software
	• Software plug-ins
	• Firewall server
	• Router
	• Internet connection
Content and Service	• Creative design and development
	• Ongoing design fees
	• Web project manager
	• Technical site manager
	• Content staff
	• Graphics staff
	• Support staff
	• Site enhancement funds
	• Fees to license outside content
	• Programming, consulting, and research
	• Training and travel
Marketing	• Direct mail
	• Launch and ongoing public relations
	• Print advertisement
	• Paid links to other websites
	• Promotions
	• Marketing staff
	• Advertising sales staff

assist in designing and delivering presentations, making an effective presentation has never been easier. Microsoft's PowerPoint has emerged as the de facto standard for creating computer-based presentations. Although this program is relatively easy to use, it can also be misused such that the "bells and whistles" added to a computer-based presentation actually detract from the presentation. Like any project, to make an effective presentation it must be planned, designed, and delivered. Planning and designing your presentation is equally important as delivering it. If your slides are poorly laid out, hard to read, or inconsistent, it won't matter how good your delivery is; your audience will think more about the poor quality of the slides than about what you are saying. Fortunately, with a little work it is easy to design a high-quality presentation if you follow a few simple steps, which are outlined in Table 5-8.

ELECTRONIC COMMERCE APPLICATIONS: INITIATING AND PLANNING SYSTEMS DEVELOPMENT PROJECTS

Initiating and planning systems development projects for an Internet-based EC application is very similar to the process followed for more traditional applications. In Chapter 4, you read how PVF's management began the WebStore project—to sell furniture products over the Internet. In this section, we highlight some of the issues that relate directly to the process of identifying and selecting systems development projects.

PINE
VALLEY
FURNITURE

TABLE 5-10   PVF WebStore: Project Benefits and Costs

Tangible Benefits	Intangible Benefits
• Lower per-transaction overhead cost • Repeat business	• First to market • Foundation for complete Web-based IS • Simplicity for customers
Tangible Costs (one-time)	**Intangible Costs**
• Internet service setup fee • Hardware • Development cost • Data entry	• No face-to-face interaction • Not all customers use Internet
Tangible Costs (recurring)	
• Internet service hosting fee • Software • Support • Maintenance • Decreased sales via traditional channels	

TABLE 5-11   PVF WebStore: Feasibility Concerns

Feasibility Concern	Description
Operational	Online store is open 24/7/365 Returns/customer support
Technical	New skill set for development, maintenance, and operation
Schedule	Must be open for business by Q3
Legal	Credit card fraud
Political	Traditional distribution channel loses business

Initiating and Planning Systems Development Projects for Pine Valley Furniture's WebStore

Given the high priority of the WebStore project, Vice President of Marketing Jackie Judson, and senior systems analyst, Jim Woo, were assigned to work on this project. Like the CTS described earlier in this chapter, their initial activity was to begin the project's initiation and planning activities.

Initiating and Planning PVF's E-Commerce System   To start the initiation and planning process, Jim and Jackie held several meetings over several days. At the first meeting they agreed that "WebStore" would be the proposed system project name. Next, they worked on identifying potential benefits, costs, and feasibility concerns. To assist in this process, Jim developed a list of potential costs from developing Web-based systems that he shared with Jackie and the other project team members (see Table 5-9).

WebStore Project Walkthrough   After meeting with the project team, Jim and Jackie established an initial list of benefits and costs (see Table 5-10) as well as several feasibility concerns (see Table 5-11). Next, Jim worked with several of PVF's technical specialists to develop an initial project schedule. Figure 5-16 shows the Gantt chart for this 84-day schedule. Finally, Jim and Jackie presented their initial project plans in a walkthrough to PVF's board of directors and senior management. All were excited about the project plan, and approval was given to move the WebStore project into the analysis phase.

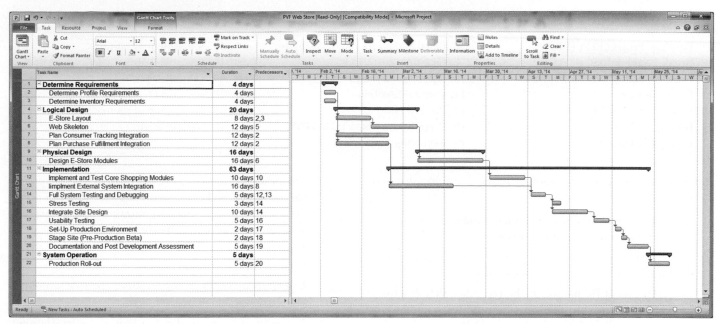

FIGURE 5-16
Schedule for WebStore project at Pine Valley Furniture
(*Source:* Microsoft Corporation.)

SUMMARY

The project initiation and planning (PIP) phase is a critical activity in the life of a project. It is at this point that projects are accepted for development, rejected as infeasible, or redirected. The objective of this process is to transform a vague system request into a tangible system description clearly outlining the objectives, feasibility issues, benefits, costs, and time schedules for the project.

Project initiation includes forming the project initiation team, establishing customer relationships, developing a plan to get the project started, setting project management procedures, and creating an overall project management environment. A key activity in project planning is the assessment of numerous feasibility issues associated with the project. Feasibilities that should be examined include economic, technical, operational, schedule, legal and contractual, and political ones. These issues are influenced by the project size, the type of system proposed, and the collective experience of the development group and potential customers of the system. High project costs and risks are not necessarily bad; rather it is more important that the organization understands the costs and risks associated with a project and with the portfolio of active projects before proceeding.

After completing all analyses, a BPP can be created. A BPP includes a high-level description of the proposed system or system change, an outline of the various feasibilities, and an overview of management issues specific to the project. Before the development of an information system can begin, the users, management, and development group must review and agree on this specification. The focus of this walkthrough review is to assess the merits of the project and to ensure that the project, if accepted for development, conforms to organizational standards and goals. An objective of this process is also to make sure that all relevant parties understand and agree with the information contained in the plan before subsequent development activities begin.

Project initiation and planning is a challenging and time-consuming activity that requires active involvement from many organizational participants. The eventual success of a development project, and the information systems function in general, hinges on the effective use of disciplined, rational approaches such as the techniques outlined in this chapter. In subsequent chapters, you will be exposed to numerous other tools that will equip you to become an effective designer and developer of information systems.

KEY TERMS

1. Baseline Project Plan (BPP)
2. Break-even analysis
3. Business case
4. Discount rate
5. Economic feasibility
6. Intangible benefit
7. Intangible cost
8. Legal and contractual feasibility
9. One-time cost
10. Operational feasibility
11. Political feasibility
12. Present value
13. Project Scope Statement (PSS)
14. Recurring cost
15. Schedule feasibility
16. Tangible benefit
17. Tangible cost
18. Technical feasibility
19. Time value of money (TVM)
20. Walkthrough

Match each of the key terms above with the definition that best fits it.

_____ The concept that money available today is worth more than the same amount tomorrow.

_____ The process of evaluating how key stakeholders within the organization view the proposed system.

_____ A document prepared for the customer that describes what the project will deliver and outlines generally at a high level all work required to complete the project.

_____ The justification for an information system, presented in terms of the tangible and intangible economic benefits and costs, and the technical and organizational feasibility of the proposed system.

_____ A process of identifying the financial benefits and costs associated with a development project.

_____ The process of assessing the degree to which a proposed system solves business problems or takes advantage of business opportunities.

_____ A cost resulting from the ongoing evolution and use of a system.

_____ The rate of return used to compute the present value of future cash flows.

_____ A benefit derived from the creation of an information system that cannot be easily measured in dollars or with certainty.

_____ The process of assessing the degree to which the potential time frame and completion dates for all major activities within a project meet organizational deadlines and constraints for affecting change.

_____ A cost associated with an information system that can be easily measured in dollars and certainty.

_____ A peer group review of any product created during the systems development process.

_____ A process of assessing the development organization's ability to construct a proposed system.

_____ A cost associated with project start-up and development or system start-up.

_____ The current value of a future cash flow.

_____ A benefit derived from the creation of an information system that can be measured in dollars and with certainty.

_____ The process of assessing potential legal and contractual ramifications due to the construction of a system.

_____ A cost associated with an information system that cannot be easily measured in terms of dollars or with certainty.

_____ This plan is the major outcome and deliverable from the project initiation and planning phase and contains the best estimate of the project's scope, benefits, costs, risks, and resource requirements.

_____ A type of cost-benefit analysis to identify at what point (if ever) benefits equal costs.

REVIEW QUESTIONS

1. Contrast the following terms:
   a. Break-even analysis; present value; net present value; return on investment
   b. Economic feasibility; legal and contractual feasibility; operational feasibility; political feasibility; schedule feasibility
   c. Intangible benefit; tangible benefit
   d. Intangible cost; tangible cost

2. List and describe the steps in the project initiation and planning process.

3. What is contained in a BPP? Are the content and format of all baseline plans the same? Why or why not?

4. Describe three commonly used methods for performing economic cost-benefit analysis.

5. List and discuss the different types of project feasibility factors. Is any factor most important? Why or why not?

6. What are the potential consequences of not assessing the technical risks associated with an information systems development project?

7. In what ways could you identify that one IS project is riskier than another?

8. What are the types or categories of benefits of an IS project?

9. What intangible benefits might an organization obtain from the development of an information system?

10. Describe the concept of the time value of money. How does the discount rate affect the value of $1 today versus one year from today?

11. Describe the structured walkthrough process. What roles need to be performed during a walkthrough?

PROBLEMS AND EXERCISES

1. Consider the purchase of a PC and laser printer for use at your home and assess the risk for this project using the project risk assessment factors in Table 5-7.

2. Consider your use of a PC at either home or work and list tangible benefits from an information system. Based on this list, does your use of a PC seem to be beneficial? Why or why not? Now do the same using Table 5-3, the intangible benefits from an information system. Does this analysis support or contradict your previous analysis? Based on both analyses, does your use of a PC seem to be beneficial?

3. Assume you are put in charge of launching a new website for a local nonprofit organization. What costs would you need to account for? Make a list of expected costs and benefits for the project. You don't need to list values, just sources of expense. Consider both one-time and recurring costs.

4. Consider the situation you addressed in Problem and Exercise 3. Create numeric cost estimates for each of the costs you listed. Calculate the net present value and return on investment. Include a break-even analysis. Assume a 10 percent discount rate and a five-year time horizon.

5. Consider the situation you addressed in Problem and Exercise 3. Create a sample Project Scope Statement, following the structure shown in Figure 5-10.

6. Assuming monetary benefits of an information system at $85,000 per year, one-time costs of $75,000, recurring costs of $35,000 per year, a discount rate of 12 percent, and a five-year time horizon, calculate the net present value of these costs and benefits of an information system. Also calculate the overall return on investment of the project and then present a break-even analysis. At what point does breakeven occur?

7. Use the outline for the BPP provided in Figure 5-9 to present the system specifications for the information system you chose for Problems and Exercises 3.

8. Change the discount rate for Problem and Exercise 3 to 10 percent and redo the analysis.

9. Change the recurring costs in Problem and Exercise 3 to $40,000 and redo the analysis.

10. Change the time horizon in Problem and Exercise 3 to three years and redo the analysis.

11. Assume monetary benefits of an information system of $40,000 the first year and increasing benefits of $10,000 a year for the next five years (year 1 = $50,000, year 2 = $60,000, year 3 = $70,000, year 4 = $80,000, year 5 = $90,000). One-time development costs were $80,000 and recurring costs were $45,000 over the duration of the system's life. The discount rate for the company was 11 percent. Using a six-year time horizon, calculate the net present value of these costs and benefits. Also calculate the overall return on investment and then present a break-even analysis. At what point does break-even occur?

12. Change the discount rate for Problem and Exercise 11 to 12 percent and redo the analysis.

13. Change the recurring costs in Problem and Exercise 11 to $40,000 and redo the analysis.

14. For the system you chose for Problems and Exercises 3, complete section 1.0, A, Project Overview, of the BPP Report. How important is it that this initial section of the BPP Report is done well? What could go wrong if this section is incomplete or incorrect?

15. For the system you chose for Problems and Exercises 3, complete section 2.0, A, Alternatives, of the BPP Report. Without conducting a full-blown feasibility analysis, what is your gut feeling as to the feasibility of this system?

16. For the system you chose for Problems and Exercises 3, complete section 3.0, A–F, Feasibility Analysis, of the BPP Report. How does this feasibility analysis compare with your gut feeling from the previous question? What might go wrong if you rely on your gut feeling in determining system feasibility?

17. For the system you chose for Problems and Exercises 3, complete section 4.0, A–C, Management Issues, of the BPP Report. Why might people sometimes feel that these additional steps in the project plan are a waste of time? What would you say to them to convince them that these steps are important?

FIELD EXERCISES

1. Describe several projects you are involved in or plan to undertake, whether they are related to your education or to your professional or personal life. Some examples are purchasing a new vehicle, learning a new language, renovating a home, and so on. For each, sketch out a BPP like that outlined in Figure 5-9. Focus your efforts on item numbers 1.0 (Introduction) and 2.0 (System Description).

2. For each project from the previous question, assess the feasibility in terms of economic, operational, technical, scheduling, legal and contractual, as well as political aspects.

3. Network with a contact you have in some organization that conducts projects (these might be information systems projects, but they could be construction, product development, or any type of project). Interview a project manager and find out what type of BPP is constructed. For a typical project, in what ways are baseline plans modified during the life of a project? Why are plans modified after the project begins? What does this tell you about project planning?

4. Through a contact you have in some organization that uses packaged software, interview an IS manager responsible for systems in an area that uses packaged application software. What contractual limitations, if any, has the organization encountered with using the package? If possible, review the license agreement for the software and make a list of all the restrictions placed on a user of this software.

5. Choose an organization that you are familiar with and determine what is done to initiate information systems projects. Who is responsible for initiating projects? Is this process formal or informal? Does this appear to be a top-down or bottom-up process? How could this process be improved?

6. Find an organization that does not use BPP for their IS projects. Why doesn't this organization use this method? What are the advantages and disadvantages of not using this method? What benefits could be gained from implementing the use of BPP? What barriers are there to implementing this method?

REFERENCES

Applegate, L. M., Austin, R. D., and Soule, D. L. 2009. *Corporate Information Strategy and Management*, 8th ed.. New York: McGraw-Hill.

Brynjolfsson, E., and S. Yang. 1997. *The Intangible Benefits and Costs of Investments: Evidence from Financial Markets.* Proceedings of the International Conference on Information Systems, pp. 147–66. Available at *http://portal.acm.org.* Accessed July 27, 2012.

Cresswell, A. M. 2004. *Return on Investment in Information Technology: A Guide for Managers Center for Technology in Government.* University at Albany, SUNY. Available at *www.ctg.albany.edu.* Accessed July 27, 2012.

DeGiglio, M. 2002. "Measure for Measure: The Value of IT." Cio. com, June 17. Available at *www.cio.com.* Accessed July 27, 2012.

Hubbard, D. 2007. "The IT Measurement Inversion." Cio.com, June 13. Available at *www.cio.com.* Accessed February 10, 2009.

Keen, J. 2003. "Intangible Benefits Can Play Key Role in Business Case." Cio.com, September 1. Available at *www.cio.com.* Accessed July 27, 2012.

King, J. L., and E. Schrems. 1978. "Cost Benefit Analysis in Information Systems Development and Operation." *ACM Computing Surveys* 10(1): 19–34.

Kirsch, L. J. 2000. "Software Project Management: An Integrated Perspective for an Emerging Paradigm." In R. W. Zmud (ed.), *Framing the Domains of IT Management: Projecting the Future from the Past,* 285–304. Cincinnati, OH: Pinnaflex Educational Resources.

Laplante, P. A. 2006. "Software Return on Investment (ROI)." In P. A. Laplante and T. Costello (eds.), *CIO Wisdom II,* 163–76. Upper Saddle River, NJ: Prentice Hall.

Lederer, A. L., and J. Prasad. 1992. "Nine Management Guidelines for Better Cost Estimating." *Communications of the ACM* 35(2): 51–59.

Morris, R., and B. M. Sember. 2008. *Project Management That Works.* AMACOM Division of American Management Association, New York City, NY.

Nash, K. S. 2008. "TCO Versus ROI." Cio.com, April 9. Available at *www.cio.com.* Accessed July 27, 2012.

Parker, M. M., and R. J. Benson. 1988. *Information Economics.* Upper Saddle River, NJ: Prentice Hall.

Pressman, R. S. 2005. *Software Engineering,* 6th ed. New York: McGraw-Hill.

Sonje, R. 2008. "Improving Project Estimation Effectiveness." Available at *www.projectperfect.com.au.* Accessed July 27, 2012.

Tech Republic. 2005. "Project Risk Factors Checklist." Version 2.0. Available at *http://articles.techrepublic.com.com/5100 -10878_11-1041706.html.* Accessed July 27, 2012.

White, S., and S. Lui. 2005. "Distinguishing Costs of Cooperation and Control in Alliances." *Strategic Management Journal* 26(10): 913–32.

Yourdon, E. 1989. *Structured Walkthroughs,* 4th ed. Upper Saddle River, NJ: Prentice Hall.

PETRIE ELECTRONICS

Chapter 5: Initiating and Planning Systems Development Projects

Now that the "No Customer Escapes" project team has been formed and that a plan had been developed for distributing project information, Jim began working on the project's scope statement, workbook, and Baseline Project Plan. He first drafted the project's scope statement and posted it on the project's intranet (see PE Figure 5-1). Once posted on the intranet, he sent a short e-mail message to all team members requesting feedback.

Minutes after posting the project charter, Jim's office phone rang.

"Jim, it's Sally. I just looked over the scope statement and have a few comments."

"Great," replied Jim, "It's just a draft. What do you think?"

"Well, I think that we need to explain more about how the system will work and why we think this new system will more than pay for itself."

"Those are good suggestions; I am sure many others will also want to know that information. However, the scope statement is a pretty high-level document and doesn't get into too much detail. Basically, its purpose is to just formally announce the project, providing a very high-level description as well as briefly listing the objectives, key assumptions, and stakeholders. The other documents that I am working on, the workbook and the Baseline Project Plan, are intended to provide more details on specific deliverables, costs, benefits, and so on. So, anyway, that type of more detailed information will be coming next."

"Oh, OK, that makes sense. I have never been on a project like this, so this is all new to me," said Sally.

"Don't worry," replied Jim, "Getting that kind of feedback from you and the rest of the team will be key for us doing a thorough feasibility analysis. I am going to need a lot of your help in identifying possible costs and benefits of the system. When we develop the Baseline Project Plan, we do a very thorough feasibility analysis—we examine financial, technical, operational, schedule, legal and contractual feasibility, as well as potential political issues arising through the development of the system."

"Wow, we have to do all that? Why can't we just build the system? I think we all know what we want," replied Sally.

"That is another great question," replied Jim. "I used to think exactly the same way, but what I learned in my last job was that there are great benefits to following a fairly formal project management process with a new system. By moving forward with care, we are much more likely to have the right system, on time and on budget"

"So," asked Sally, "What is the next step?"

"Well, we need to do the feasibility analyses I just mentioned, which becomes part of the project's Baseline Project Plan. Once this is completed, we will have a walkthrough presentation to management to make sure they agree with and understand the scope, risks, and costs associated with making 'No Customer Escapes' a reality," said Jim.

"This is going to be a lot of work, but I am sure I am going to learn a lot," replied Sally.

"So, let me get to work on the feasibility analyses," said Jim. "I will be sending requests out to all the team members to get their ideas. I should have this e-mail ready within an hour or so."

"Great, I'll look for it and respond as soon as I can," answered Sally.

"Thanks, the faster we get this background work done, the sooner we will be able to move on to what the system will do," replied Jim.

"Sounds good, talk to you later. Bye," Sally said.

"Bye, Sally, and thanks for your quick feedback," answered Jim.

Case Questions

1. Look over the scope statement (PE Figure 5-1) If you were an employee at Petrie Electronics, would you want to work on this project? Why or why not?

2. If you were part of the management team at Petrie Electronics, would you approve the project outlined in the scope statement in PE Figure 5-1? What changes, if any, need to be made to the document?

3. Identify a preliminary set of tangible and intangible costs you think would occur for this project and the system it describes. What intangible benefits do you anticipate for the system?

4. What do you consider to be the risks of the project as you currently understand it? Is this a low-, medium-, or high-risk project? Justify your answer. Assuming you were part of Jim's team, would you have any particular risks?

5. If you were assigned to help Jim with this project, how would you utilize the concept of incremental commitment in the design of the Baseline Project Plan?

6. If you were assigned to Jim's team for this project, when in the project schedule (in what phase or after which activities are completed) do you think you could develop an economic analysis of the proposed system? What economic feasibility factors do you think would be relevant?

7. If you were assigned to Jim's team for this project, what activities would you conduct in order to prepare the details for the Baseline Project Plan? Explain the purpose of each activity and show a timeline or schedule for these activities.

8. In Case Question 4, you analyze the risks associated with this project. Once deployed, what are the potential operational risks of the proposed system? How do you factor operation risks into a systems development plan?

PART THREE

Analysis

Analysis is the first systems development life cycle (SDLC) phase where you begin to understand, in depth, the need for system changes. Systems analysis involves a substantial amount of effort and cost, and is therefore undertaken only after management has decided that the systems development project under consideration has merit and should be pursued through this phase. The analysis team should not take the analysis process for granted or attempt to speed through it. Most observers would agree that many of the errors in developed systems are directly traceable to inadequate efforts in the analysis and design phases of the life cycle. Because analysis is a large and involved process, we divide it into two main activities to make the overall process easier to understand:

- *Requirements determination.* This is primarily a fact-finding activity.
- *Requirements structuring.* This activity creates a thorough and clear description of current business operations and new information processing services.

The purpose of analysis is to determine what information and information processing services are needed to support selected objectives and functions of the organization. Gathering this information is called *requirements determination*, the subject of Chapter 6. The fact-finding techniques in Chapter 6 are used to learn about the current system, the organization that the replacement system will support, and user requirements or expectations for the replacement system.

In Chapter 6, we also discuss a major source of new systems, Business Process Reengineering (BPR). In contrast to the incremental improvements that drive many systems development projects, BPR results in a radical redesign of the processes that information systems are designed to support. We show how BPR relates to information systems analysis in Chapter 7, where we use data flow diagrams to support the reengineering process. In Chapter 6, you will also learn about new requirements determination techniques sometimes used as part of Agile Methodologies. These include the Planning Game, from eXtreme Programming, and Usage-Centered Design.

Information about current operations and requirements for a replacement system must be organized for analysis and design. Organizing, or structuring, system requirements results in diagrams and descriptions (models) that can be analyzed to show deficiencies, inefficiencies, missing elements, and illogical components of the current business operation and information systems. Along with user requirements, they are used to determine the strategy for the replacement system.

The results of the requirements determination can be structured according to three essential views of the current and replacement information systems:

- *Process.* The sequence of data movement and handling operations within the system
- *Logic and timing.* The rules by which data are transformed and manipulated and an indication of what triggers data transformation
- *Data.* The inherent structure of data independent of how or when they are processed

The *process* view of a system can be represented by data flow diagrams, the primary subject of Chapter 7. The chapter also includes a section on decision tables, one of the ways to describe the *logic and timing* of what goes on inside the process boxes in data flow diagrams. Chapter 7 ends with four appendices. The first three are each dedicated to one of three techniques from the object-oriented view of development. The first appendix introduces you to use case modeling, an object-oriented method used to map a system's functionality. The second appendix introduces you to activity diagrams, while the third features sequence diagrams. These object-oriented models focus on system logic and timing. The fourth appendix covers business process modeling, which is not part of the object-oriented approach. Finally, the *data view* of a system, discussed in Chapter 8, shows the rules that govern the structure and integrity of data and concentrates on what data about business entities and relationships among these entities must be accessed within the system. Chapter 8 features entity relationship techniques in the body of the chapter and class diagramming techniques for modeling data in a special object-oriented section at the end of the chapter. Petrie's case installments follow Chapters 7 and 8 to illustrate the processes, logic, and data models that describe a new system. The cases also illustrate how diagrams and models for each of these three views of a system relate to one another to form a consistent and thorough structured description of a proposed system.

CHAPTER SIX

Determining System Requirements

Learning Objectives

After studying this chapter, you should be able to:

- Describe options for designing and conducting interviews and develop a plan for conducting an interview to determine system requirements.

- Explain the advantages and pitfalls of observing workers and analyzing business documents to determine system requirements.

- Explain how computing can provide support for requirements determination.

- Participate in and help plan a Joint Application Design session.

- Use prototyping during requirements determination.

- Describe contemporary approaches to requirements determination.

- Understand how requirements determination techniques apply to the development of electronic commerce applications.

Introduction

Systems analysis is the part of the systems development life cycle in which you determine how the current information system functions and assess what users would like to see in a new system. Analysis has two subphases: requirements determination and requirements structuring. In this chapter, you will learn about determining system requirements. Techniques used in requirements determination have evolved over time to become more structured and increasingly rely on computer support. We will first study the more traditional requirements determination methods, including interviewing, observing users in their work environment, and collecting procedures and other written documents. We will then discuss more current methods for collecting system requirements. The first of these methods is Joint Application Design (JAD). Next, you will read about how analysts rely more and more on information systems to help them perform analysis. As you will see, CASE tools, discussed in Chapter 1, are useful in requirements determination, and prototyping has become a key tool for some requirements determination efforts. Finally, you will learn how requirements analysis continues to be an important part of systems analysis and design, whether the approach involves business process redesign or new agile

177

techniques such as constant user involvement or usage-centered design, or focuses on developing Internet applications.

PERFORMING REQUIREMENTS DETERMINATION

As stated earlier and shown in Figure 6-1, there are two subphases to systems analysis: requirements determination and requirements structuring. We will address these as separate steps, but you should consider the steps as parallel and iterative. For example, as you determine some aspects of the current and desired system(s), you begin to structure these requirements or build prototypes to show users how a system might behave. Inconsistencies and deficiencies discovered through structuring and prototyping lead you to explore further the operation of current system(s) and the future needs of the organization. Eventually, your ideas and discoveries converge on a thorough and accurate depiction of current operations and what the requirements are for the new system. As you think about beginning the analysis phase, you are probably wondering what exactly is involved in requirements determination. We discuss this process in the next section.

The Process of Determining Requirements

Once management has granted permission to pursue development of a new system (this was done at the end of the project identification and selection phase of the SDLC) and a project is initiated and planned (see Chapter 5), you begin determining what the new system should do. During requirements determination, you and other analysts gather information on what the system should do from as many sources as possible: from users of the current system; from observing users; and from reports, forms, and procedures. All of the system requirements are carefully documented and made ready for structuring, the subject of Chapters 7 and 8.

In many ways, gathering system requirements is like conducting any investigation. Have you read any of the Sherlock Holmes or similar mystery stories? Do you enjoy solving puzzles? From these experiences, we can detect some similar characteristics for a good systems analyst during the requirements determination subphase. These characteristics include the following:

- *Impertinence.* You should question everything. You need to ask questions such as: Are all transactions processed the same way? Could anyone be charged something other than the standard price? Might we someday want to allow and encourage employees to work for more than one department?

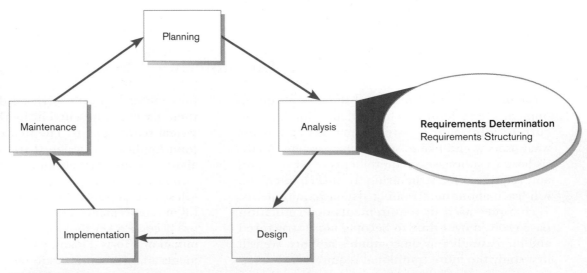

FIGURE 6-1
Systems development life cycle with analysis phase highlighted

- *Impartiality.* Your role is to find the best solution to a business problem or opportunity. It is not, for example, to find a way to justify the purchase of new hardware or to insist on incorporating what users think they want into the new system requirements. You must consider issues raised by all parties and try to find the best organizational solution.
- *Relax constraints.* Assume that anything is possible and eliminate the infeasible. For example, do not accept this statement: "We've always done it that way, so we have to continue the practice." Traditions are different from rules and policies. Traditions probably started for a good reason but, as the organization and its environment change, they may turn into habits rather than sensible procedures.
- *Attention to details.* Every fact must fit with every other fact. One element out of place means that even the best system will fail at some time. For example, an imprecise definition of who a customer is may mean that you purge customer data when a customer has no active orders, yet these past customers may be vital contacts for future sales.
- *Reframing.* Analysis is, in part, a creative process. You must challenge yourself to look at the organization in new ways. You must consider how each user views his or her requirements. You must be careful not to jump to the following conclusion: "I worked on a system like that once—this new system must work the same way as the one I built before."

Deliverables and Outcomes

The primary deliverables from requirements determination are the various forms of information gathered during the determination process: transcripts of interviews; notes from observation and analysis of documents; sets of forms, reports, job descriptions, and other documents; and computer-generated output such as system prototypes. In short, anything that the analysis team collects as part of determining system requirements is included in the deliverables resulting from this subphase of the systems development life cycle. Table 6-1 lists examples of some specific information that might be gathered during requirements determination.

These deliverables contain the information you need for systems analysis within the scope of the system you are developing. In addition, you need to understand the following components of an organization:

- The business objectives that drive what and how work is done
- The information people need to do their jobs
- The data (definition, volume, size, etc.) handled within the organization to support the jobs
- When, how, and by whom or what the data are moved, transformed, and stored
- The sequence and other dependencies among different data-handling activities
- The rules governing how data are handled and processed
- Policies and guidelines that describe the nature of the business and the market and environment in which it operates
- Key events affecting data values and when these events occur

TABLE 6-1   Deliverables for Requirements Determination

1. Information collected from conversations with or observations of users: interview transcripts, notes from observation, meeting minutes
2. Existing written information: business mission and strategy statements, sample business forms and reports and computer displays, procedure manuals, job descriptions, training manuals, flowcharts and documentation of existing systems, consultant reports
3. Computer-based information: results from JAD sessions, CASE repository contents and reports of existing systems, and displays and reports from system prototypes

As should be obvious, such a large amount of information must be organized in order to be useful. This is the purpose of the next subphase—requirements structuring.

From just this subphase of analysis, you have probably already realized that the amount of information to be gathered could be huge, especially if the scope of the system under development is broad. The time required to collect and structure a great deal of information can be extensive and, because it involves so much human effort, quite expensive. Too much analysis is not productive, and the term *analysis paralysis* has been coined to describe a systems development project that has become bogged down in an abundance of analysis work. Because of the dangers of excessive analysis, today's systems analysts focus more on the system to be developed than on the current system. The techniques you will learn about later in this chapter, JAD and prototyping, were developed to keep the analysis effort at a minimum yet still keep it effective. Newer techniques have also been developed to keep requirements determination fast and flexible, including continual user involvement, usage-centered design, and the Planning Game from eXtreme Programming. Traditional fact-gathering techniques are the subject of the next section.

TRADITIONAL METHODS FOR DETERMINING REQUIREMENTS

At the core of systems analysis is the collection of information. At the outset, you must collect information about the information systems that are currently being used and how users would like to improve the current systems and organizational operations with new or replacement information systems. One of the best ways to get this information is to talk to the people who are directly or indirectly involved in the different parts of the organizations affected by the possible system changes: users, managers, funders, and so on. Another way to find out about the current system is to gather copies of documentation relevant to current systems and business processes. In this chapter, you will learn about various ways to get information directly from stakeholders: interviews, group interviews, the Nominal Group Technique, and direct observation. You will learn about collecting documentation on the current system and organizational operation in the form of written procedures, forms, reports, and other hard copy. These traditional methods of collecting system requirements are listed in Table 6-2.

Interviewing and Listening

Interviewing is one of the primary ways analysts gather information about an information systems project. Early in a project, an analyst may spend a large amount of time interviewing people about their work, the information they use to do it, and the types of information processing that might supplement their work. Other stakeholders are interviewed to understand organizational direction, policies, expectations managers have on the units they supervise, and other nonroutine aspects of

TABLE 6-2   Traditional Methods of Collecting System Requirements

- Individually interview people informed about the operation and issues of the current system and future systems needs
- Interview groups of people with diverse needs to find synergies and contrasts among system requirements
- Observe workers at selected times to see how data are handled and what information people need to do their jobs
- Study business documents to discover reported issues, policies, rules, and directions as well as concrete examples of the use of data and information in the organization

organizational operations. During interviewing you will gather facts, opinions, and speculation and observe body language, emotions, and other signs of what people want and how they assess current systems.

There are many ways to effectively interview someone, and no one method is necessarily better than another. Some guidelines you should keep in mind when you interview, summarized in Table 6-3, are discussed next.

First, you should prepare thoroughly before the interview. Set up an appointment at a time and for a duration convenient for the interviewee. The general nature of the interview should be explained to the interviewee in advance. You may ask the interviewee to think about specific questions or issues or to review certain documentation to prepare for the interview. You should spend some time thinking about what you need to find out and write down your questions. Do not assume that you can anticipate all possible questions. You want the interview to be natural, and, to some degree, you want to spontaneously direct the interview as you discover what expertise the interviewee brings to the session.

You should prepare an interview guide or checklist so that you know in which sequence you intend to ask your questions and how much time you want to spend in each area of the interview. The checklist might include some probing questions to ask as follow-up if you receive certain anticipated responses. You can, to some degree, integrate your interview guide with the notes you take during the interview, as depicted in a sample guide in Figure 6-2. This same guide can serve as an outline for a summary of what you discover during an interview.

The first page of the sample interview guide contains a general outline of the interview. Besides basic information on who is being interviewed and when, you list major objectives for the interview. These objectives typically cover the most important data you need to collect, a list of issues on which you need to seek agreement (e.g., content for certain system reports), and which areas you need to explore, not necessarily with specific questions. You also include reminder notes to yourself on key information about the interviewee (e.g., job history, known positions taken on issues, and role with current system). This information helps you to be personal, shows that you consider the interviewee to be important, and may assist you in interpreting some answers. Also included is an agenda for the interview with approximate time limits for different sections of the interview. You may not follow the time limits precisely, but the schedule helps you cover all areas during the time the interviewee is available. Space is also allotted for general observations that do not fit under specific questions and for notes taken during the interview about topics skipped or issues raised that could not be resolved.

On subsequent pages you list specific questions; the sample form in Figure 6-2 includes space for taking notes on these questions. Because unanticipated information arises, you will not strictly follow the guide in sequence. You can, however, check off the questions you have asked and write reminders to yourself to return to or skip certain questions as the dynamics of the interview unfold.

Choosing Interview Questions   You need to decide what mix and sequence of open-ended and closed-ended questions you will use. **Open-ended questions** are usually used to probe for information for which you cannot anticipate all possible responses or for which you do not know the precise question to ask. The person being interviewed is encouraged to talk about whatever interests him or her within the general bounds of the question. An example is, "What would you say is the best thing about the information system you currently use to do your job?" or "List the three most frequently used menu options." You must react quickly to answers and determine whether or not any follow-up questions are needed for clarification or elaboration. Sometimes body language will suggest that a user has given an incomplete answer or is reluctant to divulge some information; a follow-up question might yield additional insight. One advantage of open-ended questions in an interview is that previously unknown information can surface. You can then continue exploring along unexpected lines of inquiry to reveal even more new information. Open-ended questions also

TABLE 6-3   Guidelines for Effective Interviewing

Plan the Interview
- Prepare interviewee: appointment, priming questions
- Prepare checklist, agenda, and questions

Listen carefully and take notes (record if permitted)

Review notes within 48 hours of interview

Be neutral

Seek diverse views

Open-ended questions
Questions in interviews that have no prespecified answers.

FIGURE 6-2
Typical interview guide

Interview Outline

Interviewee: *Name of person being interviewed*	Interviewer: *Name of person leading interview*
Location/Medium: *Office, conference room,* *or phone number*	Appointment Date: Start Time: End Time:
Objectives: *What data to collect* *On what to gain agreement* *What areas to explore*	Reminders: *Background/experience of interviewee* *Known opinions of interviewee*

Agenda:	Approximate Time:
Introduction	1 minute
Background on Project	2 minutes
Overview of Interview	
Topics to Be Covered	1 minute
Permission to Record	
Topic 1 Questions	5 minutes
Topic 2 Questions	7 minutes
...	...
Summary of Major Points	2 minutes
Questions from Interviewee	5 minutes
Closing	1 minute

General Observations:
Interviewee seemed busy probably need to call in a few days for follow-up questions because he gave only short answers. PC was turned off—probably not a regular PC user.

Unresolved Issues, Topics Not Covered:
He needs to look up sales figures from 1999. He raised the issue of how to handle returned goods, but we did not have time to discuss.

Interviewee:	Date:
Questions:	Notes:

When to ask question, if conditional
Question: 1
   Have you used the current sales
   tracking system? If so, how often ?
	Answer
	   Yes, I ask for a report on my
	   product line weekly.
	
	Observations
	   Seemed anxious—may be
	   overestimating usage frequency.

If yes, go to Question 2	

Question: 2
   What do you like least about the
system?
	Answer
	   Sales are shown in units, not
	   dollars.
	
	Observations
	   System can show sales in dollars,
	   but user does not know this.

often put the interviewees at ease because they are able to respond in their own words using their own structure; open-ended questions give interviewees more of a sense of involvement and control in the interview. A major disadvantage of open-ended questions is the length of time it can take for the questions to be answered. In addition, open-ended questions can be difficult to summarize.

Closed-ended questions provide a range of answers from which the interviewee may choose. Here is an example:

> *Which of the following would you say is the one best thing about the information system you currently use to do your job (pick only one)?*

> a. *Having easy access to all of the data you need*
> b. *The system's response time*
> c. *The ability to access the system from remote locations*

Closed-ended questions work well when the major answers to questions are well known. Another plus is that interviews based on closed-ended questions do not necessarily require a large time commitment—more topics can be covered. You can see body language and hear voice tone, which can aid in interpreting the interviewee's responses. Closed-ended questions can also be an easy way to begin an interview and to determine which line of open-ended questions to pursue. You can include an "other" option to encourage the interviewee to add unanticipated responses. A major disadvantage of closed-ended questions is that useful information that does not quite fit into the defined answers may be overlooked as the respondent tries to make a choice instead of providing his or her best answer.

Closed-ended questions, like objective questions on an examination, can follow several forms, including the following choices:

- True or false.
- Multiple choice (with only one response or selecting all relevant choices).
- Rating a response or idea on a scale, say from bad to good or strongly agree to strongly disagree. Each point on the scale should have a clear and consistent meaning to each person, and there is usually a neutral point in the middle of the scale.
- Ranking items in order of importance.

Interview Guidelines   First, with either open- or closed-ended questions, do not phrase a question in a way that implies a right or wrong answer. The respondent must feel that he or she can state his or her true opinion and perspective and that his or her idea will be considered equally with those of others. Questions such as "Should the system continue to provide the ability to override the default value, even though most users now do not like the feature?" should be avoided because such wording predefines a socially acceptable answer.

The second guideline to remember about interviews is to listen very carefully to what is being said. Take careful notes or, if possible, record the interview (be sure to ask permission first!). The answers may contain extremely important information for the project. Also, this may be the only chance you have to get information from this particular person. If you run out of time and still need to get information from the person you are talking to, ask to schedule a follow-up interview.

Third, once the interview is over, go back to your office and type up your notes within 48 hours. If you recorded the interview, use the recording to verify the material in your notes. After 48 hours, your memory of the interview will fade quickly. As you type and organize your notes, write down any additional questions that might arise from lapses in your notes or from ambiguous information. Separate facts from your opinions and interpretations. Make a list of unclear points that need clarification. Call the person you interviewed and get answers to these new questions. Use the phone call as an opportunity to verify the accuracy of your notes. You may also want to send a written copy of your notes to the person you interviewed so the person

Closed-ended questions
Questions in interviews that ask those responding to choose from among a set of specified responses.

can check your notes for accuracy. Finally, make sure you thank the person for his or her time. You may need to talk to your respondent again. If the interviewee will be a user of your system or is involved in another way in the system's success, you want to leave a good impression.

Fourth, be careful during the interview not to set expectations about the new or replacement system unless you are sure these features will be part of the delivered system. Let the interviewee know that there are many steps to the project and the perspectives of many people need to be considered, along with what is technically possible. Let respondents know that their ideas will be carefully considered, but that due to the iterative nature of the systems development process, it is premature to say now exactly what the ultimate system will or will not do.

Fifth, seek a variety of perspectives from the interviews. Find out what potential users of the system, users of other systems that might be affected by changes, managers and superiors, information systems staff who have experience with the current system, and others think the current problems and opportunities are and what new information services might better serve the organization. You want to understand all possible perspectives so that in a later approval step you will have information on which to base a recommendation or design decision that all stakeholders can accept.

Interviewing Groups

One drawback to using interviews to collect systems requirements is the need for the analyst to reconcile apparent contradictions in the information collected. A series of interviews may turn up inconsistent information about the current system or its replacement. You must work through all of these inconsistencies to figure out what might be the most accurate representation of current and future systems. Such a process requires several follow-up phone calls and additional interviews. Catching important people in their offices is often difficult and frustrating, and scheduling new interviews may become very time consuming. In addition, new interviews may reveal new questions that in turn require additional interviews with those interviewed earlier. Clearly, gathering information about an information system through a series of individual interviews and follow-up calls is not an efficient process.

Another option available to you is the group interview. In a group interview, several key people are interviewed at once. To make sure all of the important information is collected, you may conduct the interview with one or more analysts. In the case of multiple interviewers, one analyst may ask questions while another takes notes, or different analysts might concentrate on different kinds of information. For example, one analyst may listen for data requirements while another notes the timing and triggering of key events. The number of interviewees involved in the process may range from two to however many you believe can be comfortably accommodated.

A group interview has a few advantages. One, it is a much more effective use of your time than a series of interviews with individuals (although the time commitment of the interviewees may be more of a concern). Two, interviewing several people together allows them to hear the opinions of other key people and gives them the opportunity to agree or disagree with their peers. Synergies also often occur. For example, the comments of one person might cause another person to say, "That reminds me of" or "I didn't know that was a problem." You can benefit from such a discussion as it helps you identify issues on which there is general agreement and areas where views diverge widely.

The primary disadvantage of a group interview is the difficulty in scheduling it. The more people who are involved, the more difficult it will be finding a convenient time and place for everyone. Modern videoconferencing technology can minimize the geographical dispersion factors that make scheduling meetings so difficult. Group interviews are at the core of the JAD process, which we discuss in a later section in this chapter. A specific technique for working with groups, Nominal Group Technique, is discussed next.

Nominal Group Technique Many different techniques have been developed over the years to improve the process of working with groups. One of the more popular techniques for generating ideas among group members is called **Nominal Group Technique (NGT)**. NGT is exactly what the name indicates—the individuals working together to solve a problem are a group in name only, or nominally. Group members may be gathered in the same room for NGT, but they all work alone for a period of time. Typically, group members make a written list of their ideas. At the end of the idea-generation time, group members pool their individual ideas under the guidance of a trained facilitator. Pooling usually involves having the facilitator ask each person in turn for an idea that has not been presented before. As the person reads the idea aloud, someone else writes down the idea on a blackboard or flip chart. After all of the ideas have been introduced, the facilitator will then ask for the group to openly discuss each idea, primarily for clarification.

Once all of the ideas are understood by all of the participants, the facilitator will try to reduce the number of ideas the group will carry forward for additional consideration. There are many ways to reduce the number of ideas. The facilitator may ask participants to choose only a subset of ideas that they believe are important. Then the facilitator will go around the room, asking each person to read aloud an idea that is important to him or her that has not yet been identified by someone else. Or the facilitator may work with the group to identify and either eliminate or combine ideas that are very similar to others. At some point, the facilitator and the group end up with a tractable set of ideas, which can be further prioritized.

In a requirements determination context, the ideas being sought in an NGT exercise would typically apply to problems with the existing system or ideas for new features in the system being developed. The end result would be a list of either problems or features that group members themselves had generated and prioritized. There should be a high level of ownership of such a list, at least for the group that took part in the NGT exercise.

There is some evidence to support the use of NGT to help focus and refine the work of a group in that the number and quality of ideas that result from an NGT may be higher than what would normally be obtained from an unfacilitated group meeting. An NGT exercise could be used to complement the work done in a typical group interview or as part of a Joint Application Design effort, described in more detail later in this chapter.

> **Nominal Group Technique (NGT)**
> A facilitated process that supports idea generation by groups. At the beginning of the process, group members work alone to generate ideas, which are then pooled under the guidance of a trained facilitator.

Directly Observing Users

All the methods of collecting information that we have been discussing up until now involve getting people to recall and convey information they have about an organizational area and the information systems that support these processes. People, however, are not always very reliable informants, even when they try to be reliable and tell what they think is the truth. As odd as it may sound, people often do not have a completely accurate appreciation of what they do or how they do it. This is especially true concerning infrequent events, issues from the past, or issues for which people have considerable passion. Because people cannot always be trusted to reliably interpret and report their own actions, you can supplement and corroborate what people tell you by watching what they do or by obtaining relatively objective measures of how people behave in work situations. (See the box "Lost Soft Drink Sales" for an example of the importance of systems analysts learning firsthand about the business for which they are designing systems.)

For example, one possible view of how a hypothetical manager does her job is that a manager carefully plans her activities, works long and consistently on solving problems, and controls the pace of her work. A manager might tell you that is how she spends her day. When Mintzberg (1973) observed how managers work, however, he found that a manager's day is actually punctuated by many, many interruptions. Managers work in a fragmented manner, focusing on a problem or on a

Lost Soft Drink Sales

A systems analyst was quite surprised to read that sales of all soft-drink products were lower, instead of higher, after a new delivery truck routing system was installed. The software was designed to reduce stock-outs at customer sites by allowing drivers to visit each customer more often using more efficient delivery routes.

Confused by the results, management asked the analyst to delay a scheduled vacation, but he insisted that he could look afresh at the system only after a few overdue days of rest and relaxation.

Instead of taking a vacation, however, the analyst called a delivery dispatcher he had interviewed during the design of the system and asked to be given a route for a few days. The analyst drove a route (for a regular driver who was actually on vacation), following the schedule developed from the new system. What the analyst discovered was that the route was very efficient, as expected; so at first the analyst could not see any reason for lost sales.

During the third and last day of his "vacation," the analyst stayed overtime at one store to ask the manager if she had any ideas why sales might have dropped off in recent weeks. The manager had no explanation but did make a seemingly unrelated observation that the regular route driver appeared to have less time to spend in the store. He did not seem to take as much interest in where the products were displayed and did not ask for promotional signs to be displayed, as he had often done in the past.

From this conversation, the analyst concluded that the new delivery truck routing system was, in one sense, too good. It placed the driver on such a tight schedule that he had no time left for the "schmoozing" required to get special treatment, which gave the company's products an edge over the competition.

Without firsthand observation of the system in action gained by participating as a system user, the analyst might never have discovered the true problem with the system design. Once time was allotted for not only stocking new products but also for necessary marketing work, product sales returned to and exceeded levels achieved before the new system had been introduced.

---

communication for only a short time before they are interrupted by phone calls or visits from their subordinates and other managers. An information system designed to fit the work environment described by our hypothetical manager would not effectively support the actual work environment in which that manager finds herself.

As another example, consider the difference between what another employee might tell you about how much he uses e-mail and how much e-mail use you might discover through more objective means. An employee might tell you he is swamped with e-mail messages and that he spends a significant proportion of his time responding to e-mail messages. However, if you were able to check electronic mail records, you might find that this employee receives only 3 e-mail messages per day on average, and that the most messages he has ever received during one eight-hour period is 10. In this case, you were able to obtain an accurate behavioral measure of how much e-mail this employee copes with without having to watch him read his e-mail.

The intent behind obtaining system records and direct observation is the same, however, and that is to obtain more firsthand and objective measures of employee interaction with information systems. In some cases, behavioral measures will be a more accurate reflection of reality than what employees believe. In other cases, the behavioral information will substantiate what employees have told you directly. Although observation and obtaining objective measures are desirable ways to collect pertinent information, such methods are not always possible in real organizational settings. Thus, these methods are not totally unbiased, just as no other one data-gathering method is unbiased.

For example, observation can cause people to change their normal operating behavior. Employees who know they are being observed may be nervous and make more mistakes than normal, may be careful to follow exact procedures they do not typically follow, and may work faster or slower than normal. Moreover, because observation typically cannot be continuous, you receive only a snapshot image of the person or task you observe, which may not include important events or activities. Because observation is very time consuming, you will not only observe for a limited time, but also a limited number of people and a limited number of sites. Again, observation yields only a small segment of data from a possibly vast variety of data sources. Exactly which people or sites to observe is a difficult selection problem. You want to pick both typical and atypical people and sites, and observe during normal and abnormal conditions and times to receive the richest possible data from observation.

Analyzing Procedures and Other Documents

As noted earlier, asking questions of the people who use a system every day or who have an interest in a system is an effective way to gather information about current and future systems. Observing current system users is a more direct way of seeing how an existing system operates, but even this method provides limited exposure to all aspects of current operations. These methods of determining system requirements can be enhanced by examining system and organizational documentation to discover more details about current systems and the organization these systems support.

Although we discuss here several important types of documents that are useful in understanding possible future system requirements, our discussion does not exhaust all possibilities. You should attempt to find all written documents about the organizational areas relevant to the systems under redesign. Besides the few specific documents we discuss, organizational mission statements, business plans, organization charts, business policy manuals, job descriptions, internal and external correspondence, and reports from prior organizational studies can all provide valuable insight.

What can the analysis of documents tell you about the requirements for a new system? In documents you can find information about the following:

- Problems with existing systems (e.g., missing information or redundant steps)
- Opportunities to meet new needs if only certain information or information processing were available (e.g., analysis of sales based on customer type)
- Organizational direction that can influence information system requirements (e.g., trying to link customers and suppliers more closely to the organization)
- Titles and names of key individuals who have an interest in relevant existing systems (e.g., the name of a sales manager who led a study of buying behavior of key customers)
- Values of the organization or individuals who can help determine priorities for different capabilities desired by different users (e.g., maintaining market share even if it means lower short-term profits)
- Special information processing circumstances that occur irregularly that may not be identified by any other requirements determination technique (e.g., special handling needed for a few very large-volume customers and that requires use of customized customer ordering procedures)
- The reason why current systems are designed as they are, which can suggest features left out of current software, which may now be feasible and more desirable (e.g., data about a customer's purchase of competitors' products were not available when the current system was designed; these data are now available from several sources)
- Data, rules for processing data, and principles by which the organization operates that must be enforced by the information system (e.g., each customer is assigned exactly one sales department staff member as a primary contact if the customer has any questions)

One type of useful document is a written work procedure for an individual or a work group. The procedure describes how a particular job or task is performed, including data and information that are used and created in the process of performing the job. For example, the procedure shown in Figure 6-3 includes the data (list of features and advantages, drawings, inventor name, and witness names) that are required to prepare an invention disclosure. It also indicates that besides the inventor, the vice president for research and department head and dean must review the material, and that a witness is required for any filing of an invention disclosure. These insights clearly affect what data must be kept, to whom information must be sent, and the rules that govern valid forms.

FIGURE 6-3
Example of a procedure

GUIDE FOR PREPARATION OF INVENTION DISCLOSURE
(See FACULTY and STAFF MANUALS for Detailed
Patent Policy and Routing Procedures.)

(1) DISCLOSE ONLY ONE INVENTION PER FORM.

(2) PREPARE COMPLETE DISCLOSURE.

The disclosure of your invention is adequate for patent purposes ONLY if it enables a person skilled in the art to understand the invention.

(3) CONSIDER THE FOLLOWING IN PREPARING A COMPLETE DISCLOSURE:

(a) All essential elements of the invention, their relationship to one another, and their mode of operation.

(b) Equivalents that can be substituted for any elements.

(c) List of features believed to be new.

(d) Advantages this invention has over the prior art.

(e) Whether the invention has been built and/or tested.

(4) PROVIDE APPROPRIATE ADDITIONAL MATERIAL.

Drawings and descriptive material should be provided as needed to clarify the disclosure. Each page of this material must be signed and dated by each inventor and properly witnessed. A copy of any current and/or planned publication relating to the invention should be included.

(5) INDICATE PRIOR KNOWLEDGE AND INFORMATION.

Pertinent publications, patents or previous devices, and related research or engineering activities should be identified.

(6) HAVE DISCLOSURE WITNESSED.

Persons other than coinventors should serve as witnesses and should sign each sheet of the disclosure only after reading and understanding the disclosure.

(7) FORWARD ORIGINAL PLUS ONE COPY (two copies if supported by grant/contract) TO VICE PRESIDENT FOR RESEARCH VIA DEPARTMENT HEAD AND DEAN.

Procedures are not trouble-free sources of information, however. Sometimes your analysis of several written procedures will reveal a duplication of effort in two or more jobs. You should call such duplication to the attention of management as an issue to be resolved before system design can proceed. That is, it may be necessary to redesign the organization before the redesign of an information system can achieve its full benefits. Another problem you may encounter with a procedure occurs when the procedure is missing. Again, it is not your job to create a document for a missing procedure—that is up to management. A third and common problem with a written procedure happens when the procedure is out of date. You may realize the procedure is out of date when you interview the person responsible for performing the task described in the procedure. Once again, the decision to rewrite the procedure so that it matches reality is made by management, but you may make suggestions based upon your understanding of the organization. A fourth problem often encountered with written procedures is that the formal procedures may contradict information you collected from interviews and observation about how the organization operates and what information is required. As in the other cases, resolution rests with management.

All of these problems illustrate the difference between **formal systems** and **informal systems**. Formal systems are recognized in the official documentation of the organization; informal systems are the way in which the organization actually works. Informal systems develop because of inadequacies of formal procedures, individual work habits and preferences, resistance to control, and other factors. It is important to understand both formal and informal systems because each provides insight into information requirements and what will be required to convert from present to future information services.

A second type of document useful to systems analysts is a business form (see Figure 6-4). Forms are used for all types of business functions, from recording an order acknowledging the payment of a bill to indicating what goods have been shipped. Forms are important for understanding a system because they explicitly

Formal system
The official way a system works as described in organizational documentation.

Informal system
The way a system actually works.

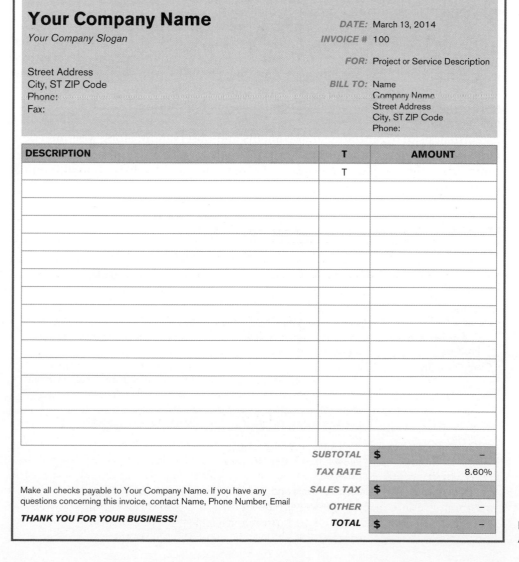

FIGURE 6-4
An invoice form from Microsoft Excel

indicate what data flow in or out of a system and which are necessary for the system to function. In the sample invoice form in Figure 6-4, we see locations for data such as the name and bill to address of the customer, the invoice number, data (quantity, description, amount) about each line item on the invoice, and calculated data such as the total.

A form gives us crucial information about the nature of the organization. For example, the company can ship and bill to different addresses; customers can have discounts applied; and the freight expense is charged to the customer. A printed form may correspond to a computer display that the system will generate for someone to enter and maintain data or to display data to online users. Forms are most useful to you when they contain actual organizational data because this allows you to determine the characteristics of the data that are actually used by the application. The ways in which people use forms change over time, and data that were needed when a form was designed may no longer be required. You can use the systems analysis techniques presented in Chapters 7 and 8 to help you determine which data are no longer required.

A third type of useful document is a report generated by current systems. As the primary output for some types of systems, a report enables you to work backward from the information on the report to the data that must have been necessary to generate them. Figure 6-5 presents an example of a typical financial report, a statement of cash flows. You would analyze such reports to determine which data need to be captured over what time period and what manipulation of these raw data would be necessary to produce each field on the report.

If the current system is computer-based, a fourth set of useful documents are those that describe the current information systems—how they were designed and how they work. A lot of different types of documents fit this description, everything from flowcharts to data dictionaries and CASE tool reports, to user manuals. An analyst who has access to such documents is lucky; many information systems developed in-house lack complete documentation (unless a CASE tool has been used).

Analysis of organizational documents and observation, along with interviewing, are the methods most often used for gathering system requirements. Table 6-4 summarizes the comparative features of observation and analysis of organizational documents.

TABLE 6-4   Comparison of Observation and Document Analysis

Characteristic	Observation	Document Analysis
Information Richness	High (many channels)	Low (passive) and old
Time Required	Can be extensive	Low to moderate
Expense	Can be high	Low to moderate
Chance for Follow-Up and Probing	Good: probing and clarification questions can be asked during or after observation	Limited: probing possible only if original author is available
Confidentiality	Observee is known to interviewer; observee may change behavior when observed	Depends on nature of document; does not change simply by being read
Involvement of Subject	Interviewees may or may not be involved and committed depending on whether they know if they are being observed	None, no clear commitment
Potential Audience	Limited numbers and limited time (snapshot) of each	Potentially biased by which documents were kept or because document was not created for this purpose

FIGURE 6-5
An example of a report: a statement of cash flows

Mellankamp Industries
Statement of Cash Flows
October 1 through December 31, 2014

	Oct 1–Dec 31, 2014
OPERATING ACTIVITIES	
Net Income	$38,239.15
Adjustments to reconcile Net Income to net cash provided by operations:	
Accounts Receivable	−$46,571.69
Employee Loans	−62.00
Inventory Asset	−18,827.16
Retainage	−2,461.80
Accounts Payable	29,189.66
Business Credit Card	70.00
BigOil Card	−18.86
Sales Tax Payable	687.65
Net cash provided by Operating Activities	$244.95
INVESTING ACTIVITIES	
Equipment	−$44,500.00
Prepaid Insurance	2,322.66
Net cash provided by Investing Activities	−$42,177.34
FINANCING ACTIVITIES	
Bank Loan	−$868.42
Emergency Loan	3,911.32
Note Payable	−17,059.17
Equipment Loan	43,013.06
Opening Balance Equity	−11,697.50
Owner's Equity: Owner's Draw	−6,000.00
Retained Earnings	8,863.39
Net cash provided by Financing Activities	$20,162.68
Net cash increase for period	−$21,769.71
Cash at beginning of period	−$21,818.48
Cash at end of period	**−$43,588.19**

CONTEMPORARY METHODS FOR DETERMINING SYSTEM REQUIREMENTS

Even though we called interviews, observation, and document analysis traditional methods for determining a system's requirements, all of these methods are still very much used by analysts to collect important information. Today, however, there are additional techniques to collect information about the current system, the organizational area requesting the new system, and what the new system should be like. In this section, you will learn about several contemporary information-gathering techniques for analysis (listed in Table 6-5): JAD, CASE tools to support JAD, and prototyping. As we said earlier, these techniques can support effective information collection and structuring while reducing the amount of time required for analysis.

Joint Application Design

Joint Application Design (JAD) started in the late 1970s at IBM and since then the practice of JAD has spread throughout many companies and industries. For example, it is quite popular in the insurance industry in Connecticut, where a JAD users' group has been formed. In fact, several generic approaches to JAD have been documented and popularized (see Wood and Silver, 1995, for an example). The main idea behind JAD is to bring together the key users, managers, and systems analysts involved in the analysis of a current system. In that respect, JAD is similar to a group interview; a JAD, however, follows a particular structure of roles and agenda that is quite different from a group interview during which analysts control the sequence of questions answered by users. The primary purpose of using JAD in the analysis phase is to collect systems requirements simultaneously from the key people involved with the system. The result is an intense and structured, but highly effective, process. As with a group interview, having all the key people together in one place at one time allows analysts to see where there are areas of agreement and where there are conflicts. Meeting with all of these important people for over a week of intense sessions allows you the opportunity to resolve conflicts, or at least to understand why a conflict may not be simple to resolve.

JAD sessions are usually conducted at a location other than the place where the people involved normally work. The idea behind such a practice is to keep participants away from as many distractions as possible so that they can concentrate on systems analysis. A JAD may last anywhere from four hours to an entire week and may consist of several sessions. A JAD employs thousands of dollars of corporate resources, the most expensive of which is the time of the people involved. Other expenses include the costs associated with flying people to a remote site and putting them up in hotels and feeding them for several days.

The typical participants in a JAD are listed below:

- *JAD session leader.* The **JAD session leader** organizes and runs the JAD. This person has been trained in group management and facilitation as well as in systems analysis. The JAD leader sets the agenda and sees that it is met; he or she remains neutral on issues and does not contribute ideas or opinions, but rather concentrates on keeping the group on the agenda, resolving conflicts and disagreements, and soliciting all ideas.

Joint Application Design (JAD)
A structured process in which users, managers, and analysts work together for several days in a series of intensive meetings to specify or review system requirements.

JAD session leader
The trained individual who plans and leads Joint Application Design sessions.

TABLE 6-5 **Contemporary Methods for Collecting System Requirements**

- Bringing together in a *JAD* session users, sponsors, analysts, and others to discuss and review system requirements
- Using *CASE tools* during a JAD to analyze current systems to discover requirements to meet changing business conditions
- Iteratively developing system *prototypes* that refine the understanding of system requirements in concrete terms by showing working versions of system features

- *Users.* The key users of the system under consideration are vital participants in a JAD. They are the only ones who have a clear understanding of what it means to use the system on a daily basis.

- *Managers.* Managers of the work groups who use the system in question provide insight into new organizational directions, motivations for and organizational impacts of systems, and support for requirements determined during the JAD.

- *Sponsor.* As a major undertaking due to its expense, a JAD must be sponsored by someone at a relatively high level in the company. If the sponsor attends any sessions, it is usually only at the very beginning or the end.

- *Systems analysts.* Members of the systems analysis team attend the JAD, although their actual participation may be limited. Analysts are there to learn from users and managers, not to run or dominate the process.

- *Scribe.* The **scribe** takes notes during the JAD sessions. This is usually done on a laptop. Notes may be taken using a word processor, or notes and diagrams may be entered directly into a CASE tool.

- *IS staff.* Besides systems analysts, other information systems (IS) staff, such as programmers, database analysts, IS planners, and data center personnel, may attend to learn from the discussion and possibly to contribute their ideas on the technical feasibility of proposed ideas or on technical limitations of current systems.

Scribe
The person who makes detailed notes of the happenings at a Joint Application Design session.

JAD sessions are usually held in special-purpose rooms where participants sit around horseshoe-shaped tables, as shown in Figure 6-6. These rooms are typically equipped with whiteboards. Other audiovisual tools may be used, such as magnetic symbols that can be easily rearranged on a whiteboard, flip charts, and computer-generated displays. Flip-chart paper is typically used for keeping track of issues that cannot be resolved during the JAD or for those issues requiring additional information that can be gathered during breaks in the proceedings. Computers may be used to create and display form or report designs, diagram existing or replacement systems, or create prototypes.

FIGURE 6-6
Illustration of the typical room layout for a JAD
(*Source:* Based on Wood and Silver, 1995.)

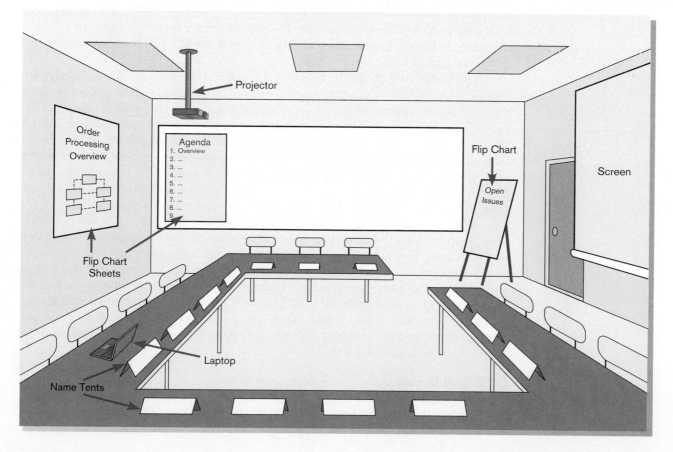

When a JAD is completed, the end result is a set of documents that detail the workings of the current system related to the study of a replacement system. Depending on the exact purpose of the JAD, analysts may also walk away from the JAD with some detailed information on what is desired of the replacement system.

Taking Part in a JAD   Imagine that you are a systems analyst taking part in your first JAD. What might participating in a JAD be like? Typically, JADs are held off-site at comfortable conference facilities. On the first morning of the JAD, you and your fellow analysts walk into a room that looks much like the one depicted in Figure 6-6. The JAD facilitator is already there; she is finishing writing the day's agenda on a flip chart. The scribe is seated in a corner with his laptop, preparing to take notes on the day's activities. Users and managers begin to enter in groups and seat themselves around the U-shaped table. You and the other analysts review your notes that describe what you have learned so far about the information system you are all here to discuss. The session leader opens the meeting with a welcome and a brief rundown of the agenda. The first day will be devoted to a general overview of the current system and major problems associated with it. The next two days will be devoted to an analysis of current system screens. The last two days will be devoted to analysis of reports.

The session leader introduces the corporate sponsor, who talks about the organizational unit and current system related to the systems analysis study and the importance of upgrading the current system to meet changing business conditions. He leaves, and the JAD session leader takes over. She yields the floor to the senior analyst, who begins a presentation on key problems with the system that have already been identified. After the presentation, the session leader opens the discussion to the users and managers in the room.

After a few minutes of talk, a heated discussion begins between two users from different corporate locations. One user, who represents the office that served as the model for the original systems design, argues that the system's perceived lack of flexibility is really an asset, not a problem. The other user, who represents an office that was part of another company before a merger, argues that the current system is so inflexible as to be virtually unusable. The session leader intervenes and tries to help the users isolate particular aspects of the system that may contribute to the system's perceived lack of flexibility.

Questions arise about the intent of the original developers. The session leader asks the analysis team about their impressions of the original system design. Because these questions cannot be answered during this meeting (none of the original designers are present and none of the original design documents are readily available), the session leader assigns the question about intent to the "to do" list. This question becomes the first one on a flip-chart sheet of "to do" items, and the session leader gives you the assignment of finding out about the intent of the original designers. She writes your name next to the "to do" item on the list and continues with the session. Before the end of the JAD, you must get an answer to this question.

The JAD will continue like this for its duration. Analysts will make presentations, help lead discussions on form and report design, answer questions from users and managers, and take notes on what is being said. After each meeting, the analysis team will meet, usually informally, to discuss what has occurred that day and to consolidate what they have learned. Users will continue to contribute during the meetings, and the session leader will facilitate, intervening in conflicts and seeing that the group follows the agenda. When the JAD is over, the session leader and her assistants must prepare a report that documents the findings in the JAD and is circulated among users and analysts.

CASE Tools During JAD   For requirements determination and structuring, the most useful CASE tools are for diagramming and for form and report generation. The more interaction analysts have with users during this phase, the more useful this set of tools is. The analyst can use diagramming and prototyping tools to give graphic form to system requirements, show the tools to users, and make changes based on the users' reactions. The same tools are very valuable for requirements structuring as well. Using common CASE tools during requirements determination and structuring makes the transition between these two subphases easier and reduces the total time spent. In structuring, CASE tools that analyze requirements information for correctness, completeness, and consistency are also useful. Finally, for alternative generation and selection, diagramming and prototyping tools are key to presenting users with graphic illustrations of what the alternative systems will look like. Such a practice provides users and analysts with better information to select the most desirable alternative system.

Some observers advocate using CASE tools during JADs (Lucas, 1993). Running a CASE tool during a JAD enables analysts to enter system models directly into a CASE tool, providing consistency and reliability in the joint model-building process. The CASE tool captures system requirements in a more flexible and useful way than can a scribe or an analysis team making notes. Further, the CASE tool can be used to project menu, display, and report designs, so users can directly observe old and new designs and evaluate their usefulness for the analysis team.

Using Prototyping During Requirements Determination

Prototyping is an iterative process involving analysts and users whereby a rudimentary version of an information system is built and rebuilt according to user feedback. Prototyping can replace the systems development life cycle or augment it. What we are interested in here is how prototyping can augment the requirements determination process.

In order to gather an initial basic set of requirements, you will still have to interview users and collect documentation. Prototyping, however, will enable you to quickly convert basic requirements into a working, though limited, version of the desired information system. The prototype will then be viewed and tested by the user. Typically, seeing verbal descriptions of requirements converted into a physical system will prompt the user to modify existing requirements and generate new ones. For example, in the initial interviews, a user might have said that he wanted all relevant utility billing information such as the client's name and address, the service record, and payment history, on a single computer display form. Once the same user sees how crowded and confusing such a design would be in the prototype, he might change his mind and instead ask to have the information organized on several screens, but with easy transitions from one screen to another. He might also be reminded of some important requirements (data, calculations, etc.) that had not surfaced during the initial interviews.

You would then redesign the prototype to incorporate the suggested changes (Figure 6-7). Once modified, users would again view and test the prototype.

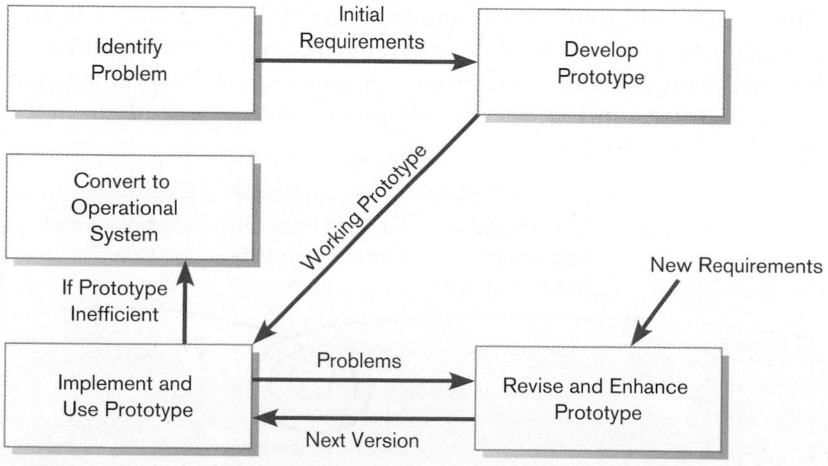

Prototyping
An iterative process of systems development in which requirements are converted to a working system that is continually revised through close collaboration between an analyst and users.

FIGURE 6-7
The prototyping methodology
(*Source:* Based on "Prototyping: The New Paradigm for Systems Development," by J. D. Naumann and A. M. Jenkins, *MIS Quarterly* 6(3): 29–44.)

And, once again, you would incorporate their suggestions for change. Through such an iterative process, the chances are good that you will be able to better capture a system's requirements.

As the prototype changes through each iteration, more and more of the design specifications for the system are captured in the prototype. The prototype can then serve as the basis for the production system, in a process called *evolutionary prototyping*. Alternatively, the prototype can serve only as a model, which is then used as a reference for the construction of the actual system. In this process, called *throwaway prototyping*, the prototype is discarded after it has been used.

Evolutionary Prototyping In evolutionary prototyping, you begin by modeling parts of the target system and, if the prototyping process is successful, you evolve the rest of the system from those parts (McConnell, 1996). A life-cycle model of evolutionary prototyping illustrates the iterative nature of the process and the tendency to refine the prototype until it is ready to release (Figure 6-8). One key aspect of this approach is that the prototype becomes the actual production system. Because of this, you often start with those parts of the system that are most difficult and uncertain.

Although a prototype system may do a great job of representing easy-to-see aspects of a system, such as the user interface, the production system itself will perform many more functions, several of which are transparent or invisible to the users. Any given system must be designed to facilitate database access, database integrity, system security, and networking. Systems also must be designed to support scalability, multiuser support, and multiplatform support. Few of these design specifications will be coded into a prototype. Further, as much as 90 percent of a system's functioning is devoted to handling exceptional cases (McConnell, 1996). Prototypes are designed to handle only the typical cases, so exception handling must be added to the prototype as it is converted to the production system. Clearly, the prototype captures only part of the system requirements.

Throwaway Prototyping Unlike evolutionary prototyping, throwaway prototyping does not preserve the prototype that has been developed. With throwaway prototyping, there is never any intention to convert the prototype into a working system. Instead, the prototype is developed quickly to demonstrate some aspect of a system design that is unclear or to help users decide among different features or interface characteristics. Once the uncertainty the prototype was created to address has been reduced, the prototype can be discarded, and the principles learned from its creation and testing can then become part of the requirements determination.

Prototyping is most useful for requirements determination when:

- User requirements are not clear or well understood, which is often the case for totally new systems or systems that support decision making
- One or a few users and other stakeholders are involved with the system
- Possible designs are complex and require concrete form to fully evaluate
- Communication problems have existed in the past between users and analysts and both parties want to be sure that system requirements are as specific as possible
- Tools (such as form and report generators) and data are readily available to rapidly build working systems

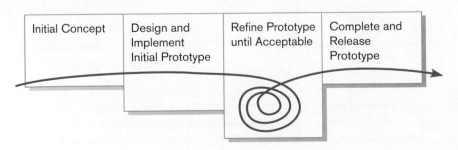

FIGURE 6-8
McConnell's evolutionary prototyping model

Prototyping also has some drawbacks as a tool for requirements determination. These include the following:

- Prototypes have a tendency to avoid creating formal documentation of system requirements, which can then make the system more difficult to develop into a fully working system.
- Prototypes can become very idiosyncratic to the initial user and difficult to diffuse or adapt to other potential users.
- Prototypes are often built as stand-alone systems, thus ignoring issues of sharing data and interactions with other existing systems, as well as issues with scaling up applications.
- Checks in the SDLC are bypassed so that some more subtle, but still important, system requirements might be forgotten (e.g., security, some data entry controls, or standardization of data across systems).

RADICAL METHODS FOR DETERMINING SYSTEM REQUIREMENTS

Whether traditional or contemporary, the methods for determining system requirements that you have read about in this chapter apply to any requirements determination effort, regardless of its motivation. But most of what you have learned has traditionally been applied to systems development projects that involve automating existing processes. Analysts use system requirements determination to understand current problems and opportunities, as well as to determine what is needed and desired in future systems. Typically, the current way of doing things has a large impact on the new system. In some organizations, though, management is looking for new ways to perform current tasks. These new ways may be radically different from how things are done now, but the payoffs may be enormous: Fewer people may be needed to do the same work, relationships with customers may improve dramatically, and processes may become much more efficient and effective, all of which can result in increased profits. The overall process by which current methods are replaced with radically new methods is generally referred to as **business process reengineering (BPR)**. Although the term *BPR* is usually associated with a management *fad* that occurred in the 1990s, businesses remain vitally interested in business processes and how to improve them (Sharp and McDermott, 2001). Even if the term *business process reengineering* may seem dated to some, process orientation remains a lasting legacy of the BPR movement.

Business process reengineering (BPR)
The search for, and implementation of, radical change in business processes to achieve breakthrough improvements in products and services.

To better understand BPR, consider the following analogy. Suppose you are a successful European golfer who has tuned your game to fit the style of golf courses and weather in Europe. You have learned how to control the flight of the ball in heavy winds, roll the ball on wide open greens, putt on large and undulating greens, and aim at a target without the aid of the landscaping common on North American courses. When you come to the United States to make your fortune on the U.S. tour, you discover that incrementally improving your putting, driving accuracy, and sand shots will help, but the new competitive environment is simply not suited to your style of the game. You need to reengineer your whole approach, learning how to aim at targets, spin and stop a ball on the green, and manage the distractions of crowds and the press. If you are good enough, you may survive, but without reengineering, you will never be a winner.

Just as the competitiveness of golf forces good players to adapt their games to changing conditions, the competitiveness of our global economy has driven most companies into a mode of continuously improving the quality of their products and services (Dobyns and Crawford-Mason, 1991). Organizations realize that creatively using information technologies can yield significant improvements in most business processes. The idea behind BPR is not just to improve each business process, but, in a systems modeling sense, to reorganize the complete flow of data in major sections of an organization to eliminate unnecessary steps, achieve synergies among previously separate steps, and become more responsive to future changes.

Companies such as IBM, Procter & Gamble, Walmart, and Ford are actively pursuing BPR efforts and have had great success. Yet many other companies have found difficulty in applying BPR principles (Moad, 1994). Nonetheless, BPR concepts are actively applied in both corporate strategic planning and information systems planning as a way to radically improve business processes (as described in Chapter 4).

BPR advocates suggest that radical increases in the quality of business processes can be achieved through creative application of information technologies. BPR advocates also suggest that radical improvement cannot be achieved by tweaking existing processes but rather by using a clean sheet of paper and asking, "If we were a new organization, how would we accomplish this activity?" Changing the way work is performed also changes the way information is shared and stored, which means that the results of many BPR efforts are the development of information system maintenance requests or requests for system replacement. It is likely that you will encounter or have encountered BPR initiatives in your own organization.

Identifying Processes to Reengineer

A first step in any BPR effort relates to understanding what processes to change. To do this, you must first understand which processes represent the key business processes for the organization. **Key business processes** are the structured set of measurable activities designed to produce a specific output for a particular customer or market. The important aspect of this definition is that key processes are focused on some type of organizational outcome, such as the creation of a product or the delivery of a service. Key business processes are also customer focused. In other words, key business processes would include all activities used to design, build, deliver, support, and service a particular product for a particular customer. BPR efforts, therefore, first try to understand those activities that are part of the organization's key business processes and then alter the sequence and structure of activities to achieve radical improvements in speed, quality, and customer satisfaction. The same techniques you learned to use for systems requirement determination can be used to discover and understand key business processes. Interviewing key individuals, observing activities, reading and studying organizational documents, and conducting JADs can all be used to find and fathom key business processes.

After identifying key business processes, the next step is to identify specific activities that can be radically improved through reengineering. Hammer and Champy (1993), who are most closely identified with the term *BPR* and the process itself suggest that three questions be asked to identify activities for radical change:

1. How important is the activity to delivering an outcome?
2. How feasible is changing the activity?
3. How dysfunctional is the activity?

The answers to these questions provide guidance for selecting which activities to change. Those activities deemed important, changeable, yet dysfunctional, are primary candidates. To identify dysfunctional activities, they suggest you look for activities where there are excessive information exchanges between individuals, where information is redundantly recorded or needs to be rekeyed, where there are excessive inventory buffers or inspections, and where there is a lot of rework or complexity. Many of the tools and techniques for modeling data, processes, events, and logic within the IS development process are also being applied to model business processes within BPR efforts (see Davenport, 1993). Thus, the skills of a systems analyst are often central to many BPR efforts.

Disruptive Technologies

Once key business processes and activities have been identified, information technologies must be applied to radically improve business processes. To do this, Hammer and Champy (1993) suggest that organizations think "inductively" about

---

Key business processes
The structured, measured set of activities designed to produce a specific output for a particular customer or market.

TABLE 6-6 **Long-Held Organizational Rules That Are Being Eliminated through Disruptive Technologies**

Rule	Disruptive Technology
Information can appear in only one place at a time.	Distributed databases allow the sharing of information.
Businesses must choose between centralization and decentralization.	Advanced telecommunications networks can support dynamic organizational structures.
Managers must make all decisions.	Decision-support tools can aid nonmanagers.
Field personnel need offices where they can receive, store, retrieve, and transmit information.	Wireless data communication and portable computers provide a "virtual" office for workers.
The best contact with a potential buyer is personal contact.	Interactive communication technologies allow complex messaging capabilities.
You have to find out where things are.	Automatic identification and tracking technology knows where things are.
Plans get revised periodically.	High-performance computing can provide real-time updating.

information technology. Induction is the process of reasoning from the specific to the general, which means that managers must learn the power of new technologies and think of innovative ways to alter the way work is done. This is contrary to deductive thinking, where problems are first identified and solutions are then formulated.

Hammer and Champy suggest that managers especially consider disruptive technologies when applying deductive thinking. **Disruptive technologies** are those that enable the breaking of long-held business rules that inhibit organizations from making radical business changes. For example, Procter & Gamble (P&G), the huge consumer products company, uses information technology to "innovate innovation" (Teresko, 2004). Technology helps different organizational units work together seamlessly on new products. P&G also uses computer simulations to quicken product design and to test potential products with consumers early in the design process. Table 6-6 shows several long-held business rules and beliefs that constrain organizations from making radical process improvements. For example, the first rule suggests that information can only appear in one place at a time. However, the advent of distributed databases (see Chapter 12) and pervasive wireless networking have "disrupted" this long-held business belief.

Disruptive technologies
Technologies that enable the breaking of long-held business rules that inhibit organizations from making radical business changes.

REQUIREMENTS DETERMINATION USING AGILE METHODOLOGIES

You've already learned about many different ways to determine the requirements for a system. Yet new methods and techniques are constantly being developed. Three more requirements determination techniques are presented in this section. The first is continual user involvement in the development process, a technique that works especially well with small and dedicated development teams. The second approach is a JAD-like process called Agile Usage-Centered Design. The third approach is the Planning Game, which was developed as part of eXtreme Programming.

Continual User Involvement

In Chapter 1, you read about the criticisms of the traditional waterfall SDLC. One of those criticisms was that the waterfall SDLC allowed users to be involved in the development process only in the early stages of analysis. Once requirements had been gathered from them, the users were not involved again in the process until the system was being installed and they were asked to sign off on it. Typically, by the time the users saw the system again, it was nothing like what they had imagined. Also, given how their business processes had changed since analysis had ended, the system most likely did not adequately address user needs. This view of the traditional

FIGURE 6-9
The iterative analysis–design–code–test cycle

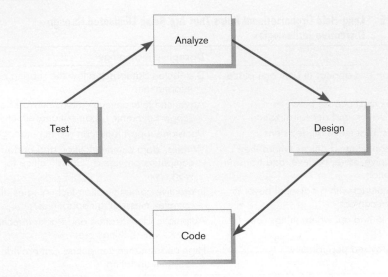

waterfall SDLC and user involvement is a stereotype of the process, and it does not describe every systems development project that used the waterfall model. However, limited user involvement has been common enough to be perceived as a real and serious problem in systems development.

One approach to the problem of limited user involvement is to involve the users continually, throughout the entire analysis and design process. Such an approach works best when development can follow the analysis–design–code–test cycle favored by the Agile Methodologies (Figure 6-9), because the user can provide information on requirements and then watch and evaluate as those requirements are designed, coded, and tested. This iterative process can continue through several cycles, until most of the major functionality of the system has been developed. Extensive involvement of users in the analysis and design process is a key part of many Agile approaches, but it was also a key part of Rapid Application Development (see Chapter 1).

Continual user involvement was a key aspect of the success of Boeing's Wire Design and Wire Install system for the 757 aircraft (Bedoll, 2003). The system was intended to support engineers who customize plane configurations for customers, allowing them to analyze all 50,000 wires that can possibly be installed on a 757. A previous attempt at building a similar system took over three years, and the resulting system was never used. The second attempt, relying on Agile Methodologies, resulted in a system that was in production after only six weeks. One of the keys to success was a user liaison who spent half of his time with the small development team and half with the other end users. In addition to following the analysis–design–code–test cycle, the team also had weekly production releases. The user liaison was involved every step of the way. Obviously, for such a requirements determination to succeed, the user who works with the development team must be very knowledgeable, but he or she must also be in a position to give up his or her normal business responsibilities in order to become so heavily involved in a system's development.

Agile Usage-Centered Design

Continual user involvement in systems development is an excellent way to ensure that requirements are captured accurately and immediately implemented in system design. However, such constant interaction works best when the development team is small, as was the case in the Boeing example. Also, it is not always possible to have continual access to users for the duration of a development project. Thus, Agile developers have come up with other means for effectively involving users in the requirements determination process. One such method is called *Agile Usage-Centered Design*, originally developed by Larry Constantine (2002) and adapted for Agile

TABLE 6-7   Steps in the Agile Usage-Centered Design Method for Requirements Determination

1. Gather a group of people, including analysts, users, programmers, and testing staff, and sequester them in a room to collaborate on this design. Include a facilitator who knows this process.
2. Give everyone a chance to vent about the current system and to talk about the features everyone wants in the new system. Record all of the complaints and suggestions for change on whiteboards or flip charts for everyone to see.
3. Determine what the most important user roles would be. Determine who will be using the system and what their goals are for using the system. Write the roles on 3 × 5 cards. Sort the cards so that similar roles are close to each other. Patton (2002) calls this a *role model.*
4. Determine what tasks user roles will have to complete in order to achieve their goals. Write these down on 3 × 5 cards. Order tasks by importance and then by frequency. Place the cards together based on how similar the tasks are to each other. Patton calls this a *task model.*
5. Task cards will be grouped together on the table based on their similarity. Grab a stack of cards. This is called an *interaction context.*
6. For each task card in the interaction context, write a description of the task directly on the task card. List the steps that are necessary to complete the task. Keep the descriptions conversational to make them easy to read. Simplify.
7. Treat each stack as a tentative set of tasks to be supported by a single aspect of the user interface, such as a screen, page, or dialog, and create a paper-and-pencil prototype for that part of the interface. Show the basic size and placement of the screen components.
8. Take on a user role and step through each task in the interaction context as modeled in the paper-and-pencil prototype. Make sure the user role can achieve its goals by using the prototype. Refine the prototype accordingly.

Methodologies by Jeff Patton (2002). Patton describes the process in nine steps, which we have adapted and presented as eight steps in Table 6-7.

Notice how similar the overall process is to a JAD meeting. All of the experts are gathered together and work with the help of the facilitator. What is unique about the Agile Usage-Centered Design is the process that supports it, which focuses on user roles, user goals, and the tasks necessary to achieve those goals. Then, tasks are grouped and turned into paper-and-pencil prototypes before the meeting is over. Requirements captured from users and developers are captured as prototyped system screens. Patton (2002) believes that the two most effective aspects of this approach are the venting session, which lets everyone get their complaints out in the open, and the use of 3 × 5 cards, which serve as very effective communication tools. As with any analysis and design process or technique, however, Agile Usage-Centered Design will not work for every project or every company.

The Planning Game from eXtreme Programming

You read about eXtreme Programming in Chapter 1, and you know that it is an approach to software development put together by Kent Beck (Beck and Andres, 2004). You also know that it is distinguished by its short cycles, its incremental planning approach, its focus on automated tests written by programmers and customers to monitor the process of development, and its reliance on an evolutionary approach to development that lasts throughout the lifetime of the system. One of the key emphases of eXtreme Programming is its use of two-person programming teams and having a customer on-site during the development process. The relevant parts of eXtreme Programming that relate to requirements determination are (1) how planning, analysis, design, and construction are all fused together into a single phase of activity and (2) its unique way of capturing and presenting system requirements and design specifications. All phases of the life cycle converge into series of activities based on the basic processes of coding, testing, listening, and designing.

What is of interest here, however, is the way requirements and specifications are dealt with. Both of these activities take place in what Beck calls the "Planning Game." The Planning Game is really just a stylized approach to development that seeks to maximize fruitful interaction between those who need a new system and those who build it. The players in the Planning Game, then, are Business and Development. Business is the customer and is ideally represented by someone who knows the processes to be supported by the system being developed. Development is represented by those actually designing and constructing the system. The game pieces are what Beck calls "Story Cards." These cards are created by Business and contain a description of a procedure or feature to be included in the system. Each card is dated and numbered and has space on it for tracking its status throughout the development effort.

The Planning Game has three phases: exploration, commitment, and steering (Figure 6-10). In exploration, Business creates a Story Card for something it wants the new system to do. Development responds with an estimation of how long it would take to implement the procedure. At this point, it may make sense to split a Story Card into multiple Story Cards, as the scope of features and procedures becomes more clear during discussion. In the commitment phase, Business sorts Story Cards into three stacks: one for essential features, one for features that are not essential but would still add value, and one for features that would be nice to have. Development then sorts Story Cards according to risk, based on how well they can estimate the time needed to develop each feature. Business then selects the cards that will be included in the next release of the product. In the final phase, steering, Business has a chance to see how the development process is progressing and to work with Development to adjust the plan accordingly. Steering can take place as often as once every three weeks.

The Planning Game between Business and Development is followed by the Iteration Planning Game, played only by programmers. Instead of Story Cards, programmers write Task Cards, which are based on Story Cards. Typically, several Task Cards are generated for each Story Card. The Iteration Planning Game has the same three phases as the Planning Game: exploration, commitment, and steering. During exploration, programmers convert Story Cards into Task Cards. During commitment, they accept responsibility for tasks and balance their workloads. During steering, the programmers write the code for the feature, test it, and if it works, they integrate the

FIGURE 6-10
eXtreme Programming's Planning Game

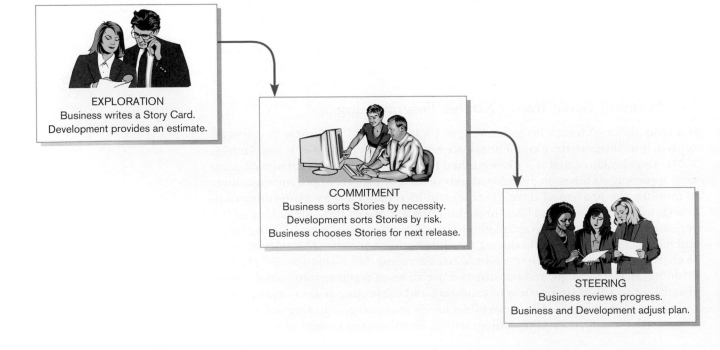

EXPLORATION
Business writes a Story Card.
Development provides an estimate.

COMMITMENT
Business sorts Stories by necessity.
Development sorts Stories by risk.
Business chooses Stories for next release.

STEERING
Business reviews progress.
Business and Development adjust plan.

feature into the product being developed. The Iteration Planning Game takes place during the time intervals between steering phase meetings in the Planning Game.

You can see how the Planning Game is similar in some ways to Agile Usage-Centered Design. Both rely on participation by users, rely on cards as communication devices, and focus on tasks the system being designed is supposed to perform. Although these approaches differ from some of the more traditional ways of determining requirements, such as interviews and prototyping, many of the core principles are the same. Customers, or users, remain the source for what the system is supposed to do. Requirements are still captured and negotiated. The overall process is still documented, although the extent and formality of the documentation may differ. Given the way requirements are identified and recorded and broken down from stories to tasks, design specifications can easily incorporate the characteristics of quality requirements: completeness, consistency, modifiability, and traceability.

ELECTRONIC COMMERCE APPLICATIONS: DETERMINING SYSTEM REQUIREMENTS

Determining system requirements for an Internet-based electronic commerce application is no different than the process followed for other applications. In the last chapter, you read how PVF's management began the WebStore project, a project to sell furniture products over the Internet. In this section, we examine the process followed by PVF to determine system requirements and highlight some of the issues and capabilities that you may want to consider when developing your own Internet-based application.

Determining System Requirements for Pine Valley Furniture's WebStore

To collect system requirements as quickly as possible, Jim and Jackie decided to hold a three-day JAD session. In order to get the most out of these sessions, they invited a broad range of people, including representatives from Sales and Marketing, Operations, and Information Systems. Additionally, they asked an experienced JAD facilitator, Cheri Morris, to conduct the session. Together with Cheri, Jim and Jackie developed a very ambitious and detailed agenda for the session. Their goal was to collect requirements on the following items:

- System Layout and Navigation Characteristics
- WebStore and Site Management System Capabilities
- Customer and Inventory Information
- System Prototype Evolution

In the remainder of this section, we briefly highlight the outcomes of the JAD session.

System Layout and Navigation Characteristics As part of the process of preparing for the JAD session, all participants were asked to visit several established retail websites, including *www.amazon.com, www.landsend.com, www.sony.com,* and *www.pier1.com.* At the JAD session, participants were asked to identify characteristics of these sites that they found appealing and those characteristics that they found cumbersome. This allowed participants to identify and discuss those features that they wanted the WebStore to possess. The outcomes of this activity are summarized in Table 6-8.

WebStore and Site Management System Capabilities After agreeing to the general layout and navigational characteristics of the WebStore, the session then turned its focus to the basic system capabilities. To assist in this process, systems analysts from the Information Systems Department developed a draft skeleton of the WebStore.

TABLE 6-8   Desired Layout and Navigation Feature of WebStore

Layout and Design	• Navigation menu and logo placement should remain consistent throughout the entire site (this allows users to maintain familiarity while using the site and minimizes users who get "lost" in the site) • Graphics should be lightweight to allow for quick page display • Text should be used over graphics whenever possible
Navigation	• Any section of the store should be accessible from any other section via the navigation menu • Users should always be aware of what section they are currently in

This skeleton was based on the types of screens common to and capabilities of popular retail websites. For example, many retail websites have a "shopping cart" feature that allows customers to accumulate multiple items before checking out rather than buying a single item at a time. After some discussion, the participants agreed that the system structure shown in Table 6-9 would form the foundation for the WebStore system.

In addition to the WebStore capabilities, members of the Marketing and Sales Department described several reports that would be necessary to effectively manage customer accounts and sales transactions. In addition, the department wants to be able to conduct detailed analyses of site visitors, sales tracking, and so on. Members of the Operations Department expressed a need to easily update the product catalog. These collective requests and activities were organized into a system design structure, called the *Site Management System*, which is summarized in Table 6-9. The structures of both the WebStore and Site Management Systems will be given to the Information Systems Department as the baseline for further analysis and design activities.

Customer and Inventory Information   The WebStore will be designed to support the furniture purchases of three distinct types of customers:

- Corporate customers
- Home office customers
- Student customers

To effectively track the sales to these different types of customers, distinct information must be captured and stored by the system. Table 6-10 summarizes this information for each customer type that was identified during the JAD session. In addition to

TABLE 6-9   System Structure of the WebStore and Site Management Systems

WebStore System	Site Management System

❏ Main Page
  • Product Line (catalog)
    ✓ Desks
    ✓ Chairs
    ✓ Tables
    ✓ File Cabinets
  • Shopping Cart
  • Checkout
  • Account Profile
  • Order Status/History
  • Customer Comments
❏ Company Info
❏ Feedback
❏ Contact Information
	❏ User Profile Manager
	❏ Order maintenance Manager
	❏ Content (catalog) Manager
	❏ Reports
	  • Total Hits
	  • Most Frequent Page Views
	  • Users/Time of Day
	  • Users/Day of Week
	  • Shoppers Not Purchasing (used shopping cart—did not checkout)
	  • Feedback Analysis

TABLE 6-10 **Customer and Inventory Information for WebStore**

Corporate Customer	Home Office Customer	Student Customer
• Company Name	• Name	• Name
• Company Address	• Doing Business as (company name)	• School
• Company Phone	• Address	• Address
• Company Fax	• Phone	• Phone
• Company Preferred Shipping Method	• Fax	• E-Mail
• Buyer Name	• E-Mail	
• Buyer Phone		
• Buyer E-Mail		

Inventory Information		
• SKU	• Finished Product Size	• Available Colors
• Name	• Finished Product Weight	• Price
• Description	• Available Materials	• Lead Time

the customer information, information about the products ordered must also be captured and stored. Orders reflect the range of product information that must be specified to execute a sales transaction. Thus, in addition to capturing the customer information, product and sales data must also be captured and stored. Table 6-10 lists the results of this analysis.

System Prototype Evolution As a final activity, the JAD participants, benefiting from extensive input from the Information Systems staff, discussed how the system implementation should evolve. After completing analysis and design activities, it was agreed that the system implementation should progress in three main stages so that changes to the requirements could more easily be identified and implemented. Table 6-11 summarizes these stages and the functionality that would be incorporated at each stage of the implementation.

At the conclusion of the JAD session there was a good feeling among the participants. All felt that a lot of progress had been made and that clear requirements had been identified. With these requirements in hand, Jim and the Information Systems staff could now begin to turn these lists of requirements into formal analysis and design specifications. To show how information flows through the WebStore, data flow diagrams (Chapter 7) will be produced. To show a conceptual model of the data used within WebStore, an entity-relationship diagram (Chapter 8) will be produced. Both of these analysis documents will become part of the foundation for detailed system design and implementation.

TABLE 6-11 **Stages of System Implementation of WebStore**

Stage 1—Basic Functionality:
• Simple catalog navigation; two products per section—limited attributes set
• 25 sample users
• Simulated credit card transaction
• Full shopping cart functionality
Stage 2—Look and Feel:
• Full product attribute set and media (images, video)—commonly referred to as the "product data catalog"
• Full site layout
• Simulated integration with Purchasing Fulfillment and Customer Tracking systems
Stage 3—Staging/Preproduction:
• Full integration with Purchasing Fulfillment and Customer Tracking systems
• Full credit card processing integration
• Full product data catalog

SUMMARY

As we saw in Chapter 1, there are two subphases in the systems analysis phase of the systems development life cycle: requirements determination and requirements structuring. Chapter 6 has focused on requirements determination, the gathering of information about current systems and the need for replacement systems. Chapters 7 and 8 will address techniques for structuring the requirements elicited during requirements determination.

For requirements determination, the traditional sources of information about a system include interviews; observation; group interviews; and procedures, forms, and other useful documents. Often many or even all of these sources are used to gather perspectives on the adequacy of current systems and the requirements for replacement systems. Each form of information collection has its advantages and disadvantages. Selecting the methods to use depends on the need for rich or thorough information, the time and budget available, the need to probe deeper once initial information is collected, the need for confidentiality for those providing assessments of system requirements, the desire to get people involved and committed to a project, and the potential audience from which requirements should be collected.

Both open- and closed-ended questions can be posed during interviews. In either case, you must be very precise in formulating a question in order to avoid ambiguity and to ensure a proper response. During observation you must try not to intrude or interfere with normal business activities so that the people being observed do not modify their activities from normal processes. The results of all requirements-gathering methods should be compared because there may be differences between the formal or official system and the way people actually work, the informal system.

You also learned about contemporary methods to collect requirements information, many of which make use of information systems. JAD begins with the idea of the group interview and adds structure and a JAD session leader to it. Typical JAD participants include the session leader, a scribe, key users, managers, a sponsor, and systems analysts. JAD sessions are usually held off-site and may last as long as one week.

Systems analysis is increasingly performed with computer assistance, as is the case in using CASE tools and prototyping to support requirements determination. As part of the prototyping process, users and analysts work closely together to determine requirements that the analyst then builds into a model. The analyst and user then work together on revising the model until it is close to what the user desires.

BPR is an approach to radically changing business processes. BPR efforts are a source of new information requirements. Information systems and technologies often enable BPR by allowing an organization to eliminate or relax constraints on traditional business rules. Agile requirements determination techniques are another contemporary approach to figuring out what a new or improved system is supposed to do. Continual customer involvement relies on high levels of user participation. Agile Usage-Centered Design and the Planning Game rely on novel interactions between users and developers to uncover basic tasks and features the new system should include.

Most of the same techniques used for requirements determination for traditional systems can also be fruitfully applied to the development of Internet applications. Accurately capturing requirements in a timely manner for Internet applications is just as important as for more traditional systems.

The result of requirements determination is a thorough set of information, including some charts, that describes the current systems being studied and the need for new and different capabilities to be included in the replacement systems. This information, however, is not in a form that makes analysis of true problems and clear statements of new features possible. Thus, you and other analysts will study this information and structure it into standard formats suitable for identifying problems and unambiguously describing the specifications for new systems. We discuss a variety of popular techniques for structuring requirements in the next two chapters.

KEY TERMS

1. Business process reengineering (BPR)
2. Closed-ended questions
3. Disruptive technologies
4. Formal system
5. Informal system
6. JAD session leader
7. Joint Application Design (JAD)
8. Key business processes
9. Nominal Group Technique (NGT)
10. Open-ended questions
11. Prototyping
12. Scribe

Match each of the key terms above with the definition that best fits it.

_____ Questions in interviews that ask those responding to choose from among a set of specified responses.

_____ Technologies that enable the breaking of long-held business rules that inhibit organizations from making radical business changes.

_____ A facilitated process that supports idea generation by groups. At the beginning of the process, group members work alone to generate ideas, which are then pooled under the guidance of a trained facilitator.

_____ The structured, measured set of activities designed to produce a specific output for a particular customer or market.

_____ An iterative process in which requirements are converted to a working system that is continually revised through organized user collaboration.

_____ The official way a system works as described in organizational documentation.

_____ The search for, and implementation of, radical change in business processes to achieve breakthrough improvements in products and services.

_____ The way a system actually works.

_____ The person who makes detailed notes of the happenings at a JAD session.

_____ Questions in interviews that have no prespecified answers.

_____ The trained individual who plans and leads JAD sessions.

_____ A structured process in which users, managers, and analysts work together for several days in a series of meetings to clarify or review requirements.

REVIEW QUESTIONS

1. Describe systems analysis and the major activities that occur during this phase of the systems development life cycle.

2. Describe four traditional techniques for collecting information during analysis. When might one be better than another?

3. What is JAD? How is it better than traditional information-gathering techniques? What are its weaknesses?

4. How has computing been used to support requirements determination?

5. How can NGT be used for requirements determination?

6. How can CASE tools be used to support requirements determination? Which type of CASE tools is appropriate for use during requirements determination?

7. Describe how prototyping can be used during requirements determination. How is it better or worse than traditional methods?

8. When conducting a business process reengineering study, what should you look for when trying to identify business processes to change? Why?

9. What are disruptive technologies and how do they enable organizations to radically change their business processes?

10. Why is continual user involvement a useful way to discover system requirements? Under what conditions might it be used? Under what conditions might it not be used?

11. Describe Agile Usage-Centered Design. Describe the Planning Game. Compare and contrast these two requirements determination techniques.

PROBLEMS AND EXERCISES

1. Choose either CASE or prototyping as a topic and review a related article from the popular press and from the academic research literature. Summarize the two articles and, based on your reading, prepare a list of arguments for why this type of system would be useful in a JAD session. Also address the limits for applying this type of system in a JAD setting.

2. One of the potential problems with gathering information requirements by observing potential system users mentioned in the chapter is that people may change their behavior when they are being observed. What could you do to overcome this potential confounding factor in accurately determining information requirements?

3. Summarize the problems with the reliability and usefulness of analyzing business documents as a method for gathering information requirements. How could you cope with these problems to effectively use business documents as a source of insights on system requirements?

4. Suppose you were asked to lead a JAD session. List 10 guidelines you would follow to assist you in playing the proper role of a JAD session leader.

5. Prepare a plan, similar to Figure 6-2, for an interview with your academic advisor to determine which courses you should take to develop the skills you need to be hired as a programmer/analyst.

6. Write at least three closed-ended questions that you might use in an interview of users of a word-processing package in order to develop ideas for the next version of the package. Test these questions by asking a friend to answer the questions; then interview your friend to determine why she responded as she did. From this interview, determine if she misunderstood any of your questions and, if so, rewrite the questions to be less ambiguous.

7. Figure 6-2 shows part of a guide for an interview. How might an interview guide differ when a group interview is to be conducted?

8. Group interviews and JADs are very powerful ways to collect system requirements, but special problems arise during group requirements collection sessions. Summarize the special interviewing and group problems that arise in such group sessions and suggest ways that you, as a group interviewer or group facilitator, might deal with these problems.

9. Review the material in Chapter 4 on corporate and information systems strategic planning. How are these processes different from BPR? What new perspectives might BPR bring that classical strategic planning methods may not have?

10. Research other Agile methodologies and write a report about how they handle systems requirements determination.

FIELD EXERCISES

1. Effective interviewing is not something that you can learn from just reading about it. You must first do some interviewing, preferably a lot of it, because interviewing skills improve only with experience. To get an idea of what interviewing is like, try the following: Find three friends or classmates to help you complete this exercise. Organize yourselves into pairs. Write down a series of questions you can use to find out about a job your partner now has or once held. You decide what questions to use, but at a minimum, you must find out the following: (1) the job's title; (2) the job's responsibilities; (3) who your partner reported to; (4) who reported to your partner, if anyone did; and (5) what information your partner used to do his or her job. At the same time, your partner should be preparing questions to ask you about a job you have had. Now conduct the interview. Take careful notes. Organize what you find into a clear form that another person could understand. Now repeat the process, but this time, your partner interviews you.

   While the two of you have been interviewing each other, your two other friends should have been doing the same thing. When all four of you are done, switch partners and repeat the entire process. When you are all done, each of you should have interviewed two people, and each of you should have been interviewed by two people. Now, you and the person who interviewed your original partner should compare your findings. Most likely, your findings will not be identical to what the other person found. If your findings differ, discover why. Did you use the same questions? Did the other person do a more thorough job of interviewing your first partner because it was the second time he or she had conducted an interview? Did you both ask follow-up questions? Did you both spend about the same amount of time on the interview? Prepare a report with this person about why your findings differed. Now find both of the people who interviewed you. Does one set of findings differ from the other? If so, try to figure out why. Did one of them (or both of them) misrepresent or misunderstand what you told them? Each person should now write a report on their experience, using it to explain why interviews are sometimes inconsistent and inaccurate and why having two people interview someone on a topic is better than having just one person do the interview. Explain the implications of what you have learned for the requirements determination subphase of the systems development life cycle.

2. Choose a work team at your work or university and interview them in a group setting. Ask them about their current system (whether computer-based or not) for performing their work. Ask each of them what information they use and/or need and from where/whom they get it. Was this a useful method for you to learn about their work? Why or why not? What comparative advantages does this method provide as compared to one-on-one interviews with each team member? What comparative disadvantages?

3. For the same work team you used in Field Exercise 2, examine copies of any relevant written documentation (e.g., written procedures, forms, reports, system documentation). Are any of these forms of written documentation missing? Why? With what consequences? To what extent does this written documentation fit with the information you received in the group interview?

4. Interview systems analysts, users, and managers who have been involved in JAD sessions. Determine the location, structure, and outcomes of each of their JAD sessions. Elicit their evaluations of their sessions. Were they productive? Why or why not?

5. Survey the literature on JAD in the academic and popular press and determine the "state of the art." How is JAD being used to help determine system requirements? Is using JAD for this process beneficial? Why or why not? Present your analysis to the IS manager at your work or at your university. Does your analysis of JAD fit with his or her perception? Why or why not? Is he or she currently using JAD, or a JAD-like method, for determining system requirements? Why or why not?

6. With the help of other students or your instructor, contact someone in an organization who has carried out a BPR study. What effects did this study have on information systems? In what ways did information technology, especially disruptive technologies, facilitate making the radical changes discovered in the BPR study?

7. Find an organization that uses Agile techniques for requirements determination. What techniques do they use? How did they discover them? What did they use before? What is their evaluation of the Agile techniques they use?

REFERENCES

Beck, K., and C. Andres. 2004. *eXtreme Programming eXplained.* Upper Saddle River, NJ: Addison-Wesley.

Bedoll, R. 2003. "A Tale of Two Projects: How 'Agile' Methods Succeeded After 'Traditional' Methods Had Failed in a Critical System-Development Project." Proceedings of 2003 XP/Agile Universe Conference. New Orleans, LA, August. Berlin: Springer-Verlag, 25–34.

Constantine, L. 2002. "Process Agility and Software Usability: Toward Lightweight Usage-Centered Design." *Information Age* August/September. Available at *www.infoage.idg.com.au /index.php?id=244792583.* Accessed February 12, 2004.

Davenport, T. H. 1993. *Process Innovation: Reengineering Work Through Information Technology.* Boston: Harvard Business School Press.

Dobyns, L., and C. Crawford-Mason. 1991. *Quality or Else.* Boston: Houghton-Mifflin.

Hammer, M. 1996. *Beyond Reengineering.* New York: Harper Business.

Hammer, M., and J. Champy. 1993. *Reengineering the Corporation.* New York: Harper Business.

Lucas, M. A. 1993. "The Way of JAD." *Database Programming & Design.* 6 (July): 42–49.

McConnell, S. 1996. *Rapid Development.* Redmond, WA: Microsoft Press.

Mintzberg, H. 1973. *The Nature of Managerial Work.* New York: Harper & Row.

Moad, J. 1994. "After Reengineering: Taking Care of Business." *Datamation.* 40 (20): 40–44.

Naumann, J. D., and A. M. Jenkins. 1982. "Prototyping: The New Paradigm for Systems Development." *MIS Quarterly* 6(3): 29–44.

Patton, J. 2002. "Designing Requirements: Incorporating Usage-Centered Design into an Agile SW Development Process." In D. Wells and L. Williams (eds.), *Extreme Programming and Agile Methods – XP/Agile Universe 2002, LNCS 2418,* 1–12. Berlin: Springer-Verlag.

Sharp, A., and P. McDermott. 2001. *Workflow Modeling: Tools for Process Improvement and Application Development.* Norwood, MA: Artech House Inc.

Teresko, J. 2004, "P&G's Secret: Innovating Innovation." *Industry Week* 253(12), 27–34.

Wood, J., and D. Silver. 1995. *Joint Application Development,* 2nd ed. New York: John Wiley & Sons.

PETRIE ELECTRONICS

Chapter 6: Determining System Requirements

Although the customer loyalty project at Petrie Electronics had gone slowly at first, the past few weeks had been fast-paced and busy, Jim Watanabe, the project manager, thought to himself. He had spent much of his time planning and conducting interviews with key stakeholders inside the company. He had also worked with the marketing group to put together some focus groups made up of loyal customers, to get some ideas about what they would value in a customer loyalty program. Jim had also spent some time studying customer loyalty programs at other big retail chains and those in other industries as well, such as the airlines, known for their extensive customer loyalty programs. As project manager, he had also supervised the efforts of his team members. Together, they had collected a great deal of data. Jim had just finished creating a high-level summary of the information into a table he could send to his team members (PE Table 6-1).

PE TABLE 6-1   Requirements and Constraints for Petrie's Customer Loyalty Project

Requirements:

- Effective customer incentives – System should be able to effectively store customer activity and convert to rewards and other incentives
- Easy for customers to use – Interface should be intuitive for customer use
- Proven performance – System as proposed should have been used successfully by other clients
- Easy to implement – Implementation should not require outside consultants or extraordinary skills on the part of our staff or require specialized hardware
- Scalable – System should be easily expandable as number of participating customers grows
- Vendor support – Vendor should have proven track record of reliable support and infrastructure in place to provide it

Constraints:

- Cost to buy – Licenses for one year should be under $500,000
- Cost to operate – Total operating costs should be no more than $1 million per year
- Time to implement – Duration of implementation should not exceed three months
- Staff to implement – Implementation should be successful with the staff we have and with the skills they already possess

From the list of requirements, it was clear that he and his team did not favor building a system from scratch in-house. Jim was glad that the team felt that way. Not only was building a system like this in-house an antiquated practice, it was expensive and time consuming. As nice as it might have been to develop a unique system just for Petrie, there was little point in reinventing the wheel. The IT staff would customize the system interface, and there would be lots of work for Sanjay's staff in integrating the new system and its related components with Petrie's existing systems, but the core of the system would have already been developed by someone else.

Just as he was finishing the e-mail he would send to his team about the new system's requirements and constraints, he received a new message from Sanjay. He had asked Sanjay to take the lead in scouting out existing customer loyalty systems that Petrie could license. Sanjay had conducted a preliminary investigation that was now complete. His e-mail contained the descriptions of three of the systems he had found and studied (PE Table 6-2). Obviously, Jim and his team would need to have a lot more information about these alternatives, but Jim was intrigued by the possibilities. He sent a reply to Sanjay, asking him to pass the alternatives on to the team, and also asking him to prepare a briefing for the team that would include more detailed information about each alternative.

PE TABLE 6-2   Alternatives for Petrie's Customer Loyalty Project

Alternative A:

Data warehousing-centered system designed and licensed by Standard Basic Systems Inc. (SBSI). The data warehousing tools at the heart of the system were designed and developed by SBSI and work with standard relational DBMS and relational/OO hybrid DBMS. The SBSI tools and approach have been used for many years and are well known in the industry, but SBSI-certified staff are essential for implementation, operation, and maintenance. The license is relatively expensive. The customer loyalty application using the SBSI data warehousing tools is an established application, used by many retail businesses in other industries.

Alternative B:

Customer Relationship Management-centered system designed and licensed by XRA Corporation. XRA is a pioneer in CRM systems, so its CRM is widely recognized as an industry leader. The system includes tools that support customer loyalty programs. The CRM system itself is large and complex, but pricing in this proposal is based only on modules used for the customer loyalty application.

Alternative C:

Proprietary system designed and licensed by Nova Innovation Group, Inc. The system is relatively new and leading edge, so it has only been implemented in a few sites. The vendor is truly innovative but small and inexperienced. The customer interface, designed for a standard web browser, is stunning in its design and is extremely easy for customers to use to check on their loyalty program status. The software runs remotely, in the "cloud," and data related to the customer loyalty program would be stored in the cloud too.

Case Questions

1. What do you think are the sources of the information Jim and his team collected? How do you think they collected all of that information?
2. Examine PE Table 6-1. Are there any requirements or constraints that you can think of that were overlooked? List them.
3. If you were looking for alternative approaches for Petrie's customer loyalty program, where would you look for information? Where would you start? How would you know when you were done?
4. Using the web, find three customizable customer loyalty program systems being sold by vendors. Create a table like PE Table 6-2 that compares them.
5. Why shouldn't Petrie's staff build their own unique system in-house?

Structuring System Process Requirements

Learning Objectives

After studying this chapter, you should be able to:

- Understand the logical modeling of processes by studying examples of data flow diagrams.

- Draw data flow diagrams following specific rules and guidelines that lead to accurate and well-structured process models.

- Decompose data flow diagrams into lower-level diagrams.

- Balance higher-level and lower-level data flow diagrams.

- Use data flow diagrams as a tool to support the analysis of information systems.

- Discuss process modeling for electronic commerce applications.

- Use decision tables to represent the logic of choice in conditional statements.

Introduction

In the last chapter, you learned of various methods that systems analysts use to collect the information necessary to determine information systems requirements. In this chapter, our focus will be on one tool that is used to coherently represent the information gathered as part of requirements determination—data flow diagrams. Data flow diagrams enable you to model how data flow through an information system, the relationships among the data flows, and how data come to be stored at specific locations. Data flow diagrams also show the processes that change or transform data. Because data flow diagrams concentrate on the movement of data between processes, these diagrams are called *process models*.

As its name indicates, a data flow diagram is a graphical tool that allows analysts (and users, for that matter) to depict the flow of data in an information system. The system can be physical or logical, manual or computer-based. In this chapter, you will learn how to draw and revise data flow diagrams. We present the basic symbols used in such diagrams and a set of rules that govern how these diagrams are drawn. You will also learn about what to do and what *not* to do when drawing data flow diagrams. Two important concepts related to data flow diagrams are also presented: balancing and decomposition. Toward the end of the chapter, we present the use of data flow diagrams as part of the analysis of an information system and as a tool for supporting business process reengineering. You will also learn how process modeling is important for the analysis of electronic commerce applications. Also in this chapter, you will learn about decision tables. Decision tables allow you to represent the conditional logic that is part of some data flow diagram processes. Finally, at the end of the chapter, we have included special sections on an object-oriented development approach to process and logic modeling. These sections cover use cases, activity diagrams, and sequence diagrams. We have also included an appendix on business process modeling.

PROCESS MODELING

Process modeling involves graphically representing the functions, or processes, that capture, manipulate, store, and distribute data between a system and its environment and between components within a system. A common form of a process model is a **data flow diagram (DFD)**. Over the years, several different tools have been developed for process modeling. In this chapter, we focus on DFDs, the traditional process modeling technique of structured analysis and design and one of the techniques most frequently used today for process modeling. We also introduce you to decision tables, a well-known way to model the conditional logic contained in many DFD processes.

Data flow diagram (DFD)
A picture of the movement of data between external entities and the processes and data stores within a system.

Modeling a System's Process for Structured Analysis

As Figure 7-1 shows, the analysis phase of the systems development life cycle has two subphases: requirements determination and requirements structuring. The analysis team enters the requirements structuring phase with an abundance of information gathered during the requirements determination phase. During requirements structuring, you and the other team members must organize the information into a meaningful representation of the information system that currently exists and of the requirements desired in a replacement system. In addition to modeling the processing elements of an information system and how data are transformed in the system, you must also model the processing logic (decision tables) and the structure of data within the system (Chapter 8). For traditional structured analysis, a process model is only one of three major complementary views of an information system. Together, process, logic, and data models provide a thorough specification of an information system and, with the proper supporting tools, also provide the basis for the automatic generation of many working information system components.

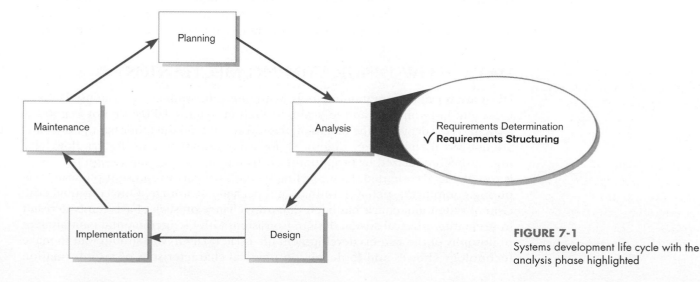

FIGURE 7-1
Systems development life cycle with the analysis phase highlighted

TABLE 7-1 **Deliverables for Process Modeling**

1. Context DFD
2. DFDs of the system (adequately decomposed)
3. Thorough descriptions of each DFD component

Deliverables and Outcomes

In structured analysis, the primary deliverables from process modeling are a set of co-herent, interrelated DFDs. Table 7-1 provides a more detailed list of the deliverables that result when DFDs are used to study and document a system's processes. First, a context diagram shows the scope of the system, indicating which elements are inside and which are outside the system. Second, DFDs of the system specify which pro-cesses move and transform data, accepting inputs and producing outputs. These dia-grams are developed with sufficient detail to understand the current system and to eventually determine how to convert the current system into its replacement. Finally, entries for all of the objects included in all of the diagrams are included in the proj-ect dictionary or CASE repository. This logical progression of deliverables allows you to understand the existing system. You can then abstract this system into its essential elements to show how the new system should meet the information-processing re-quirements identified during requirements determination. Remember, the deliver-ables of process modeling are simply stating what you learned during requirements determination; in later steps in the systems development life cycle, you and other project team members will make decisions on exactly how the new system will de-liver these new requirements in specific manual and automated functions. Because requirements determination and structuring are often parallel steps, DFDs evolve from the more general to the more detailed as current and replacement systems are better understood.

Even though DFDs remain popular tools for process modeling and can signifi-cantly increase software development productivity, DFDs are not used in all systems development methodologies. Some organizations, such as EDS, have developed their own diagrams to model processes. Other organizations rely on process model-ing tools in CASE tool sets. Some methodologies, such as RAD and object-oriented analysis and design methodologies (see Chapter 1), do not model processes sepa-rately at all.

DFDs provide notation as well as illustrate important concepts about the move-ment of data between manual and automated steps, and offer a way to depict work flow in an organization. DFDs continue to be beneficial to information systems pro-fessionals as tools for both analysis and communication. For that reason, we devote almost an entire chapter to DFDs, but we complement our coverage of DFDs with an introduction to use cases and use case diagrams in the chapter appendix on use case.

DATA FLOW DIAGRAMMING MECHANICS

DFDs are versatile diagramming tools. With only four symbols, you can use DFDs to represent both physical and logical information systems. DFDs are not as good as flowcharts for depicting the details of physical systems; on the other hand, flowcharts are not very useful for depicting purely logical information flows. In fact, flowchart-ing has been criticized by proponents of structured analysis and design because it is too physically oriented. Flowcharting symbols primarily represent physical com-puting equipment, such as terminals and permanent storage. One continual criti-cism of system flowcharts has been that overreliance on such charts tends to result in premature physical system design. Consistent with the incremental commitment philosophy of the systems development life cycle (SDLC), you should wait to make technology choices and to decide on physical characteristics of an information system until you are sure all functional requirements are correct and accepted by users and other stakeholders.

DFDs do not share this problem of premature physical design because they do not rely on any symbols to represent specific physical computing equipment. They are also easier to use than flowcharts because they involve only four different symbols.

Definitions and Symbols

There are two different standard sets of DFD symbols (see Figure 7-2); each set consists of four symbols that represent the same things: data flows, data stores, processes, and sources/sinks (or external entities). The set of symbols we will use in this book was devised by Gane and Sarson (1979). The other standard set was developed by DeMarco (1979) and Yourdon (Yourdon and Constantine, 1979).

A *data flow* can be best understood as data in motion, moving from one place in a system to another. A data flow could represent data on a customer order form or a payroll check; it could also represent the results of a query to a database, the contents of a printed report, or data on a data entry computer display form. A data flow is data that move together, so it can be composed of many individual pieces of data that are generated at the same time and that flow together to common destinations. A **data store** is data at rest. A data store may represent one of many different physical locations for data; for example, a file folder, one or more computer-based file(s), or a notebook. To understand data movement and handling in a system, it is not important to understand the system's physical configuration. A data store might contain data about customers, students, customer orders, or supplier invoices. A **process** is the work or actions performed on data so that they are transformed, stored, or distributed. When modeling the data processing of a system, it does not matter whether a process is performed manually or by a computer. Finally, a **source/sink** is the origin and/or destination of the data. Sources/sinks are sometimes referred to as external entities because they are outside the system. Once processed, data or information leave the system and go to some other place. Because sources and sinks are outside the system we are studying, many of the characteristics of sources and sinks are of no interest to us. In particular, we do not consider the following:

- Interactions that occur between sources and sinks
- What a source or sink does with information or how it operates (i.e., a source or sink is a "black box")

Data store
Data at rest, which may take the form of many different physical representations.

Process
The work or actions performed on data so that they are transformed, stored, or distributed.

Source/sink
The origin and/or destination of data; sometimes referred to as external entities.

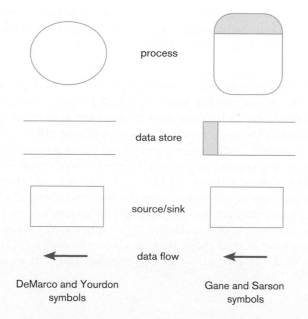

	process	
	data store	
	source/sink	
	data flow	
DeMarco and Yourdon symbols		Gane and Sarson symbols

FIGURE 7-2
Comparison of DeMarco and Yourdon and Gane and Sarson DFD symbol sets

- How to control or redesign a source or sink because, from the perspective of the system we are studying, the data a sink receives and often what data a source provides are fixed
- How to provide sources and sinks direct access to stored data because, as external agents, they cannot directly access or manipulate data stored within the system; that is, processes within the system must receive or distribute data between the system and its environment

The symbols for each set of DFD conventions are presented in Figure 7-2. In both conventions, a data flow is depicted as an arrow. The arrow is labeled with a meaningful name for the data in motion; for example, Customer Order, Sales Receipt, or Paycheck. The name represents the aggregation of all the individual elements of data moving as part of one packet, that is, all the data moving together at the same time. A square is used in both conventions for sources/sinks and has a name that states what the external agent is, such as Customer, Teller, EPA Office, or Inventory Control System. The Gane and Sarson symbol for a process is a rectangle with rounded corners; it is a circle for DeMarco and Yourdon. The Gane and Sarson rounded rectangle has a line drawn through the top. The upper portion is used to indicate the number of the process. Inside the lower portion is a name for the process, such as Generate Paycheck, Calculate Overtime Pay, or Compute Grade Point Average. The Gane and Sarson symbol for a data store is a rectangle that is missing its right vertical side. At the left end is a small box used to number the data store, and inside the main part of the rectangle is a meaningful label for the data store, such as Student File, Transcripts, or Roster of Classes. The DeMarco and Yourdon data store symbol consists of two parallel lines, which may be depicted horizontally or vertically.

As stated earlier, sources/sinks are always outside the information system and define the boundaries of the system. Data must originate outside a system from one or more sources, and the system must produce information to one or more sinks (these are principles of open systems, and almost every information system is an open system). If any data processing takes place inside the source/sink, it is of no interest because this processing takes place outside the system we are diagramming. A source/sink might consist of the following:

- Another organization or organization unit that sends data to or receives information from the system you are analyzing (e.g., a supplier or an academic department—in either case, the organization is external to the system you are studying)
- A person inside or outside the business unit supported by the system you are analyzing who interacts with the system (e.g., a customer or loan officer)
- Another information system with which the system you are analyzing exchanges information

Many times students who are just learning how to use DFDs will become confused as to whether something is a source/sink or a process within a system. This dilemma occurs most often when the data flows in a system cross office or departmental boundaries so that some processing occurs in one office and the processed data are moved to another office where additional processing occurs. Students are tempted to identify the second office as a source/sink to emphasize the fact that the data have been moved from one physical location to another (Figure 7-3a). However, we are not concerned with where the data are physically located. We are more interested in how they are moving through the system and how they are being processed. If the processing of data in the other office may be automated by your system or the handling of data there may be subject to redesign, then you should represent the second office as one or more processes rather than as a source/sink (Figure 7.3b).

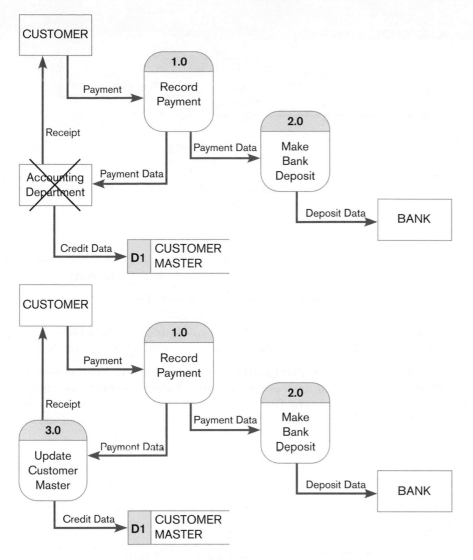

FIGURE 7-3
Differences between sources/sinks and processes
(a) An improperly drawn DFD showing a process as a source/sink

(b) A DFD showing proper use of a process

Developing DFDs: An Example

To illustrate how DFDs are used to model the logic of data flows in information systems, we will present and work through an example. Consider Hoosier Burger, a fictional restaurant in Bloomington, Indiana, owned by Bob and Thelma Mellankamp. Some are convinced that its hamburgers are the best in Bloomington, maybe even in southern Indiana. Many people, especially Indiana University students and faculty, frequently eat at Hoosier Burger. The restaurant uses an information system that takes customer orders, sends the orders to the kitchen, monitors goods sold and inventory, and generates reports for management.

The information system is depicted as a DFD in Figure 7-4. The highest-level view of this system, shown in the figure, is called a **context diagram**. You will notice that this context diagram contains only one process, no data stores, four data flows, and three sources/sinks. The single process, labeled 0, represents the entire system; all context diagrams have only one process, labeled 0. The sources/sinks represent the environmental boundaries of the system. Because the data stores of the system are conceptually inside one process, data stores do not appear on a context diagram.

The analyst must determine which processes are represented by the single process in the context diagram. As you can see in Figure 7-5, we have identified four

Context diagram

An overview of an organizational system that shows the system boundaries, external entities that interact with the system, and the major information flows between the entities and the system.

FIGURE 7-4
Context diagram of Hoosier Burger's
food-ordering system

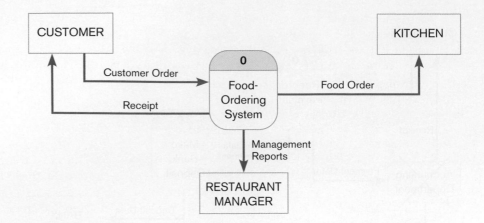

separate processes. The main processes represent the major functions of the system, and these major functions correspond to actions such as the following:

1. Capturing data from different sources (e.g., Process 1.0)
2. Maintaining data stores (e.g., Processes 2.0 and 3.0)
3. Producing and distributing data to different sinks (e.g., Process 4.0)
4. High-level descriptions of data transformation operations (e.g., Process 1.0)

These major functions often correspond to the activities on the main system menu.

We see that the system begins with an order from a customer, as was the case with the context diagram. In the first process, labeled 1.0, we see that the customer order is processed. The result is four streams, or flows, of data: (1) the food order is transmitted to the kitchen, (2) the customer order is transformed into a list of goods

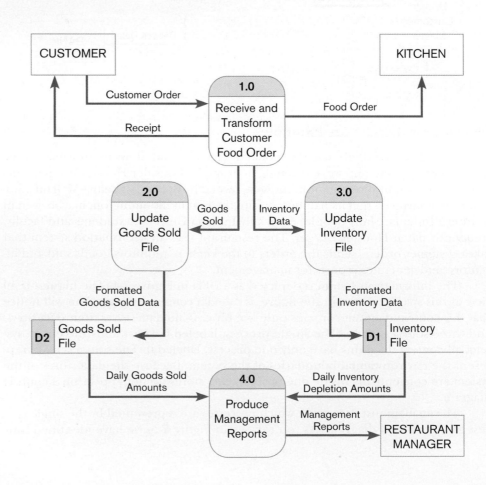

FIGURE 7-5
Level-0 DFD of Hoosier Burger's
food-ordering system

sold, (3) the customer order is transformed into inventory data, and (4) the process generates a receipt for the customer.

Notice that the sources/sinks are the same in the context diagram and in this diagram: the customer, the kitchen, and the restaurant's manager. This diagram is called a **level-0 diagram** because it represents the primary individual processes in the system at the highest possible level. Each process has a number that ends in .0 (corresponding to the level number of the DFD).

Two of the data flows generated by the first process, Receive and Transform Customer Food Order, go to external entities, so we no longer have to worry about them. We are not concerned about what happens outside our system. Let's trace the flow of the data represented in the other two data flows. First, the data labeled Goods Sold go to Process 2.0, Update Goods Sold File. The output for this process is labeled Formatted Goods Sold Data. This output updates a data store labeled Goods Sold File. If the customer order was for two cheeseburgers, one order of fries, and a large soft drink, each of these categories of goods sold in the data store would be incremented appropriately. The Daily Goods Sold Amounts are then used as input to Process 4.0, Produce Management Reports. Similarly, the remaining data flow generated by Process 1.0, Inventory Data, serves as input for Process 3.0, Update Inventory File. This process updates the Inventory File data store, based on the inventory that would have been used to create the customer order. For example, an order of two cheeseburgers would mean that Hoosier Burger now has two fewer hamburger patties, two fewer burger buns, and four fewer slices of American cheese. The Daily Inventory Depletion Amounts are then used as input to Process 4.0. The data flow leaving Process 4.0, Management Reports, goes to the sink Restaurant Manager.

Figure 7-5 illustrates several important concepts about information movement. Consider the data flow Inventory Data moving from Process 1.0 to Process 3.0. We know from this diagram that Process 1.0 produces this data flow and that Process 3.0 receives it. However, we do not know the timing of when this data flow is produced, how frequently it is produced, or what volume of data is sent. Thus, this DFD hides many physical characteristics of the system it describes. We do know, however, that this data flow is needed by Process 3.0 and that Process 1.0 provides these needed data.

Also implied by the Inventory Data data flow is that whenever Process 1.0 produces this flow, Process 3.0 must be ready to accept it. Thus, Processes 1.0 and 3.0 are coupled with each other. In contrast, consider the link between Process 2.0 and Process 4.0. The output from Process 2.0, Formatted Goods Sold Data, is placed in a data store and, later, when Process 4.0 needs such data, it reads Daily Goods Sold Amounts from this data store. In this case, Processes 2.0 and 4.0 are decoupled by placing a buffer, a data store, between them. Now, each of these processes can work at their own pace, and Process 4.0 does not have to be ready to accept input at any time. Further, the Goods Sold File becomes a data resource that other processes could potentially draw upon for data.

Data Flow Diagramming Rules

You must follow a set of rules when drawing DFDs. Unlike system flowcharts, these rules allow you (or a CASE tool) to evaluate DFDs for correctness. The rules for DFDs are listed in Table 7-2. Figure 7-6 illustrates incorrect ways to draw DFDs and the corresponding correct application of the rules. The rules that prescribe naming conventions (rules C, G, I, and P) and those that explain how to interpret data flows in and out of data stores (rules N and O) are not illustrated in Figure 7-6.

In addition to the rules in Table 7-2, there are two DFD guidelines that often apply:

1. *The inputs to a process are different from the outputs of that process.* The reason is that processes, because they have a purpose, typically transform inputs into outputs, rather than simply pass the data through without some manipulation. What may

Level-0 diagram
A DFD that represents a system's major processes, data flows, and data stores at a high level of detail.

TABLE 7-2   Rules Governing Data Flow Diagramming

Process:

A. No process can have only outputs. It is making data from nothing (a miracle). If an object has only outputs, then it must be a source.

B. No process can have only inputs (a black hole). If an object has only inputs, then it must be a sink.

C. A process has a verb phrase label.

Data Store:

D. Data cannot move directly from one data store to another data store. Data must be moved by a process.

E. Data cannot move directly from an outside source to a data store. Data must be moved by a process that receives data from the source and places the data into the data store.

F. Data cannot move directly to an outside sink from a data store. Data must be moved by a process.

G. A data store has a noun phrase label.

Source/Sink:

H. Data cannot move directly from a source to a sink. It must be moved by a process if the data are of any concern to our system. Otherwise, the data flow is not shown on the DFD.

I. A source/sink has a noun phrase label.

Data Flow:

J. A data flow has only one direction of flow between symbols. It may flow in both directions between a process and a data store to show a read before an update. The latter is usually indicated, however, by two separate arrows because these happen at different times.

K. A fork in a data flow means that exactly the same data goes from a common location to two or more different processes, data stores, or sources/sinks (this usually indicates different copies of the same data going to different locations).

L. A join in a data flow means that exactly the same data come from any of two or more different processes, data stores, or sources/sinks to a common location.

M. A data flow cannot go directly back to the same process it leaves. There must be at least one other process that handles the data flow, produces some other data flow, and returns the original data flow to the beginning process.

N. A data flow to a data store means update (delete or change).

O. A data flow from a data store means retrieve or use.

P. A data flow has a noun phrase label. More than one data flow noun phrase can appear on a single arrow as long as all of the flows on the same arrow move together as one package.

(*Source:* Adapted from Celko, 1987.)

happen is that the same input goes in and out of a process but the process also produces other new data flows that are the result of manipulating the inputs.

2. *Objects on a DFD have unique names.* Every process has a unique name. There is no reason for two processes to have the same name. To keep a DFD uncluttered, however, you may repeat data stores and sources/sinks. When two arrows have the same data flow name, you must be careful that these flows are exactly the same. It is easy to reuse the same data flow name when two packets of data are almost the same, but not identical. A data flow name represents a specific set of data, and another data flow that has even one more or one less piece of data must be given a different, unique name.

Decomposition of DFDs

In the earlier example of Hoosier Burger's food-ordering system, we started with a high-level context diagram. Upon thinking more about the system, we saw that the larger system consisted of four processes. The act of going from a single system to four component processes is called *(functional) decomposition*. **Functional decomposition** is an iterative process of breaking the description or perspective of a system down into finer and finer detail. This process creates a set of hierarchically related

Functional decomposition
An iterative process of breaking the description of a system down into finer and finer detail, which creates a set of charts in which one process on a given chart is explained in greater detail on another chart.

Rule	Incorrect	Correct

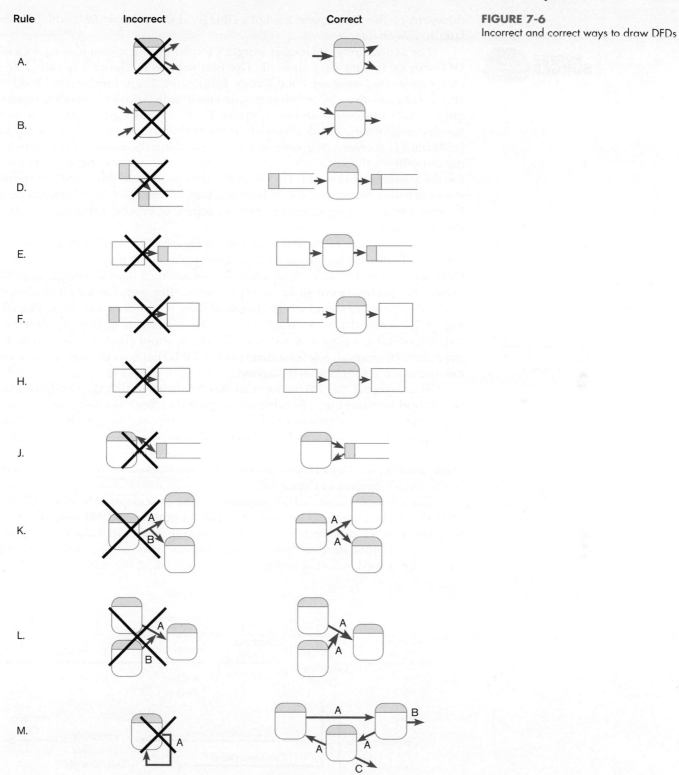

FIGURE 7-6
Incorrect and correct ways to draw DFDs

charts in which one process on a given chart is explained in greater detail on another chart. For the Hoosier Burger system, we broke down, or decomposed, the larger system into four processes. Each resulting process (or subsystem) is also a candidate for decomposition. Each process may consist of several subprocesses. Each subprocess may also be broken down into smaller units. Decomposition continues until you have reached the point at which no subprocess can logically be broken

down any further. The lowest level of a DFD is called a *primitive DFD*, which we define later in this chapter.

Let's continue with Hoosier Burger's food-ordering system to see how a level-0 DFD can be further decomposed. The first process in Figure 7-5, called Receive and Transform Customer Food Order, transforms a customer's verbal food order (e.g., "Give me two cheeseburgers, one small order of fries, and one regular orange soda") into four different outputs. Process 1.0 is a good candidate process for decomposition. Think about all of the different tasks that Process 1.0 has to perform: (1) receive a customer order, (2) transform the entered order into a form meaningful for the kitchen's system, (3) transform the order into a printed receipt for the customer, (4) transform the order into goods sold data, and (5) transform the order into inventory data. At least five logically separate functions can occur in Process 1.0. We can represent the decomposition of Process 1.0 as another DFD, as shown in Figure 7-7.

Note that each of the five processes in Figure 7-7 is labeled as a subprocess of Process 1.0: Process 1.1, Process 1.2, and so on. Also note that, just as with the other DFDs we have looked at, each of the processes and data flows is named. You will also notice that no sources or sinks are represented. Although you may include sources and sinks, the context and level-0 diagrams show the sources and sinks. The DFD in Figure 7-7 is called a level-1 diagram. If we should decide to decompose Processes 2.0, 3.0, or 4.0 in a similar manner, the DFDs we would create would also be level-1 diagrams. In general, a **level-*n* diagram** is a DFD that is generated from *n* nested decompositions from a level-0 diagram.

Processes 2.0 and 3.0 perform similar functions in that they both use data input to update data stores. Because updating a data store is a singular logical function, neither of these processes needs to be decomposed further. We can, however, decompose Process 4.0, Produce Management Reports, into at least three subprocesses: Access Goods Sold and Inventory Data, Aggregate Goods Sold and Inventory Data, and Prepare Management Reports. The decomposition of Process 4.0 is shown in the level-1 diagram of Figure 7-8.

Each level-1, -2, or -*n* DFD represents one process on a level-*n*-1 DFD; each DFD should be on a separate page. As a rule of thumb, no DFD should have more than about seven processes because too many processes will make the diagram too crowded and difficult to understand. To continue with the decomposition of Hoosier Burger's food-ordering system, we examine each of the subprocesses identified

---

Level-*n* diagram
A DFD that is the result of *n* nested decompositions from a process on a level-0 diagram.

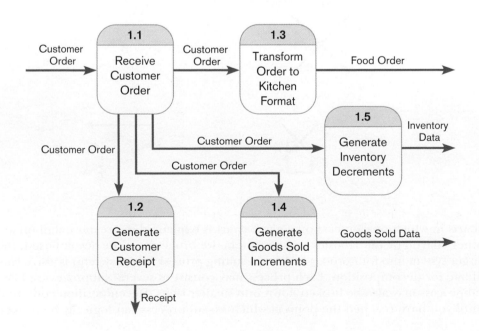

FIGURE 7-7
Level-1 diagram showing the decomposition of Process 1.0 from the level-0 diagram for Hoosier Burger's food-ordering system

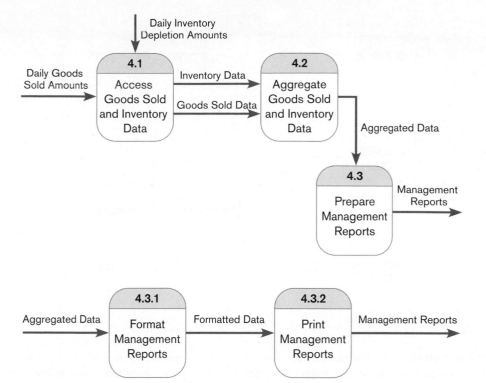

FIGURE 7-8
Level-1 diagram showing the
decomposition of Process 4.0 from the
level-0 diagram for Hoosier Burger's
food-ordering system

FIGURE 7-9
Level-2 diagram showing the
decomposition of Process 4.3 from
the level-1 diagram for Process 4.0 for
Hoosier Burger's food-ordering system

in the two level-1 diagrams we have produced, one for Process 1.0 and one for Process 4.0. Should we decide that any of these subprocesses should be further decomposed, we would create a level-2 diagram showing that decomposition. For example, if we decided that Process 4.3 in Figure 7-8 should be further decomposed, we would create a diagram that looks something like Figure 7-9. Again, notice how the subprocesses are labeled.

Just as the labels for processes must follow numbering rules for clear communication, process names should also be clear yet concise. Typically, process names begin with an action verb, such as Receive, Calculate, Transform, Generate, or Produce. Process names often are the same as the verbs used in many computer programming languages. Example process names include Merge, Sort, Read, Write, and Print. Process names should capture the essential action of the process in just a few words, yet be descriptive enough of the process's action so that anyone reading the name gets a good idea of what the process does. Many times, students just learning DFDs will use the names of people who perform the process or the department in which the process is performed as the process name. This practice is not very useful because we are more interested in the action the process represents than the person performing it or the place where it occurs.

Balancing DFDs

When you decompose a DFD from one level to the next, there is a conservation principle at work. You must conserve inputs and outputs to a process at the next level of decomposition. In other words, Process 1.0, which appears in a level-0 diagram, must have the same inputs and outputs when decomposed into a level-1 diagram. This conservation of inputs and outputs is called **balancing**.

Let's look at an example of balancing a set of DFDs. Look back at Figure 7-4. This is the context diagram for Hoosier Burger's food-ordering system. Notice that there is one input to the system, the customer order, which originates with the customer. Notice also that there are three outputs: the customer receipt, the food order intended for the kitchen, and management reports. Now look at Figure 7-5. This is

Balancing
The conservation of inputs and outputs to a DFD process when that process is decomposed to a lower level.

FIGURE 7-10
An unbalanced set of DFDs
(a) Context diagram

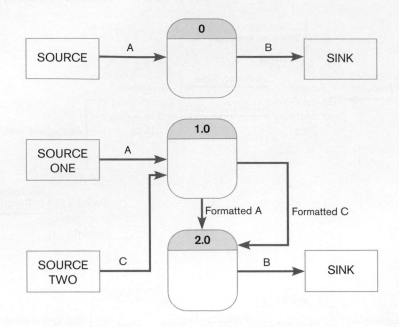

(b) Level-0 diagram

the level-0 diagram for the food-ordering system. Remember that all data stores and flows to or from them are internal to the system. Notice that the same single input to the system and the same three outputs represented in the context diagram also appear at level 0. Further, no new inputs to or outputs from the system have been introduced. Therefore, we can say that the context diagram and level-0 DFDs are balanced.

Now look at Figure 7-7, where Process 1.0 from the level-0 DFD has been decomposed. As we have seen before, Process 1.0 has one input and four outputs. The single input and multiple outputs all appear on the level-1 diagram in Figure 7-7. No new inputs or outputs have been added. Compare Process 4.0 in Figure 7-5 with its decomposition in Figure 7-8. You see the same conservation of inputs and outputs.

Figure 7-10 shows one example of what an unbalanced DFD could look like. The context diagram shows one input to the system, A, and one output, B. Yet in the level-0 diagram, there is an additional input, C, and flows A and C come from different sources. These two DFDs are not balanced. If an input appears on a level-0 diagram, it must also appear on the context diagram. What happened with this example? Perhaps, when drawing the level-0 DFD, the analyst realized that the system also needed C in order to compute B. A and C were both drawn in the level-0 DFD, but the analyst forgot to update the context diagram. When making corrections, the analyst should also include "SOURCE ONE" and "SOURCE TWO" on the context diagram. It is very important to keep DFDs balanced from the context diagram all the way through each level of diagram you create.

A data flow consisting of several subflows on a level-*n* diagram can be split apart on a level-*n* diagram for a process that accepts this composite data flow as input. For example, consider the partial DFDs from Hoosier Burger, illustrated in Figure 7-11. In Figure 7-11a, we see that a composite, or package, data flow, Payment and Coupon, is input to the process. That is, the payment and coupon always flow together and are input to the process at the same time. In Figure 7-11b, the process is decomposed (sometimes referred to as exploded or nested) into two subprocesses, and each subprocess receives one of the components of the composite data flow from the higher-level DFD. These diagrams are still balanced because exactly the same data are included in each diagram.

The principle of balancing and the goal of keeping a DFD as simple as possible led to four additional, advanced rules for drawing DFDs. These advanced rules are summarized in Table 7-3. Rule Q covers the situation illustrated in Figure 7-11. Rule R

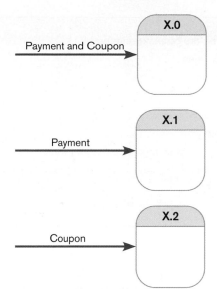

FIGURE 7-11
Example of data flow splitting
(a) Composite data flow

(b) Disaggregated data flows

TABLE 7-3   Advanced Rules Governing Data Flow Diagramming

Q. A composite data flow on one level can be split into component data flows at the next level, but no new data can be added and all data in the composite must be accounted for in one or more subflows.

R. The inputs to a process must be sufficient to produce the outputs (including data placed in data stores) from the process. Thus, all outputs can be produced, and all data in inputs move somewhere: to another process or to a data store outside the process or onto a more detailed DFD showing a decomposition of that process.

S. At the lowest level of DFDs, new data flows may be added to represent data that are transmitted under exceptional conditions; these data flows typically represent error messages (e.g., "Customer not known; do you want to create a new customer?") or confirmation notices (e.g., "Do you want to delete this record?").

T. To avoid having data flow lines cross each other, you may repeat data stores or sources/ sinks on a DFD. Use an additional symbol, like a double line on the middle vertical line of a data store symbol or a diagonal line in a corner of a sink/source square, to indicate a repeated symbol.

(*Source:* Adapted from Celko, 1987.)

covers a conservation principle about process inputs and outputs. Rule S addresses one exception to balancing. Rule T tells you how you can minimize clutter on a DFD.

AN EXAMPLE DFD

To illustrate the creation and refinement of DFDs, we will look at another example from Hoosier Burger. We saw that the food-ordering system generates two types of usage data—for goods sold and for inventory. At the end of each day, the manager, Bob Mellankamp, generates the inventory report that tells him how much inventory should have been used for each item associated with a sale. The amounts shown on the inventory report are just one input to a largely manual inventory control system Bob uses every day. Figure 7-12 lists the steps involved in Bob's inventory control system.

In the Hoosier Burger inventory system, three sources of data come from outside: suppliers, the food-ordering system inventory report, and stock on hand. Suppliers provide invoices as input, and the system returns payments and orders as outputs to the suppliers. Both the inventory report and the stock-on-hand amounts provide inventory counts as system inputs. When Bob receives invoices from suppliers, he records their receipt on an invoice log sheet and files the actual invoices in

FIGURE 7-12

List of activities involved in Bob Mellankamp's inventory control system for Hoosier Burger

1. Meet delivery trucks before opening restaurant.
2. Unload and store deliveries.
3. Log invoices and file in accordion file.
4. Manually add amounts received to stock logs.
5. After closing, print inventory report.
6. Count physical inventory amounts.
7. Compare inventory report totals to physical count totals.
8. Compare physical count totals to minimum order quantities. If the amount is less, make order; if not, do nothing.
9. Pay bills that are due and record them as paid.

his accordion file. Using the invoices, Bob records the amount of stock delivered on the stock logs, which are paper forms posted near the point of storage for each inventory item. Figure 7-13 gives a partial example of Hoosier Burger's stock log. Notice that the minimum order quantities—the stock level at which orders must be placed in order to avoid running out of an item—appear on the log form. The stock log also has spaces for entering the starting amount, amount delivered, and the amount used for each item. Amounts delivered are entered on the sheet when Bob logs stock deliveries; amounts used are entered after Bob has compared the amounts of stock used according to a physical count and according to the numbers on the inventory report generated by the food-ordering system. We should note that Hoosier Burger has standing daily delivery orders for some perishable items that are used every day, such as burger buns, meats, and vegetables. Bob uses the minimum order quantities and the amount of stock on hand to determine which orders need to be placed. He uses the invoices to determine which bills need to be paid, and he carefully records each payment.

To create the DFD, we need to identify the essence of the inventory system Bob has established. What are the key data necessary to keep track of inventory and to pay bills? What are the key processes involved? At least four key processes make up Hoosier Burger's inventory system: (1) account for anything added to inventory, (2) account for anything taken from inventory, (3) place orders, and (4) pay bills. Key data used by the system include inventories and stock-on-hand counts, however they are determined. Major outputs from the system continue to be orders and

FIGURE 7-13

Hoosier Burger's stock log form

Stock Log					
Date:			Jan 1		Jan 2
Item	Reorder Quantity	Starting Amount	Amount Delivered	Amount Used	Starting Amount
Hamburger buns	50 dozen	5	50	43	12
Hot dog buns	25 dozen	0	25	22	3
English muffins	10 dozen	6	10	12	4
Napkins	2 cases	10	0	2	8
Straws	1 case	1	0	1	0

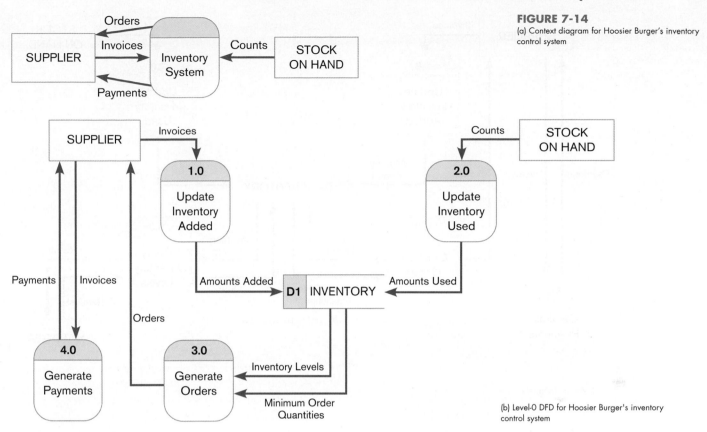

FIGURE 7-14
(a) Context diagram for Hoosier Burger's inventory control system

(b) Level-0 DFD for Hoosier Burger's inventory control system

payments. If we focus on the essential elements of the system, we obtain the context diagram and the level-0 DFD shown in Figure 7-14.

At this point, we can revise the DFD based on any new functionality desired for the system. For Hoosier Burger's inventory system, Bob Mellankamp would like to add three additional functions. First, Bob would like data on new shipments to be entered into an automated system, thus doing away with paper stock log sheets. Bob would like shipment data to be as current as possible because it will be entered into the system as soon as the new stock arrives at the restaurant. Second, Bob would like the system to determine automatically whether a new order should be placed. Automatic ordering would relieve Bob of worrying about whether Hoosier Burger has enough of everything in stock at all times. Finally, Bob would like to be able to know, at any time, the approximate inventory level for each good in stock. For some goods, such as hamburger buns, Bob can visually inspect the amount in stock and determine approximately how much is left and how much more is needed before closing time. For other items, however, Bob may need a rough estimate of what is in stock more quickly than he can estimate via a visual inspection.

The revised DFD for Hoosier Burger's inventory system is shown in Figure 7-15. The main difference between the DFD in Figure 7-14b and the revised DFD in Figure 7-15 is the new Process 5.0, which allows for querying the inventory data to get an estimate of how much of an item is in stock. Bob's two other requests for change can both be handled within the existing logical view of the inventory system. Process 1.0, Update Inventory Added, does not indicate whether the updates are in real time or batched, or whether the updates occur on paper or as part of an automated system. Therefore, immediately entering shipment data into an automated system is encompassed by Process 1.0. Similarly, Process 2.0, Generate Orders, does not indicate whether Bob or a computer generates orders or whether the orders are generated on a real-time or batch basis, so Bob's request that orders be generated automatically by the system is already represented by Process 3.0.

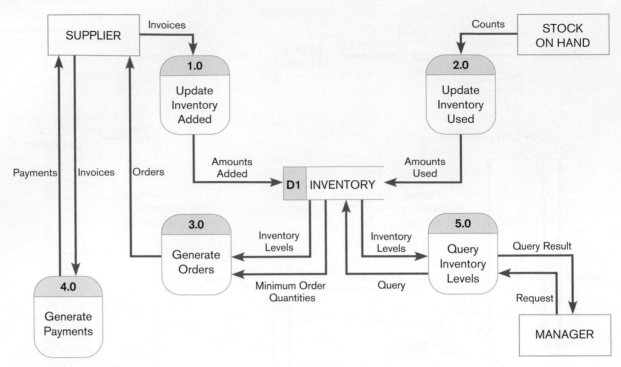

FIGURE 7-15
Revised level-0 DFD for Hoosier Burger's
inventory control system

USING DATA FLOW DIAGRAMMING IN THE ANALYSIS PROCESS

Learning the mechanics of drawing DFDs is important because DFDs have proven to be essential tools for the structured analysis process. Beyond the issue of drawing mechanically correct DFDs, there are other issues related to process modeling with which you, as an analyst, must be concerned. Such issues, including whether the DFDs are complete and consistent across all levels, are dealt with in the next section, which covers guidelines for drawing DFDs. Another issue to consider is how you can use DFDs as a useful tool for analysis. In these final sections, we also illustrate how DFDs can be used to support business process reengineering.

Guidelines for Drawing DFDs

In this section, we will consider additional guidelines for drawing DFDs that extend beyond the simple mechanics of drawing diagrams and making sure that the rules listed in Tables 7-2 and 7-3 are followed. These guidelines include (1) completeness, (2) consistency, (3) timing considerations, (4) the iterative nature of drawing DFDs, and (5) primitive DFDs.

DFD completeness

The extent to which all necessary components of a DFD have been included and fully described.

Completeness   The concept of **DFD completeness** refers to whether you have included in your DFDs all of the components necessary for the system you are modeling. If your DFD contains data flows that do not lead anywhere or data stores, processes, or external entities that are not connected to anything else, your DFD is not complete. Most CASE tools have built-in facilities that you can run to help you determine if your DFD is incomplete. When you draw many DFDs for a system, it is not uncommon to make errors. CASE tool analysis functions or walk-throughs with other analysts can help you identify such problems.

Not only must all necessary elements of a DFD be present, each of the components must be fully described in the project dictionary. With most CASE tools, the project dictionary is linked with the diagram. That is, when you define a process, data flow, source/sink, or data store on a DFD, an entry is automatically created in the repository for that element. You must then enter the repository and complete the element's description. Different descriptive information can be kept about each of the four types of elements on a DFD, and each CASE tool or project dictionary standard an organization adopts has different entry information. Data flow repository entries typically include the following:

- The label or name for the data flow as entered on the DFDs (*Note:* Case and punctuation of the label matter, but if exactly the same label is used on multiple DFDs, whether nested or not, then the same repository entry applies to each reference.)
- A short description defining the data flow
- A list of other repository objects grouped into categories by type of object
- The composition or list of data elements contained in the data flow
- Notes supplementing the limited space for the description that go beyond defining the data flow to explaining the context and nature of this repository object
- A list of locations (the names of the DFDs) on which this data flow appears and the names of the sources and destinations on each of these DFDs for the data flow

By the way, it is this tight linkage between diagrams and the CASE repository that creates much of the value of a CASE tool. Although very sophisticated drawing tools, as well as forms and word-processing systems, exist, these stand-alone tools do not integrate graphical objects with their textual descriptions as CASE tools do.

Consistency The concept of **DFD consistency** refers to whether or not the depiction of the system shown at one level of a nested set of DFDs is compatible with the depictions of the system shown at other levels. A gross violation of consistency would be a level-1 diagram with no level-0 diagram. Another example of inconsistency would be a data flow that appears on a higher-level DFD but not on lower levels (also a violation of balancing). Yet another example of inconsistency is a data flow attached to one object on a lower-level diagram but also attached to another object at a higher level; for example, a data flow named Payment, which serves as input to Process 1 on a level-0 DFD, appears as input to Process 2.1 on a level-1 diagram for Process 2.

CASE tools also have analysis facilities that you can use to detect such inconsistencies across nested DFD. For example, when you draw a DFD using a CASE tool, most tools will automatically place the inflows and outflows of a process on the DFD you create when you inform the tool to decompose that process. In manipulating the lower-level diagram, you could accidentally delete or change a data flow that would cause the diagrams to be out of balance; thus, a consistency-check facility with a CASE tool is quite helpful.

Timing You may have noticed in some of the DFD examples we have presented that DFDs do not do a very good job of representing time. On a given DFD, there is no indication of whether a data flow occurs constantly in real time, once per week, or once per year. There is also no indication of when a system would run. For example, many large, transaction-based systems may run several large, computing-intensive jobs in batch mode at night, when demands on the computer system are lighter. A DFD has no way of indicating such overnight batch processing. When you draw DFDs, then, draw them as if the system you are modeling has never started and will never stop.

Iterative Development The first DFD you draw will rarely capture perfectly the system you are modeling. You should count on drawing the same diagram over and over again, in an iterative fashion. With each attempt, you will come closer to a good

DFD consistency
The extent to which information contained on one level of a set of nested DFDs is also included on other levels.

approximation of the system or aspect of the system you are modeling. Iterative DFD development recognizes that requirements determination and requirements structuring are interacting, not sequential, subphases of the analysis phase of the SDLC.

One rule of thumb is that it should take you about three revisions for each DFD you draw. Fortunately, CASE tools make revising drawings a lot easier than it would be if you had to draw each revision with a pencil and a template.

Primitive DFDs One of the more difficult decisions you need to make when drawing DFDs is when to stop decomposing processes. One rule is to stop drawing when you have reached the lowest logical level; however, it is not always easy to know what the lowest logical level is. Other, more concrete rules for when to stop decomposing include the following:

- When you have reduced each process to a single decision or calculation or to a single database operation, such as retrieve, update, create, delete, or read
- When each data store represents data about a single entity, such as a customer, employee, product, or order
- When the system user does not care to see any more detail or when you and other analysts have documented sufficient detail to do subsequent systems development tasks
- When every data flow does not need to be split further to show that different data are handled in different ways
- When you believe that you have shown each business form or transaction, computer online display, and report as a single data flow (this often means, for example, that each system display and report title corresponds to the name of an individual data flow)
- When you believe there is a separate process for each choice on all lowest-level menu options for the system

Obviously, the iteration guideline discussed earlier and the various feedback loops in the SDLC (see Figure 7-1) suggest that when you think you have met the rules for stopping, you may later discover nuances to the system that require you to further decompose a set of DFDs.

By the time you stop decomposing a DFD, it may be quite detailed. Seemingly simple actions, such as generating an invoice, may pull information from several entities and may also return different results depending on the specific situation. For example, the final form of an invoice may be based on the type of customer (which would determine such things as discount rate), where the customer lives (which would determine such things as sales tax), and how the goods are shipped (which would determine such things as the shipping and handling charges). At the lowest-level DFD, called a **primitive DFD**, all of these conditions would have to be met. Given the amount of detail called for in a primitive DFD, perhaps you can see why many experts believe analysts should not spend their time completely diagramming the current physical information system because much of the detail will be discarded when the current logical DFD is created.

Using the guidelines presented in this section will help you to create DFDs that are more than just mechanically correct. Your DFDs will also be robust and accurate representations of the information system you are modeling. Primitive DFDs also facilitate consistency checks with the documentation produced from other requirements structuring techniques and also makes it easy for you to transition to system design steps. Having mastered the skills of drawing good DFDs, you can now use them to support the analysis process, the subject of the next section.

Primitive DFD
The lowest level of decomposition for a DFD.

Using DFDs as Analysis Tools

We have seen that DFDs are versatile tools for process modeling and that they can be used to model systems that are either physical or logical, current or new. DFDs can also be used in a process called **gap analysis**. Analysts can use gap analysis to discover discrepancies between two or more sets of DFDs, representing two or more states of an information system, or discrepancies within a single DFD.

Once the DFDs are complete, you can examine the details of individual DFDs for such problems as redundant data flows, data that are captured but that are not used by the system, and data that are updated identically in more than one location. These problems may not have been evident to members of the analysis team or to other participants in the analysis process when the DFDs were created. For example, redundant data flows may have been labeled with different names when the DFDs were created. Now that the analysis team knows more about the system it is modeling, such redundancies can be detected. Such redundancies can be detected most easily from CASE tool repository reports. For example, many CASE tools can generate a report that lists all of the processes that accept a given data element as input (remember, a list of data elements is likely part of the description of each data flow). From the labels of these processes, you can determine whether the data are captured redundantly or if more than one process is maintaining the same data stores. In such cases, the DFDs may well accurately mirror the activities occurring in the organization. Because the business processes being modeled took many years to develop, sometimes with participants in one part of the organization adapting procedures in isolation from other participants, redundancies and overlapping responsibilities may well have resulted. The careful study of the DFDs created as part of analysis can reveal these procedural redundancies and allow them to be corrected as part of system design.

Inefficiencies can also be identified by studying DFDs, and there are a wide variety of inefficiencies that might exist. Some inefficiencies relate to violations of DFD drawing rules. For example, a violation of rule R from Table 7-3 could occur because obsolete data are captured but never used within a system. Other inefficiencies are due to excessive processing steps. For example, consider the correct DFD in item M of Figure 7-1. Although this flow is mechanically correct, such a loop may indicate potential delays in processing data or unnecessary approval operations.

Similarly, a set of DFDs that models the current logical system can be compared with DFDs that model the new logical system to better determine which processes systems developers need to add or revise when building the new system. Processes for which inputs, outputs, and internal steps have not changed can possibly be reused in the construction of the new system. You can compare alternative logical DFDs to identify those few elements that must be discussed in evaluating competing opinions on system requirements. The logical DFDs for the new system can also serve as the basis for developing alternative design strategies for the new physical system. As we saw with the Hoosier Burger example, a process on a DFD can be implemented in several different physical ways.

Using DFDs in Business Process Reengineering

DFDs are also useful for modeling processes in business process reengineering (BPR), which you read about in Chapter 6. To illustrate the usefulness of DFDs for BPR, let's look at an example from Hammer and Champy (1993). Hammer and Champy use IBM Credit Corporation as an example of a firm that successfully reengineered its primary business process. IBM Credit Corporation provides financing for customers making large purchases of IBM computer equipment. Its job is to analyze deals proposed by salespeople and write the final contracts governing those deals.

According to Hammer and Champy, IBM Credit Corporation typically took six business days to process each financing deal. The process worked like this: First, the salesperson called in with a proposed deal. The call was taken by one of a half dozen people sitting around a conference table. Whoever received the call logged it and wrote the details on a piece of paper. A clerk then carried the paper to a second person, who initiated the next step in the process by entering the data into a computer system and checking the client's creditworthiness. This person then wrote the details on a piece of paper and carried the paper, along with the original documentation, to a loan officer. Then, in a third step, the loan officer modified the standard IBM loan agreement for the customer. This step involved a separate computer system from the one used in step two.

In the fourth step, details of the modified loan agreement, along with the other documentation, were sent on to the next station in the process, where a different clerk determined the appropriate interest rate for the loan. This step also involved its own information system. In step five, the resulting interest rate and all of the paper generated up to this point were then carried to the next stop, where the quote letter was created. Once complete, the quote letter was sent via overnight mail back to the salesperson.

Only reading about this process makes it seem complicated. We can use DFDs to illustrate the overall process (see Figure 7-16). DFDs help us see that the process is not as complicated as it is tedious and wasteful, especially when you consider that so many different people and computer systems were used to support the work at each step.

According to Hammer and Champy, two IBM managers decided to see if they could improve the overall process at IBM Credit Corporation. They took a call from a salesperson and walked it through the system. These managers found that the actual work being done on a contract only took 90 minutes. For much of the rest of the six days it took to process the deal, the various bits of documentation were sitting in someone's in-basket waiting to be processed.

IBM Credit Corporation management decided to reengineer its entire process. The five sets of task specialists were replaced with generalists. Now each call from the field goes to a single clerk, who does all the work necessary to process the contract. Instead of having different people check for creditworthiness, modify the basic loan agreement, and determine the appropriate interest rate, one person does it all. IBM Credit Corporation still has specialists for the few cases that are significantly different from what the firm routinely encounters. In addition, the process is now supported by a single computer system. The new process is modeled by the DFD in Figure 7-17. The most striking difference between the DFD in Figure 7-16 and the DFD in Figure 7-17, other than the number of process boxes in each one, is the lack of documentation flow in Figure 7-17. The resulting process is much simpler and cuts down dramatically on any chance of documentation getting lost between steps.

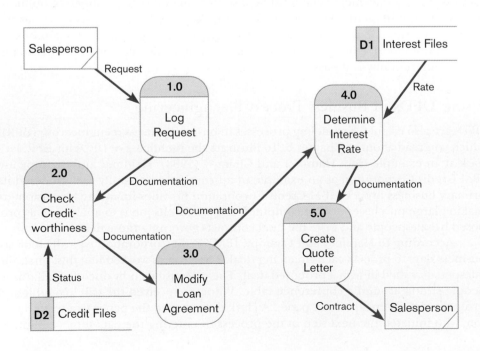

FIGURE 7-16
IBM Credit Corporation's primary work process before BPR

(*Source:* Based on Hammer and Champy, 1993.)

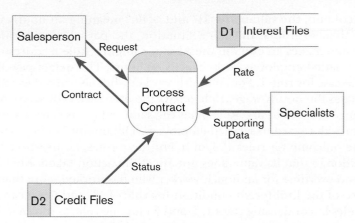

FIGURE 7-17
IBM Credit Corporation's primary work process after BPR
(*Source:* Based on Hammer and Champy, 1993.)

Redesigning the process from beginning to end allowed IBM Credit Corporation to increase the number of contracts it can handle by 100-fold—not 100 percent, which would only be doubling the amount of work. BPR allowed IBM Credit Corporation to handle 100 times more work in the same amount of time and with fewer people!

MODELING LOGIC WITH DECISION TABLES

A **decision table** is a diagram of process logic where the logic is reasonably complicated. All of the possible choices and the conditions the choices depend on are represented in tabular form, as illustrated in the decision table in Figure 7-18.

The decision table in Figure 7-18 models the logic of a generic payroll system. The table has three parts: the **condition stubs**, the **action stubs**, and the **rules**. The condition stubs contain the various conditions that apply to the situation the table is modeling. In Figure 7-18, there are two condition stubs for employee type and hours worked. Employee type has two values: "S," which stands for salaried, and "H," which stands for hourly. Hours worked has three values: less than 40, exactly 40, and more than 40. The action stubs contain all the possible courses of action that result from combining values of the condition stubs. There are four possible courses of action in this table: Pay Base Salary, Calculate Hourly Wage, Calculate Overtime, and Produce Absence Report. You can see that not all actions are triggered by all combinations of conditions. Instead, specific combinations trigger specific actions. The part of the table that links conditions to actions is the section that contains the rules.

To read the rules, start by reading the values of the conditions as specified in the first column: Employee type is "S," or salaried, and hours worked is less than 40. When both of these conditions occur, the payroll system is to pay the base salary.

Decision table
A matrix representation of the logic of a decision, which specifies the possible conditions for the decision and the resulting actions.

Condition stubs
The part of a decision table that lists the conditions relevant to the decision.

Action stubs
The part of a decision table that lists the actions that result for a given set of conditions.

Rules
The part of a decision table that specifies which actions are to be followed for a given set of conditions.

	Conditions/ Courses of Action	Rules					
		1	2	3	4	5	6
Condition Stubs	Employee type	S	H	S	H	S	H
	Hours worked	<40	<40	40	40	>40	>40
Action Stubs	Pay base salary	X		X		X	
	Calculate hourly wage		X		X		X
	Calculate overtime						X
	Produce absence report		X				

FIGURE 7-18
Complete decision table for payroll system example

In the next column, the values are "H" and "<40," meaning an hourly worker who worked less than 40 hours. In such a situation, the payroll system calculates the hourly wage and makes an entry in the Absence Report. Rule 3 addresses the situation when a salaried employee works exactly 40 hours. The system pays the base salary, as was the case for rule 1. For an hourly worker who has worked exactly 40 hours, rule 4 calculates the hourly wage. Rule 5 pays the base salary for salaried employees who work more than 40 hours. Rule 5 has the same action as rules 1 and 3 and governs behavior with regard to salaried employees. The number of hours worked does not affect the outcome for rules 1, 3, or 5. For these rules, hours worked is an **indifferent condition** in that its value does not affect the action taken. Rule 6 calculates hourly pay and overtime for an hourly worker who has worked more than 40 hours.

Because of the indifferent condition for rules 1, 3, and 5, we can reduce the number of rules by condensing rules 1, 3, and 5 into one rule, as shown in Figure 7-19. The indifferent condition is represented with a dash. Whereas we started with a decision table with six rules, we now have a simpler table that conveys the same information with only four rules.

In constructing these decision tables, we have actually followed a set of basic procedures:

1. *Name the conditions and the values that each condition can assume.* Determine all of the conditions that are relevant to your problem and then determine all of the values each condition can take. For some conditions, the values will be simply "yes" or "no" (called a *limited entry*). For others, such as the conditions in Figures 7-18 and 7-19, the conditions may have more values (called an *extended entry*).
2. *Name all possible actions that can occur.* The purpose of creating decision tables is to determine the proper course of action given a particular set of conditions.
3. *List all possible rules.* When you first create a decision table, you have to create an exhaustive set of rules. Every possible combination of conditions must be represented. It may turn out that some of the resulting rules are redundant or make no sense, but these determinations should be made only after you have listed every rule so that no possibility is overlooked. To determine the number of rules, multiply the number of values for each condition by the number of values for every other condition. In Figure 7-18, we have two conditions, one with two values and one with three, so we need 2 × 3 or 6, rules. If we added a third condition with three values, we would need 2 × 3 × 3, or 18, rules.

When creating the table, alternate the values for the first condition, as we did in Figure 7-18 for type of employee. For the second condition, alternate the values but repeat the first value for all values of the first condition, then repeat the second value for all values of the first condition, and so on. You essentially follow this procedure for all subsequent conditions. Notice how we alternated

Indifferent condition

In a decision table, a condition whose value does not affect which actions are taken for two or more rules.

Conditions/ Courses of Action	Rules			
	1	2	3	4
Employee type	S	H	H	H
Hours worked	–	<40	40	>40
Pay base salary	X			
Calculate hourly wage		X	X	X
Calculate overtime				X
Produce absence report		X		

FIGURE 7-19
Reduced decision table for payroll system example

the values of hours worked in Figure 7-18. We repeated "<40" for both values of type of employee, "S" and "H." Then we repeated "40," and then ">40."

4. *Define the actions for each rule.* Now that all possible rules have been identified, provide an action for each rule. In our example, we were able to figure out what each action should be and whether all of the actions made sense. If an action doesn't make sense, you may want to create an "impossible" row in the action stubs in the table to keep track of impossible actions. If you can't tell what the system ought to do in that situation, place question marks in the action stub spaces for that particular rule.

5. *Simplify the decision table.* Make the decision table as simple as possible by removing any rules with impossible actions. Consult users on the rules where system actions aren't clear and either decide on an action or remove the rule. Look for patterns in the rules, especially for indifferent conditions. We were able to reduce the number of rules in the payroll example from six to four, but greater reductions are often possible.

Let's look at an example from Hoosier Burger. The Mellankamps are trying to determine how they reorder food and other items they use in the restaurant. If they are going to automate the inventory control functions at Hoosier Burger, they need to articulate their reordering process. In thinking through the problem, the Mellankamps realize that how they reorder depends on whether the item is perishable. If an item is perishable, such as meat, vegetables, or bread, the Mellankamps have a standing order with a local supplier stating that a prespecified amount of food is delivered each weekday for that day's use and each Saturday for weekend use. If the item is not perishable, such as straws, cups, and napkins, an order is placed when the stock on hand reaches a certain predetermined minimum reorder quantity. The Mellankamps also realize the importance of the seasonality of their work. Hoosier Burger's business is not as good during the summer months when the students are off campus as it is during the academic year. They also note that business falls off during Christmas and spring break. Their standing orders with all their suppliers are reduced by specific amounts during the summer and holiday breaks. Given this set of conditions and actions, the Mellankamps put together an initial decision table (see Figure 7-20).

Conditions/ Courses of Action	Rules											
	1	2	3	4	5	6	7	8	9	10	11	12
Type of item	P	N	P	N	P	N	P	N	P	N	P	N
Time of week	D	D	W	W	D	D	W	W	D	D	W	W
Season of year	A	A	A	A	S	S	S	S	H	H	H	H
Standing daily order	X				X				X			
Standing weekend order			X				X				X	
Minimum order quantity		X		X		X		X		X		X
Holiday reduction									X		X	
Summer reduction					X		X					

Type of item:
P = perishable
N = nonperishable

Time of week:
D = weekday
W = weekend

Season of year:
A = academic year
S = summer
H = holiday

FIGURE 7-20
Complete decision table for Hoosier Burger's inventory reordering

FIGURE 7-21
Reduced decision table for Hoosier
Burger's inventory reordering

Conditions/ Courses of Action	Rules						
	1	2	3	4	5	6	7
Type of item	P	P	P	P	P	P	N
Time of week	D	W	D	W	D	W	–
Season of year	A	A	S	S	H	H	–
Standing daily order	X		X		X		
Standing weekend order		X		X		X	
Minimum order quantity							X
Holiday reduction					X	X	
Summer reduction			X	X			

Notice three things about Figure 7-20. First, notice how the values for the third condition have been repeated, providing a distinctive pattern for relating the values for all three conditions to each other. Every possible rule is clearly provided in this table. Second, notice that we have 12 rules. Two values for the first condition (type of item) times 2 values for the second condition (time of week) times 3 values for the third condition (season of year) equals 12 possible rules. Third, notice how the action for nonperishable items is the same, regardless of day of week or time of year. For nonperishable goods, both time-related conditions are indifferent. Collapsing the decision table accordingly gives us the decision table in Figure 7-21. Now there are only 7 rules instead of 12.

You have now learned how to draw and simplify decision tables. You can also use decision tables to specify additional decision-related information. For example, if the actions that should be taken for a specific rule are more complicated than one or two lines of text can convey or if some conditions need to be checked only when other conditions are met (nested conditions), you may want to use separate, linked decision tables. In your original decision table, you can specify an action in the action stub that says "Perform Table B." Table B could contain an action stub that returns to the original table, and the return would be the action for one or more rules in Table B. Another way to convey more information in a decision table is to use numbers that indicate sequence rather than Xs where rules and action stubs intersect. For example, for rules 3 and 4 in Figure 7-21, it would be important for the Mellankamps to account for the summer reduction to modify the existing standing order for supplies. "Summer reduction" would be marked with a "1" for rules 3 and 4, whereas "Standing daily order" would be marked with a "2" for rule 3, and "Standing weekend order" would be marked with a "2" for rule 4.

You have seen how decision tables can model the relatively complicated logic of a process. As such, decision tables are compact; you can pack a lot of information into a small table. Decision tables also allow you to check for the extent to which your logic is complete, consistent, and not redundant.

ELECTRONIC COMMERCE APPLICATION: PROCESS MODELING USING DATA FLOW DIAGRAMS

Process modeling for an Internet-based electronic commerce application is no different than the process followed for other applications. In chapter 6, you read how Pine Valley Furniture (PVF) determined the system requirements for their Web-Store project, a project to sell furniture products over the Internet. In this section, we analyze the WebStore's high-level system structure and develop a level-0 DFD for those requirements.

Process Modeling for Pine Valley Furniture's WebStore

After completing the Joint Application Design (JAD) session, senior systems analyst Jim Woo went to work on translating the WebStore system structure into a DFD. His first step was to identify the level-0—major system—processes. To begin, he carefully examined the outcomes of the JAD session that focused on defining the system structure of the WebStore system. From this analysis, he identified six high-level processes that would become the foundation of the level-0 DFD. These processes, listed in Table 7-4, were the "work" or "action" parts of the website. Note that each of these processes corresponds to the major processing items listed in the system structure.

Next, Jim determined that it would be most efficient if the WebStore system exchanged information with existing PVF systems rather than capture and store redundant information. This analysis concluded that the WebStore should exchange information with the Purchasing Fulfillment System—a system for tracking orders (see Chapter 3)—and the Customer Tracking System—a system for managing customer information. These two existing systems will be "sources" (providers) and "sinks" (receivers) of information for the WebStore system. When a customer opens an account, his or her information will be passed from the WebStore system to the Customer Tracking System. When an order is placed (or when a customer requests status information on a prior order), information will be stored (retrieved) in (from) the Purchasing Fulfillment System.

Finally, Jim found that the system would need to access two additional data sources. First, in order to produce an online product catalog, the system would need to access the inventory database. Second, to store the items a customer wanted to purchase in the Webstore's shopping cart, a temporary database would need to be created. Once a transaction is completed, the shopping cart data can be deleted. With this information, Jim was able to develop the level-0 DFD for the Webstore system, which is shown in Figure 7-22. He then felt that he had a good understanding of how information would flow through the Webstore, of how a customer would interact with the system, and of how the Webstore would share information with existing PVF systems. Each of these high-level processes would eventually need to be further decomposed before system design could proceed. Yet, before doing that, he wanted to get a clear picture of exactly what data were needed throughout the entire system. We will discover the outcomes of this analysis activity—conceptual data modeling—in Chapter 8.

TABLE 7-4 System Structure of the WebStore and Corresponding Level-0 Processes

WebStore System	Processes
❑ Main Page	Information Display (minor/no processes)
• Product Line (Catalog)	1.0 Browse Catalog
✓ Desks	2.0 Select Item for Purchase
✓ Chairs	
✓ Tables	
✓ File Cabinets	
• Shopping Cart	3.0 Display Shopping Cart
• Checkout	4.0 Check Out Process Order
• Account Profile	5.0 Add/Modify Account Profile
• Order Status/History	6.0 Order Status Request
• Customer Comments	Information Display (minor/no processes)
❑ Company Information	
❑ Feedback	
❑ Contact Information	

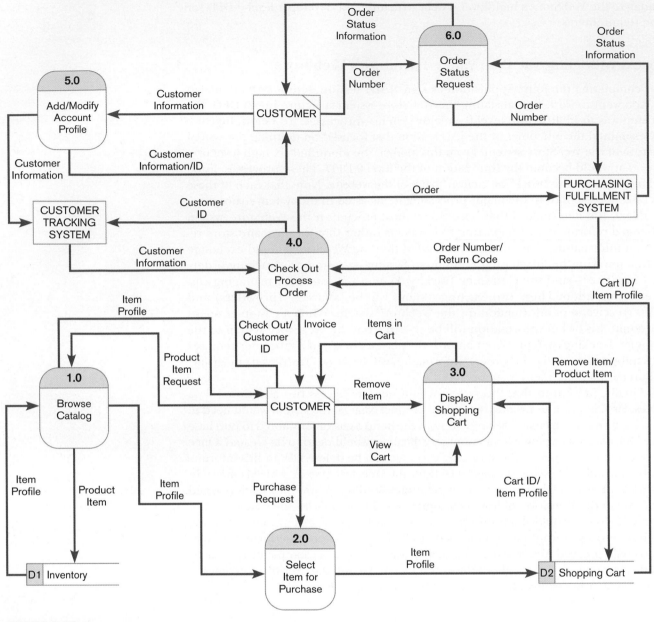

FIGURE 7-22
Level-0 DFD for the WebStore

SUMMARY

Processes can be modeled in many different ways, but this chapter has focused on data flow diagrams, or DFDs. DFDs are very useful for representing the overall data flows into, through, and out of an information system. DFDs rely on four symbols to represent the four conceptual components of a process model: data flows, data stores, processes, and sources/sinks. DFDs are hierarchical in nature, and each level of a DFD can be decomposed into smaller, simpler units on a lower-level diagram. You begin process modeling by constructing a context diagram, which shows the entire system as a single process. The next step is to generate a

level-0 diagram, which shows the most important high-level processes in the system. You then decompose each process in the level-0 diagram, as warranted, until it makes no logical sense to go any further. When decomposing DFDs from one level to the next, it is important that the diagrams be balanced; that is, inputs and outputs on one level must be conserved on the next level.

DFDs should be mechanically correct, but they should also accurately reflect the information system being modeled. To that end, you need to check DFDs for completeness and consistency and draw them as if the system being modeled were timeless. You should be willing to revise DFDs several times. Complete sets of DFDs should extend to the primitive level, where every component reflects certain irreducible properties; for example, a process represents a single database operation and every data store represents data about a single entity. Following these guidelines, you can produce DFDs that can be used to analyze the gaps between existing and desired procedures and between current and new systems.

Decision tables are a graphical method for representing process logic. In decision tables, conditions are listed in the conditions stubs, possible actions are listed in the action stubs, and rules link combinations of conditions

into the actions that should result. Analysts reduce the complexity of decision tables by eliminating rules that do not make sense and by combining rules with different conditions.

Although analysts have been modeling processes for information systems for over 30 years, dating back at least to the beginnings of the philosophy of structured analysis and design, it is just as important for electronic commerce applications as it is for more traditional systems. Future chapters will show, as this one did, how traditional tools and techniques developed for structured analysis and design provide powerful assistance for the electronic commerce development.

KEY TERMS

1. Action stubs
2. Balancing
3. Condition stubs
4. Context diagram
5. Data flow diagram (DFD)
6. Data store

7. Decision table
8. DFD completeness
9. DFD consistency
10. Functional decomposition
11. Gap analysis
12. Indifferent condition

13. Level-0 diagram
14. Level-n diagram
15. Primitive DFD
16. Process
17. Rules
18. Source/sink

Match each of the key terms above with the definition that best fits it.

_____ A picture of the movement of data between external entities and the processes and data stores within a system.

_____ The part of a decision table that lists the actions that result for a given set of conditions.

_____ The conservation of inputs and outputs to a DFD process when that process is decomposed to a lower level.

_____ A DFD that represents a system's major processes, data flows, and data stores at a high level of detail.

_____ The origin and/or destination of data; sometimes referred to as external entities.

_____ In a decision table, a condition whose value does not affect which actions are taken for two or more rules.

_____ An overview of an organizational system that shows the system boundary, external entities that interact with the system, and the major information flows between the entities and the system.

_____ The lowest level of decomposition for a DFD.

_____ The extent to which all necessary components of a DFD have been included and fully described.

_____ A matrix representation of the logic of a decision; it specifies the possible conditions for the decision and the resulting actions.

_____ The extent to which information contained on one level of a set of nested DFDs is also included on other levels.

_____ A DFD that is the result of n nested decompositions of a series of subprocesses from a process on a level-0 diagram.

_____ The part of a decision table that lists the conditions relevant to the decision.

_____ The work or actions performed on data so that they are transformed, stored, or distributed.

_____ Data at rest, which may take the form of many different physical representations.

_____ The process of discovering discrepancies between two or more sets of DFDs or discrepancies within a single DFD.

_____ The part of a decision table that specifies which actions are to be followed for a given set of conditions.

_____ An iterative process of breaking the description of a system down into finer and finer detail, which creates a set of charts in which one process on a given chart is explained in greater detail on another chart.

REVIEW QUESTIONS

1. What is a DFD? Why do systems analysts use DFDs?
2. Explain the rules for drawing good DFDs.
3. What is decomposition? What is balancing? How can you determine if DFDs are not balanced?
4. Explain the convention for naming different levels of DFDs.
5. Why do analysts draw multiple sets of DFDs?
6. How can DFDs be used as analysis tools?
7. Explain the guidelines for deciding when to stop decomposing DFDs.
8. How do you decide if a system component should be represented as a source/sink or as a process?
9. What unique rules apply to drawing context diagrams?
10. What are the steps in creating a decision table? How do you reduce the size and complexity of a decision table?
11. What does the term *limited entry* mean in a decision table?
12. What is the formula that is used to calculate the number of rules a decision table must cover?

PROBLEMS AND EXERCISES

1. Using the example of a retail clothing store in a mall, list relevant data flows, data stores, processes, and sources/sinks. Observe several sales transactions. Draw a context diagram and a level-0 diagram that represent the selling system at the store. Explain why you chose certain elements as processes versus sources/sinks.

2. Choose a transaction that you are likely to encounter, perhaps ordering a cap and gown for graduation, and develop a high-level DFD or a context diagram. Decompose this to a level-0 diagram.

3. Evaluate your level-0 DFD from Problem and Exercise 2 using the rules for drawing DFDs in this chapter. Edit your DFD so that it does not break any of these rules.

4. Choose an example like the one in Problem and Exercise 2 and draw a context diagram. Decompose this diagram until it does not make sense to continue. Be sure that your diagrams are balanced.

5. Refer to Figure 7-23, which contains drafts of a context and level-0 DFD for a university class registration system. Identify and explain potential violations of rules and guidelines on these diagrams.

6. What is the relationship between DFDs and entries in the project dictionary or CASE repository?

7. Consider the DFD in Figure 7-24. List three errors (rule violations) on this DFD.

8. Consider the three DFDs in Figure 7-25. List three errors (rule violations) on these DFDs.

9. Starting with a context diagram, draw as many nested DFDs as you consider necessary to represent all the details of the employee hiring system described in the following narrative. You must draw at least a context and a level-0 diagram. If you discover while drawing these diagrams that the narrative is incomplete, make up reasonable explanations to complete the story. Supply these extra explanations along with the diagrams.

   Projects, Inc., is an engineering firm with approximately 500 engineers of different types. The company keeps records on all employees, their skills, projects assigned, and departments worked in. New employees are hired by the personnel manager based on data in an application form and evaluations collected from other managers who interview the job candidates. Prospective employees may apply at any time. Engineering managers notify the personnel manager when a job opens and list the characteristics necessary to be eligible for the job. The personnel manager compares the qualifications of the available pool of applicants with the characteristics of an open job and then schedules interviews between the manager in charge of the open position and the three best candidates from the pool. After receiving evaluations on each interview from the manager, the personnel manager makes the hiring decision based upon the evaluations and applications of the candidates and the characteristics of the job and then notifies the interviewees and the manager about the decision. Applications of rejected applicants are retained for one year, after which time the application is purged. When hired, a new engineer completes a nondisclosure agreement, which is filed with other information about the employee.

10. Starting with a context diagram, draw as many nested DFDs as you consider necessary to represent all of the details of the system described in the following narrative. In drawing these diagrams, if you discover that the narrative is incomplete, make up reasonable explanations to make the story complete. Supply these extra explanations along with the diagrams.

   Maximum Software is a developer and supplier of software products to individuals and businesses. As part of their operations, Maximum provides a 1-800 help desk line for clients who have questions about software purchased from Maximum. When a call comes in, an operator inquires about the nature of the call. For calls that are not truly help desk functions, the operator redirects the call to another unit of the company (such as Order Processing or Billing). Because many customer questions require in-depth knowledge of a product, help desk consultants are organized by product. The operator directs the call to a consultant skilled on the software that the caller needs help with. Because a consultant is not always immediately available, some calls must be put into a queue for the next available consultant. Once a consultant answers the call, the consultant determines if this is the first call from this customer about a particular problem. If it is,

Context Diagram

FIGURE 7-23
Class registration system for Problem and Exercise 5

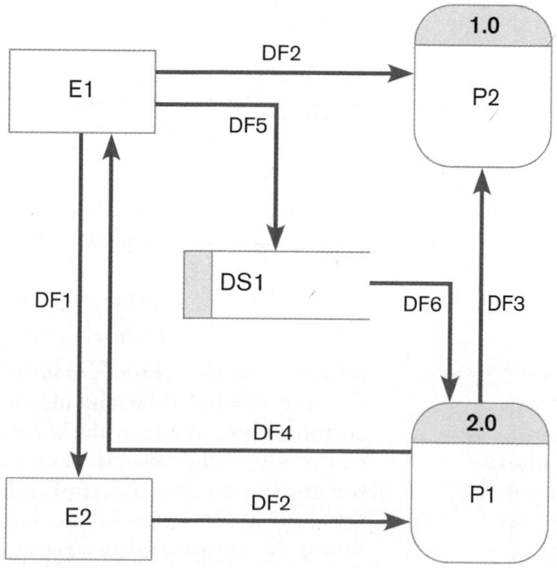

Student

Class Schedule

0
Class
Registration
System

Course Request

List of Courses

Department

Possible Classes

Scheduled
Classes

D1 Roster
of Classes

Level-0 Diagram

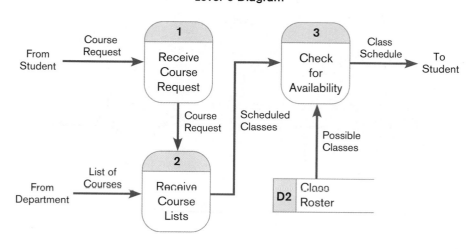

From
Student

Course
Request

1
Receive
Course
Request

3
Check
for
Availability

Class
Schedule

To
Student

Course
Request

Scheduled
Classes

Possible
Classes

From
Department

List of
Courses

2
Receive
Course
Lists

D2 Class
Roster

FIGURE 7-24
DFD for Problem and Exercise 7

E1

DF2

1.0
P2

DF5

DF1

DS1

DF6

DF3

DF4

2.0
P1

E2

DF2

the consultant creates a new call report to keep track of all information about the problem. If it is not the first call about a problem, the consultant asks the customer for a call report number and retrieves the open call report to determine the status of the inquiry. If the caller does not know the call report number, the consultant collects other identifying information such as the caller's name, the software involved, or the name of the consultant who has handled the previous calls on the problem in order to conduct a search for the

FIGURE 7-25
DFD for Problem and Exercise 8

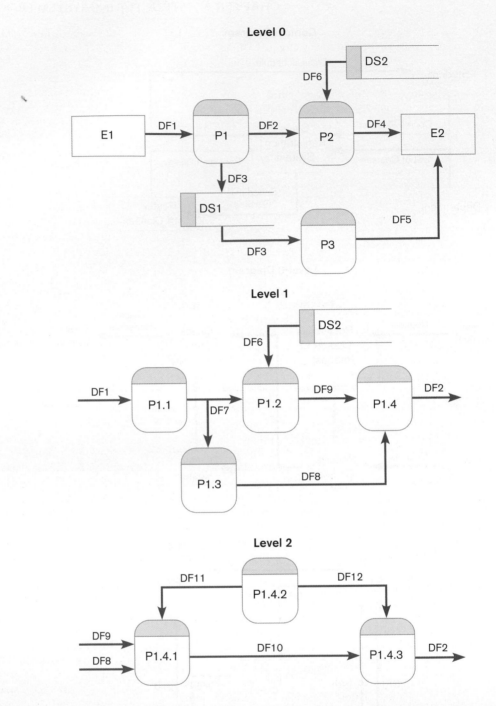

appropriate call report. If a resolution of the customer's problem has been found, the consultant informs the client as to what that resolution is, indicates on the report that the customer has been notified, and closes out the report. If resolution has not been discovered, the consultant finds out if the consultant handling this problem is on duty. If so, he or she transfers the call to the other consultant (or puts the call into the queue of calls waiting to be handled by that consultant). Once the proper consultant receives the call, that consultant records any new details the customer may have. For continuing problems and for new call reports, the consultant tries to discover an answer to the problem by using the relevant software and looking up

information in reference manuals. If the consultant can now resolve the problem, the consultant tells the customer how to deal with the problem and closes the call report. Otherwise, the consultant files the report for continued research and tells the customer that someone at Maximum will get back to him or her, and that if the customer discovers new information about the problem, he or she should call Maximum with the information, identifying the problem with a specified call report number.

Analyze the DFDs you created in the first part of this question. What recommendations for improvements in the help desk system at Maximum can you make based on this analysis? Draw new logical DFDs that represent the requirements you would suggest for an improved help desk system. Remember, these are to be logical DFDs, so consider improvements independent of technology that can be used to support the help desk.

11. Develop a context diagram and level-0 diagram for the hospital pharmacy system described in the following narrative. If you discover that the narrative is incomplete, make up reasonable explanations to complete the story. Supply these extra explanations along with the diagrams.

> The pharmacy at Mercy Hospital fills medical prescriptions for all hospital patients and distributes these medications to the nurse stations responsible for the patients' care. Prescriptions are written by doctors and sent to the pharmacy. A pharmacy technician reviews each prescription and sends it to the appropriate pharmacy station. Prescriptions for drugs that must be formulated (made on-site) are sent to the lab station, prescriptions for off-the-shelf drugs are sent to the shelving station, and prescriptions for narcotics are sent to the secure station. At each station, a pharmacist reviews the order, checks the patient's file to determine the appropriateness of the prescription, and fills the order if the dosage is at a safe level and it will not negatively interact with the other medications or allergies indicated in the patient's file. If the pharmacist does not fill the order, the prescribing doctor is contacted to discuss the situation. In this case, the order may ultimately be filled, or the doctor may write another prescription depending on the outcome of the discussion. Once filled, a prescription label is generated listing the patient's name, the drug type and dosage, an expiration date, and any special instructions. The label is placed on the drug container, and the order is sent to the appropriate nurse station. The patient's admission number, the drug type and amount dispensed, and the cost of the prescription are then sent to the Billing department.

12. Develop a context diagram and a level-0 diagram for the contracting system described in the following narrative. If you discover that the narrative is incomplete, make up reasonable explanations to complete the story. Supply these extra explanations along with the diagrams.

> Government Solutions Company (GSC) sells computer equipment to federal government agencies. Whenever a federal agency needs to purchase equipment from GSC, it issues a purchase order against a standing contract previously negotiated with the company. GSC holds several standing contracts with various federal agencies. When a purchase order is received by GSC's contracting officer, the contract number referenced on the purchase order is entered into the contract database. Using information from the database, the contracting officer reviews the terms and conditions of the contract and determines whether the purchase order is valid. The purchase order is valid if the contract has not expired, the type of equipment ordered is listed on the original contract, and the total cost of the equipment does not exceed a predetermined limit. If the purchase order is not valid, the contracting officer sends the purchase order back to the requesting agency with a letter stating why the purchase order cannot be filled, and a copy of the letter is filed. If the purchase order is valid, the contracting officer enters the purchase order number into the contract database and flags the order as outstanding. The purchase order is then sent to the Order Fulfillment department. Here the inventory is checked for each item ordered. If any items are not in stock, the Order Fulfillment department creates a report listing the items not in stock and attaches it to the purchase order. All purchase orders are forwarded to the warehouse, where the items in stock are pulled from the shelves and shipped to the customer. The warehouse then attaches to the purchase order a copy of the shipping bill listing the items shipped and sends it to the contracting officer. If all items were shipped, the contracting officer closes the outstanding purchase order record in the database. The purchase order, shipping bill, and exception report (if attached) are then filed in the contracts office.

13. Develop a context diagram and as many nested DFDs as you consider necessary to represent all the details of the training logistics system described in the following narrative. If you discover that the narrative is incomplete, make up reasonable explanations to complete the story. Supply these extra explanations along with the diagrams.

> Training, Inc., conducts training seminars in major U.S. cities. For each seminar, the Logistics department must make arrangements for the meeting facilities, the training consultant's travel, and the shipment of any seminar materials. For each scheduled seminar, the Bookings department notifies the logistics coordinator of the type of seminar, the dates and city location, and the name of the consultant who will conduct the training. To arrange for meeting facilities, the logistics coordinator gathers information on possible meeting sites in the scheduled city. The meeting site location decision is made based on date availability, cost, type of meeting space available, and convenience of the location. Once the site decision is made, the coordinator speaks with the sales manager of the meeting facility to reserve the meeting room(s), plan the seating arrangement(s), and reserve any necessary audiovisual equipment. The coordinator estimates the number and size of meeting rooms, the type of seating arrangements, and the audiovisual equipment needed for each seminar from the information kept in a logistics database on each type of seminar offered and the number of anticipated registrants for a particular booking. After negotiations are conducted by the logistics coordinator and the sales manager of the meeting facility, the sales manager creates a contract agreement specifying the negotiated arrangements and sends two copies of it to the logistics coordinator. The coordinator reviews the agreement and approves it if no changes are needed. One copy of the agreement is filed and the other copy is sent back to the sales manager. If changes are needed, the agreement copies are changed and returned to the sales manager for approval. This approval process continues

until both parties have approved the agreement. The coordinator must also contact the training consultant to make travel arrangements. First, the coordinator reviews the consultant's travel information in the logistics database and researches flight schedules. Then the consultant is contacted to discuss possible travel arrangements; subsequently, the coordinator books a flight for the consultant with a travel agency. Once the consultant's travel arrangements have been completed, a written confirmation and itinerary are sent to the consultant. Two weeks before the date of the seminar, the coordinator determines what, if any, seminar materials (e.g., transparencies, training guides, pamphlets, etc.) need to be sent to the meeting facility. Each type of seminar has a specific set of materials assigned to it. For some materials, the coordinator must know how many participants have registered for the seminar in order to determine how many to send. A request for materials is sent to the Materials-handling department, where the materials are gathered, boxed, and sent to the meeting address listed on the request. Once the requested materials have been shipped, a notification is sent to the logistics coordinator.

14. Look at the set of DFDs created in this chapter for Hoosier Burger's food-ordering system. Represent the decision logic of one or more of the processes as decision tables.

15. What types of questions need to be asked during requirements determination in order to gather the information needed for logic modeling? Give examples.

16. In one company, the rules for buying personal computers specify that if the purchase is over $15,000, it has to go out for bid, and the Request for Proposals (RFP) must be approved by the Purchasing department. If the purchase is under $15,000, the personal computers can simply be bought from any approved vendor; however, the Purchase Order must still be approved by the Purchasing department. If the purchase goes out for bid, there must be at least three proposals received for the bid. If not, the RFP must go out again. If there are still not enough proposals, then the process can continue with the one or two vendors that have submitted proposals. The winner of the bid must be on an approved list of vendors for the company and, in addition, must not have any violations against them for affirmative action or environmental matters. At this point, if the proposal is complete, the Purchasing department can issue a Purchase Order. Draw a decision table to represent the logic in this process. Notice the similarities between the text in this question and the format of your answer.

17. In a relatively small company that sells thin, electronic keypads and switches, the rules for selling products specify that sales representatives are assigned to unique regions of the country. Sales come either from cold calling, referrals, or current customers with new orders. A sizable portion of their business comes from referrals from larger competitors who send their excess and/or "difficult" projects to this company. The company tracks these referrals and returns the favors to these competitors by sending business their way. The sales reps receive a 10 percent commission on actual purchases, not on orders, in their region. They can collaborate on a sale with reps in other regions and share the commissions, with 8 percent going to the "home" rep and 2 percent going to the "visiting" rep. For any sales beyond the rep's previously stated and approved individual annual sales goals, he or she receives an additional 5 percent commission, an additional end-of-the-year bonus determined by management, and a special vacation for his or her family. Customers receive a 10 percent discount for any purchases over $100,000 per year, which are factored into the rep's commissions. In addition, the company focuses on customer satisfaction with the product and service, so there is an annual survey of customers in which they rate the sales rep. These ratings are factored into the bonuses such that a high rating increases the bonus amount, a moderate rating does nothing, and a low rating can lower the bonus amount. The company also wants to ensure that the reps close all sales. Any differences between the amount of orders and actual purchases are also factored into the rep's bonus amount. As best you can, present the logic of this business process using a decision table. Write down any assumptions you have to make.

18. The following is an example that demonstrates the rules of the tenure process for faculty at many universities. Present the logic of this business using a decision table. Write down any assumptions you have to make.

A faculty member applies for tenure in his or her sixth year by submitting a portfolio summarizing his or her work. In rare circumstances, a faculty member can come up for tenure earlier than the sixth year, but only if the faculty member has the permission of the department chair and college dean. New professors who have worked at other universities before taking their current jobs rarely, if ever, start their new jobs with tenure. They are usually asked to undergo one probationary year during which they are evaluated; only then can they be granted tenure. Top administrators coming to a new university job, however, can often negotiate for retreat rights that enable them to become a tenured faculty member should their administrative post end. These retreat arrangements generally have to be approved by faculty. The tenure review process begins with an evaluation of the candidate's portfolio by a committee of faculty within the candidate's department. The committee then writes a recommendation on tenure and sends it to the department's chairperson, who then makes a recommendation and passes the portfolio and recommendation to the next level, a college-wide faculty committee. This committee does the same as the department committee and passes its recommendation, the department's recommendation, and the portfolio to the next level, a university-wide faculty committee. This committee does the same as the other two committees and passes everything to the provost (or sometimes the academic vice president). The provost then writes his or her own recommendation and passes everything to the president, the final decision maker. This process, from the time the candidate creates his or her portfolio until the time the president makes a decision, can take an entire academic year. The focus of the evaluation is on research, which could be grants, presentations, and publications, though preference is given for empirical research that has been published in top-ranked, refereed journals and where the publication makes a contribution to the field. The candidate must also do well in teaching and service (i.e., to the university, the community, or the discipline), but the primary emphasis is on research.

19. An organization is in the process of upgrading microcomputer hardware and software for all employees. Hardware will be allocated to each employee in one of three packages. The first hardware package includes a standard microcomputer with a color monitor of moderate resolution and moderate storage capabilities. The second package includes a high-end microcomputer with a high-resolution color monitor and a great deal of RAM and ROM. The third package is a high-end notebook-sized microcomputer. Each computer comes with a network interface card so that it can be connected to the network for printing and e-mail. The notebook computers come with a modem for the same purpose. All new and existing employees will be evaluated in terms of their computing needs (e.g., the types of tasks they perform, how much and in what ways they can use the computer). Light users receive the first hardware package. Heavy users receive the second package. Some moderate users will receive the first package and some will receive the second package, depending on their needs. Any employee who is deemed to be primarily mobile (e.g., most of the sales force) will receive the third package. Each employee will also be considered for additional hardware. For example, those who need scanners and/or printers will receive them. A determination will be made regarding whether the user receives a color or black-and-white scanner and whether they receive a slow or fast or color or black-and-white printer. In addition, each employee will receive a suite of software that includes a word processor, a spreadsheet, and a presentation maker. All employees will be evaluated for additional software needs. Depending on their needs, some will receive a desktop publishing package, some will receive a database management system (and some will also receive a developer's kit for the DBMS), and some will receive a programming language. Every 18 months, those employees with the high-end systems will receive new hardware, and their old systems will be passed on to those who previously had the standard systems. All those employees with the portable systems will receive new notebook computers. Present the logic of this business process using a decision table. Write down any assumptions you have to make.

20. Read the narratives below and follow the directions for each. If you discover that the narrative is incomplete, make up reasonable explanations to complete the story. Supply these extra explanations along with your answers.

    a. Samantha must decide which courses to register for this semester. She has a part-time job, and she is waiting to find out how many hours per week she will be working during the semester. If she works 10 hours or less per week, she will register for three classes, but if she works more than 10 hours per week, she will register for only two classes. If she registers for two classes, she will take one class in her major area and one elective. If she registers for three classes, she will take two classes in her major area and one elective. Use a decision table to represent this logic.

    b. Jerry plans on registering for five classes this semester: English Composition, Physics, Physics Lab, Java, and Music Appreciation. However, he is not sure if these classes are being offered this semester or if there will be timing conflicts. Also, two of the classes, Physics and Physics Lab, must be taken together during the same semester. Therefore, if he can register for only one of them, he will not take either class. If, for any reason, he cannot register for a class, he will identify and register for a different class to take its place and that fits his time schedule. Use a decision table that shows all rules to represent this logic.

21. Mary is trying to decide which graduate programs she will apply to. She wants to stay in the southeastern region of the United States, but if a program is considered one of the top 10 in the country, she is willing to move to another part of the United States. Mary is interested in both the MBA and Master of MIS programs. An MBA program must have at least one well-known faculty member and meet her location requirements before she will consider applying to it. Additionally, any program she applies to must offer financial aid, unless she is awarded a scholarship. Use a decision table to represent this logic.

22. At a local bank, loan officers must evaluate loan applications before approving or denying them. During this evaluation process, many factors regarding the loan request and the applicant's background are considered. If the loan is for less than $2000, the loan officer checks the applicant's credit report. If the credit report is rated good or excellent, the loan officer approves the loan. If the credit report is rated fair, the officer checks to see if the applicant has an account at the bank. If the applicant holds an account, the application is approved; otherwise, the application is denied. If the credit report is rated poor, the application is denied. Loan applications for amounts between $2000 and $200,000 are divided into four categories: car, mortgage, education, and other. For car, mortgage, and other loan requests, the applicant's credit report is reviewed and an employment check is made to verify the applicant's reported salary income. If the credit report rating is poor, the loan is denied. If the credit report rating is fair, good, or excellent and the salary income is verified, the loan is approved. If the salary income is not verifiable, the applicant is contacted and additional information is requested. In this case, the loan application, along with the additional information, is sent to the vice president for review and a final loan decision. For educational loans, the educational institution the applicant will attend is contacted to determine the estimated cost of attendance. This amount is then compared to the amount of the loan requested in the application. If the requested amount exceeds the cost of attendance, the loan is denied. Otherwise, education loan requests for amounts between $2000 and $34,999 are approved if the applicant's credit rating is fair, good, or excellent. Education loan applications requesting amounts from $35,000 to $200,000 are approved only if the credit rating is good or excellent. All loan applications for amounts greater than $200,000 are sent to the vice president for review and approval. Use a decision table to represent this logic.

FIELD EXERCISES

1. Talk with a systems analyst who works at an organization. Ask the analyst to show you a complete set of DFDs from a current project. Interview the analyst about his or her views concerning DFDs and their usefulness for analysis.

2. Interview several people in an organization about a particular system. What is the system like now and what would they like to see changed? Create a complete set of DFDs for the system. Show your DFDs to some of the people you interviewed and ask for their reactions. What kinds of comments do they make? What kinds of suggestions?

3. Talk with a systems analyst who uses a CASE tool. Investigate what capabilities the CASE tool has for automatically checking for rule violations in DFDs. What reports can the CASE tool produce with error and warning messages to help analysts correct and improve DFDs?

4. Find out which, if any, drawing packages, word processors, forms design, and database management systems your university or company supports. Research these packages to determine how they might be used in the production of a project dictionary. For example, do the drawing packages include a set of standard DFD symbols in their graphic symbol palette?

5. At an organization with which you have contact, ask one or more employees to draw a "picture" of the business process they interact with at that organization. Ask them to draw the process using whatever format suits them. Ask them to depict in their diagram each of the components of the process and the flow of information among these components at the highest level of detail possible. What type of diagram have they drawn? In what ways does it resemble (and not resemble) a DFD? Why? When they have finished, help the employees to convert their diagram to a standard DFD. In what ways is the DFD stronger and/or weaker than the original diagram?

REFERENCES

Celko, J. 1987. "I. Data Flow Diagrams." *Computer Language* 4 (January): 41–43.

DeMarco, T. 1979. *Structured Analysis and System Specification.* Upper Saddle River, NJ: Prentice Hall.

Gane C., and T. Sarson. 1979. *Structured Systems Analysis.* Upper Saddle River, NJ: Prentice Hall.

Hammer, M., and J. Champy. 1993. *Reengineering the Corporation.* New York: Harper Business.

Vessey, I., and R. Weber. 1986. "Structured Tools and Conditional Logic." *Communications of the ACM* 29(1): 48–57.

Wieringa, R. 1998. "A Survey of Structured and Object-Oriented Software Specification Methods and Techniques." *ACM Computing Surveys* 30(4): 459–527.

Yourdon, E. 1989. *Managing the Structured Techniques*, 4th ed. Upper Saddle River, NJ: Prentice Hall.

Yourdon, E., and L. L. Constantine. 1979. *Structured Design.* Upper Saddle River, NJ: Prentice Hall.

APPENDIX 7A

Object-Oriented Analysis and Design

Use Cases*

Learning Objectives

After studying this section, you should be able to:

- Explain use cases and use case diagrams and how they can be used to model system functionality.

- Present the basic aspects of how to create written use cases.

- Discuss process modeling with use cases for electronic commerce applications.

Introduction

Here we will introduce you to use cases and use case diagrams. Use cases are a different way to model the functionality of a business process that facilitates the development of information systems to support that process. Although common in object-oriented systems analysis and design, use case modeling can also be used with more traditional methods for modeling business processes. After learning the basics about use cases, including use case diagrams and written use cases, you will also learn how process modeling can be done with use cases for the analysis of electronic commerce applications.

*The original version of this appendix was written by Professor Atish P. Sinha.

USE CASES

As Chapter 7 has shown, DFDs are powerful modeling tools that you can use to show a system's functionality and the flow of data necessary for the system to perform its functions. DFDs are not the only way to show functionality, of course. Another way is use case modeling. Use case modeling helps analysts analyze the functional requirements of a system. Use case modeling helps developers understand the functional requirements of the system without worrying about how those requirements will be implemented. The process is inherently iterative—analysts and users work together throughout the model development process to further refine their use case models. Although use case modeling is most often associated with object-oriented systems analysis and design, the concept is flexible enough that it can also be used within more traditional approaches. In this section of the chapter, you will learn about use cases, use case diagrams and their constituent parts, and written use cases.

What Is a Use Case?

Use case
A depiction of a system's behavior or functionality under various conditions as the system responds to requests from users.

A **use case** shows the behavior or functionality of a system (see Figure 7-26). It consists of a set of possible sequences of interactions between a system and a user in a particular environment, possible sequences that are related to a particular goal. A use case describes the behavior of a system under various conditions as the system responds to requests from principal actors. A principal actor initiates a request of the system, related to a goal, and the system responds. A use case can be stated as a present-tense verb phrase, containing the verb (what the system is supposed to do) and the object of the verb (what the system is to act on). For example, use case names would include Enter Sales Data, Compute Commission, Generate Quarterly Report. As with DFDs, use cases do not reflect all of the system requirements; they must be augmented by documents that detail requirements, such as business rules, data fields and formats, and complex formulas.

Actor
An external entity that interacts with a system.

A use case model consists of actors and use cases. An **actor** is an external entity that interacts with the system. It is someone or something that exchanges information with the system. For the most part, a use case represents a sequence of related actions initiated by an actor to accomplish a specific goal; it is a specific way of using the system (Jacobson et al., 1992). Note that there is a difference between an actor and a user. A user is anyone who uses the system. An actor, on the other hand, represents a role that a user can play. The actor's name should indicate that role. An actor is a type or class of users; a user is a specific instance of an actor class playing the

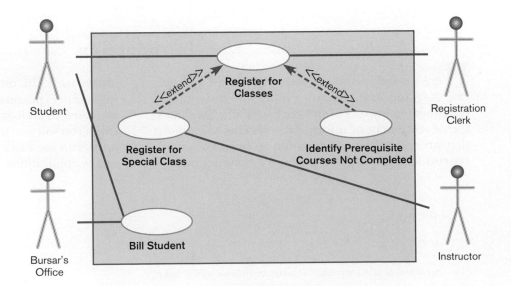

FIGURE 7-26
A use case diagram for a university registration system

actor's role. Note that the same user can play multiple roles. For example, if William Alvarez plays two roles, one as an instructor and the other as an adviser, we represent him as an instance of an actor called Instructor and as an instance of another actor called Adviser. Because actors are outside the system, you do not need to describe them in detail. The advantage of identifying actors is that it helps you to identify the use cases they carry out.

For identifying use cases, Jacobson et al. (1992) recommend that you ask the following questions:

- What are the main tasks performed by each actor?
- Will the actor read or update any information in the system?
- Will the actor have to inform the system about changes outside the system?
- Does the actor have to be informed of unexpected changes?

Use Case Diagrams

Use cases help you capture the functional requirements of a system. As you saw in Chapter 6, during the requirements analysis stage, the analyst sits down with the intended users of a system and makes a thorough analysis of what functions they desire from the system. When it comes time to structure these requirements, the identified system functions are represented as use cases. For example, a university registration system has a use case for class registration and another for student billing. These use cases, then, represent the typical interactions the system has with its users.

A **use case diagram** is depicted diagrammatically, as in Figure 7-26. It is a picture that shows system behavior, along with the key actors that interact with the system. The use case diagram in Figure 7-26 is for a university registration system, which is shown as a box. Outside the box are four actors—Student, Registration Clerk, Instructor, and Bursar's Office—that interact with the system. An actor is shown using a stick-figure symbol with its name below it. Inside the box are four use cases—Register for Classes, Register for Special Class, Identify Prereq Courses Not Completed, and Bill Student—which are shown as ellipses with their names underneath. These use cases are performed by the actors outside the system.

Typically, a use case is initiated by an actor. For example, Bill Student is initiated by the Bursar's Office. A use case can interact with actors other than the one that initiated it. The Bill Student use case, although initiated by the Bursar's Office, interacts with the Students by mailing them tuition invoices. Another use case, Register for Classes, is carried out by two actors, Student and Registration Clerk. This use case performs a series of related actions aimed at registering a student for a class. Although use cases are typically initiated by actors, in some circumstances a use case is initiated by another use case. Such use cases are called *abstract use cases*. We will discuss these in more detail later in this appendix.

A use case represents complete functionality. You should not represent an individual action that is part of an overall function as a use case. For example, although submitting a registration form and paying tuition are two actions performed by users (students) in the university registration system, we do not show them as use cases because they do not specify a complete course of events; each of these actions is executed only as part of an overall function or use case. You can think of Submit Registration Form as one of the actions of the Register for Classes use case and of Pay Tuition as one of the actions of the Bill Student use case.

Use case diagram

A picture showing system behavior, along with the key actors that interact with the system.

Definitions and Symbols

Use case diagramming is relatively simple because it involves only a few symbols. However, like DFDs and other relatively simple diagramming tools, these few symbols can be used to represent quite complex situations. Mastering use case diagramming takes lots of practice. The key symbols in a use case diagram are illustrated in Figure 7-26 and explained below:

- *Actor.* As explained earlier, an actor is a role, not an individual. Individuals are instances of actors. One particular individual may play many roles simultaneously. An actor is involved with the functioning of a system at some basic level. Actors are represented by stick figures.
- *Use case.* Each use case is represented as an ellipse. Each use case represents a single system function. The name of the use case can be listed inside the ellipse or just below it.
- *System boundary.* The system boundary is represented as a box that includes all of the relevant use cases. Note that actors are outside the system boundary.
- *Connections.* In Figure 7-26, note that the actors are connected to use cases with lines, and that use cases are connected to each other with arrows. A solid line connecting an actor to a use case shows that the actor is involved in that particular system function. The solid line does not mean that the actor is sending data to or receiving data from the use case. Note that all of the actors in a use case diagram are not involved in all the use cases in the system. The dotted-line arrows that connect use cases also have labels (there is an <<extend>> label on the arrows in Figure 7-26). These use case connections and their labels are explained next. Note that use cases do not have to be connected to other use cases. The arrows between use cases do not illustrate data or process flows.

Extend relationship
An association between two use cases where one adds new behaviors or actions to the other.

- *Extend relationship.* An **extend relationship** extends a use case by adding new behaviors or actions. It is shown as a dotted-line arrow pointing toward the use case that has been extended and labeled with the <<extend>> symbol. The dotted-line arrow does not indicate any kind of data or process flow between use cases. In Figure 7-26, for example, the Register for Special Class use case extends the Register for Classes use case by capturing the additional actions that need to be performed in registering a student for a special class. Registering for a special class requires prior permission of the instructor, in addition to the other steps carried out for a regular registration. You may think of Register for Classes as the basic course, which is always performed—independent of whether the extension is performed or not—and Register for Special Class as an alternative course, which is performed only under special circumstances.

Note also that the Instructor actor is needed for Register for Special Class. The Instructor is not needed for Register for Classes, which involves the Student and Registrar actors only. The reason for not including the Instructor for normal registration but including him or her for registering for special classes is that certain additional actions are required from the Instructor for a special class. The Instructor's approval is likely needed just to create an instance of a special class, and there may be other special requirements that need to be met for the class to be created. None of these special arrangements are necessary for normal registration, so the Instructor is not needed under normal circumstances.

Another example of an extend relationship is that between the Identify Prereq Courses Not Completed and Register for Classes use cases. The former extends the latter in situations where a student registering for a class has not taken the prerequisite courses.

Include relationship
An association between two use cases where one use case uses the functionality contained in the other.

- *Include relationship.* Another kind of relationship between use cases is an **include relationship**, which arises when one use case uses another use case. An include relationship is shown diagrammatically as a dotted-line arrow pointed toward the use case that is being used. The line is labeled with the . symbol. The dotted-line arrow does not indicate any kind of data or process flow between use cases. An include relationship implies that the use case where the arrow originates uses the use case where the arrow ends while it is executing. Typically, the use case that is "included" represents a generic function that is common to many

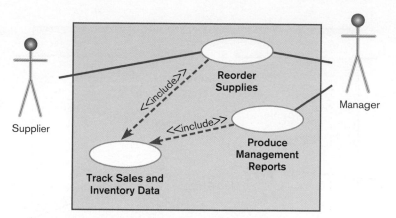

FIGURE 7-27
A use case diagram featuring an include relationship

business functions. Rather than reproduce that functionality within every use case that needs it, the functionality is factored out into a separate use case that can then be used by other use cases. An example of an include relationship is shown in Figure 7-27.

Figure 7-27 shows a generic use case diagram for any business that needs to reorder supplies on a regular basis, such as a retail establishment or a restaurant. Because this is a generic use case diagram, its use cases are high level. Three different use cases are identified in the figure: Reorder Supplies, Produce Management Reports, and Track Sales and Inventory Data. Two actors have been identified: Supplier and Manager. Reorder Supplies involves the Manager and Supplier actors. A manager initiates the use case, which then sends requests to suppliers for various items. The Produce Management Reports use case involves only the Manager actor. In Figure 7-27, the include relationship between the Reorder Supplies and Track Sales and Inventory Data use cases implies that the former uses the latter while executing. Simply put, when a manager reorders supplies, the sales and inventory data are tracked. The same data are also tracked when management reports are produced, so there is another include relationship between the Produce Management Reports and Track Sales and Inventory Data use cases.

The Track Sales and Inventory Data is a generalized use case, representing the common behavior among the specialized use cases Reorder Supplies and Produce Management Reports. When Reorder Supplies or Produce Management Reports is performed, the entire Track Sales and Inventory Data use case is used. Note, however, that it is used only when one of the specialized use cases is performed. Such a use case, which is never performed by itself, is called an abstract use case (Eriksson and Penker, 1998; Jacobson et al., 1992). An abstract case does not interact directly with an actor.

Figure 7-28 shows a use case diagram for Hoosier Burger. Several actors and use cases can be identified. The actor that first comes to mind is Customer, which represents the class of all customers who order food at Hoosier Burger; Order food is therefore represented as a use case. The other actor that is involved in this use case is Service Person. A specific scenario would represent a customer (an instance of Customer) placing an order with a service person (an instance of Service Person). At the end of each day, the manager of Hoosier Burger reorders supplies by calling suppliers. We represent this by a use case called Reorder Supplies, which involves the Manager and Supplier actors. A manager initiates the use case, which then sends requests to suppliers for various items.

Hoosier Burger also hires employees from time to time. Therefore, we have identified a use case, called Hire employee, in which two actors, Manager and Applicant, are involved. When a person applies for a job at Hoosier Burger, the manager makes a hiring decision.

FIGURE 7-28
Use case diagram for Hoosier Burger

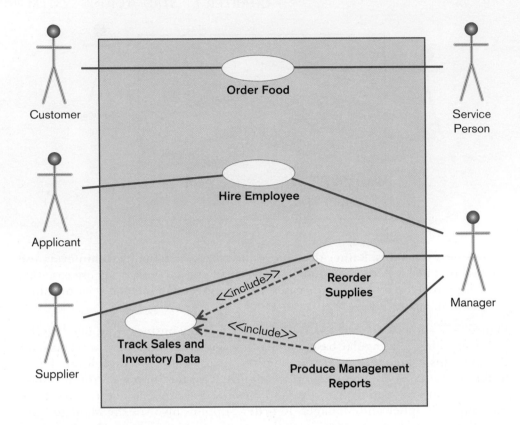

Figure 7-28 provides another example of an include relationship, shown diagrammatically as a dashed line pointing toward the use case that is being used; the line is labeled with the <<include>> symbol. In Figure 7-28, for example, the include relationship between the Reorder supplies and Track sales and inventory data use cases implies that the former uses the latter while executing. When a manager reorders supplies, the sales and inventory data are tracked. The same data are also tracked when management reports are produced, so there is another include relationship between the Produce management reports and Track sales and inventory data use cases.

Track sales and inventory data is a generalized use case, representing the common behavior among the specialized use cases Reorder supplies and Produce management reports. When a task like Reorder supplies or Produce management reports is performed, the entire Track sales and inventory data is used. Note, however, that it is used only when one of the specialized use cases is performed. As you will recall, such a use case, which is never performed by itself, is called an abstract use case (Eriksson and Penker, 1998; Jacobson et al., 1992).

WRITTEN USE CASES

Use case diagrams can represent the functionality of a system by showing use case names and the actors who are involved with them. The names of the use cases alone do not provide much of the information that is necessary to continue with analysis and to move on to the design phase. We also need to know what goes on inside each use case. The contents of a use case can be written in simple text, as was explained before for the Register for Classes use case in Figure 7-26. Others recommend templates that force consideration of all of the important information one needs to have about use cases.

Cockburn (2001) recommends a specific template for writing use cases (Figure 7-29). Templates can be simpler than the one Cockburn recommends or more

| Use Case Title: |
| Primary Actor: |
| Level: |
| Stakeholders: |
| Precondition: |
| Minimal Guarantee: |
| Success Guarantee: |
| Trigger: |
| Main Success Scenario: |
| Extensions: |

FIGURE 7-29
A template for writing use cases
(*Source:* Cockburn, Alistair, *Writing Effective Use Cases*, 1st ed., © 2001. Reprinted and Electronically reproduced by permission of Pearson Education, Inc. Upper Saddle River, New Jersey.)

complicated. The point is not the format of the template so much as it is how the template encourages analysts to write complete use cases. Each heading reminds the analyst of the information that needs to be provided. In the template in Figure 7-29, it should be clear what information is being sought. The use case title and the name of the primary actor role, both of which were featured in the discussion of use case diagrams, can be found on the use case diagram. The other information asked for in the template is new and will be discussed in more detail. The next section will deal exclusively with an important concept, the *level* of the use case. The following section will deal with the rest of the terms in the template.

Level

Level has to do with the level of detail at which the use case is being described. Level can range from *high* to low, where high is general and abstract, and low is detailed. Cockburn suggests five different levels of detail:

- White: As seen from the clouds, as if flying in a plane at 35,000 feet
- Kite: You're still in the air, but more detail than at cloud level.
- Blue: Also known as sea level.
- Fish: This is below sea level with a lot of detail. The detail increases deeper down, just like air pressure.
- Black: This is the bottom of the sea where the maximum amount of detail is provided.

Both the white and kite levels provide a summary of the use case goals. These goals are at a very high level. Goals at the white level are enterprise-wide, whereas at the kite level, the goals are those of a single business unit. Use cases at the white and kite levels are sometime called summary use cases. Summary use cases do not include functional requirements. Use cases written at the blue level, or sea level, focus on user goals: What is the user trying to achieve in interacting with the system? Use cases written at the fish and black levels (sometimes called the *clam level*) are much more detailed and focus on subfunction goals. To see how the levels relate to each other, think about the view of the Caribbean Sea you would get if you were flying over it in a big plane like a 757. You can't see the bottom of the sea, and at this altitude, you can't even see much detail about the surface of the water. This would be the white level. Then think about how the same stretch of the Caribbean would look from about 100 feet up. This is the kite level. From the kite level, you would be able to see a lot more detail on the surface, compared to being in the 757 jet, but you still can't see a lot of detail on the sea bottom, even with the water as clear as it is in a lot of the Caribbean. Now imagine the view of the same place from a rowboat. This is the user goal or sea level view. You can see the bottom much more clearly now, but it's still not completely clear. Now dive under

Level
Perspective from which a use case description is written, typically ranging from high level to extremely detailed.

FIGURE 7-30
Use case levels and detail when moving
from top to bottom
(*Source:* George, Hoffer, Valacich, Batra,
2006. *Object-Oriented Systems Analysis and
Design*, 2nd ed. Upper Saddle River, NJ:
Prentice Hall.)

 ☁	Buy parts to build cars
 🪁	Buy parts to build Escorts
 ▬	Order Escort parts from suppliers
 🐟	Choose supplier for part
 🦪	Encrypt data for secure transmission

the water and go down about 50 feet. You are a lot closer to the bottom—the fish level—and so now you can see a lot more detail at the bottom of the sea. But you don't see the most detail possible until you are sitting on the bottom itself—the black or clam level.

To put all this into a business function perspective, let's imagine five levels of use cases written for the Ford Motor Company. The white level use cases would serve an enterprise-wide goal ("Buy parts to build cars"), whereas a kite level use case would serve one business unit ("Buy parts to build Escorts"). If a system user has the role of procurement manager for the Escort model, the user goals at sea level might be "Order Escort parts from suppliers" and "Pay bills." A fish level goal for the procurement system might include "Choose supplier for a part." A black or clam level goal for the same system might include "Establish a secure connection." Figure 7-30 shows the relationships among the levels.

The Rest of the Template

Stakeholder
People who have a vested interest in the system being developed.

Next in the use case template is the list of **stakeholders**: those people who have some key interest in the development of the system. They would include the system's users as well as the manager, other managers in the company, customers, stockholders, the vendors that supply the company, and so on. Stakeholders are important to identify because they typically have some impact on what the system does and on how it is designed. It should be obvious that some stakeholders have more of a stake than others, and the most involved stakeholders are the ones that probably should be listened to first.

Preconditions
Things that must be true before a use case can start.

The next term in Cockburn's template (Figure 7-29) is **preconditions**, which are those things the system must ensure are true before the use case can start. For example, in Figure 7-26, for the use case Register for Classes, students would not be allowed to register if they had any outstanding debts due to the university. No outstanding debts would be listed under preconditions for Register for Classes in its written use case template.

Minimal guarantee
The least amount promised to the stakeholder by a use case.

Next is **minimal guarantee**. According to Cockburn, the minimal guarantee is the least amount promised by the use case to the stakeholder. One way to determine what this should be is to ask, "What would make the stakeholder unhappy?" For some use cases, the minimal guarantee might be simply Nothing happens. The stakeholder would be unhappy because the system does not do what it is supposed to. However, no detrimental effects result either; no bad data are entered into the system, no data are lost, and the system does not crash. For many use cases, the best thing to offer for a minimal guarantee is to roll back the transaction to its original starting place; nothing is gained but no harm is done either.

Success guarantee
What a use case must do effectively in order to satisfy stakeholders.

A **success guarantee** lists what it takes to satisfy stakeholders if the use case is completed successfully. For example, in Figure 7-26, for the use case Bill Student, a success guarantee would involve the successful compilation of charges due from the student and the successful creation of an accurate invoice that reflects those charges.

This does not imply that the student is *happy* with the result; he or she might think the charges are too high or too low (although rarely the latter). What is important is that the use case functioned correctly and achieved its goals.

Next is the slot in the template for **trigger**, the thing that initiates the use case. A trigger could be a phone call, a letter, or even a call from another use case. In the example of Bill Student, the trigger could be a message indicating that the class registration process was complete.

The last item in Cockburn's written use case template is **extensions**. Maybe the best way to think about an extension is as the "else statement" that follows an "if statement." An extension is invoked only if its associated condition is encountered. In a written use case, the conditions that invoke extensions usually refer to some type of system failure. For example, if a use case involves access through the Internet and a network failure occurs so that the Internet connection is lost, what happens? If the system requires a login and the user provides the wrong account name, what happens? If the user provides the wrong password, what happens? All of the actions that would follow these conditions would be listed in the use case template as extensions.

Figure 7-31 shows a use case diagram for a reservation system. Figure 7-32 shows a finished, written use case, based on the reservation use case diagram. This use case description is written at the kite, or summary, level, which means that it shows only the user goals rather than the functional requirements. You'll notice that five user goals are described, four of which are carried out by the customer, and this reflects the content of the use case diagram in Figure 7-31. Although Figure 7-31 is generic to any system that handles reservations, the written use case in Figure 7-32 is specific to hotel reservations. For hotel reservations made on the web, certain simplifying assumptions particular to hotel reservations have been made, such as customers being required to provide a one night's deposit in order to hold the reservation. You'll also notice that there is a list of extensions at the end of the written use case. There is at least one extension for each user goal, although the first function, searching for a room for a desired time period at a specific hotel, has two extensions. There is no set number of extensions required for a user goal. In fact, there is no requirement that a user goal has an extension at all.

Trigger
Event that initiates a use case.

Extension
The set of behaviors or functions in a use case that follow exceptions to the main success scenario.

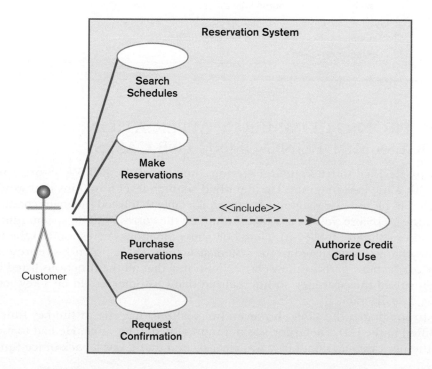

FIGURE 7-31
A use case diagram for a reservation system

(*Source:* George, Hoffer, Valacich, Batra, 2006. *Object-Oriented Systems Analysis and Design*, 2nd ed. Upper Saddle River, NJ: Prentice Hall.)

FIGURE 7-32

Kite level written use case, for making a hotel room reservation

(*Source:* George, Hoffer, Valacich, Batra, 2006. *Object-Oriented Systems Analysis and Design*, 2nd ed. Upper Saddle River, NJ: Prentice Hall.)

Use Case Title: Making a hotel room reservation
Primary Actor: Customer
Level: Kite (summary)
Stakeholders: Customer, credit bureau
Precondition: Customer accesses the hotel website
Minimal Guarantee: Rollback of any uncompleted transaction
Success Guarantees: Reservation held with one night's deposit
Trigger: Customer accesses hotel homepage

Main Success Scenario:
1. Customer searches for hotel location and room availability for desired time period.
2. Customer makes reservation for desired room for desired time period.
3. Customer holds reservation by authorizing a deposit for one night's stay.
4. Credit bureau verifies that customer has necessary credit for deposit.
5. Customer requests confirmation of reservations.

Extensions:
1a. Hotel property search function is not available.
  1a1. Customer quits site.
1b. Specific hotel room not available for desired time period.
  1b1. Customer quits site.
  1b2. Customer searches for different hotel for desired time period.
  1b3. Customer searches for same hotel for different time period.
2a. Making reservation transaction is interrupted.
  2a1. Transaction rolled back. Customer starts again.
  2a2. Transaction rolled back. Customer quits site.
3a. Holding reservation transaction is interrupted.
  3a1. Transaction rolled back. Customer starts again.
  3a2. Transaction rolled back. Customer quits site.
4a. Credit bureau cannot verify that customer has necessary credit.
  4a1. Customer notified of issue. Transaction rolled back. Customer quits site.
  4a2. Customer notified of issue. Transaction rolled back. Customer begins reservation process again with different credit card.
5a. Confirmation of transaction is interrupted.
  5a1. Customer seeks other means of confirmation.
  5a2. Customer quits site.

ELECTRONIC COMMERCE APPLICATION: PROCESS MODELING USING USE CASES

Jim Woo decided to try to model the functionality of the PVF WebStore application with a use case diagram. He identified six high-level functions that would be included in his use case diagram. Jim created a table that listed the main characteristics of the WebStore website in one column and the corresponding system functions in another column (Table 7-5). Note how these functions correspond to the major website characteristics listed in the system structure. These functions represent the "work" or "action" parts of the website. Jim noted that all the functions listed in his table involved the customer, so Jim realized that Customer would be a key actor in his use case diagram.

In looking at the table, however, Jim realized that one of the key functions identified in the JAD, Fill Order, was not represented in his table. He had to include it in the use case diagram, but it was clear to him that it was a back-office function

TABLE 7-5 **System Structure of WebStore and Corresponding Functions**

WebStore System	Functions
❏ Main Page	Browse Catalog
• Product Line (catalog)	
✓ Desks	
✓ Chairs	
✓ Tables	
✓ File Cabinets	
• Shopping Cart	Place Order
• Checkout	Place Order
• Account Profile	Maintain Account
• Order Status/History	Check Order
• Customer Comments	
❏ Company Information	
❏ Feedback	
❏ Contact Information	

and that it required adding another actor to the use case diagram. This actor would be the Shipping Clerk. Jim added Shipping Clerk to the right-hand side of his use case diagram. The finished diagram is shown in Figure 7-33.

WRITING USE CASES FOR PINE VALLEY FURNITURE'S WEBSTORE

Jim Woo was pleased with the use case diagram he created for WebStore (Figure 7-33). Now that he had identified all of the use cases necessary (he thought), he was ready to go back and start writing the use cases. The management in Pine Valley's Information Systems department had mandated that analysts use a standard template for writing use cases. Given his use case diagram, Jim decided to create two types of written use cases. The first would deal with the entire process of buying a PVF product on

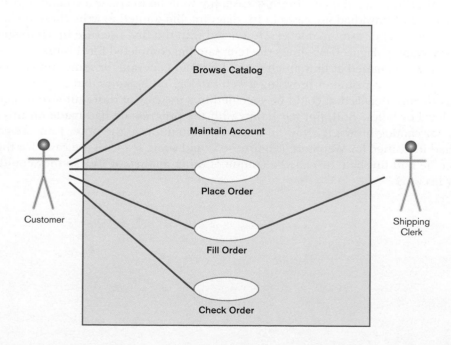

FIGURE 7-33
WebStore use case diagram

FIGURE 7-34
Jim Woo's kite level written use case for buying a product at PVF's WebStore

(*Source:* George, Hoffer, Valacich, Batra, 2006. *Object-Oriented Systems Analysis and Design,* 2nd ed. Upper Saddle River, NJ: Prentice Hall.)

Use Case Title: Buying a PVF Product at WebStore
Primary Actor: Customer
Level: Kite (summary)
Stakeholders: Customer, shipping clerk
Precondition: Customer accesses the WebStore website
Minimal Guarantee: Rollback of any uncompleted transaction
Success Guarantees: Order filled
Trigger: Customer accesses WebStore homepage
Main Success Scenario: 1. Customer browses catalog. 2. Customer places order for desired product(s). 3. Shipping clerk fills order. 4. Customer checks status of order.

Extensions:
1a. Catalog is not available.
    1a1. Customer quits site.
    1a2. Customer takes action to gain access to catalog.
2a. Order transaction is interrupted.
    2a1. Transaction rolled back. Customer starts again.
    2a2. Transaction rolled back. Customer quits site.
3a. Item is out of stock.
    3a1. Shipping clerk notifies customer. Customer waits for stock to be replenished.
    3a2. Shipping clerk notifies customer. Customer cancels order.
4a. Order status is not available.
    4a1. Customer quits site.
    4a2. Customer takes action to gain access to order status.

WebStore, as depicted in his use case diagram. This written use case would be at the kite level. It would be a summary use case and would not include functional requirements. The finished product is shown in Figure 7-34.

After finishing the kite level use case, Jim went on to create a couple of written use cases for individual use cases in his diagram. Jim wanted to write these use cases at the sea level or user goal level. He started with the first use case in his diagram, Browse catalog. Figure 7-35 shows the template Jim completed for this first use case.

Jim was amazed at how much detail he could generate for something as seemingly simple as a customer browsing a web catalog. Yet he knew that he had left out many details, details that could be specified in a use case at different levels, such as fish level or below. Still, Jim was happy with the progress he had made on this use case for catalog browsing. Now he turned his attention to the other four use cases he had identified for WebStore (Figure 7-33) and wrote sea level use cases for them. Once he had finished, he called a couple of other analysts at PVF so they could review his work.

Use Case Title: Browse catalog
Primary Actor: Customer
Level: Sea level (user goal)
Stakeholders: Customer
Precondition: Customer must be online with Web access
Minimal Guarantee: Rollback of any uncompleted transaction; system logs progress until failure
Success Guarantees: Files customer desires load correctly
Trigger: Customer accesses WebStore homepage
Main Success Scenario: 1. Cookie created on customer hard drive. 2. Customer selects category of item to view from list (e.g., home, office, patio). 3. Customer selects subcategory of item to view from list (e.g., home is subdivided into kitchen, dining room, bedroom, living room, den, etc.). 4. Customer selects specific item from list in subcategory to view (e.g., TV stand in den). 5. Customer selects specific item from list of products (e.g., Smith & Wesson TV stand). 6. Customer clicks on thumbnail photo of item to get regular-sized photo to view. 7. Customer selects "Product Specifications" to get detailed information on product. 8. Customer uses Web browser "Back" button to go back to see other products or other rooms or other types of furniture. 9. Customer selects from choices on menu bar to go elsewhere, either "Other Types of Furniture," "WebStore Home," or "PVF Home."

Extensions:
 	
1.a. Cookie cannot be created.
    1.a.1. Message created indicates to customer that browsing is not possible because his or her Web browser does not allow for the creation of cookies.
    1.a.2. Customer either adjusts the browser's cookie settings and tries again or leaves the site.
6.a. Full-sized photo does not load.
    6.a.1. Customer gets a broken-link symbol.
    6.a.2. Customer hits the refresh button and the photo loads successfully.
    6.a.3. Customer hits the refresh button and the photo does not load successfully; customer leaves the site.
2-7.a. The requested Web page does not load or cannot be found.
    2-7.a.1. Customer gets a "page not found" error page in browser.
    2-7.a.2. Customer hits the refresh button and the requested page loads successfully.
    2-7.a.3. Customer hits the refresh button and the requested page does not load successfully; customer leaves the site.

FIGURE 7-35
Jim Woo's completed template for PVF's Browse catalog use case
(*Source:* George, Hoffer, Valacich, Batra, 2006. *Object-Oriented Systems Analysis and Design*, 2nd ed. Upper Saddle River, NJ: Prentice Hall.)

SUMMARY

Use case modeling, featuring use case diagrams and written use cases, is another method you can use to model business processes. Use cases focus on system functionality and business processes, and they provide little, if any, information about how data flow through a system. In many ways, use case modeling complements DFD modeling. The use case approach provides another tool for analysts to use in structuring system requirements.

KEY TERMS

1. Actor	5. Level	9. Success guarantee
2. Extend relationship	6. Minimal guarantee	10. Trigger
3. Extension	7. Preconditions	11. Use case
4. Include relationship	8. Stakeholder	12. Use case diagram

Match each of the key terms above with the definition that best fits it.

_____ People who have a vested interest in the system being developed

_____ The least amount promised to the stakeholder by a use case.

_____ An association between two use cases where one adds new behaviors or actions to the other.

_____ An external entity that interacts with a system.

_____ Event that initiates a use case.

_____ A depiction of a system's behavior or functionality under various conditions as the system responds to requests from users.

_____ The set of behaviors or functions in a use case that follow exceptions to the main success scenario.

_____ A picture showing system behavior, along with the key actors that interact with the system.

_____ What a use case must do effectively in order to satisfy stakeholders.

_____ An association between two use cases where one use case uses the functionality contained in the other.

_____ Things that must be true before a use case can start.

_____ Perspective from which a use case description is written, typically ranging from high level to extremely detailed.

REVIEW QUESTIONS

1. What are use cases?
2. What is use case modeling?
3. What is a use case diagram?
4. What is a written use case and how does it compare to a use case diagram?
5. Explain an include relationship.
6. Explain an extend relationship.
7. Compare DFDs with use case diagrams.

8. What can a written description of a use case provide that a use case diagram cannot?
9. Describe Cockburn's template for a written use case.
10. List and explain the five levels from which use case descriptions can be written.
11. What is the difference between a minimal guarantee and a success guarantee?
12. What are extensions?

PROBLEMS AND EXERCISES

1. Draw a use case diagram for the situation described in Problem and Exercise 9, page 240.
2. Draw a use case diagram for the situation described in Problem and Exercise 10, page 240.
3. Draw a use case diagram for the situation described in Problem and Exercise 11 page 243.
4. Draw a use case diagram for the situation described in Problem and Exercise 12, page 243.
5. Draw a use case diagram for the situation described in Problem and Exercise 13, page 243.
6. Draw a use case diagram based on the level-0 diagram in Figure 7-23. How does your use case diagram for Figure 7-23 differ from the one in Figure 7-26, which is also about registering for classes? To what do you attribute the differences?
7. Develop a use case diagram for using an ATM machine to withdraw cash.

8. Develop a written use case for using an ATM machine to withdraw cash.
9. Choose a transaction that you are likely to encounter, perhaps ordering a cap and gown for graduation, and develop a use case diagram for it.
10. Choose a transaction that you are likely to encounter and develop a written use case for it.
11. The diagram in Figure 7-33 includes five use cases. In this chapter, Jim Woo wrote descriptions for one of them, Browse catalog. Prepare written descriptions for the other use cases in Figure 7-33.
12. An auto rental company wants to develop an automated system that can handle car reservations, customer billing, and car auctions. Usually a customer reserves a car, picks it up, and then returns it after a certain period of time. At the time of pickup, the customer has the option to buy or waive

collision insurance on the car. When the car is returned, the customer receives a bill and pays the specified amount. In addition to renting cars, every six months or so, the auto rental company auctions the cars that have accumulated over 20,000 miles. Draw a use case diagram for capturing the requirements of the system to be developed. Include an abstract use case for capturing the common behavior among any two use cases. Extend the diagram to capture corporate billing, where corporate customers are not billed directly; rather, the corporations they work for are billed and payments are made sometime later.

FIELD EXERCISE

1. At an organization with which you have contact, find an analyst who uses use case modeling. Find out how long he or she has been writing use cases and how he or she feels about use cases compared with DFDs.

REFERENCES

Cockburn, A. 2001. *Writing Effective Use Cases.* Reading, MA: Addison-Wesley.

Eriksson, H., and M. Penker. 1998. *UML Toolkit.* New York: Wiley.

Jacobson, I., M. Christerson, P. Jonsson, and G. Overgaard. 1992. *Object-Oriented Software Engineering: A Use-Case-Driven Approach.* Reading, MA: Addison-Wesley.

Object-Oriented Analysis and Design

Activity Diagrams*

Learning Objectives

After studying this section, you should be able to:

- Understand how to represent system logic with activity diagrams.

Introduction

An **activity diagram** shows the conditional logic for the sequence of system activities needed to accomplish a business process. An individual activity may be manual or automated. Further, each activity is the responsibility of a particular organizational unit.

Figure 7-36 illustrates a typical customer ordering process for a stock-to-order business, such as a catalog or an Internet sales company. Interactions with other business processes, such as replenishing inventory, forecasting sales, or analyzing profitability, are not shown.

In Figure 7-36, each column, called a swimlane, represents the organizational unit responsible for certain activities. The vertical axis is time, but without a time scale (i.e., the distance between symbols implies nothing about the absolute amount of time passing). The process starts at the large dot. Activities are represented by rectangles with rounded corners. A fork means that several parallel, independent sequences of activities are initiated (such as after the Receive Order activity), and a join (such as before the Send Invoice activity) signifies that independent streams of activities now must all reach completion to move on to the next step.

*The original version of this appendix was written by Professor Atish P. Sinha.

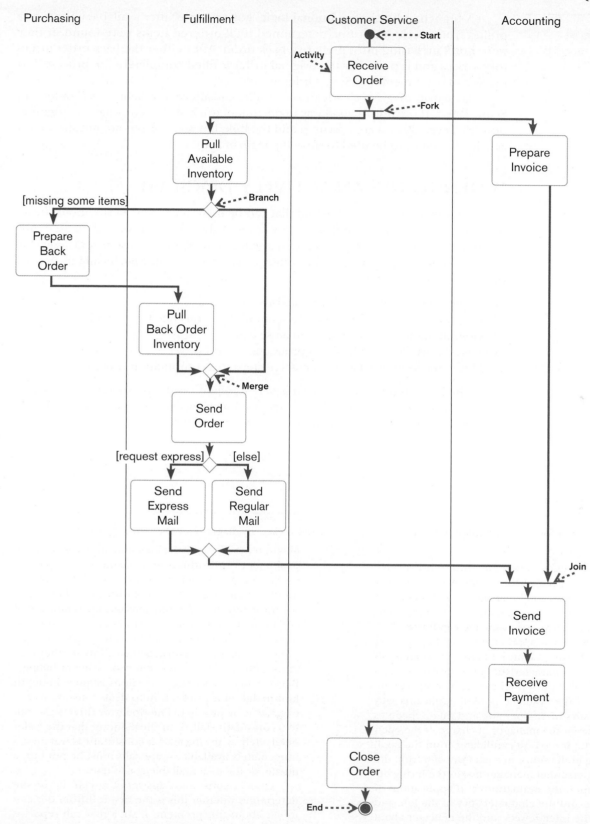

FIGURE 7-36
Activity diagram for a customer order process

A branch indicates conditional logic. For example, after available inventory is pulled from stock, it must be determined if all ordered items were found. If they were not, Purchasing must prepare a back order. After either the back order inventory arrives and is pulled or the original order is filled completely, the process flow merges to continue to the Send Order activity.

An activity diagram clearly shows parallel and alternative behaviors (Fowler and Scott, 1999). It provides a good way to document work or flows through an organization. However, objects are obscured and the links between objects are not shown. An activity diagram can be used to show the logic of a use case.

WHEN TO USE AN ACTIVITY DIAGRAM

An activity diagram is a flexible tool that can be used in a variety of situations. It can be used at a high level as well as at a low level of abstraction. It should be used only when it adds value to the project. Our recommendation is to use it sparingly. Ask the following question: Does it add value, or is it redundant? Specifically, an activity diagram can be used to:

1. Depict the flow of control from activity to activity.
2. Help in use case analysis to understand what actions need to take place.
3. Help in identifying extensions in a use case.
4. Model work flow and business processes.
5. Model the sequential and concurrent steps in a computation process.

The interpretation of the term *activity* depends on the perspective from which one is drawing the diagram. At a conceptual level, an activity is a task that needs to be done, whether by a human or a computer (Fowler and Scott, 1999). At an implementation level, an activity is a method or a class.

PROBLEMS AND EXERCISES

1. Draw an activity diagram for the following employee hiring process.

Projects, Inc., is an engineering firm with approximately 500 engineers in different specialties. New employees are hired by the personnel manager, based on data in an application form and evaluations collected from other managers who interview the job candidates. Prospective employees may apply at any time. Engineering managers notify the personnel manager when a job opens and list the characteristics necessary to be eligible for the job. The personnel manager compares the qualifications of the available pool of applicants with the characteristics of an open job and then schedules interviews between the manager in charge of the open position and the three best candidates from the pool. After receiving evaluations on each interview from the manager, the personnel manager makes the hiring decision based upon the evaluations and applications of the candidates and the characteristics of the job, and then notifies the interviewees and the manager about the decision. Applications of rejected applicants are retained for 1 year, after which time the application is purged. When hired, a new engineer completes a nondisclosure agreement, which is filed with other information about the employee.

2. Draw an activity diagram for the following case.

Maximum Software develops and supplies software products to individuals and businesses. As part of its operations, Maximum provides an 800 telephone number help desk for clients with questions about software purchased from Maximum. When a call comes in, an operator inquires about the nature of the call. For calls that are not truly help desk functions, the operator redirects the call to another unit of the company (such as order processing or billing). Because many customer questions require in-depth knowledge of a product, help desk consultants are organized by product. The operator directs the call to a consultant skilled on the software that the caller needs help with. Because a consultant is not always immediately available, some calls must be put into a queue for the next available consultant.

Once a consultant answers the call, he or she determines whether this is the first call from this customer about this problem. If so, a new call report is created to keep track of all information about the problem. If not, the customer is asked for a call report number so the consultant can retrieve the open call report to determine the status of the inquiry. If the caller does not know the call report number, the consultant collects other identifying information such as the caller's name, the software involved, or the name of the consultant who has handled the previous calls on the problem in order to conduct a search for the appropriate call report. If a resolution of the customer's problem has been found, the consultant informs the client what that resolution is, indicates on the report that the customer has been notified, and closes out the report. If a resolution has not been discovered, the consultant finds out whether the consultant handling this problem is on duty. If so, the call is transferred to the other consultant (or the call is put into the queue of calls waiting to be handled by that consultant).

Once the proper consultant receives the call, any new details the customer may have are recorded. For continuing problems and for new call reports, the consultant tries to discover an answer to the problem by using the relevant software and looking up information in reference manuals. If the problem can be resolved, the customer is told how to deal with the problem, and the call report is closed. Otherwise, the consultant files the report for continued research and tells the customer that someone at Maximum will be in touch, or if the customer discovers new information about the problem, he or she can call back, identifying the problem with a specified call report number.

REFERENCE

Fowler, M., and K. Scott. 1999. *UML Distilled*, 2nd ed. Reading, MA: Addison-Wesley.

Object-Oriented Analysis and Design

Sequence Diagrams*

Learning Objectives

After studying this section, you should be able to:

- Understand how to represent system logic with sequence diagrams.

Introduction

In this section on object-oriented analysis and design, we will introduce you to sequence diagrams. We will first show how to design some of the use cases we identified earlier in the analysis phase (Chapter 7), using *sequence diagrams*. A use case design describes how each use case is performed by a set of communicating objects (Jacobson et al., 1992). In UML, an interaction diagram is used to show the pattern of interactions among objects for a particular use case. There are two types of interaction diagrams: sequence diagrams and collaboration diagrams (Object Management Group, 2008). Both express similar information, but they do so in different ways. Whereas sequence diagrams show the explicit sequencing of messages, collaboration diagrams show the relationships among objects. In the next section, we will show you how to design use cases using sequence diagrams.

*The original version of this appendix was written by Professor Atish P. Sinha.

DYNAMIC MODELING: SEQUENCE DIAGRAMS

A **sequence diagram** depicts the interactions among objects during a certain period of time. Because the pattern of interactions varies from one use case to another, each sequence diagram shows only the interactions pertinent to a specific use case. It shows the participating objects by their lifelines and the interactions among those objects—arranged in time sequence—by the messages they exchange with one another.

A sequence diagram may be presented either in a generic form or in an instance form. The generic form shows all possible sequences of interactions, that is, the sequences corresponding to all the scenarios of a use case. For example, a generic sequence diagram for the Class registration use case (see Figure 7-26) would capture the sequence of interactions for every valid scenario of that use case. The instance form, on the other hand, shows the sequence for only one scenario. A scenario in UML refers to a single path, among many possible different paths, through a use case (Fowler, 2003). A path represents a specific combination of conditions within the use case. In Figure 7-37, a sequence diagram is shown, in instance form, for a scenario in which a student registers for a course that specifies one or more prerequisite courses as requirements.

The vertical axis of the diagram represents time, and the horizontal axis represents the various participating objects. Time increases as we go down the vertical axis. The diagram has six objects, from an instance of Registration Window on the left to an instance of Registration called "a New Registration" on the right. The ordering of the objects has no significance. However, you should try to arrange the objects so

Sequence diagram
Depicts the interactions among objects during a certain period of time.

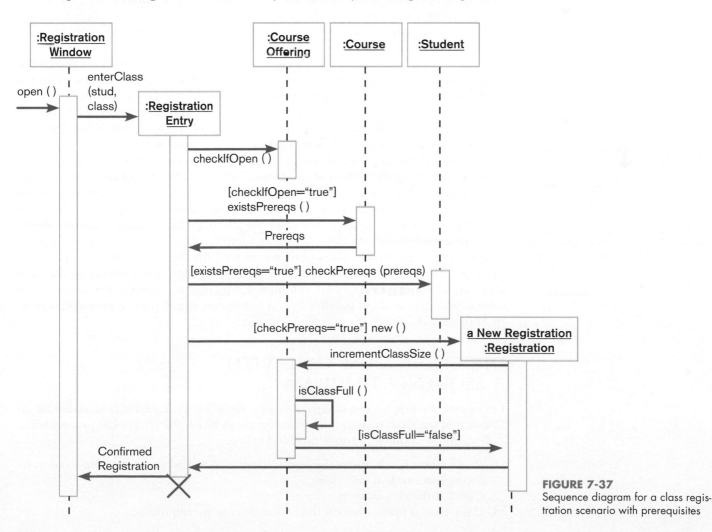

FIGURE 7-37
Sequence diagram for a class registration scenario with prerequisites

that the diagram is easy to read and understand. Each object is shown as a vertical dashed line called the lifeline; the lifeline represents the object's existence over a certain period of time. An object symbol—a box with the object's name underlined—is placed at the head of each lifeline.

A thin rectangle superimposed on the lifeline of an object represents an activation of the object. An **activation** shows the time period during which an object performs an operation, either directly or through a call to some subordinate operation. The top of the rectangle, which is at the tip of an incoming message, indicates the initiation of the activation; the bottom indicates its completion.

Objects communicate with one another by sending messages. A message is shown as a solid arrow from the sending object to the receiving object. For example, the checkIfOpen message is represented by an arrow from the Registration Entry object to the Course Offering object. When the arrow points directly into an object box, a new instance of that object is created. Normally the arrow is drawn horizontally, but in some situations (discussed later), you may have to draw a sloping message line.

Messages can be of different types (Object Management Group, 2008). Each type is indicated in a diagram by a particular type of arrowhead. A **synchronous message**, shown as a full, solid arrowhead, is one where the caller has to wait for the receiving object to complete execution of the called operation before it itself can resume execution. An example of a synchronous message is checkIfOpen. When a Registration Entry object sends this message to a Course Offering object, the latter responds by executing an operation called checkIfOpen (same name as the message). After the execution of this operation is complete, control is transferred back to the calling operation within Registration Entry with a return value of "true" or "false."

A synchronous message always has an associated return message. The message may provide the caller with some return value(s) or simply acknowledge to the caller that the operation called has been successfully completed. We have not shown the return for the checkIfOpen message; it is implicit. We have explicitly shown the return for the existsPrereqs message from Registration Entry to Course. The tail of the return message is aligned with the base of the activation rectangle for the existsPrereqs operation. The message returns the list of prerequisites, if any, for the course in question. Return messages, if shown, unnecessarily clutter the diagram; you can show only the ones that help in understanding the sequence of interactions.

A **simple message** simply transfers control from the sender to the recipient without describing the details of the communication. In a diagram, the arrowhead for a simple message is drawn as a transverse tick mark. As we have seen, the return of a synchronous message is a simple message. The "open" message in Figure 7-37 is also a simple message; it simply transfers control to the Registration Window object.

An **asynchronous message**, shown as a half arrowhead in a sequence diagram, is one where the sender does not have to wait for the recipient to handle the message. The sender can continue executing immediately after sending the message. Asynchronous messages are common in concurrent, real-time systems, in which several objects operate in parallel. We do not discuss asynchronous messages further in Appendix 7C.

DESIGNING A USE CASE WITH A SEQUENCE DIAGRAM

Let us now see how we can design use cases. We will draw a sequence diagram for an instance of the Class registration use case, one in which the course has prerequisites. A description of this scenario is provided below.

1. Registration Clerk opens the registration window and enters the registration information (student and class).
2. Check if the class is open.
3. If the class is open, check if the course has any prerequisites.

Activation
The time period during which an object performs an operation.

Synchronous message
A type of message in which the caller has to wait for the receiving object to finish executing the called operation before it can resume execution itself.

Simple message
A message that transfers control from the sender to the recipient without describing the details of the communication.

Asynchronous message
A message in which the sender does not have to wait for the recipient to handle the message.

4. If the course has prerequisites, then check if the student has taken all of those prerequisites.
5. If the student has taken those prerequisites, then register the student for the class and increment the class size by one.
6. Check if the class is full; if not, do nothing.
7. Display the confirmed registration in the registration window.

The diagram of Figure 7-37 shows the sequence of interactions for this scenario. In response to the "open" message from Registration Clerk (external actor), the registration window pops up on the screen and the registration information is entered. This creates a new Registration Entry object, which then sends a checkIfOpen message to the Course Offering object (representing the class the student wants to register for). There are two possible return values: "true" or "false." In this scenario, the assumption is that the class is open. We have therefore placed a guard condition, checkIfOpen ="true," on the message existsPrereqs. The guard condition ensures that the message will be sent only if the class is open. The return value is a list of prerequisites; the return is shown explicitly in the diagram.

For this scenario, the fact that the course has prerequisites is captured by the guard condition existsPrereqs = "true." If this condition is satisfied, the Registration Entry object sends a checkPrereqs message, with "prereqs" as an argument, to the Student object to determine if the student has taken those prerequisites. If the student has taken all the prerequisites, the Registration Entry object creates an object called "a New Registration," which denotes a new registration.

Next, "a New Registration" sends a message called incrementClassSize to Course Offering in order to increase the class size by one. The incrementClassSize operation within Course Offering then calls upon isClassFull, another operation within the same object; this is known as *self-delegation* (Fowler, 2003). Assuming that the class is not full, the isClassFull operation returns control to the calling operation with a value of "false." Next, the incrementClassSize operation completes and relinquishes control to the calling operation within "a New Registration."

Finally, on receipt of the return message from "a New Registration," the Registration Entry object destroys itself (the destruction is shown with a large X) and sends a confirmation of the registration to the registration window. Note that Registration Entry is not a persistent object; it is created on the fly to control the sequence of interactions and is deleted as soon as the registration is completed. In between, it calls several other operations within other objects by sequencing the following messages: checkIfOpen, existsPrereqs, checkPrereqs, and new. Hence, Registration Entry may be viewed as a control object (Jacobson et al., 1992).

Apart from the Registration Entry object, "a New Registration" is also created during the time period captured in the diagram. The messages that created these objects are represented by arrows pointing directly toward the object symbols. For example, the arrow representing the message called "new" is connected to the object symbol for "a New Registration." The lifeline of such an object begins when the message that creates it is received (the dashed vertical line is hidden behind the activation rectangle).

As we discussed before, the Registration Entry object is destroyed at the point marked by X. The lifeline of this object, therefore, extends from the point of creation to the point of destruction. For objects that are neither created nor destroyed during the time period captured in the diagram—for example, Course Offering, Course, and Student—the lifelines extend from the top to the bottom of the diagram.

Figure 7-38 shows the sequence diagram for a slightly different scenario—when a student registers for a course without any prerequisites. Notice that the guard condition to be satisfied for creating "a New Registration," existsPrereqs ="false," is different from that in the previous scenario. Also, because there is no need to check if the student has taken the prerequisites, there is no need to send the checkPrereqs message to Student. Thus, the Student object does not participate in this scenario.

FIGURE 7-38
Sequence diagram for a class
registration scenario without
prerequisites

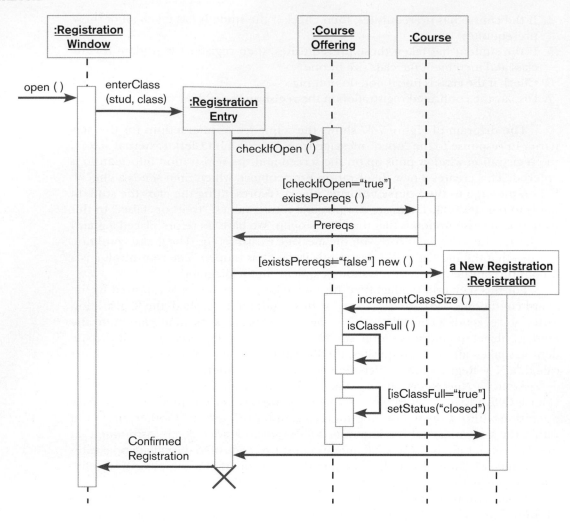

There is another difference between this scenario and the previous one. In this scenario, when the incrementClassSize operation within Course Offering calls isClassFull, the value returned is "true." Before returning control to "a New Registration," the incrementClassSize operation self-delegates again, this time calling setStatus to set the status of the class to "closed."

Both of the sequence diagrams we have seen so far are in instance form. In Figure 7-39, we present a sequence diagram in generic form. This diagram encompasses all possible combinations of conditions for the Prereq courses not completed use case (see Figure 7-26). Because this use case is an extension of the Class registration use case, we have not shown the Registration Window object. It is assumed that the Registration Entry object has already been created by the original use case. To improve understandability, we have provided textual description in the left margin. You may provide such descriptions in either the left or the right margins, but try to align the text horizontally with the corresponding element in the diagram. The contents of the use case are described as follows:

1. If the student has not taken one or more of the prerequisites for the course he or she wants to register for, check if the student has been granted a waiver for each of those prerequisites.
2. If a waiver was not granted for one or more of the prerequisites not taken, then check if the student tested out of each of those prerequisites by taking an exam.
3. If the student did not test out of any of those prerequisites, then deny registration. Otherwise, register the student for the class and provide a confirmation.

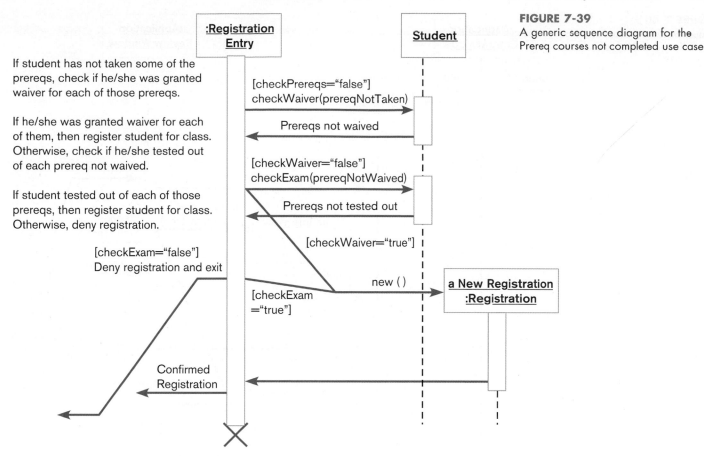

FIGURE 7-39
A generic sequence diagram for the
Prereq courses not completed use case

Because this use case extension pertains only to those registration situations where a student has not taken the prerequisite courses, we have placed a guard condition, checkPrereqs = "false," on the checkWaiver message from Registration Entry to Student. This message invokes the checkWaiver operation within Student to find out if the student has been granted waivers on all the prerequisites he or she has not taken. Note that the operation has to be applied to each of the prerequisites not taken. The iteration is described in the text on the left margin.

The diagram also exhibits branching, with multiple arrows leaving a single point. Each branch is labeled by a guard condition. The first instance of branching is based on the value returned by the checkWaiver operation. If checkWaiver ="true," the system creates "a New Registration" object, bypassing other operations. If checkWaiver ="false"—meaning that some of the prerequisites in question were not waived—Registration Entry sends another message, checkExam, to Student to check if he or she tested out of each of the prerequisite courses not waived.

There is another instance of branching at this point. If checkExam ="false," Registration Entry sends a message (to Registration Window), denying the registration and exiting the system. We have deliberately bent the message line downward to show that none of the other remaining interactions take place. If checkExam = "true," then "a New Registration" is created.

A SEQUENCE DIAGRAM FOR HOOSIER BURGER

In Figure 7-40, we show another sequence diagram, in generic form, for Hoosier Burger's Hire employee use case (see Figure 7-28). The description of the use case follows:

1. On receipt of an application for a job at Hoosier Burger, the data relating to the applicant are entered through the application entry window.
2. The manager opens the application review window and reviews the application.

FIGURE 7-40
Sequence diagram for Hoosier
Burger's Hire employee use case

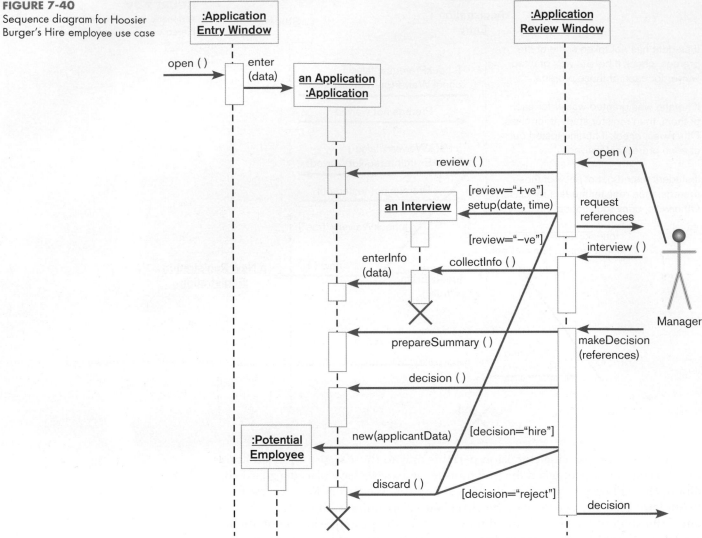

3. If the initial review is negative, the manager discards the application and conveys the rejection decision to the applicant. No further processing of the application is involved.

4. If the initial review is positive, then the manager sets up a date and time to interview the applicant. The manager also requests that the references specified in the application provide recommendation letters.

5. The manager interviews the candidate and enters the additional information gathered during the interview into the application file.

6. When the recommendation letters come in, the manager is ready to make a decision. First, he or she prepares a summary of the application. Based on the summary, he or she then makes a decision. If the decision is to reject the candidate, the application is discarded and the decision is conveyed to the applicant. The processing of the application comes to an end.

7. If the decision is to hire the candidate, a potential employee file is created and all relevant information about the candidate (e.g., name, Social Security number, birth date, address, phone number, etc.) is entered into this file. The hiring decision is conveyed to the applicant.

In the sequence diagram for this use case, we have explicitly shown Manager as an external actor. The branching after the return value from the review message is received represents the two options the Manager has. If review equals "+ve," then an object called "an Interview" is created through the setup message shown in the upper branch. We have shown the arguments to the message—date and time—because their values are required to set up an interview. Notice that if review equals "–ve" (lower branch), the discard message is sent to destroy the Application. The operations in between, for example, enterInfo, prepareSummary, and so forth, are completely bypassed.

Note that within the "an Interview" object created by the setup operation, there is another operation called collectInfo, which is invoked when the object receives the collectInfo message from the Application Review Window. The operation collects all of the relevant information during the interview and enters this information into an Application. After "an Interview" receives a successful return message (not shown) from "an Application," it self-destructs because there is no longer a need for it.

Next, Manager sends a makeDecision message, which invokes a corresponding operation within Application Review Window. This operation first sends a prepare-Summary message to "an Application," followed by another called "decision" to the same object. There is branching again at this point, depending on the return value. If decision equals "hire," then a message called "new" is sent to create an instance of Potential Employee, which stores the relevant applicant data. If decision equals "reject," the discard operation destroys "an Application." In either case, the decision is conveyed to the applicant.

SUMMARY

In this appendix, we showed you how to design use cases by drawing sequence diagrams. A sequence diagram is an invaluable tool for specifying and understanding the flow of control. When coding the system, sequence diagrams help you to effectively and easily capture the dynamic aspects of the system by implementing the operations, messages, and the sequencing of those messages in the target programming language.

KEY TERMS

1. Activation
2. Asynchronous message
3. Sequence diagram
4. Simple message
5. Synchronous message

Match each of the key terms above with the definition that best fits it.

_____ Depicts the interactions among objects during a certain period of time.

_____ The time period during which an object performs an operation.

_____ A type of message in which the caller has to wait for the receiving object to finish executing the called operation before it can resume execution.

_____ A message that transfers control from the sender to the recipient without describing the details of the communication.

_____ A message in which the sender does not have to wait for the recipient to handle the message.

REVIEW QUESTIONS

1. Contrast the following terms (you will have to use what you learned in the object-oriented sections of Chapters 7 and 8 to contrast all of these terms):

   a. Actor; use case
   b. Extends relationship; uses relationship
   c. Object class; object
   d. Attribute; operation
   e. Operation; method
   f. Query operation; update operation
   g. Abstract class; concrete class
   h. Class diagram; object diagram
   i. Association; aggregation
   j. Generalization; aggregation
   k. Aggregation; composition
   l. Generic sequence diagram; instance sequence diagram
   m. Synchronous message; asynchronous message
   n. Sequence diagram; activity diagram

2. State the activities involved in each of the following phases of the object-oriented development life cycle: object-oriented analysis, object-oriented design, and object-oriented implementation.

3. Compare and contrast the object-oriented analysis and design models with the structured analysis and design models.

PROBLEMS AND EXERCISES

You will need to consult Chapters 7 and 8 to work Problems and Exercises 1 and 2.

1. Draw a use case diagram, as well as a class diagram, for the following situation (state any assumptions you believe you have to make in order to develop a complete diagram).

> Stillwater Antiques buys and sells one-of-a-kind antiques of all kinds (e.g., furniture, jewelry, china, and clothing). Each item is uniquely identified by an item number and is also characterized by a description, an asking price, and a condition as well as open-ended comments. Stillwater works with many different individuals, called *clients*, who sell items to and buy items from the store. Some clients only sell items to Stillwater, some only buy items, and others both sell and buy. A client is identified by a client number and is also described by a client name and client address. When Stillwater sells an item in stock to a client, the owners want to record the commission paid, the actual selling price, the sales tax (tax of zero indicates a tax-exempt sale), and the date the item sold. When Stillwater buys an item from a client, the owners want to record the purchase cost, the purchase date, and the condition of the item at the time of purchase.

2. Draw a use case diagram, as well as a class diagram, for the following situation (state any assumptions you believe you have to make in order to develop a complete diagram).

> The H. I. Topi School of Business operates international business programs in ten locations throughout Europe. The school had its first class of 9000 graduates in 1965.

The school keeps track of each graduate's student number, name, country of birth, current country of citizenship, current name, current address, and the name of each major the student completed (each student has one or two majors). To maintain strong ties to its alumni, the school holds various events around the world. Events have a title, date, location, and type (e.g., reception, dinner, or seminar). The school needs to keep track of which graduates have attended which events. When a graduate attends an event, a comment is recorded about the information school officials learned from that graduate at that event. The school also keeps in contact with graduates by mail, e-mail, telephone, and fax interactions. As with events, the school records information learned from the graduate from each of these contacts. When a school official knows that he or she will be meeting or talking to a graduate, a report is produced showing the latest information about that graduate and the information learned during the past two years from that graduate from all contacts and events the graduate attended.

3. See Problem and Exercise 12 in Appendix 7A. One of the use cases for the auto rental system in this exercise is "Car reservation." Draw a sequence diagram, in instance form, to describe the sequence of interactions for each of the following scenarios of this use case:

a. Car is available during the specified time period.

b. No car in the desired category (e.g., compact, midsize, etc.) is available during the specified time period.

FIELD EXERCISE

1. Interview a systems analyst at a local company that uses object-oriented programming and system development tools. Ask to see any analysis and design diagrams he or she has drawn of their databases and applications. Compare these diagrams to the ones in this chapter. What differences do you see? What additional features and notations are used, and what are their purposes?

REFERENCES

Booch, G., R. A. Maksimchuk, M. W. Engel, B. J. Young, J. Collallen, and K. A. Houston. 2007. *Object-Oriented Analysis and Design with Applications*, 3rd ed. Redwood City, CA: Addison Wesley Professional.

Coad, P., and E. Yourdon. 1991a. *Object-Oriented Analysis*, 2nd ed. Upper Saddle River, NJ: Prentice Hall.

Coad, P., and E. Yourdon. 1991b. *Object-Oriented Design*. Upper Saddle River, NJ: Prentice Hall.

Erikson, H., M. Penker, B. Lyons, and D. Fado. 2003. *UML 2 Toolkit*. New York: John Wiley.

Fowler, M. 2003. *UML Distilled: A Brief Guide to the Standard Object Modeling Language*, 3rd ed. Reading, MA: Addison-Wesley.

Jacobson, I., M. Christerson, P. Jonsson, and G. Overgaard. 1992. *Object-Oriented Software Engineering: A Use-Case Driven Approach*. Reading, MA: Addison-Wesley.

Object Management Group. 2008. Unified Modeling Language Notation Guide. Version 2.0. Available at *www.omg.org*. Accessed on February 17, 2009.

Object Management Group. 2009. Unified Modeling Language Document Set. Version 2.2. Available at *www.omg.org/spec /UML/2.2/*. Accessed on February 17, 2009.

Rumbaugh, J., M. Blaha, W. Premerlani, F. Eddy, and W. Lorensen. 1991. *Object-Oriented Modeling and Design*. Upper Saddle River, NJ: Prentice Hall.

Business Process Modeling

Learning Objective

After studying this section, you should be able to:

- Understand how to represent business processes with business process diagrams.

Introduction

At the heart of just about any information system developed for organizations, there is a business process. A business process is a standard method for accomplishing a particular task necessary for an organization to function. A business process can come from any business function, from accounting to supply chain management to after-sales service. It can cross business functions as well. A business process can be simple or complex, but the more complex it is, the harder it is to automate. Complexity also makes a process more difficult to understand for those who are not familiar with it. Communication tools are needed to describe business processes to those who need to know about

them, such as systems analysts, but who have no first-hand knowledge of the processes. There are many ways to represent business processes, from data flow diagrams to flow charts. The Object Management Group (OMG), the same group that is responsible for standards for object-oriented programming, has established a specific modeling approach for business processes. It is called Business Process Modeling Notation (BPMN). This appendix provides a very brief introduction to BPMN. First, we will introduce you to the basic notation in BPMN, and second, we will provide an example. If you are interested in mastering BPMN, there are many materials available (see the reference list).

BASIC NOTATION

Business Process Modeling Notation is much more complicated than data flow diagrams notation, as it is made up of many more symbols, and each symbol has numerous variations. (The interested reader is referred to the BPMN standards and numerous other documents to learn about all of the various aspects of the complete BPMN standard. See the reference list at the end of the appendix.) However, there are four basic concepts in BPMN, each of which has its own basic symbol. These basic concepts are events, activities, gateways, and flows. Their symbols are as follows:

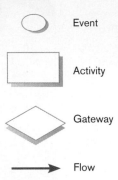

Event
In business process modeling, a trigger that initiates the start of a process.

Activity
In business process modeling, an action that must take place for a process to be completed.

Gateway
In business process modeling, a decision point.

Flow
In business process modeling, it shows the sequence of action in a process.

All business processes begin and end with an **event**. The symbol for an event is a circle. For a starting event, the walls of the circle are thin. For the ending event, the walls are thicker. A starting event can be colored green, and an ending event can be colored red. An **activity** is some action that must take place for the process to be completed. An activity can be completed by people or by a computerized system. The symbol for an activity is a rectangle with rounded edges. A **gateway**, symbolized by a diamond, is a decision point. The final primary concept is **flow**, represented by an arrow. Flow shows sequence, the order in which activities occur. A simple example, without content, of a process represented by BPMN, is as follows:

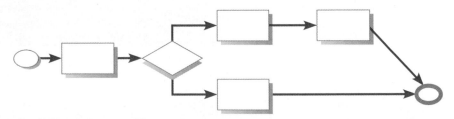

In this simple example, you see that the business process starts with some event, shown with an event symbol on the left. BPMN diagrams are always read from left to right. The event is followed by the first activity. An arrow symbolizing flow connects the event to the activity. The first activity is followed by a gateway. This is a decision point, indicated by two choices, "Yes" and "No." Some condition is associated with the gateway, and that condition can either be met (Yes) or not (No). Whether the condition is met or not determines where the flow goes next in the diagram. Both conditions lead to an additional activity. If the flow goes through the top of the diagram, there is one more activity that takes place before the process ends at its ending event. If the flow goes through the bottom part of the diagram, the process concludes after just one activity is completed. Note how the walls of the circle that represents the final event are much thicker than those of the circle representing the beginning event.

Typically gateways are exclusive, which is to say that flow must follow only one path out of the gateway and only one downstream activity can take place. However, a gateway may be inclusive, which means that more than one downstream activity can occur. If a gateway is inclusive, the downstream activities that follow it must also be followed by a merge gateway, where all the flows come back together. Such a situation would look like this:

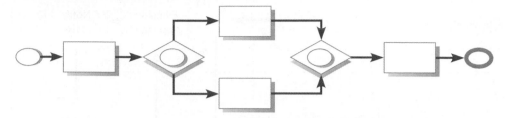

In this example, both or either of the activities that follow the gateway could occur. The merge gateway follows the activities, reuniting the two possible flows into a single flow. Note that the symbol for an inclusive gateway adds a circle inside the diamond. If it is necessary to indicate an exclusive gateway, you can add an "X" to the inside of the diamond.

Although it is beyond the scope of this appendix to introduce all of the specialized varieties of the four basic concepts in BPMN, it is useful to present a few varieties of some of the concepts. For example, here are the symbols for a couple of types of events. They both feature the basic circle as a symbol for an event, but each one has something inside. The first has an envelope inside, and the envelope stands for a message. An event shown like this, at the beginning of a process, means that the process starts with a message. A message is a basic flow of information, such as the receipt of an order or of a customer inquiry. The other event symbol has a clock inside. If a process starts with this type of event, it means the process starts at a particular time. In both cases, the starting event is triggered by an action outside the process itself, either a message or a particular time.

Another example of variations in a basic concept appears below, for flow. You have seen the basic symbol for flow, the arrow. The next symbol for flow includes a slashing line near its beginning. This indicates a default flow, and you will usually see this symbol after a gateway. It shows that flow through a gateway typically follows one path out of those available. The third flow symbol is a little different. The arrow line is dotted, and it begins with a circle. This symbol is used to signify the flow of a message rather than the flow of sequence from one activity to the next.

We have presented just a few of the many variations in BPMN available for the basic concepts. There are many more, and all of them are designed to address very specific circumstances. Having all of these variations available makes BPMN very precise and therefore very powerful. However, all of the variations also make BPMN relatively complex and harder to learn than diagramming notations that employ less variety.

Before we leave this section on notation, we need to address one other concept: swimlanes. A process diagram can be depicted with or without a **swimlane**, which is a way to visually encapsulate a process. Swimlanes can be depicted either vertically or horizontally. Whether a swimlane is used or not, the diagram shows only one process, with one actor. If more than one actor is part of the process, then the process diagram is shown in a **pool**. A pool is made up of at least two swimlanes, each of which focuses on the actions of one participant. The participant need not be a single person; it can be a team or a department that participates in a part of the process. Pools can also be depicted vertically or horizontally. When a pool is used in a business process diagram, it is called a collaboration diagram.

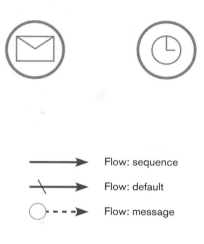

Flow: sequence

Flow: default

Flow: message

Swimlane

In business process modeling, a way to visually encapsulate a process.

Pool

In business process modeling, a way to encapsulate a process that has two or more participants.

LANE

LANE

FIGURE 7-41
Depicting a recruiting
process with BPMN

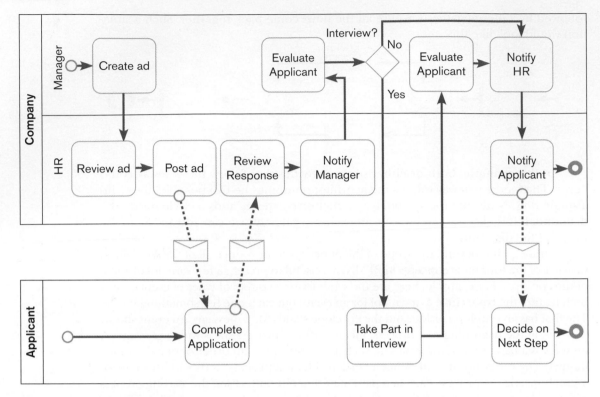

EXAMPLE

An example of a business process diagram that features both swimlanes and a pool is shown in Figure 7-41. The process depicted is recruiting. There are three participants: the job applicant (shown in a swimlane), a manager, and a Human Resources (HR) department. The manager and HR department are in the same company and so are both shown in a pool. All communication between the company (pool) and the applicant (swimlane) is done through messaging. Note the lines indicating communication between the applicant and the organization are dotted lines and feature envelopes at mid-line. The envelopes symbolize messages, or information.

To read the diagram, start at the top left, with the event symbol in the manager's lane in the company pool. Then continue to read from left to right. Follow the arrows, which indicate flow, up, down, and across. The manager needs to recruit someone for a job, so she creates an advertisement for the job. Flow control then passes to the HR department, where the ad is reviewed and then posted. At this point, a job applicant sees the ad and completes and returns an application. The application is received by HR, where it is evaluated and passed on to the manager. The manager evaluates the application and then must decide whether or not to interview the applicant. This decision is indicated by the gateway symbol. There are two possible outcomes: yes, interview the applicant, and no, don't interview the applicant. If the decision is "no," then the manager notifies HR. HR notifies the applicant, and the applicant must decide what to do next. The process ends for both the company and the applicant. If the decision is "yes," then the applicant takes part in an interview. The results are evaluated by the manager. At this point, whether the manager decides to hire the applicant or not, she notifies HR of her decision, and HR then notifies the applicant. The applicant must decide on his next step, and the process ends for all involved.

Obviously, the recruiting process has been simplified for this example. Many more activities are typically involved in recruiting, such as conducting credit and background checks of the applicant. Much of the detail was removed to make the example easier to understand and to depict.

SUMMARY

In this appendix, we introduced you to Business Process Modeling Notation (BPMN). BPMN is a standardized way of depicting business processes. It is overseen by the Object Management Group (OMG), the same group that oversees notation for object-orientation. We introduced you to the four basic concepts of BPMN—event, activity, gateway, and flow—and the symbols for each of them. We also introduced you to swimlanes and pools. BPMN is a very precise and complex modeling notation, but that makes it very powerful. Because BPMN is not technical in nature, it can be used effectively for communications between systems analysts and systems users.

KEY TERMS

1. Activity
2. Event
3. Flow
4. Gateway
5. Pool
6. Swimlane

Match each of the key terms above with the definition that best fits it.

_____ A way to visually encapsulate a process.

_____ A trigger that initiates the start of a process.

_____ A way to encapsulate a process that has two or more participants.

_____ A decision point.

_____ Shows the sequence of action in a process.

_____ An action that must take place for a process to be completed.

REVIEW QUESTIONS

1. What is a business process? Why is business process diagramming important?

2. What is BPMN? Who is responsible for it?

3. List and define the four main concepts that are part of BPMN.

4. What is the difference between a swimlane and a pool? When do you use each one?

5. BPMN includes many different varieties of its key concepts. You were introduced to three different variations of the symbol for flow. Explain each one of them.

PROBLEMS AND EXERCISES

1. BPMN includes many different varieties of its key concepts. Go to *www.bpmn.org* (and some of the other BPMN sites listed in the reference list) and look up all of the many variations that are possible for each concept. Prepare a report on six possible variations for each of the four major concepts.

2. The appendix features two BPMN examples that showed symbols but lacked content. Think of actual processes that can be described with the "empty" process diagrams in the chapter. These processes will have to be pretty simple, given how small and simple the diagrams are.

3. Use BPMN to depict Hoosier Burger's food-ordering system from Figure 7-5 as a business process model.

4. Use BPMN to depict Hoosier Burger's inventory control system from Figure 7-15 as a business process model.

FIELD EXERCISES

1. Find a company in your area that uses BPMN. Interview analysts and users about this business process modeling approach. What do they think of it? How useful is it? Ask for some examples of diagrams they have created.

2. Think of several business processes you take part in regularly as a customer. For example, think about withdrawing cash from an ATM. Think about ordering a movie and downloading a movie online. Consider purchasing something with a credit card from a big box store. Use BPMN to depict each of the processes you can think of.

REFERENCES

http://www.bpmn.org/
http://en.wikipedia.org/wiki/Business_Process_Modeling_Notation
http://www.omg.org/spec/BPMN/1.2/
http://www.omg.org/spec/BPMN/2.0/
http://www.omg.org/spec/BPMN/2.0/examples/PDF/
http://www.sparxsystems.com/platforms/business_process_modeling.html

PETRIE ELECTRONICS

Chapter 7: Structuring System Process Requirements

Jim and Sanjay chatted in Jim's office while they waited for Sally to arrive.

"Good work on researching those alternatives," Jim said.

"Thanks," replied Sanjay. "There are a lot of alternatives out there. I think we found the best three, considering what we are able to pay."

Just then Sally walked in. "Sorry I'm late. Things are getting really busy in Marketing right now. I've been putting out fires all morning."

Sally sat down at the table across from Jim.

"I understand," Jim said. "But to stay on schedule, we need to start focusing on the specifics of what we want our system to do. Remember when you wanted more details on what the system would do? Well, now we start to spend some serious energy on getting that done."

"Awesome," replied Sally, as she pulled a Red Bull out of her oversized bag and popped it open.

"I've got a list here of four core functions the system must perform," said Sanjay, pulling copies of a list from a folder on the table (PE Table 7-1). "Let's look at these."

After reviewing the list Sanjay had given them, Jim said, "Nice job, Sanjay. But we need to put this in graphical format, so that everyone can see what the inputs and outputs are for each function and how they are related to each other. We also need to see how the new system fits in with our existing data sources. We need...."

"Some data flow diagrams," Sanjay interrupted.

"Exactly," said Jim.

"They are already done," replied Sanjay, handing diagrams to both Jim and Sally. "I've already created a first draft of the context diagram (PE Figure 7-1) and a level-1 diagram (PE Figure 7-2). You can see how I've defined the boundaries of our system, and I've included our existing product and marketing databases."

"What can I say?" Jim said. "Again, a nice job on your part. These diagrams are both good places for us to start. Let's get copies of all of this to the team."

"I'll be right back," Sally said, standing up. "I need to get some coffee."

PE TABLE 7-1   Four Core Functions of Petrie's Customer Loyalty System

Function	Description
Record customer activities	When a customer makes a purchase, the transaction must be recorded in the customer loyalty system, as the rewards the system generates are driven by purchases. Similarly, when a customer uses a coupon generated by the system, it must also be recorded, so that the customer activity records can be updated to show that the coupon has been used and is now invalid.
Send promotions	Data about customer activities provide information about what types of products customers tend to buy and in what quantities. This information helps determine what sales promotion materials are best targeted at what customers. Customers who buy lots of video games should receive promotions about games, game platforms, and HD TVs, for example.
Generate point redemption coupons	Data about customer activities is used to generate coupons for future purchases. Those coupons must be made available to customers, either as paper coupons sent in the mail, or they should be made available online, in the customer's private account area. Once created, the customer activity database needs to be updated to show the creation of the coupon. The loyalty points needed to create the coupon must be deducted from the customer's total points.
Generate customer reports	From time to time, either in the mail or electronically, customers need to be send account reports that show their recent purchases, the coupons they have been issued that have not yet been redeemed, and the total points they have amassed from their purchases.

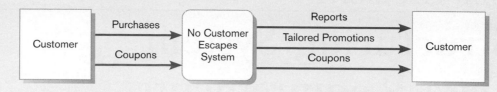

PE FIGURE 7-1
Context diagram

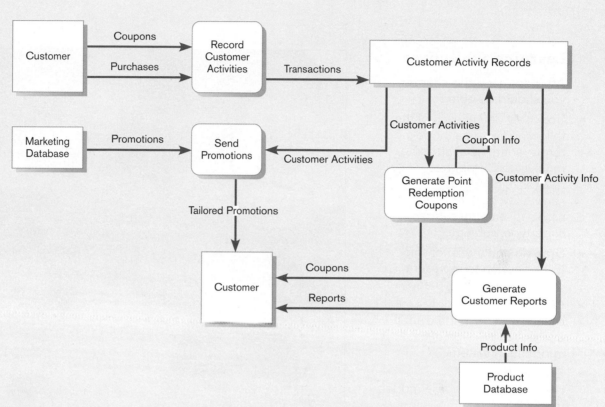

PE FIGURE 7-2 Level-1 DFD

Case Questions

1. Are the DFDs in PE Figures 7-1 and 7-2 balanced? Show that they are, or are not. If they are not balanced, how can they be fixed?
2. Decompose each of the core processes in PE Figure 7-2 and draw a new DFD for each core process.
3. Has the team overlooked any core processes in the system that should be in PE Table 7-1 and PE Figure 7-2? What would they be? Add them to PE Table 7-1 and PE Figure 7-2.

4. Redesign PE Figures 7-1 and 7-2 so that they are clearer and more efficient and more comprehensive.
5. Why is it important for the team to create DFDs if they are not going to write the actual system code themselves?

Structuring System Data Requirements

Learning Objectives

After studying this chapter, you should be able to:

- Concisely define each of the following key data modeling terms: *entity type, attribute, multivalued attribute, relationship, degree, cardinality, business rule, associative entity, trigger, supertype, subtype.*

- Draw an entity-relationship (E-R) diagram to represent common business situations.

- Explain the role of conceptual data modeling in the overall analysis and design of an information system.

- Explain the role of prepackaged database models (patterns) in data modeling.

- Distinguish among unary, binary, and ternary relationships and give an example of each.

- Define four basic types of business rules in a conceptual data model.

- Relate data modeling to process and logic modeling as different views of describing an information system.

Introduction

In Chapter 7, you learned how to model and analyze two important views of an information system: (1) the flow of data between manual or automated steps and (2) the decision logic of processing data. None of the techniques discussed so far, however, has concentrated on the data that must be retained in order to support the data flows and processing described. For example, you learned how to show data stores, or *data at rest*, in a data flow diagram (DFD). The natural *structure* of data, however, was not shown. DFDs, use

cases, and various processing logic techniques show *how, where,* and *when* data are used or changed in an information system, but these techniques do not show the *definition, structure,* and *relationships* within the data. Data modeling develops these missing, and crucial, pieces of description of a system.

In fact, some systems developers believe that a data model is the most important part of the statement of information system requirements. This belief is based on the following reasons. First, the characteristics of data captured during data modeling are crucial in the design of databases, programs, computer screens, and printed reports. For example, facts such as these—a data element is numeric, a product can be in only one product line at a time, a line item on a customer order can never be moved to another customer order, customer region name is limited to a specified set of values—are all essential pieces of information in ensuring data integrity in an information system.

Second, data, not processes, are the most complex aspects of many modern information systems and hence require a central role in structuring system requirements. Transaction processing systems can have considerable process complexity in validating data, reconciling errors, and coordinating the movement of data to various databases. Current systems development focuses more on management information systems (such as sales tracking), decision support systems (such as short-term cash investment), and business intelligence systems (such as market basket analysis). Such systems are more data intensive. The exact nature of processing is also more ad hoc than with transaction processing systems, so the details of processing steps cannot be anticipated. Thus, the goal is to provide a rich data resource that might support any type of information inquiry, analysis, and summarization.

Third, the characteristics about data (e.g., length, format, and relationships with other data) are reasonably permanent and have significant similarity for different organizations in the same business. In contrast, the paths and design of data flow are quite dynamic. A data model explains the inherent nature of the organization, not its transient form. Therefore, an information system design based on a data orientation, rather than a process or logic orientation, should have a longer useful life and should have common features for the same applications or domains in different organizations. Finally, structural information about data is essential for automatic program generation. For example, the fact that a customer order has many line items on it instead of just one line item affects the automatic design of a computer screen for entry of customer orders. Although a data model specifically documents the file and database requirements for an information system, the business meaning, or semantics, of data included in the data model has a broader effect on the design and construction of a system.

The most common format used for data modeling is *entity-relationship (E-R) diagramming*. A similar format used with object-oriented analysis and design methods is *class diagramming*, which is included in a special section at the end of this chapter on the object-oriented development approach to data modeling. Data models that use E-R and class diagram notations explain the characteristics and structure of data independent of how the data may be stored in computer memories. A data model is usually developed iteratively, either from scratch or from a purchased data model for the industry or business area to be supported. Information system (IS) planners use this preliminary data model to develop an enterprise-wide data model with very broad categories of data and little detail. Next, during the definition of a project, a specific data model is built to help explain the scope of a particular systems analysis and design effort. During requirements structuring, a data model represents conceptual data requirements for a particular system. Then, after system inputs and outputs are fully described during logical design, the data model is refined before it is translated into a logical format (typically a relational data model) from which database definition and physical database design are done. A data model represents certain types of business rules that govern the properties of data. Business rules are important statements of business policies that ideally will be enforced through the database and database management system ultimately used for the application you are designing. Thus, you will use E-R and class diagramming in many systems development project steps, and most IS project members need to know how to develop and read data model diagrams. Therefore, mastery of the requirements structuring methods and techniques addressed in this chapter is critical to your success on a systems development project team.

CONCEPTUAL DATA MODELING

Conceptual data model
A detailed model that captures the overall structure of organizational data that is independent of any database management system or other implementation considerations.

A **conceptual data model** is a representation of organizational data. The purpose of a conceptual data model is to show as many rules about the meaning and interrelationships among data as are possible.

Conceptual data modeling is typically done in parallel with other requirements analysis and structuring steps during systems analysis (see Figure 8-1), as outlined in prior chapters. On larger systems development teams, a subset of the project team concentrates on data modeling while other team members focus attention on process or logic modeling. Analysts develop (or use from prior systems development) a conceptual data model for the current system and then build or refine a purchased conceptual data model that supports the scope and requirements for the proposed or enhanced system.

The work of all team members is coordinated and shared through the project dictionary or repository. This repository is often maintained by a common Computer-Aided Software Engineering (CASE) or data modeling software tool, but some organizations still use manual documentation. Whether automated or manual, it is essential that the process, logic, and data model descriptions of a system be consistent and complete because each describes different, but complementary, views of the same information system. For example, the names of data stores on the primitive-level DFDs often correspond to the names of data entities in E-R diagrams, and the data elements associated with data flows on DFDs must be accounted for by attributes of entities and relationships in E-R diagrams.

The Conceptual Data Modeling Process

The process of conceptual data modeling begins with developing a conceptual data model for the system being replaced, if a system already exists. This is essential for planning the conversion of the current files or database into the database of the new system. Further, this is a good, but not a perfect, starting point for your understanding of the data requirements of the new system. Then, a new conceptual data model is built (or a standard one is purchased) that includes all of the data requirements for the new system. You discovered these requirements from the fact-finding methods employed during requirements determination. Today, given the popularity of rapid development methodologies, such as the use of predefined patterns, these requirements often evolve through various iterations from some starting point in a purchased application or database design. Even when developed from scratch, data modeling is an iterative process with many checkpoints.

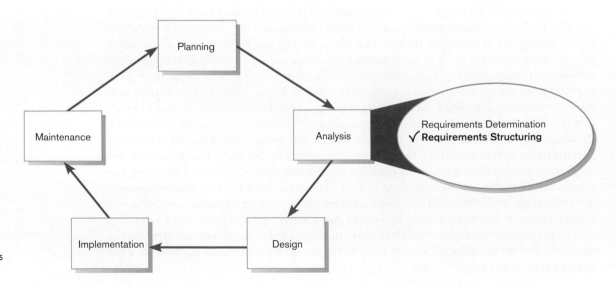

FIGURE 8-1
Systems development life cycle with analysis phase highlighted

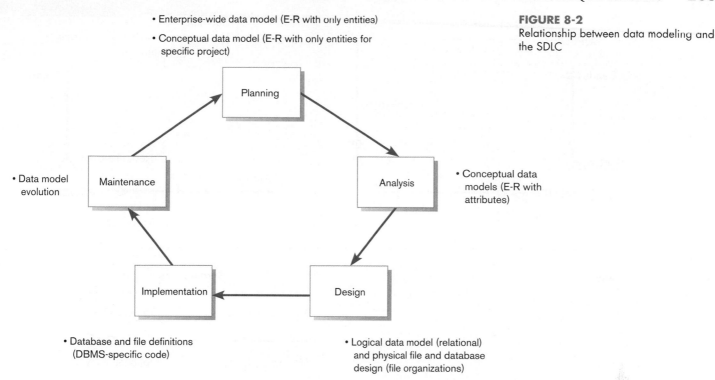

• Enterprise-wide data model (E-R with only entities)

• Conceptual data model (E-R with only entities for specific project)

• Data model evolution

• Conceptual data models (E-R with attributes)

• Database and file definitions (DBMS-specific code)

• Logical data model (relational) and physical file and database design (file organizations)

FIGURE 8-2
Relationship between data modeling and the SDLC

Conceptual data modeling is one kind of data modeling and database design carried out throughout the systems development process. Figure 8-2 shows the different kinds of data modeling and database design that go on during the whole systems development life cycle (SDLC). The conceptual data modeling methods we discuss in this chapter are suitable for the planning and analysis phases; these methods can be used with either a data model developed from scratch or based on a purchased data model. The planning phase of the SDLC addresses issues of system scope, general requirements, and content independent of technical implementation. E-R and class diagramming are suited for this phase because these diagrams can be translated into a wide variety of technical architectures for data, such as relational, network, and hierarchical architectures. A data model evolves from the early stages of planning through the analysis phase as it becomes more specific and is validated by more detailed analyses of system needs.

In the design phase, the final data model developed in analysis is matched with designs for systems inputs and outputs and is translated into a format from which physical data storage decisions can be made. After specific data storage architectures are selected, then, in implementation, files and databases are defined as the system is coded. Through the use of the project repository, a field in a physical data record can, for example, be traced back to the conceptual data attribute that represents it on a data model diagram. Thus, the data modeling and design steps in each of the SDLC phases are linked through the project repository.

Deliverables and Outcomes

Most organizations today do conceptual data modeling using E-R modeling, which uses a special notation to represent as much meaning about data as possible. Because of the rapidly increasing interest in object-oriented methods, class diagrams using unified modeling language (UML) drawing tools such as IBM's Rational products or Microsoft Visio are also popular. We will focus first on E-R diagramming and then later show how it differs from class diagramming.

The primary deliverable from the conceptual data modeling step within the analysis phase is an E-R diagram, similar to the one shown in Figure 8-3. This figure shows the major categories of data (rectangles on the diagram) and the business

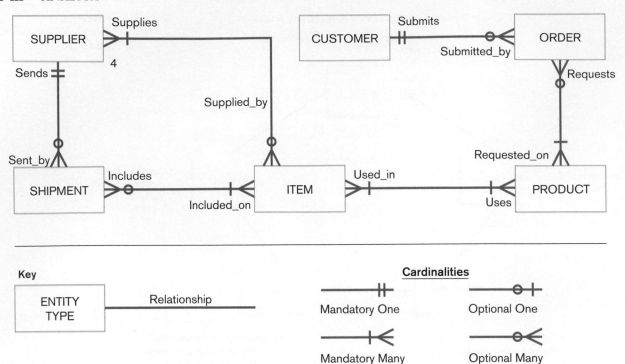

FIGURE 8-3
Sample conceptual data model

relationships between them (lines connecting rectangles). For example, Figure 8-3 shows that, for the business represented by this diagram, a SUPPLIER sometimes Supplies ITEMs to the company, and an ITEM is always Supplied by one to four SUPPLIERS. The fact that a supplier only sometimes supplies items implies that the business wants to keep track of some suppliers without designating what they can supply. This diagram includes two names on each line so that a relationship can be read in each direction. For simplicity, we will not typically include two names on lines in E-R diagrams in this book; however, this is a standard used in many organizations.

The other deliverable from conceptual data modeling is a full set of entries about data objects that will be stored in the project dictionary, repository, or data modeling software. The repository is the mechanism that links the data, processes, and logic models of an information system. For example, there are explicit links between a data model and a DFD. Some important links are explained briefly here.

- Data elements included in data flows also appear in the data model, and vice versa. You must include in the data model any raw data captured and retained in a data store, and a data model can include only data that have been captured or that have been computed from captured data. Because a data model is a general business picture of data, both manual and automated data stores will be included.
- Each data store in a process model must relate to business objects (what we will call data entities) represented in the data model. For example, in Figure 7-5, the Inventory File data store must correspond to one or several data objects on a data model.

You can use an automated repository to verify these linkages.

GATHERING INFORMATION FOR CONCEPTUAL DATA MODELING

Requirements determination methods must include questions and investigations that take a data, not only a process and logic, focus. For example, during interviews with potential system users—during Joint Application Design (JAD) sessions or through requirements interviews—you must ask specific questions in order to gain the perspective on data that you need to develop or tailor a purchased data model. In later sections of this chapter, we will introduce some specific terminology and constructs used in data modeling. Even without this specific data modeling language, you can begin to understand the kinds of questions that must be answered during requirements determination. These questions relate to understanding the rules and policies by which the area to be supported by the new information system operates. That is, a data model explains what the organization does and what rules govern how work is performed in the organization. You do not, however, need to know (and often can't fully anticipate) how or when data are processed or used to do data modeling.

You typically do data modeling from a combination of perspectives. The first perspective is generally called the *top-down approach*. This perspective derives the business rules for a data model from an intimate understanding of the nature of the business, rather than from any specific information requirements in computer displays, reports, or business forms. It is this perspective that is typically the basis for a purchased data model. Several very useful sources of typical questions elicit the business rules needed for data modeling (see Aranow, 1989; Gottesdiener, 1999; and Sandifer and von Halle, 1991a, 1991b). Table 8-1 summarizes a few key questions you should ask system users and business managers so that you can develop an accurate and complete data model tailored to the particular situation. The questions in this table are purposely posed in business terms. You can ask these questions whether you start with a clean sheet of paper or a purchased data model, but typically the questions are more obvious and thorough when you begin the data modeling project with a purchased data model for the industry or application under development. In this chapter, you will learn the more technical terms included in bold at the end of each set of questions. Of course, these technical terms do not mean much to a business manager, so you must learn how to frame your questions in business terms for your investigation.

TABLE 8-1 Requirements Determination Questions for Data Modeling

1. *What are the subjects/objects of the business?* What types of people, places, things, materials, events, etc. are used or interact in this business, about which data must be maintained? How many instances of each object might exist?—**data entities and their descriptions**

2. *What unique characteristic (or characteristics) distinguishes each object from other objects of the same type?* Might this distinguishing feature change over time or is it permanent? Might this characteristic of an object be missing even though we know the object exists?—**primary key**

3. *What characteristics describe each object?* On what basis are objects referenced, selected, qualified, sorted, and categorized? What must we know about each object in order to run the business?—**attributes and secondary keys**

4. *How do you use these data?* That is, are you the source of the data for the organization, do you refer to the data, do you modify it, and do you destroy it? Who is not permitted to use these data? Who is responsible for establishing legitimate values for these data?—**security controls and understanding who really knows the meaning of data**

5. *Over what period of time are you interested in these data?* Do you need historical trends, current "snapshot" values, and/or estimates or projections? If a characteristic of an object changes over time, must you know the obsolete values?—**cardinality and time dimensions of data**

6. *Are all instances of each object the same?* That is, are there special kinds of each object that are described or handled differently by the organization? Are some objects summaries or combinations of more detailed objects?—**supertypes, subtypes, and aggregations**

7. *What events occur that imply associations among various objects?* What natural activities or transactions of the business involve handling data about several objects of the same or a different type?—**relationships, and their cardinality and degree**

8. *Is each activity or event always handled the same way or are there special circumstances?* Can an event occur with only some of the associated objects, or must all objects be involved? Can the associations between objects change over time (for example, employees change departments)? Are values for data characteristics limited in any way?—**integrity rules, minimum and maximum cardinality, time dimensions of data**

FIGURE 8-4
Sample customer form

PVF CUSTOMER ORDER			
ORDER NO: 61384			CUSTOMER NO: 1273

NAME: ADDRESS: CITY-STATE-ZIP:

Contemporary Designs
123 Oak St.
Austin, TX 28384

ORDER DATE: 11/04/2014   PROMISED DATE: 11/21/2014

PRODUCT NO	DESCRIPTION	QUANTITY ORDERED	UNIT PRICE
M128	Bookcase	4	200.00
B381	Cabinet	2	150.00
R210	Table	1	500.00

You can also gather the information you need for data modeling by reviewing specific business documents—computer displays, reports, and business forms—handled within the system. This process of gaining an understanding of data is often called a *bottom-up approach*. These items will appear as data flows on DFDs and will show the data processed by the system and, hence, probably the data that must be maintained in the system's database. Consider, for example, Figure 8-4, which shows a customer order form used at Pine Valley Furniture (PVF). From this form, we determine that the following data must be kept in the database:

ORDER NO	CUSTOMER NO
ORDER DATE	NAME
PROMISED DATE	ADDRESS
PRODUCT NO	CITY-STATE-ZIP
DESCRIPTION	
QUANTITY ORDERED	
UNIT PRICE	

We also see that each order is from one customer and that an order can have multiple line items, each for one product. We will use this kind of understanding of an organization's operation to develop data models.

INTRODUCTION TO E-R MODELING

The basic E-R modeling notation uses three main constructs: data entities, relationships, and their associated attributes. Several different E-R notations exist, and many CASE and E-R drawing software support multiple notations. For simplicity, we have adopted one common notation for this book; this notation uses the so-called crow's foot symbols and places data attribute names within entity rectangles. This notation is very similar to that used by many E-R drawing tools, including Microsoft Visio®. If you use another notation in courses or at work, you should be able to easily translate between notations.

An **entity-relationship data model (E-R model)** is a detailed, logical representation of the data for an organization or for a business area. The E-R model is expressed in terms of entities in the business environment, the relationships or associations among those entities, and the attributes or properties of both the entities and their relationships. An E-R model is normally expressed as an **entity-relationship diagram (E-R diagram)**, which is a graphical representation of an E-R model. The notation we will use for E-R diagrams appears in Figure 8-5, and subsequent sections explain this notation.

Entity-relationship data model (E-R model)
A detailed, logical representation of the entities, associations, and data elements for an organization or business area.

Entity-relationship diagram (E-R diagram)
A graphical representation of an E-R model.

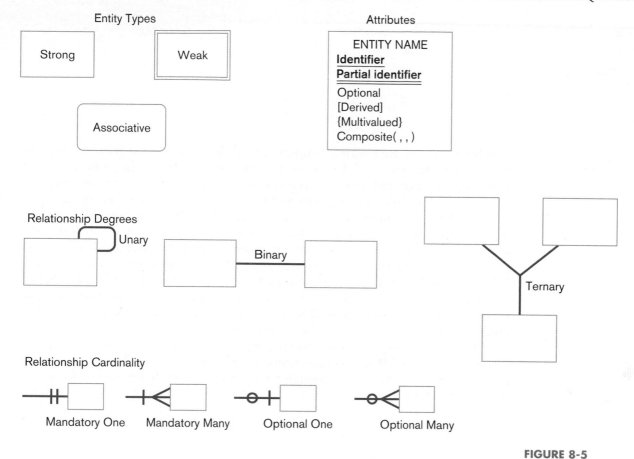

FIGURE 8-5
Basic E-R notation

Entities

An entity (see the first question in Table 8-1) is a person, place, object, event, or concept in the user environment about which the organization wishes to maintain data. An entity has its own identity that distinguishes it from each other entity. Some examples of entities follow:

- Person: EMPLOYEE, STUDENT, PATIENT
- Place: STORE, WAREHOUSE, STATE
- Object: MACHINE, BUILDING, AUTOMOBILE, PRODUCT
- Event: SALE, REGISTRATION, RENEWAL
- Concept: ACCOUNT, COURSE, WORK CENTER

There is an important distinction between entity *types* and entity *instances*. An **entity type** (sometimes called an *entity class*) is a collection of entities that share common properties or characteristics. Each entity type in an E-R model is given a name. Because the name represents a class or set, it is singular. Also, because an entity is an object, we use a simple noun to name an entity type. We use capital letters in naming an entity type and, in an E-R diagram, the name is placed inside a rectangle representing the entity, as shown in Figure 8-6a.

An **entity instance** (also known simply as an *instance*) is a single occurrence of an entity type. An entity type is described just once in a data model, whereas many instances of that entity type may be represented by data stored in the database. For example, there is one EMPLOYEE entity type in most organizations, but there may be hundreds (or even thousands) of instances of this entity type stored in the database.

A common mistake many people make when they are just learning to draw E-R diagrams, especially if they already know how to do data flow diagramming, is to confuse data entities with sources/sinks or system outputs and relationships with data flows. A simple rule to avoid such confusion is that a true data entity will have many

Entity type
A collection of entities that share common properties or characteristics.

Entity instance
A single occurrence of an entity type. Also known as an instance.

FIGURE 8-6
Representing entity types
(a) Three entity types

(b) Questionable entity types

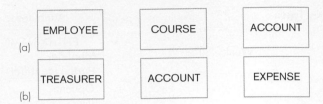

possible instances, each with a distinguishing characteristic, as well as one or more other descriptive pieces of data. Consider the entity types that might be associated with a sorority expense system, as represented in Figure 8-6b. In this situation, the sorority treasurer manages accounts and records expense transactions against each account. However, do we need to keep track of data about the treasurer and her supervision of accounts as part of this accounting system? The treasurer is the person entering data about accounts and expenses and making inquiries about account balances and expense transactions by category. Because there is only one treasurer, TREASURER data do not need to be kept. On the other hand, if each account has an account manager (e.g., a sorority officer) who is responsible for assigned accounts, then we may wish to have an ACCOUNT MANAGER entity type with pertinent attributes as well as relationships to other entity types.

In this same situation, is an expense report an entity type? Because an expense report is computed from expense transactions and account balances, it is a data flow, not an entity type. Even though there will be multiple instances of expense reports over time, the report contents are already represented by the ACCOUNT and EXPENSE entity types.

Often when we refer to entity types in subsequent sections, we will simply say *entity*. This is common among data modelers. We will clarify that we mean an entity by use of the term *entity instance*.

Naming and Defining Entity Types   Clearly naming and defining data, such as entity types, are important tasks during requirements determination and structuring. When naming and defining entity types, you should use the following guidelines:

- An entity type name is a *singular noun* (such as CUSTOMER, STUDENT, or AUTOMOBILE).
- An entity type name should be *descriptive and specific to the organization*. For example, a PURCHASE ORDER for orders placed with suppliers is distinct from CUSTOMER ORDER for orders placed by customers. Both of these entity types cannot be named ORDER.
- An entity type name should be *concise*; for example, in a university database, use REGISTRATION for the event of a student registering for a class rather than STUDENT REGISTRATION FOR CLASS.
- *Event entity types* should be named for the *result of the event*, not the activity or process of the event. For example, the event of a project manager assigning an employee to work on a project results in an ASSIGNMENT.

Some specific guidelines for defining entity types follow:

- An entity type definition should include a statement of *what the unique characteristic(s) is (are) for each instance* of the entity type.
- An entity type definition should make clear *what entity instances are included and not included* in the entity type. For example, "A customer is a person or organization that has placed an order for a product from us or that we have contacted to advertise or promote our products. A customer does not include persons or organizations that buy our products only through our customers, distributors, or agents."
- An entity type definition often includes a description of *when an instance of the entity type is created and deleted*.

- For some entity types, the definition must specify *when an instance might change into an instance of another entity type*; for example, a bid for a construction company becomes a contract once it is accepted.
- For some entity types, the definition must specify *what history is to be kept about entity instances*. Statements about keeping history may have ramifications about how we represent the entity type on an E-R diagram and eventually how we store data for the entity instances.

Attributes

Each entity type has a set of attributes (see the third question in Table 8-1) associated with it. An **attribute** is a property or characteristic of an entity that is of interest to the organization (relationships may also have attributes, as we will see in the section on relationships). Following are some typical entity types and associated attributes:

> STUDENT: Student_ID, Student_Name, Home_Address, Phone_Number, Major
> AUTOMOBILE: Vehicle_ID, Color, Weight, Horsepower
> EMPLOYEE: Employee_ID, Employee_Name, Payroll_Address, Skill

We use an initial capital letter, followed by lowercase letters, and nouns in naming an attribute; underscores may or may not be used to separate words. In E-R diagrams, we represent an attribute by placing its name inside the rectangle for the associated entity (see Figure 8-5). We use different notations for attributes to distinguish between different types of attributes, which we describe below. Our notation is similar to that used by many CASE and E-R drawing tools, such as Microsoft Visio or Oracle's Designer. Precisely how different types of attributes are shown varies by tool.

Naming and Defining Attributes   Often several attributes have approximately the same name and meaning. Thus, it is important to carefully name attributes using the following guidelines:

- An attribute name is a *noun* (such as Customer_ID, Age, or Product_Minimum_Price).
- An attribute name should be *unique*. No two attributes of the same entity type may have the same name, and it is desirable, for clarity, that no two attributes across all entity types have the same name.
- To make an attribute name unique and for clarity, *each attribute name should follow a standard format*. For example, your university may establish Student_GPA, as opposed to GPA_of_Student, as an example of the standard format for attribute naming.
- *Similar attributes of different entity types should use similar but distinguishing names*; for example, the city of residence for faculty and students should be, respectively, Faculty_Residence_City_Name and Student_Residence_City_Name.

Some specific guidelines for defining attributes follow:

- An attribute definition states *what the attribute is and possibly why it is important*.
- An attribute definition should make it clear *what is included and what is not included* in the attribute's value; for example, "Employee_Monthly_ Salary_ Amount is the amount of money paid each month in the currency of the country of residence of the employee exclusive of any benefits, bonuses, reimbursements, or special payments."
- Any *aliases*, or alternative names, for the attribute can be specified in the definition.
- It may also be desirable to state in the definition *the source of values for the attribute*. Stating the source may make the meaning of the data clearer.
- An attribute definition should indicate *if a value for the attribute is required or optional.* This business rule about an attribute is important for maintaining data integrity.

Attribute
A named property or characteristic of an entity that is of interest to the organization.

- An attribute definition may indicate *if a value for the attribute may change* once a value is provided and before the entity instance is deleted. This business rule also controls data integrity.
- An attribute definition may also indicate any *relationships that attribute has with other attributes*; for example, "Employee_Vacation_Days_Number is the number of days of paid vacation for the employee. If the employee has a value of 'Exempt' for Employee_Type, then the maximum value for Employee_ Vacation_Days_Number is determined by a formula involving the number of years of service for the employee."

Candidate Keys and Identifiers

Every entity type must have an attribute or set of attributes that distinguishes one instance from other instances of the same type (see the second question in Table 8-1). A **candidate key** is an attribute (or combination of attributes) that uniquely identifies each instance of an entity type. A candidate key for a STUDENT entity type might be Student_ID.

Sometimes a combination of attributes is required to identify a unique entity. For example, consider the entity type GAME for a basketball league. The attribute Team_Name is clearly not a candidate key because each team plays several games. If each team plays exactly one home game against each other team, then the combination of the attributes Home_Team and Visiting_Team is a composite candidate key for GAME.

Some entities may have more than one possible candidate key. One candidate key for EMPLOYEE is Employee_ID; a second is the combination of Employee_Name and Address (assuming that no two employees with the same name live at the same address). If there is more than one possible candidate key, the designer must choose one of the candidate keys as the identifier. An **identifier** is a candidate key that has been selected to be used as the unique characteristic for an entity type. We show the identifier attribute(s) by placing a solid underline below the identifier (see Figure 8-5). Bruce (1992) suggests the following criteria for selecting identifiers:

- Choose a candidate key that will not change its value over the life of each instance of the entity type. For example, the combination of Employee_Name and Payroll_Address would probably be a poor choice as an identifier for EMPLOYEE because the values of Payroll_Address and Employee_Name could easily change during an employee's term of employment.
- Choose a candidate key so that, for each instance of the entity, the attribute is guaranteed to have valid values and not be null. To ensure valid values, you may have to include special controls in data entry and maintenance routines to eliminate the possibility of errors. If the candidate key is a combination of two or more attributes, make sure that all parts of the key have valid values.
- Avoid the use of so-called intelligent identifiers, whose structure indicates classifications, locations, and so on. For example, the first two digits of a key for a PART entity may indicate the warehouse location. Such codes are often modified as conditions change, which renders the primary key values invalid.
- Consider substituting single-attribute surrogate keys for large composite keys. For example, an attribute called Game_ID could be used for the entity GAME instead of the combination of Home_Team and Visiting_Team.

Figure 8-7 shows the representation for a STUDENT entity type using our E-R notation. STUDENT has a simple identifier, Student_ID, and three other simple attributes.

Other Attribute Types

A **multivalued attribute** may take on more than one value for each entity instance. Suppose that Skill is one of the attributes of EMPLOYEE. If each employee can have more than one skill, Skill is a multivalued attribute. Two ways of showing multivalued

Candidate key
An attribute (or combination of attributes) that uniquely identifies each instance of an entity type.

Identifier
A candidate key that has been selected as the unique, identifying characteristic for an entity type.

```
┌─────────────────────────┐
│         STUDENT         │
│   Student_ID            │
│   Student_Name          │
│   Student_Campus_Address│
│   Student_Campus_Phone  │
└─────────────────────────┘
```

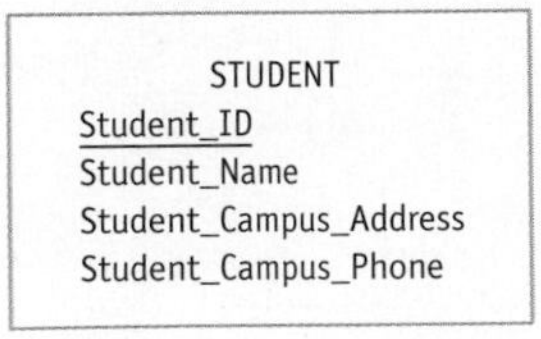

FIGURE 8-7
STUDENT entity type with attributes

Multivalued attribute
An attribute that may take on more than one value for each entity instance.

FIGURE 8-8
Multivalued attributes and repeating groups

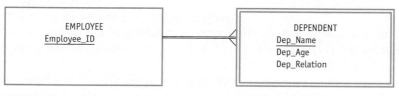

(a) Multivalued attribute skill

(b) Repeating group of dependent data

(c) Weak entity for dependent data

attributes are common. The first is to list the multivalued attribute along with other attributes, but use a special symbol to indicate that it is multivalued. This is the approach taken in Figure 8-8a, where the multivalued attribute skill is enclosed in curly brackets.

Sometimes a set of data repeats together. For example, consider Figure 8-8b for an employee entity with multivalued attributes for data about each employee's dependents. In this situation, data such as dependent name, age, and relation to employee (spouse, child, parent, etc.) are multivalued attributes about an employee, and these attributes repeat together (we show this by using one set of curly brackets around the data that repeats together). Several attributes that repeat together are called a **repeating group**.

Conceptually, dependents can also be thought of as entities. Thus, many data analysts prefer a second approach to representing a repeating group. In this approach, we separate the repeating data into another entity, called a weak (or *attributive*) entity (designated by a rectangle with a double line border), and then use a relationship (relationships are discussed in the next section) to link the weak entity to its associated regular entity (this particular relationship is also represented by a double line). We can show this in Figure 8-8c using a weak entity, DEPENDENT, and a relationship, between DEPENDENT and EMPLOYEE. The crow's foot next to DE-PENDENT means that there may be many DEPENDENTs for the same EMPLOYEE. The identifier of DEPENDENT is a combination of the dependent's name and the ID of the employee for which this person is a dependent. It is sufficient to show Dep_Name in the weak entity and use a double underline to designate it as a partial identifier.

It may be important to designate whether an attribute must have a value (**required attribute**) or may not have a value (**optional attribute**) for every entity instance. It is also common to have an attribute, such as Name or Address, which has meaningful component parts, which we call a **composite attribute**. For some applications, people may want to simply refer to the set of component attributes by a composite name, whereas in other applications, we may need to display or compute with only some of the component parts. Also in conceptual modeling, users may refer to some datum which can

Repeating group
A set of two or more multivalued attributes that are logically related.

Required attribute
An attribute that must have a value for every entity instance.

Optional attribute
An attribute that may not have a value for every entity instance.

Composite attribute
An attribute that has meaningful component parts.

Derived attribute
An attribute whose value can be computed from related attribute values.

```
            EMPLOYEE
Employee_ID
Employee_Name(First_Name, Last_Name)
Date_of_Birth
[Employee_Age]
```

FIGURE 8-9
Required, optional, composite, and derived attributes

Relationship
An association between the instance of one or more entity types that is of interest to the organization.

be computed from other data in the database, a so-called **derived attribute**. In order to represent these unique characteristics of attributes, many E-R drawing tools have special notations for each of these types of attributes. In this text, we use the notation found in Figure 8-5. Figure 8-9 illustrates an EMPLOYEE entity with each of these types of attributes using our notation. Any identifier is required, and we have designated the composite attribute Employee_Name (with atomic components First_Name and Last_Name) as also required by putting these attribute names in bold. Date_of_Birth is an optional attribute. Employee_Age, also optional, can be computed from today's date and Date_of_Birth, so it is a derived attribute.

Relationships

Relationships are the glue that holds together the various components of an E-R model (see the fifth, seventh, and eighth questions in Table 8-1). A **relationship** is an association between the instances of one or more entity types that is of interest to the organization. An association usually means that an event has occurred or that there exists some natural linkage between entity instances. For this reason, relationships are labeled with verb phrases. For example, in Figure 8-10a we represent a training department in a company that is interested in tracking which training courses each of its employees has completed. This leads to a relationship called Completes between the EMPLOYEE and COURSE entity types.

As indicated by the arrows, this is a many-to-many relationship: each employee may complete more than one course, and each course may be completed by more than one employee. More significantly, we can use the Completes relationship to determine the specific courses that a given employee has completed. Conversely, we can determine the identity of each employee who has completed a particular course. For example, consider the employees and courses shown in Figure 8-10b. In this illustration, Melton has completed three courses (C++, COBOL, and Perl) and the SQL course has been completed by Celko and Gosling.

We sometimes use two verb phrases for a relationship name so that there is an explicit name for the relationship in each direction. The standards you will follow will be determined by your organization.

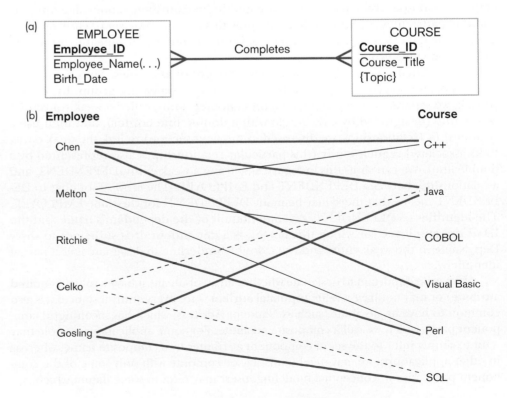

FIGURE 8-10
Relationship type and instances
(a) Relationship type (Completes)
(b) Relationship instances

CONCEPTUAL DATA MODELING AND THE E-R MODEL

The last section introduced the fundamentals of the E-R data modeling notation—entities, attributes, and relationships. The goal of conceptual data modeling is to capture as much of the meaning of data as is possible. The more details (business rules) about data that we can model, the better the system we can design and build. Further, if we can include all these details in a CASE repository, and if a CASE tool can generate code for data definitions and programs, then the more we know about data, the more code we can generate automatically. This will make system building more accurate and faster. More important, if we can keep a thorough repository of data descriptions, we can regenerate the system as needed as the business rules change. Because maintenance is the largest expense with any information system, the efficiencies gained by maintaining systems at the rule rather than the code level drastically reduce the cost.

In this section, we explore more advanced concepts needed to model data more thoroughly and learn how the E-R notation represents these concepts.

Degree of a Relationship

The **degree** of a relationship (see the seventh question in Table 8-1) is the number of entity types that participate in that relationship. Thus, the relationship Completes illustrated in Figure 8-10a is of degree two because there are two entity types: EMPLOYEE and COURSE. The three most common relationships in E-R models are unary (degree one), binary (degree two), and ternary (degree three). Higher-degree relationships are possible, but they are rarely encountered in practice, so we restrict our discussion to these three cases. Examples of unary, binary, and ternary relationships appear in Figure 8-11.

Degree
The number of entity types that participate in a relationship.

Unary Relationships Also called a *recursive relationship*, a **unary relationship** is a relationship between the instances of one entity type. Three examples are shown in Figure 8-11. In the first example, Is_married_to is shown as a one-to-one relationship between instances of the PERSON entity type. That is, each person may be currently married to one other person. In the second example, Manages is shown as a one-to-many relationship between instances of the EMPLOYEE entity type. Using this relationship, we can identify, for example, the employees who report to a particular manager; reading the Manages relationship in the opposite direction, we can identify who the manager is for a given employee. In the third example, Stands_after is shown as a one-to-one relationship between instance of the TEAM entity type. This relationship represents the sequential order of teams in a league; this sequential ordering could be based on any criteria, such as winning percentage.

Unary relationship
A relationship between instances of one entity type; also called *recursive relationship*.

Figure 8-12 shows an example of another common unary relationship, called a *bill-of-materials structure*. Many manufactured products are made of subassemblies, which in turn are composed of other subassemblies and parts, and so on. As shown in Figure 8-12a, we can represent this structure as a many-to-many unary relationship. In this figure, we use Has_components for the relationship name. The attribute Quantity, which is a property of the relationship, indicates the number of each component that is contained in a given assembly.

Two occurrences of this structure are shown in Figure 8-12b. Each of these diagrams shows the immediate components of each item as well as the quantities of that component. For example, item TX100 consists of item BR450 (quantity 2) and item DX500 (quantity 1). You can easily verify that the associations are in fact many-to-many. Several of the items have more than one component type (e.g., item MX300 has three immediate component types: HX100, TX100, and WX240). Also, some of the components are used in several higher-level assemblies. For example, item WX240 is used in both item MX300 and item WX340, even at different levels of the

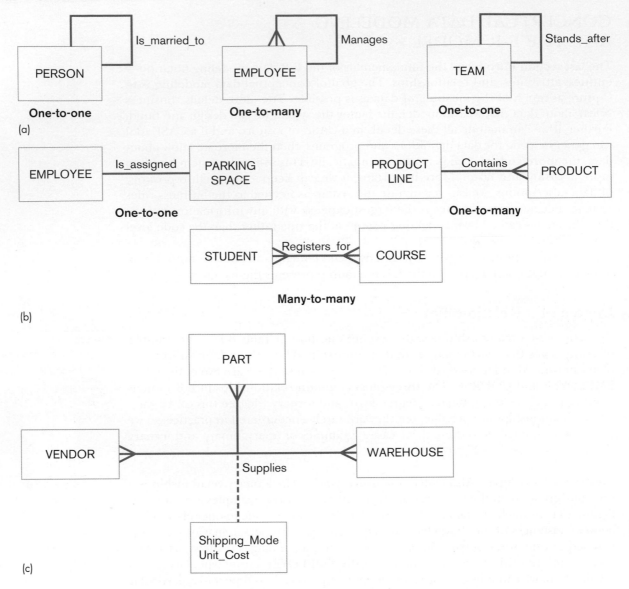

FIGURE 8-11

Examples of relationships of different degrees
(a) Unary relationships
(b) Binary relationships
(c) Ternary relationship

Binary relationship

A relationship between instances of two entity types. This is the most common type of relationship encountered in data modeling.

Ternary relationship

A simultaneous relationship among instance of three entity types.

bill of materials. The many-to-many relationship guarantees that, for example, the same subassembly structure of WX240 (not shown) is used each time item WX240 goes into making some other item.

Binary Relationships   A **binary relationship** is a relationship between instances of two entity types and is the most common type of relationship encountered in data modeling. Figure 8-11b shows three examples. The first (one-to-one) indicates that an employee is assigned one parking place, and each parking place is assigned to one employee. The second (one-to-many) indicates that a product line may contain several products, and each product belongs to only one product line. The third (many-to-many) shows that a student may register for more than one course, and that each course may have many student registrants.

Ternary Relationships   A **ternary relationship** is a *simultaneous* relationship among instances of three entity types. In the example shown in Figure 8-11c, the relationship Supplies tracks the quantity of a given part that is shipped by a particular vendor to a selected warehouse. Each entity may be a one or a many participant in a ternary relationship (in Figure 8-11, all three entities are many participants).

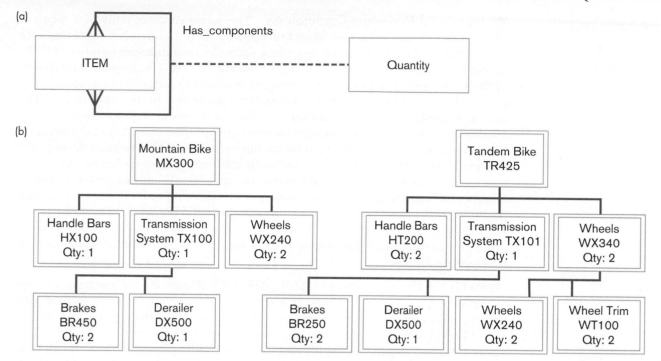

FIGURE 8-12
Representing a bill-of-materials structure
(a) Many-to-many relationship
(b) Two ITEM bill-of-materials structure instances

Note that a ternary relationship is not the same as three binary relationships. For example, Shipping_Mode is an attribute of the Supplies relationship in Figure 8-11c. Shipping_Mode cannot be properly associated with any of the three possible binary relationships among the three entity types (such as that between PART and VENDOR) because Shipping_Mode is the type of shipping carrier used for a particular PART shipped from a particular VENDOR to a particular WAREHOUSE. We strongly recommend that all ternary (and higher) relationships be represented as associative entities (described later). We examine the cardinality of relationships next.

Cardinalities in Relationships

Suppose there are two entity types, A and B, connected by a relationship. The **cardinality** of a relationship (see the fifth, seventh, and eighth questions in Table 8-1) is the number of instances of entity B that can (or must) be associated with each instance of entity A. For example, consider the relationship for DVDs at a video store shown in Figure 8-13a.

Clearly, a video store may stock more than one DVD of a given movie. In the terminology we have used so far, this example is intuitively a "many" relationship. Yet it is also true that the store may not have a single copy of a particular movie in stock. We need a more precise notation to indicate the range of cardinalities for a relationship. This notation was introduced in Figure 8-5, which you may want to review at this point.

Cardinality
The number of instances of entity B that can (or must) be associated with each instance of entity A.

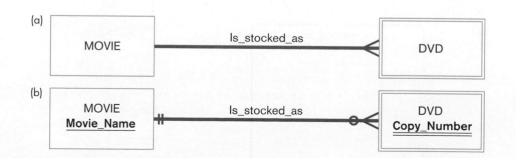

FIGURE 8-13
Introducing cardinality constraints
(a) Basic relationship
(b) Relationship with cardinality constraints

Minimum and Maximum Cardinalities   The *minimum* cardinality of a relationship is the minimum number of instances of entity B that may be associated with each instance of entity A. In the preceding example, the minimum number of DVDs available for a movie is zero, in which case we say that DVD is an *optional participant* in the Is_stocked_as relationship. When the minimum cardinality of a relationship is one, then we say that entity B is a *mandatory participant* in the relationship. The *maximum* cardinality is the maximum number of instances. For our example, this maximum is "many" (an unspecified number greater than one). Using the notation from Figure 8-5, we diagram this relationship in Figure 8-13b. The zero through the line near the DVD entity means a minimum cardinality of zero, whereas the crow's foot notation means a "many" maximum cardinality. The double underline of Copy_Number indicates that this attribute is part of the identifier of DVD, but the full composite identifier must also include the identifier of MOVIE, Movie_Name.

Examples of three relationships that show all possible combinations of minimum and maximum cardinalities appear in Figure 8-14. A brief description of each relationship follows:

1. PATIENT Has_recorded PATIENT_HISTORY (Figure 8-14a). Each patient has recorded one or more patient histories (we assume that the initial patient visit is always recorded as an instance of PATIENT HISTORY). Each instance of PATIENT HISTORY is a record for exactly one PATIENT.
2. EMPLOYEE Is_assigned_to PROJECT (Figure 8-14b). Each PROJECT has at least one assigned EMPLOYEE (some projects have more than one). Each EMPLOYEE may or (optionally) may not be assigned to any existing PROJECT, or may be assigned to several PROJECTs.
3. PERSON Is_married_to PERSON (Figure 8-14c). This is an optional zero or one cardinality in both directions because a person may or may not be married.

It is possible for the maximum cardinality to be a fixed number, not an arbitrary "many" value. For example, suppose corporate policy states that an employee may work on at most five projects at the same time. We could show this business rule by placing a "5" above or below the crow's foot next to the PROJECT entity in Figure 8-14b.

FIGURE 8-14
Examples of cardinality constraints
(a) Mandatory cardinalities

(b) One optional, one mandatory cardinality

(c) Optional cardinalities

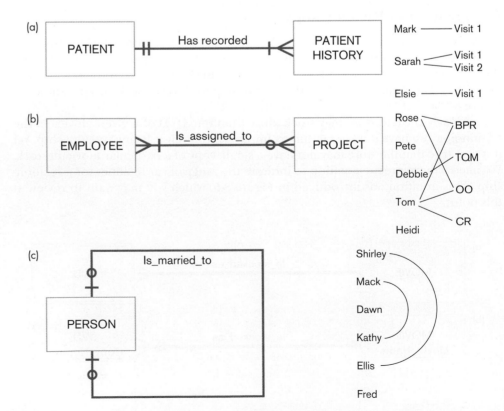

Naming and Defining Relationships

Relationships may be the most difficult component of an E-R diagram to understand. Thus, you should use a few special guidelines for naming relationships, such as the following:

- A relationship name is a *verb phrase* (such as Assigned_to, Supplies, or Teaches). Relationships represent actions, usually in the present tense. A relationship name states the action taken, not the result of the action (e.g., use Assigned_to, not Assignment).
- You should *avoid vague names,* such as Has or Is_related_to. Use descriptive verb phrases taken from the action verbs found in the definition of the relationship.

    Specific guidelines for defining relationships follow:

- A relationship definition explains *what action is being taken and possibly why it is important.* It may be important to state who or what does the action, but it is not important to explain how the action is taken.
- It may be important to *give examples to clarify the action.* For example, for a relationship Registered_for between student and course, it may be useful to explain that this covers both on-site and online registration and registrations made during the drop/add period.
- The definition should explain any *optional participation.* You should explain what conditions lead to zero associated instances, whether this can happen only when an entity instance is first created or whether this can happen at any time.
- A relationship definition should also *explain the reason for any explicit maximum cardinality* other than many.
- A relationship definition should *explain any restrictions on participation in the relationship.* For example, "Supervised_by links an employee with the other employees he or she supervises and links an employee with the other employee who supervises him or her. An employee cannot supervise him- or herself, and an employee cannot supervise other employees if his or her job classification level is below 4."
- A relationship definition should *explain the extent of history that is kept in the relationship.*
- A relationship definition should *explain whether an entity instance* involved in a relationship instance *can transfer participation to another relationship instance.* For example, "Places links a customer with the orders they have placed with our company. An order is not transferable to another customer."

Associative Entities

As seen in the examples of the Supplies relationship in Figure 8-11 and the Has_ components relationship of Figure 8-12, attributes may be associated with a many-to-many relationship as well as with an entity. For example, suppose that the organization wishes to record the date (month and year) that an employee completes each course. Some sample data follow:

Employee_ID	Course_Name	Date_Completed
549-23-1948	Basic Algebra	March 2014
629-16-8407	Software Quality	June 2014
816-30-0458	Software Quality	February 2014
549-23-1948	C Programming	May 2014

From these limited data, you can conclude that the attribute Date_Completed is not a property of the entity EMPLOYEE because a given employee, 549-23-1948, has completed courses on different dates. Nor is Date_Completed a property of COURSE because a particular course (Software Quality) may be completed on different dates. Instead, Date_Completed is a property of the relationship between EMPLOYEE and COURSE. The attribute is associated with the relationship and diagrammed in Figure 8-15.

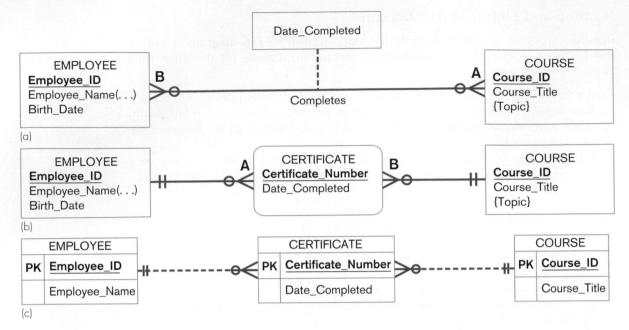

FIGURE 8-15
An associative entity
(a) Attribute on a relationship
(b) An associative entity (CERTIFICATE)
(c) An associative entity using Microsoft Visio®

Associative entity

An entity type that associates the instances of one or more entity types and contains attributes that are peculiar to the relationship between those entity instances; also called a *gerund*.

Because many-to-many and one-to-one relationships may have associated attributes, the E-R data model poses an interesting dilemma: Is a many-to-many relationship actually an entity in disguise? Often the distinction between entity and relationship is simply a matter of how you view the data. An **associative entity** (sometimes called a *gerund*) is a relationship that the data modeler chooses to model as an entity type. Figure 8-15b shows the E-R notation for representing the Completes relationship as an associative entity and Figure 8-15c shows how this would be modeled using Microsoft Visio. The lines from CERTIFICATE to the two entities are not two separate binary relationships, so they do not have labels. Note that EMPLOYEE and COURSE have mandatory one cardinality because an instance of Completes must have an associated EMPLOYEE and COURSE. The labels A and B show where the cardinalities from the Completes relation now appear. We have created an identifier for CERTIFICATE of Certificate_Number, rather than use the implied combination of the identifiers of EMPLOYEE and COURSE, Employee_ID and Course_Name, respectively.

An example of the use of an associative entity for a ternary relationship appears in Figure 8-16. This figure shows an alternative (and more explicit) representation of the ternary Supplies relationship shown in Figure 8-11. In Figure 8-16, the entity type (associative entity) SHIPMENT SCHEDULE replaces the Supplies relationship from Figure 8-11. Each instance of SHIPMENT SCHEDULE represents a real-world shipment by a given vendor of a particular part to a selected warehouse. The Shipment_Mode and Unit_Cost are attributes of SHIPMENT SCHEDULE. We have not designated an identifier for SHIPMENT SCHEDULE, so implicitly it would be a composite identifier of the identifiers of the three related entities. Business rules about participation of vendors, parts, and warehouses in supplies relationships are shown via the cardinalities next to SUPPLY SCHEDULE. Remember, these are not three separate relationships, as with any associative entity.

One situation in which a relationship *must be* turned into an associative entity is when the associative entity has other relationships with entities besides the relationship that caused its creation. For example, consider the E-R diagram in Figure 8-17a that represents price quotes from different vendors for purchased parts stocked by PVF. Now, suppose that we also need to know which price quote is in effect for each part shipment received. This additional data requirement *necessitates* that the Quotes_price relationship be transformed into an associative entity, as shown in Figure 8-17b.

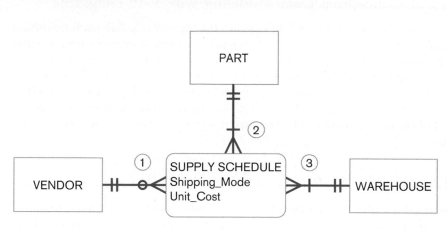

(1) Each vendor can supply many parts to any number of warehouses, but need not supply any parts.

(2) Each part can be supplied by any number of vendors to more than one warehouse, but each part must be supplied by at least one vendor to a warehouse.

(3) Each warehouse can be supplied with any number of parts from more than one vendor, but each warehouse must be supplied with at least one part.

FIGURE 8-16
Cardinality constraints in a ternary relationship

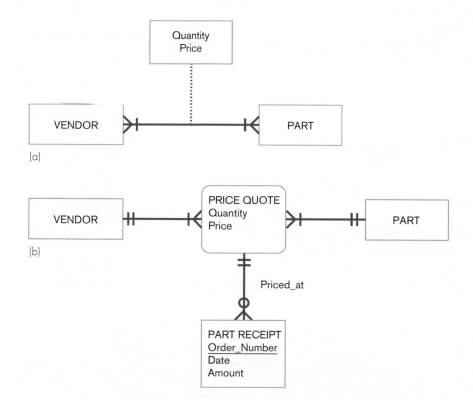

(a)

(b)

FIGURE 8-17
Situation requiring an associative entity
(a) Many-to-many relationship with attributes
(b) Associative entity with separate relationship

In this case, PRICE QUOTE is not a ternary relationship. Rather, PRICE QUOTE is a binary many-to-many relationship (associative entity) between VENDOR and PART. In addition, each PART RECEIPT, based on Amount, has an applicable, negotiated Price. Each PART RECEIPT is for a given PART from a specific VENDOR, and the Amount of the receipt dictates the purchase price in effect by matching with the Quantity attribute. Because the PRICE QUOTE pertains to a given PART and a given VENDOR, PART RECEIPT does not need direct relationships with these entities.

Summary of Conceptual Data Modeling with E-R Diagrams

The purpose of E-R diagramming is to capture the richest possible understanding of the meaning of data necessary for an information system or organization. Besides the aspects shown in this chapter, there are many other semantics about data that E-R diagramming can represent. Some of these more advanced capabilities are explained in Hoffer et al. (2011). You can also find some general guidelines for effective conceptual data modeling in Moody (1996) . The following section presents one final aspect of conceptual data modeling: capturing the relationship between similar entity types.

REPRESENTING SUPERTYPES AND SUBTYPES

Often two or more entity types seem very similar (maybe they have almost the same name), but there are a few differences. That is, these entity types share common properties but also have one or more distinct attributes or relationships. To address this situation, the E-R model has been extended to include supertype/subtype relationships. A **subtype** is a subgrouping of the entities in an entity type that is meaningful to the organization. For example, STUDENT is an entity type in a university. Two subtypes of STUDENT are GRADUATE STUDENT and UNDERGRADUATE STUDENT. A **supertype** is a generic entity type that has a relationship with one or more subtypes.

An example illustrating the basic notation used for supertype/subtype relationships appears in Figure 8-18. The supertype PATIENT is connected with a line to a circle, which in turn is connected by a line to each of the two subtypes, OUTPATIENT and RESIDENT PATIENT. Attributes that are shared by all patients (including the identifier) are associated with the supertype; attributes that are unique to a particular subtype (e.g., Checkback_Date for OUTPATIENT) are associated with that subtype. Relationships in which all types of patients participate (Is_cared_for) are associated with the supertype; relationships in which only a subtype participates (Is_assigned for RESIDENT PATIENTs) are associated only with the relevant subtype.

Several important business rules govern supertype/subtype relationships. The **total specialization rule** specifies that each entity instance of the supertype must be a member of some subtype in the relationship. The **partial specialization rule** specifies that an entity instance of the supertype does not have to belong to any subtype. Total specialization is shown on an E-R diagram by a double line from the supertype to the circle, and partial specialization is shown by a single line. The **disjoint rule** specifies that if an entity instance of the supertype is a member of one subtype, it cannot

Subtype
A subgrouping of the entities in an entity type that is meaningful to the organization and that shares common attributes or relationships distinct from other subgroupings.

Supertype
A generic entity type that has a relationship with one or more subtypes.

Total specialization rule
Specifies that each entity instance of the supertype must be a member of some subtype of the relationship.

Partial specialization rule
Specifies that an entity instance of the supertype does not have to belong to any subtype.

Disjoint rule
Specifies that if an entity instance of the supertype is a member of one subtype, it cannot simultaneously be a member of any other subtype.

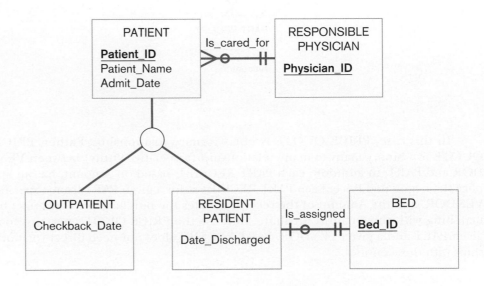

FIGURE 8-18
Supertype/subtype relationships in a hospital

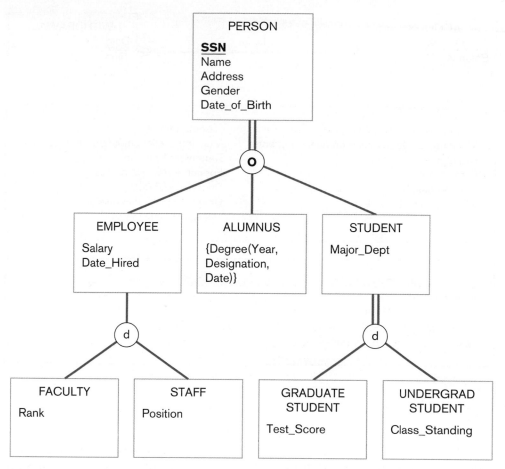

FIGURE 8-19
Example of supertype/subtype hierarchy

simultaneously be a member of any other subtype. The **overlap rule** specifies that an entity instance can simultaneously be a member of two (or more) subtypes. Disjoint versus overlap is shown by a "d" or an "o" in the circle.

Figure 8-19 illustrates several combinations of these rules for a hierarchy of supertypes and subtypes in a university database. In this example,

- A PERSON must be (total specialization) an EMPLOYEE, an ALUMNUS, or a STUDENT, or any combination of these subtypes (overlap).
- An EMPLOYEE must be a FACULTY or a STAFF (disjoint), or may be just an EMPLOYEE (partial specialization).
- A STUDENT can be only a GRADUATE STUDENT or an UNDERGRADUATE STUDENT and nothing else (total specialization and disjoint).

Overlap rule
Specifies that an entity instance can simultaneously be a member of two (or more) subtypes.

BUSINESS RULES

Conceptual data modeling is a step-by-step process for documenting information requirements, and it is concerned with both the structure of data and with rules about the integrity of those data (see the eighth question in Table 8-1). **Business rules** are specifications that preserve the integrity of the logical data model. Four basic types of business rules are as follows:

Business rules
Specifications that preserve the integrity of the logical data model.

1. *Entity integrity.* Each instance of an entity type must have a unique identifier that is not null.
2. *Referential integrity constraints.* Rules concerning the relationships between entity types.
3. *Domains.* Constraints on valid values for attributes.
4. *Triggering operations.* Other business rules that protect the validity of attribute values.

FIGURE 8-20
Examples of business rules
(a) Simple banking relationship

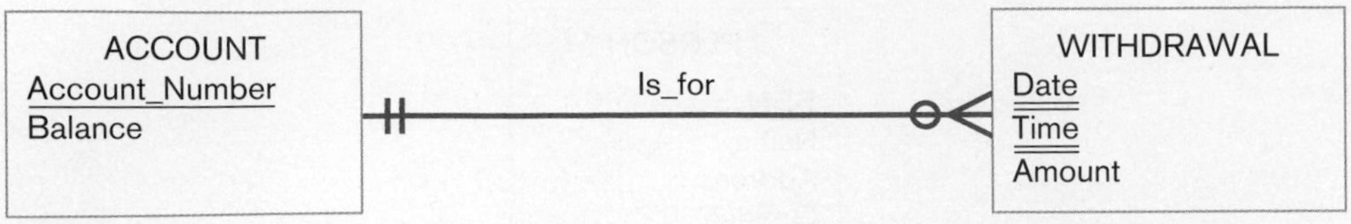

(b) Typical domain definitions

Name: Account_Number
Meaning: Customer account number in bank
Data type: Character
Format: nnn-nnnn
Uniqueness: Must be unique
Null support: Non-null

Name: Amount
Meaning: Dollar amount of transaction
Data type: Numeric
Format: 2 decimal places
Range: 0–10,000
Uniqueness: Nonunique
Null support: Non-null

(c) Typical triggering operation

User rule: WITHDRAWAL Amount may not exceed ACCOUNT Balance
Event: Insert
Entity Name: WITHDRAWAL
Condition: WITHDRAWAL Amount > ACCOUNT Balance
Action: Reject the insert transaction

The E-R model that we have described in this chapter is concerned primarily with the structure of data rather than with expressing business rules (although some elementary rules are implied in the E-R model). Generally, the business rules are captured during requirements determination and stored in the CASE repository as they are documented. Entity integrity was described earlier in this chapter, and referential integrity is described in Chapter 9 because it applies to database design. In this section, we briefly describe two types of rules: domains and triggering operations. These rules are illustrated with a simple example from a banking environment, shown in Figure 8-20a. In this example, an ACCOUNT entity has a relationship (Is_for) with a WITHDRAWAL entity.

Domains

Domain
The set of all data types and values that an attribute can assume.

A **domain** is the set of all data types and ranges of values that attributes may assume (Fleming and von Halle, 1990). Domain definitions typically specify some (or all) of the following characteristics of attributes: data type, length, format, range, allowable values, meaning, uniqueness, and null support (whether an attribute value may or may not be null).

Figure 8-20b shows two domain definitions for the banking example. The first definition is for Account_Number. Because Account_Number is an identifier attribute, the definition specifies that Account_Number must be unique and also must not be null (these specifications are true of all identifiers). The definition specifies that the attribute data type is character and that the format is nnn-nnnn. Thus, any attempt to enter a value for this attribute that does not conform to its character type or format will be rejected, and an error message will be displayed.

The domain definition for the Amount attribute (dollar amount of the requested withdrawal) also may not be null, but is not unique. The format allows for two decimal places to accommodate a currency field. The range of values has a lower limit of zero (to prevent negative values) and an upper limit of 10,000. The latter is an arbitrary upper limit for a single withdrawal transaction.

The use of domains offers several advantages:

- Domains verify that the values for an attribute (stored by insert or update operations) are valid.
- Domains ensure that various data manipulation operations (such as joins or unions in a relational database system) are logical.
- Domains help conserve effort in describing attribute characteristics.

Domains can conserve effort because we can define domains and then associate each attribute in the data model with an appropriate domain. To illustrate, suppose that a bank has three types of accounts, with the following identifiers:

Account Type	Identifier
CHECKING	Checking_Account_Number
SAVINGS	Savings_Account_Number
LOAN	Loan_Account_Number

If domains are not used, the characteristics for each of the three identifier attributes must be described separately. Suppose, however, that the characteristics for all three of the attributes are identical. Having defined the domain Account_Number once (as shown in Figure 8-13b), we simply associate each of these three attributes with Account_Number. Other common domains such as Date, Social_Security_Number, and Telephone_Number also need to be defined just once in the model.

Triggering Operations

A **triggering operation** (also called a **trigger**) is an assertion or rule that governs the validity of data manipulation operations such as insert, update, and delete. The scope of triggering operations may be limited to attributes within one entity or it may extend to attributes in two or more entities. Complex business rules may often be stated as triggering operations.

A triggering operation normally includes the following components:

1. *User rule.* A concise statement of the business rule to be enforced by the triggering operation
2. *Event.* The data manipulation operation (insert, delete, or update) that initiates the operation
3. *Entity name.* The name of the entity being accessed and/or modified
4. *Condition.* The condition that causes the operation to be triggered
5. *Action.* The action taken when the operation is triggered

Figure 8-20c shows an example of a triggering operation for the banking situation. The business rule is a simple (and familiar) one: the amount of an attempted withdrawal may not exceed the current account balance. The event of interest is an attempted insert of an instance of the WITHDRAWAL entity type (perhaps from an automated teller machine). The condition is

Amount (of the withdrawal) . ACCOUNT Balance

When this condition is triggered, the action taken is to reject the transaction. You should note two things about this triggering operation: first, it spans two entity types; second, the business rule could not be enforced through the use of domains.

The use of triggering operations is an increasingly important component of database strategy. With triggering operations, the responsibility for data integrity lies within the scope of the database management system rather than with application programs or human operators. In the banking example, tellers could conceivably check the account balance before processing each withdrawal. Human operators would be subject to human error and, in any event, manual processing would not

Triggering operation (trigger)
An assertion or rule that governs the validity of data manipulation operations such as insert, update, and delete; also called a *trigger*.

work with automated teller machines. Alternatively, the logic of integrity checks could be built into the appropriate application programs, but integrity checks would require duplicating the logic in each program. There is no assurance that the logic would be consistent (because the application programs may have been developed at different times by different people) or that the application programs will be kept up to date as conditions change.

As stated earlier, business rules should be documented in the CASE repository. Ideally, these rules will then be checked automatically by database software. Removing business rules from application programs and incorporating them in the repository (in the form of domains, referential integrity constraints, and triggering operations) has several important advantages; specifically, incorporating business rules in the repository.

1. Provides for faster application development with fewer errors because these rules can be generated into programs or enforced by the database management system
2. Reduces maintenance effort and expenditures
3. Provides for faster response to business changes
4. Facilitates end-user involvement in developing new systems and manipulating data
5. Provides for consistent application of integrity constraints
6. Reduces the time and effort required to train application programmers
7. Promotes ease of use of a database

For a more thorough treatment of business rules, see Hoffer et al. (2011) .

ROLE OF PACKAGED CONCEPTUAL DATA MODELS—DATABASE PATTERNS

Fortunately, the art and science of data modeling has progressed to the point where it is seldom necessary for an organization to develop its data models internally in their entirety. Instead, common database patterns for different business situations are available in packaged data models (or model components) that can be purchased at comparatively low cost and, after suitable customization, assembled into full-scale data models. These generic data models are developed by industry specialists, consultants, and database technology vendors based on their expertise and experience in dozens of organizations across multiple industry types. The models are typically provided as the contents of a data modeling software package, such as ERWin from Computer Associates. The software is able to produce E-R diagrams, maintain all metadata about the data model, and produce a variety of reports that help in the process of tailoring the data model to the specific situation, such as customizing data names, changing relationship characteristics, or adding data unique to your environment. The software can then generate the computer code to define the database to a database management system once the design is fully customized to the local situation. Some simple and limited generic data models can be found in books or on the Internet.

There are two principal types of packaged data models: universal data models applicable to nearly any business or organization and industry-specific data models. We discuss each of these types briefly and provide references for each type.

Universal Data Models

Numerous core subject areas are common to many (or even most) organizations, such as customers, products, accounts, documents, and projects. Although they differ in detail, the underlying data structures are often quite similar for these subjects. Further, there are core business functions such as purchasing, accounting, receiving, and project management that follow common patterns. Universal data models are templates for one or more of these subject areas and/or functions. All of the expected components of data models are generally included: entities, relationships, attributes, primary and foreign keys, and even sample data. Two examples of universal data model sets are provided by Hay (1996) and Silverston (2001a).

Industry-Specific Data Models

Industry-specific data models are generic data models that are designed to be used by organizations within specific industries. Data models are available for nearly every major industry group, including health care, telecommunications, discrete manufacturing, process manufacturing, banking, insurance, and higher education. These models are based on the premise that data model patterns for organizations are very similar within a particular industry ("a bank is a bank"). However, the data models for one industry (such as banking) are quite different from those for another (such as hospitals). Prominent examples of industry-specific data models are provided by Silverston (2001b), Kimball and Ross (2002), and Inmon (2000) at *www.billinmon.com.*

Benefits of Database Patterns and Packaged Data Models

Most people in the data modeling field refer to a purchased universal or industry-specific database pattern as a logical data model (LDM). Technically, the term *logical data model* means a conceptual data model with some additional properties associated with the most popular type of database technology—relational databases. The type of data planning and analysis we have covered in this chapter can, in fact, be done using either a conceptual or a logical data model. The process is the same; only the starting point is different.

LDMs are the database version of patterns, components, and prepackaged applications that have been discussed in prior chapters as ways to more quickly and reliably build a new application. An advantage of LDMs is that a packaged data model now exists for almost every industry and application area, for specific operational systems to enterprise systems, such as Enterprise Resource Planning (ERP) and data warehouses. They are available from database software vendors, application software providers, and consulting firms. The use of a prepackaged data model does not eliminate the need for the methods and techniques we discuss in this chapter; they only change the context in which these methods and techniques are used.

It is now important that you consider purchasing a prepackaged data model even when an application is to be built from scratch. Consider the following benefits of starting with and then tailoring a purchased data model:

- *Validated.* Purchased models are proven from extensive experience.
- *Cost reduction.* Projects with purchased models take less time and cost less because the initial discovery steps are no longer necessary, leaving only iterative tailoring and refinement to the local situation.
- *Anticipate the future, not just initial requirements.* Purchased models anticipate future needs, not just those recognized during the first version of an application. Thus, their benefits are recurring, not one-time, because the database design does not require structural change, which can have costly ramifications for reprogramming the applications using the database.
- *Facilitates systems analysis.* The purchased model actually facilitates database planning and analysis by providing a first data model, which you can use to generate specific analysis questions and concrete, not hypothetical or abstract examples of what might be in the appropriate database.
- *Consistent and complete.* The purchased data models are very general, covering almost all options employed by the associated functional area or industry. Thus, they provide a structure that, when tailored, will be consistent and complete.

See Hoffer, et al. (2011) for more details on the use of packaged data models in data modeling and database development. Of course, packaged data models are no substitute for sound database analysis and design. Skilled analysts and designers are still required to determine database requirements and to select, modify, install, and integrate any packaged systems that are used.

ELECTRONIC COMMERCE APPLICATION: CONCEPTUAL DATA MODELING

Conceptual data modeling for an Internet-based electronic commerce application is no different than the process followed when analyzing the data needs for other types of applications. In the preceding chapters, you read how Jim Woo analyzed the flow of information within the WebStore and developed a DFD. In this section, we examine the process he followed when developing the WebStore's conceptual data model.

Conceptual Data Modeling for Pine Valley Furniture's WebStore

To better understand what data would be needed within the WebStore, Jim Woo carefully reviewed the information from the JAD session and his previously developed DFD. Table 8-2 shows a summary of the customer and inventory information identified during the JAD session. Jim wasn't sure if this information was complete, but he knew that it was a good starting place for identifying what information the WebStore needed to capture, store, and process. To identify additional information, he carefully studied the DFD shown in Figure 8-21. In this diagram, two data stores—Inventory and Shopping Cart—are clearly identified; both were strong candidates to become entities within the conceptual data model. Finally, Jim examined the data flows from the DFD as additional possible sources for entities. This analysis resulted in the identification of five general categories of information that he needed to consider:

- Customer
- Inventory
- Order
- Shopping Cart
- Temporary User/System Messages

After identifying these multiple categories of data, Jim's next step was to carefully define each item. To do this, he again examined all data flows within the DFD and recorded the source and destination of all data flows. By carefully listing these flows, he could more easily move through the DFD and more thoroughly understand what information needed to move from point to point. This activity resulted in the creation of two tables that documented his growing understanding of the

TABLE 8-2   Customer and Inventory Information for WebStore

Home Office Customer	Student Customer	Inventory Information
Name	Name	SKU
Doing Business as (company's name)	School	Name
	Address	Description
Address	Phone	Finished Product Size
Phone	E-Mail	Finished Product Weight
Fax		Available Materials
E-Mail		Available Colors
		Price
		Lead Time

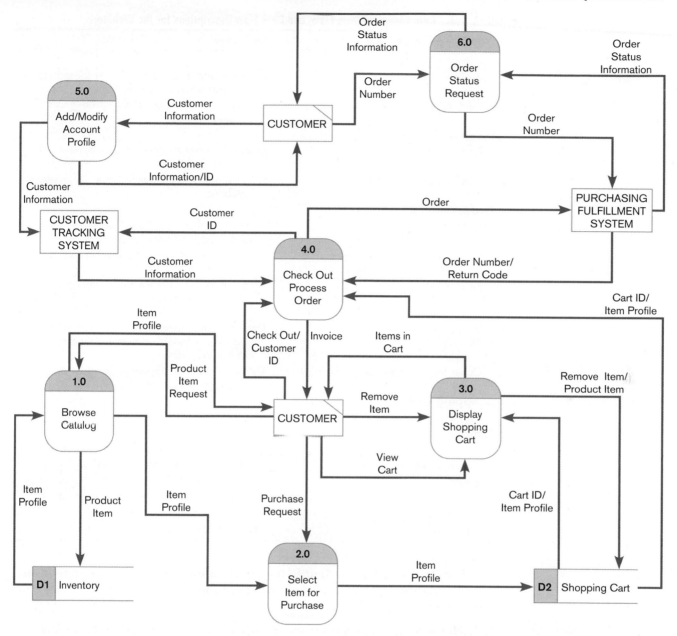

FIGURE 8-21
Level-0 DFD for the WebStore

WebStore's requirements. The first, Table 8-3, lists each of the data flows within each data category and its corresponding description. The second, Table 8-4, lists each of the unique data flows within each data category. He now felt ready to construct an E-R diagram for the WebStore.

Jim concluded that Customer, Inventory, and Order were each a unique entity and would be part of his E-R diagram. Recall that an entity is a person, place, or object; all three of these items meet this criteria. Because the Temporary User/System Messages data were not permanently stored items—nor were they a person, place, or object—he concluded that this should not be an entity in the conceptual data model. Alternatively, although the shopping cart was also a temporarily stored item, its contents needed to be stored for at least the duration of a customer's visit to the WebStore and should be considered an object. As shown in Figure 8-21, Process 4.0, Check Out Process Order, moves the Shopping Cart contents to the Purchasing Fulfillment System, where the order details are stored. Thus, he concluded that Shopping Cart—along with Customer, Inventory, and Order—would be entities in his E-R diagram.

TABLE 8-3   Data Category, Data Flow, and Data Flow Descriptions for the WebStore

Data Category/Data Flow	Description
Customer-Related	
Customer ID	Unique identifier for each customer (generated by Customer Tracking System)
Customer Information	Detailed customer information (stored in Customer Tracking System)
Inventory-Related	
Product Item	Unique identifier for each product item (stored in Inventory Database)
Item Profile	Detailed product information (stored in Inventory Database)
Order-Related	
Order Number	Unique identifier for an order (generated by Purchasing Fulfillment System)
Order	Detailed order information (stored in Purchasing Fulfillment System)
Return Code	Unique code for processing customer returns (generated by/stored in Purchasing Fulfillment System)
Invoice	Detailed order summary statement (generated from order information stored in Purchasing Fulfillment System)
Order Status Information	Detailed summary information on order status (stored/generated by)
Shopping Cart	
Cart ID	Unique identifier for shopping cart
Temporary User/System Messages	
Product Item Request	Request to view information on a catalog item
Purchase Request	Request to move an item into the shopping cart
View Cart	Request to view the contents of the shopping cart
Items in Cart	Summary report of all shopping cart items
Remove Item	Request to remove item from shopping cart
Check Out	Request to check out and process order

The final step was to identify the interrelationships among these four entities. After carefully studying all the related information, Jim came to the following conclusions:

1. Each Customer *owns* zero-or-one Shopping Cart instances; each Shopping Cart instance *is owned by* one-and-only-one Customer.
2. Each Shopping Cart instance *contains* one-and-only-one Inventory item; each Inventory item *is contained in* zero-or-many Shopping Cart instances.
3. Each Customer *places* zero-to-many Orders; each Order *is placed by* one-and-only-one Customer.
4. Each Order *contains* one-to-many Shopping Cart instances; each Shopping Cart instance *is contained in* one-and-only-one Order.

With these relationships defined, Jim drew the E-R diagram shown in Figure 8-22. He now had a very good understanding of the requirements, the flow of information within the WebStore, the flow of information between the WebStore and existing PVF systems, and now the conceptual data model. Over the next few hours, Jim planned to further refine his understanding by listing the specific attributes for each entity and then comparing these lists with the existing inventory, customer, and order database tables. Making sure that all attributes were accounted for would be the final conceptual data modeling activity before beginning the process of selecting a final design strategy.

TABLE 8-4 Data Category, Data Flow, and the Source/Destination of Data Flows within the WebStore DFD

Data Flow	From/To
Customer-Related	
Customer ID	From Customer to Process 4.0
	From Process 4.0 to Customer Tracking System
	From Process 5.0 to Customer
Customer Information	From Customer to Process 5.0
	From Process 5.0 to Customer
	From Process 5.0 to Customer Tracking System
	From Customer Tracking System to Process 4.0
Inventory-Related	
Product Item	From Process 1.0 to Data Store D1
	From Process 3.0 to Data Store D2
Item Profile	From Data Store D1 to Process 1.0
	From Process 1.0 to Customer
	From Process 1.0 to Process 2.0
	From Process 2.0 to Data Store D2
	From Data Store D2 to Process 3.0
	From Data Store D2 to Process 4.0
Order-Related	
Order Number	From Purchasing Fulfillment System to Process 4.0
	From Customer to Process 6.0
	From Process 6.0 to Purchasing Fulfillment System
Order	From Process 4.0 to Purchasing Fulfillment System
Return Code	From Purchasing Fulfillment System to Process 4.0
Invoice	From Process 4.0 to Customer
Order Status	From Process 6.0 to Customer
	From Purchasing Fulfillment System to Process 6.0
Shopping Cart	
Cart ID	From Data Store D2 to Process 3.0
	From Data Store D2 to Process 4.0
Temporary User/System Messages	
Product Item Request	From Customer to Process 1.0
Purchase Request	From Customer to Process 2.0
View Cart	From Customer to Process 3.0
Items in Cart	From Process 3.0 to Customer
Remove Item	From Customer to Process 3.0
	From Process 3.0 to Data Store D2
Check Out	From Customer to Process 4.0

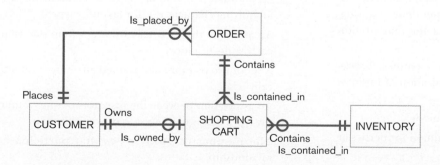

FIGURE 8-22
E-R diagram for the WebStore system

SUMMARY

We have presented the process and basic notation used to model the data requirements of an information system. We outlined the structuring of conceptual data models using the E-R notation and also discussed how the components of a conceptual data model relate to data flows and data stores.

Conceptual data modeling is based on certain constructs about the structure, not use, of data. These constructs include entity, relationship, degree, and cardinality. A data model shows the relatively permanent business rules that define the nature of an organization. Rules define characteristics of data such as the legitimate domain of values for data attributes, the unique characteristics (identifier) of entities, the relationships between different entities, and the triggering operations that protect the validity of attributes during data maintenance.

A data model shows major categories of data, called *entities* for the E-R notation; the associations or relationships between entities; and the attributes of both entities and relationships. A special type of entity called an *associative entity* is often necessary to represent a many-to-many relationship between entities. Entity types are

distinct from entity instances. Each entity instance is distinguished from other instances of the same type by an identifier attribute.

Relationships are the glue that holds a data model together. Three common relationship types are unary, binary, and ternary. The minimum and maximum number of entity instances that participate in a relationship represent important rules about the nature of the organization, as captured during requirements determination. Supertype/subtype relationships can be used to show a hierarchy of more general to more specific, related entity types that share common attributes and relationships. Rules for total and partial specialization between the supertype and subtypes and disjoint and overlap among the subtypes clarify the meaning of the related entity types.

Modern systems analysis is based on reuse, and one form of reuse is prepackaged conceptual data models. These data models can be purchased from various vendors and are very helpful in learning best practices from other organizations in the same industry or for the same business function. They save considerable time over building complex data models from scratch.

KEY TERMS

1. Associative entity
2. Attribute
3. Binary relationship
4. Business rules
5. Candidate key
6. Cardinality
7. Composite attribute
8. Conceptual data model
9. Degree
10. Derived attribute
11. Disjoint rule

12. Domain
13. Entity instance
14. Entity-relationship data model (E-R model)
15. Entity-relationship diagram (E-R diagram)
16. Entity type
17. Identifier
18. Multivalued attribute
19. Optional attribute
20. Overlap rule

21. Partial specialization rule
22. Relationship
23. Repeating group
24. Required attribute
25. Subtype
26. Supertype
27. Ternary relationship
28. Total specialization rule
29. Triggering operation (trigger)
30. Unary relationship

Match each of the key terms above with the definition that best fits it.

_____ A detailed model that captures the overall structure of organizational data and that is independent of any database management system or other implementation considerations.

_____ A detailed, logical representation of the entities, associations, and data elements for an organization or business area.

_____ A graphical representation of an E-R model.

_____ A collection of entities that share common properties or characteristics.

_____ A single occurrence of an entity type.

_____ A named property or characteristic of an entity that is of interest to the organization.

_____ An attribute (or combination of attributes) that uniquely identifies each instance of an entity type.

_____ A candidate key that has been selected as the unique, identifying characteristic for an entity type.

_____ An attribute that may take on more than one value for each entity instance.

_____ A set of two or more multivalued attributes that are logically related.

_____ An association between the instances of one or more entity types that is of interest to the organization.

_____ The number of entity types that participate in a relationship.

_____ A relationship between the instances of one entity type.

_____ A relationship between instances of two entity types.

_____ A simultaneous relationship among instances of three entity types.

_____ The number of instances of entity B that can (or must) be associated with each instance of entity A.

_____ An entity type that associates the instances of one or more entity types and contains attributes that are peculiar to the relationship between those entity instances.

_____ A subgrouping of the entities in an entity type that is meaningful to the organization.

_____ A generic entity type that has a relationship with one or more subtypes.

_____ Specifies that each entity instance of the supertype must be a member of some subtype in the relationship.

_____ Specifies that an entity instance of the supertype does not have to belong to any subtype.

_____ Specifies that if an entity instance of the supertype is a member of one subtype, it cannot simultaneously be a member of any other subtype.

_____ Specifies that an entity instance can simultaneously be a member of two (or more) subtypes.

_____ Specifications that preserve the integrity of the logical data model.

_____ The set of all data types and values that an attribute can assume.

_____ An assertion or rule that governs the validity of data manipulation operations such as insert, update, and delete.

_____ An attribute that must have a value for every entity instance.

_____ An attribute that may not have a value for every entity instance.

_____ An attribute that has meaningful component parts.

_____ An attribute whose value can be computed from related attribute values.

REVIEW QUESTIONS

1. Discuss why some systems developers believe that a data model is one of the most important parts of the statement of information system requirements.

2. Distinguish between the data modeling done during information systems planning, project initiation and planning, and the analysis phases of the SDLC.

3. What elements of a DFD should be analyzed as part of data modeling?

4. Explain why a ternary relationship is not the same as three binary relationships.

5. When must a many-to-many relationship be modeled as an associative entity?

6. What is the significance of triggering operations and business rules in the analysis and design of an information system?

7. Which of the following types of relationships—one-to-one, one-to-many, many-to-many—can have attributes associated with them?

8. What are the linkages among DFDs, decision tables, and E-R diagrams?

9. What is the degree of a relationship? Give an example of each of the relationship degrees illustrated in this chapter.

10. Give an example (different from any example in this chapter) of a ternary relationship.

11. List the deliverables from the conceptual data modeling part of the analysis phase of the systems development process.

12. Explain the relationship between minimum cardinality and optional and mandatory participation.

13. List the ideal characteristics of an entity identifier attribute.

14. Explain how conceptual data modeling is different when you start with a prepackaged data model rather than a clean sheet of paper.

15. Contrast the following terms:
    a. Subtype; supertype
    b. Total specialization rule; partial specialization rule
    c. Disjoint rule; overlap rule
    d. Attribute; operation

PROBLEMS AND EXERCISES

1. Assume that at PVF, each product (described by Product No., Description, and Cost) is composed of at least three components (described by Component No., Description, and Unit of Measure), and components are used to make one or many products (i.e., must be used in at least one product). In addition, assume that components are used to make other components and that raw materials are also considered to be components. In both cases of components being used to make products and components being used to make other components, we need to keep track of how many components go into making something else. Draw an E-R diagram for this situation and place minimum and maximum cardinalities on the diagram.

2. Much like PVF's sale of products, stock brokerages sell stocks, and the prices are continually changing. Draw an E-R diagram that takes into account the changing nature of stock prices.

3. If you were going to develop a computer-based tool to help an analyst interview users and quickly and easily create and edit E-R diagrams, what type of tool would you build? What features would it have? How would it work?

4. A software training program is divided into training modules, and each module is described by module name and the approximate practice time. Each module sometimes has prerequisite modules. Model this situation of training programs and modules with an E-R diagram.

5. Each semester, each student must be assigned an adviser who counsels students about degree requirements and helps students register for classes. Students must register for classes with the help of an adviser, but if their assigned adviser is not available, they may register with any adviser. We must keep track of students, their assigned adviser, and with whom the student registered for the current term. Represent this situation of students and advisers with an E-R diagram.

6. Assume that entity PART has attributes Part_Number, Drawing_Number, Weight, Description, Storage_Location, and Cost. Which attributes are candidate keys? Why? Which attribute would you select for the identifier of PART? Why? Or do you think that you should create another attribute to be the identifier? Why or why not?

7. Consider the E-R diagram in Figure 8-15b.

   a. What would be the identifier for the CERTIFICATE associative entity if Certificate_Number were not included?

   b. Now assume that the same employee may take the same course multiple times, on different dates. Does this change your answer to Problem and Exercise 7a? Why or why not?

8. Study the E-R diagram in Figure 8-23. Based on this E-R diagram, answer the following questions:

   a. How many PROJECTs can an employee work on?

   b. What is the degree of the Includes relationship?

   c. Are there any associative entities on this diagram? If so, name them.

   d. How else could the attribute Skill be modeled?

   e. Is it possible to attach any attributes to the Includes relationship?

   f. Could TASK be modeled as an associative entity?

   g. Employees earnings are calculated based on a different hourly pay rate for each project. Where on the E-R diagram would you represent the new attribute Hourly pay rate?

9. For the E-R diagram provided in Figure 8-24, draw in the relationship cardinalities and describe them. Describe any assumptions you must make about relevant business rules. Are there any changes or additions you would make to this diagram to make it better? Why or why not?

10. For the E-R diagram provided in Figure 8-24, assume that this company decided to assign each sales representative to a small, unique set of customers; some customers can now become "members" and receive unique benefits; small manufacturing teams will be formed and each will be assigned to the production of a small, unique set of products; and each purchasing agent will be assigned to a small, unique set of vendors. Make the necessary changes to the E-R diagram and draw and describe the new relationship cardinalities.

11. Obtain a copy of an invoice, order form, or bill used in one of your recent business transactions. Create an E-R diagram to describe your sample document.

12. Using Table 8-1 as a guide, develop the complete script (questions and possible answers) of an interview between analysts and users within the order entry function at PVF.

13. A concert ticket reservation is an association among a patron, a concert, and a seat. Select a few pertinent attributes for each of these entity types and represent a reservation in an E-R diagram.

14. Choose from your own experiences with organizations and draw an E-R diagram for a situation that has a ternary relationship.

15. Consider the E-R diagram in Figure 8-25. Are all three relationships—Holds, Goes_on, and Transports—necessary (i.e., can one of these be deduced from the other two)? Are there reasonable assumptions that make all three relationships necessary? If so, what are these assumptions?

16. Draw an E-R diagram to represent the sample customer order in Figure 8-4.

17. In a real estate database, there is an entity called PROPERTY, which is a property for sale by the agency. Each time a potential property buyer makes a purchase offer on a

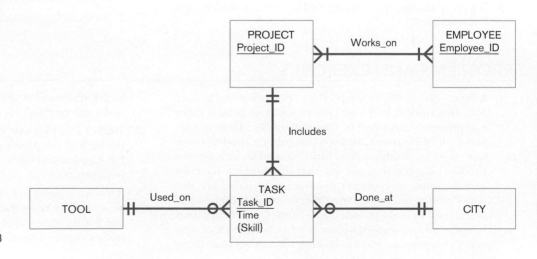

FIGURE 8-23
E-R diagram for Problem and Exercise 8

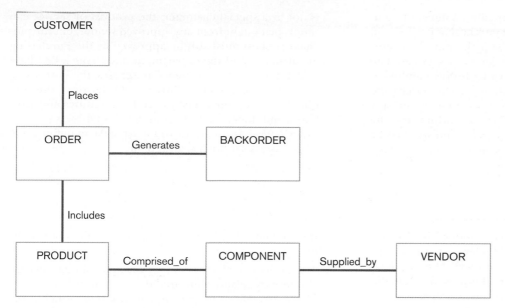

FIGURE 8-24
E-R diagram for Problem and Exercises
9 and 10

property, the agency records the date, offering price, and name of the person making the offer.

a. Represent the PROPERTY entity and its purchase offer attributes using the notation for multivalued attributes.

b. Represent the PROPERTY entity and its purchase offer attributes using two entity types.

c. Assume the agency decides to also keep data about buyers and potential buyers, including their name, phone number, and address. Buyers often have multiple phone numbers and addresses, which are not necessarily related to each other. Augment your answer to Problem and Exercise 17b to accommodate this new entity type.

d. Finally, assume that, for each purchase offer, we need to know which buyer phone number and address to associate with that offer. Augment your answer to Problem and Exercise 17c to accommodate this new requirement.

18. Consider the Is_married_to unary relationship in Figure 8-14c.

a. Assume that we want to know the date on which a marriage occurred. Augment this E-R diagram to include a Date_married attribute.

b. Because persons sometimes remarry after the death of a spouse or a divorce, redraw this E-R diagram to show the whole history of marriages (not just the current marriage) for persons. Show the Date_married attribute on this diagram.

c. In your answer to Problem and Exercise 18b, is it possible to represent a situation in which the same two people marry each other a multiple number of times? Explain.

19. Consider Figure 8-20.

a. Write a domain integrity rule for Balance.

b. Write a triggering operation for the Balance attribute for the event of inserting a new ACCOUNT.

20. How are E-R diagrams similar to and different from decision trees? In what ways are data and logic modeling techniques complementary? What problems might be encountered if either data or logic modeling techniques were not performed well or not performed at all as part of the systems development process?

21. In the Purchasing department at one company, a purchase request may be assigned an expediter within the

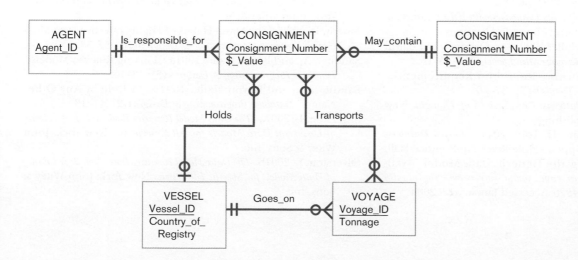

FIGURE 8-25
E-R diagram for Problem and
Exercise 15

Purchasing department. This expediter follows the purchase request through the entire purchasing process and acts as the sole contact person with the person or unit within the company buying the goods or services. The Purchasing department refers to its fellow employees buying goods and services as customers. The purchasing process is such that purchase requests from certain special customers must go out for bid to vendors, and the associated Request for Bids for these requests must be approved by the Purchasing department. If the purchase is not by a special customer, the product or service can simply be bought from any approved vendor, but the purchase request must still be approved by the Purchasing department, and the department must issue a Purchase Order. For "special customer" purchases, the Purchasing department can issue a Purchase Order once the winning bid is accepted. List the relevant entities and attributes, and draw an E-R diagram for this business process. List whatever assumptions you must make to define identifiers, assess cardinality, and so on.

FIELD EXERCISES

1. Interview a friend or family member about each of the entities, attributes, relationships, and relevant business rules he or she comes into contact with at work. Use this information to construct and present to this person an E-R diagram. Revise the diagram until it seems appropriate to you and to your friend or family member.

2. Visit an organization that provides primarily a service, such as a dry cleaner, and a company that manufactures a more tangible product. Interview employees from these organizations about the entities, attributes, relationships, and relevant business rules that are commonly encountered by their company. Use this information to construct E-R diagrams. What differences and similarities are there between the diagrams for the service- and the product-oriented companies? Does the E-R diagramming technique handle both situations equally well? Why or why not? What differences, if any, might there be in the use of this technique for a public agency?

3. Discuss with a systems analyst the role of conceptual data modeling in the overall systems analysis and design of information systems at his or her company. How, and by whom, is conceptual data modeling performed? What training in this technique is given? At what point(s) is this done in the development process? Why?

4. Ask a systems analyst to give you examples of unary, binary, and ternary relationships that they have heard of or dealt with personally at their company. Ask them which is the most common. Why?

5. Talk to MIS professionals at a variety of organizations and determine their interest in using prepackaged data models, rather than doing data modeling from scratch. If they have adopted any prepackaged data models, document how they did the customization for their local requirements.

6. Ask a systems analyst to give you a copy of the standard notation he or she uses to draw E-R diagrams. In what ways is this notation different from the notation in this text? Which notation do you prefer and why? What is the meaning of any additional notation?

7. Ask a systems analyst in a manufacturing company to show you an E-R diagram for a database in that organization that contains bill-of-materials data. Compare that E-R diagram to the one in Figure 8-7. What are the differences between these diagrams?

REFERENCES

Aranow, E. B. 1989. "Developing Good Data Definitions." *Database Programming & Design* 2(8): 36–39.

Bruce, T. A. 1992. *Designing Quality Databases with IDEF1X Information Models.* New York: Dorset House Publications.

Fleming, C. C., and B. von Halle. 1990. "An Overview of Logical Data Modeling." *Data Resource Management* 1(1): 5–15.

Gottesdiener, E. 1999. "Turning Rules into Requirements." *Application Development Trends* 6(7): 37–50.

Hay, D. 1996. *Data Model Patterns: Conventions of Thought.* New York: Dorset House Publishing.

Hoffer, J. A., V. Ramesh, and H. Topi. 2011. *Modern Database Management,* 10th ed. Upper Saddle River, NJ: Prentice Hall.

Inmon, W. H. 2000. Using the Generic Data Model. Available at *www.billinmon.com; www.dmreview.com/master .cfm?NavID=55&EdID=4820.* Accessed January 12, 2004.

Kimball, R., and M. Ross. 2002. *The Data Warehouse Toolkit: The Complete Guide to Dimensional Data Modeling,* 2nd ed. New York: John Wiley & Sons, Inc.

Moody, D. 1996. "The Seven Habits of Highly Effective Data Modelers." *Database Programming & Design* 9(10): 57, 58, 60–62, 64.

Sandifer, A., and B. von Halle. 1991a. "Linking Rules to Models." *Database Programming & Design* 4(3): 13–16.

Sandifer, A., and B. von Halle. 1991b. "A Rule by Any Other Name." *Database Programming & Design* 4(2): 11–13.

Silverston, L. 2001a. *The Data Model Resource Book, Vol. 1: A Library of Universal Data Models for All Enterprises.* New York: John Wiley & Sons, Inc.

Silverston, L. 2001b. *The Data Model Resource Book, Vol. 2: A Library of Data Models for Specific Industries.* New York: John Wiley & Sons, Inc.

Object-Oriented Analysis and Design

Object Modeling—Class Diagrams

Learning Objectives

After studying this section, you should be able to:

- Concisely define each of the following key data modeling terms: *object, state, behavior, object class, class diagram, operation, encapsulation, association role, abstract class, polymorphism, aggregation,* and *composition.*

- Draw a class diagram to represent common business situations.

- Explain the unique capabilities of class diagrams compared with E-R diagrams for modeling data.

Introduction

In this section, we show how to develop class diagrams, the object-oriented data modeling notation. We describe the main concepts and techniques involved in object modeling, including objects and classes, encapsulation of attributes and operations, aggregation relationships, polymorphism, and inheritance. We show how you can develop class diagrams, using the UML notation, to provide a conceptual view of the system being modeled. For a more thorough coverage of object modeling, see George et al. (2007).

REPRESENTING OBJECTS AND CLASSES

With the object-oriented approach, we model the world in objects. Before applying this approach to a real-world problem, we need to understand what an

Object

An entity that has a well-defined role in the application domain, and it has state, behavior, and identity characteristics.

State

Encompasses an object's properties (attributes and relationships) and the values of those properties.

Behavior

Represents how an object acts and reacts.

Object class

A logical grouping of objects that have the same (or similar) attributes, relationships, and behaviors; also called *class*.

Class diagram

Shows the static structure of an object-oriented model; the object classes, their internal structure, and the relationships in which they participate.

object really is. Similar to an entity instance, an **object** has a well-defined role in the application domain, and it has state (data), behavior, and identity characteristics. An object is a single occurrence of a class, which we define below.

An object has a state and exhibits behavior through operations that can examine or affect its state. The **state** of an object encompasses its properties (attributes and relationships) and the values of those properties. Its **behavior** represents how an object acts and reacts (Booch, 1994). An object's state is determined by its attribute values and links to other objects. An object's behavior depends on its state and the operation being performed. An operation is simply an action that one object performs upon another in order to get a response. You can think of an operation as a service provided by an object (supplier) to its clients. A client sends a message to a supplier, which delivers the desired service by executing the corresponding operation.

Consider the example of a student, Mary Jones, represented as an object. The state of this object is characterized by its attributes, let's say, name, date of birth, year, address, and phone, and the values these attributes currently have; for example, name is "Mary Jones," year is "junior," and so on. Its behavior is expressed through operations such as calc-gpa, which is used to calculate a student's current grade point average. The Mary Jones object, therefore, packages both its state and its behavior together.

All objects have an identity; that is, no two objects are the same. For example, if there are two Student instances with the same name and date of birth (or even all attributes), they are essentially two different objects. An object maintains its own identity over its life. For example, if Mary Jones gets married and changes her name, address, and phone, she will still be represented by the same object. This concept of an inherent identity is different from the identifier concept we saw earlier for E-R modeling.

We use the term **object class** (or simply class) to refer to a logical grouping of objects that have the same (or similar) attributes, relationships, and behaviors (methods) (just as we used entity type and entity instance earlier in this chapter). In our example, therefore, Mary Jones is an object instance, whereas Student is an object class (as Student was an entity type in E-R diagramming).

Classes can be depicted graphically in a class diagram, as shown in Figure 8-26. A **class diagram** shows the static structure of an object-oriented model: the object classes, their internal structure, and the relationships in which they participate. In UML, a class is represented by a rectangle with three compartments separated by horizontal lines. The class name appears in the top compartment, the list of attributes in the middle compartment, and the list of operations in the bottom compartment of the rectangle. The diagram in Figure 8-26 shows two classes, Student and Course, along with their attributes and operations.

The Student class is a group of Student objects that share a common structure and a common behavior. Each object knows to which class it belongs; for example, the Mary Jones object knows that it belongs to the Student class. Objects belonging to the same class may also participate in similar relationships with other objects; for

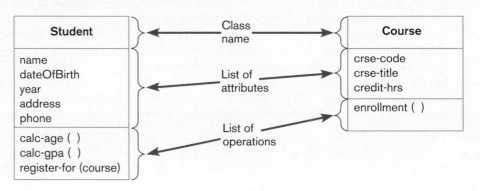

FIGURE 8-26
UML class diagram showing two classes

example, all students register for courses and, therefore, the Student class can participate in a relationship called "registers-for" with another class called Course (see the section Representing Associations).

An **operation**, such as calc-gpa in Student (see Figure 8-26), is a function or a service that is provided by all the instances of a class to invoke behavior in an object by passing a message. It is only through such operations that other objects can access or manipulate the information stored in an object. The operations, therefore, provide an external interface to a class; the interface presents the outside view of the class without showing its internal structure or how its operations are implemented. This technique of hiding the internal implementation details of an object from its external view is known as **encapsulation**, or information hiding (Booch, 1994; Rumbaugh et al., 1991). So while we provide the abstraction of the behavior common to all instances of a class in its interface, we encapsulate within the class its structure and the secrets of the desired behavior.

TYPES OF OPERATIONS

Operations can be classified into three types, depending on the kind of service requested by clients: (1) constructor, (2) query, and (3) update (*UML Notation Guide*, 1997). A **constructor operation** creates a new instance of a class. For example, you can have an operation called create-student within Student that creates a new student and initializes its state. Such constructor operations are available to all classes and, therefore, are not explicitly shown in the class diagram.

A **query operation** is an operation that does not have any side effects; it accesses the state of an object but does not alter the state (Fowler, 2000; Rumbaugh et al., 1991). For example, the Student class can have an operation called get-year (not shown) that simply retrieves the year (freshman, sophomore, junior, or senior) of the Student object specified in the query. Note that there is no need to show explicitly a query such as get-year in the class diagram because it retrieves the value of an independent base attribute. Consider, however, the calc-age operation within Student. This is also a query operation because it does not have any side effects. Note that the only argument for this query is the target Student object. Such a query can be represented as a derived attribute (Rumbaugh et al., 1991); for example, we can represent "age" as a derived attribute of Student. Because the target object is always an implicit argument of an operation, there is no need to show it explicitly in the operation declaration.

An **update operation** has side effects; it alters the state of an object. For example, consider an operation of Student called promote-student (not shown). The operation promotes a student to a new year, say, from junior to senior, thereby changing the Student object's state (value of the year attribute). Another example of an update operation is register-for(course), which, when invoked, has the effect of establishing a connection from a Student object to a specific Course object. Note that, in addition to having the target Student object as an implicit argument, the operation has an explicit argument called "course" that specifies the course for which the student wants to register. Explicit arguments are shown within parentheses.

A **class-scope operation** is an operation that applies to a class rather than an object instance. For example, avg-gpa for the Student class (not shown with the other operations for this class in Figure 8-26) calculates the average gpa across all students (the operation name is underlined to indicate that it is a scope operation).

REPRESENTING ASSOCIATIONS

Parallel to the definition of a relationship for the E-R model, an **association** is a relationship among instances of object classes. As in the E-R model, the degree of an association relationship may be one (unary), two (binary), three (ternary), or higher

Operation
A function or a service that is provided by all the instances of a class.

Encapsulation
The technique of hiding the internal implementation details of an object from its external view.

Constructor operation
An operation that creates a new instance of a class.

Query operation
An operation that accesses the state of an object but does not alter the state.

Update operation
An operation that alters the state of an object.

Class-scope operation
An operation that applies to a class rather than an object instance.

Association
A named relationship between or among object classes.

Association role
The end of an association where it connects to a class.

(*n*-ary). In Figure 8-27, we illustrate how the object-oriented model can be used to represent association relationships of different degrees. An association is shown as a solid line between the participating classes. The name given to the end of an association where it connects to a class is called an **association role** (*UML Notation Guide*, 1997). Each association has two or more roles. A role may be explicitly named with a label near the end of an association (see the "manager" role in Figure 8-27). The role name indicates the role played by the class attached to the end near which the name appears. Use of role names is optional. You can specify role names in place of or in addition to an association name. You may show the direction of an association explicitly by using a solid triangle next to the association name.

Figure 8-27 shows two unary relationships, Is-married-to and Manages. At one end of the Manages relationship, we have named the role as "manager," implying that an employee can play the role of a manager. We have not named the other roles, but we have named the associations. When the role name does not appear, you may think of the role name as being that of the class attached to that end (Fowler, 2000). For example, you may call the role for the right end of the Is-assigned relationship in Figure 8-27 "Parking Place."

Multiplicity
A specification that indicates how many objects participate in a given relationship.

Each role has a **multiplicity**, which indicates how many objects participate in a given association relationship. For example, a multiplicity of 2..5 denotes that a minimum of two and a maximum of five objects can participate in a given relationship. Multiplicities, therefore, are nothing but cardinality constraints, which you saw in E-R diagrams. In addition to integer values, the upper bound of a multiplicity can be an * (asterisk), which denotes an infinite upper bound. If a single integer value is specified, it means that the range includes only that value.

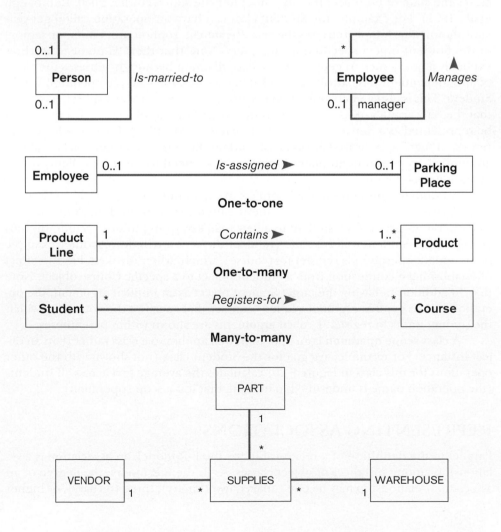

FIGURE 8-27
Examples of association relationships of different degrees

FIGURE 8-28
Example of binary associations

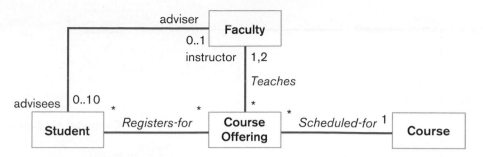

The multiplicities for both roles in the Is-married-to relationship in Figure 8-27 are 0..1, indicating that a person may be single or married to one person. The multiplicity for the manager role in the Manages relationship is 0..1 and that for the other role is 0..*, implying that an employee may be managed by only one manager, but a manager may manage zero to many employees.

In Figure 8-27, we show a ternary relationship called Supplies among Vendor, Part, and Warehouse. As is preferred in an E-R diagram, we represent a ternary relationship using a class and place the name of the relationship there.

The class diagram in Figure 8-28 shows several binary associations. The diagram shows that a student may have an adviser, while a faculty member may advise up to a maximum of ten students. Also, while a course may have multiple offerings, a given course offering is scheduled for exactly one course. UML allows you to specify numerically any multiplicity. For example, the diagram shows that a course offering may be taught by one or two instructors (1,2). You can specify a single number (e.g., 2 for the members of a bridge team), a range (e.g., 11–14 for the players of a soccer team who participated in a particular game), or a discrete set of numbers and ranges (e.g., 3, 5, 7 for the number of committee members and 20–32, 35–40 for the workload in hours per week of a company's employees).

Figure 8-28 also shows that a faculty member plays the role of an instructor, as well as that of an adviser. While the adviser role identifies the Faculty object associated with a Student object, the advisee's role identifies the set of Student objects associated with a Faculty object. We could have named the association, say, Advises, but, in this case, the role names are sufficiently meaningful to convey the semantics of the relationship.

REPRESENTING ASSOCIATIVE CLASSES

When an association itself has attributes or operations of its own or when it participates in relationships with other classes, it is useful to model the association as an **associative class** (just as we used an associative entity in E-R diagramming). For example, in Figure 8-29, the attributes term and grade really belong to the many-to-many association between Student and Course. The grade of a student for a course cannot be determined unless both the student and the course are known. Similarly, to find the term(s) in which the student took the course, both student and course must be known. The check Eligibility operation, which determines if a student is eligible to register for a given course, also belongs to the association, rather than to any of the two participating classes. We have also captured the fact that, for some course registrations, a computer account is issued to a student. For these reasons, we model Registration as an association class, with its own set of features and an association with another class (Computer Account). Similarly, for the unary Tutors association, beginDate and numberOfHrs (number of hours tutored) really belong to the association, and therefore appear in a separate association class.

You have the option of showing the name of an association class on the association path, on the class symbol, or both. When an association has only attributes but does not have any operations or does not participate in other associations,

Associative class
An association that has attributes or operations of its own or that participates in relationships with other classes.

FIGURE 8-29
Class diagram showing associative classes

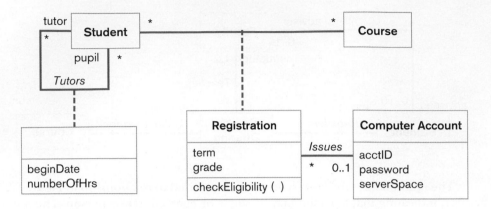

the recommended option is to show the name on the association path, but to omit it from the association class symbol, to emphasize its "association nature" (*UML Notation Guide,* 1997). That is how we have shown the Tutors association. On the other hand, we have displayed the name of the Registration association—which has two attributes and one operation of its own, as well as an association called Issues with Computer Account—within the class rectangle to emphasize its "class nature."

Figure 8-30 shows a ternary relationship among the Student, Software, and Course classes. It captures the fact that students use various software tools for different courses. For example, we could store the information that Mary Jones used Microsoft Access and Oracle for the Database Management course, Rational Rose and Visual C++ for the Object-Oriented Modeling course, and Level 5 Object for the Expert Systems Course. Now suppose we want to estimate the number of hours per week Mary will spend using Oracle for the Database Management course. This process really belongs to the ternary association, and not to any of the individual classes. Hence, we have created an associative class called Log, within which we have declared an operation called estimate Usage. In addition to this operation, we have specified three attributes that belong to the association: beginDate, expiryDate, and hoursLogged. Alternatively, the associative class Log can be placed at the intersection of the association lines, as shown in Figure 8-16; in this case, multiplicities are required on all the lines next to the Log class.

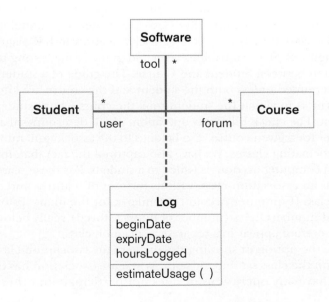

FIGURE 8-30
Ternary relationship with association classes

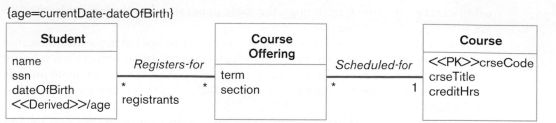

FIGURE 8-31
Stereotypes

REPRESENTING STEREOTYPES FOR ATTRIBUTES

In E-R diagrams, we designated attributes as being primary keys, and we designated them as multivalued, derived, and other types. This can also be done in a class diagram by placing a stereotype next to the attribute. Stereotypes simply extend the common UML vocabulary. For instance, in Figure 8-31, age is a derived attribute of Student because it can be calculated from the date of birth and the current date. Because the calculation is a constraint on the object class, the calculation is shown on this diagram within braces near the Student object class. Also, the crseCode is a primary key for the Course class. Other properties of attributes can be similarly shown.

REPRESENTING GENERALIZATION

In the object-oriented approach, you can abstract the common features (attributes and operations) among multiple classes, as well as the relationships they participate in, into a more general class, as we saw with supertype/subtype relationships in E-R diagramming. The classes that are generalized are called subclasses, and the class they are generalized into is called a superclass.

Consider the example shown in Figure 8-32, which is the class diagramming equivalent of the E-R diagram in Figure 8-18. There are two types of patients: outpatients and resident patients. The features that are shared by all patients—patientId, patientName, and admitDate—are stored in the Patient superclass, whereas the features that are peculiar to a particular patient type are stored in the corresponding subclass (e.g., checkbackDate of Outpatient). A generalization path is shown as a solid line from the subclass to the superclass, with an arrow at the end, and pointing toward the superclass. We also specify that this generalization is dynamic, meaning that an object may change subtypes. This generalization is also complete (there are no other subclasses) and disjoint (subclasses are not overlapping). Although not

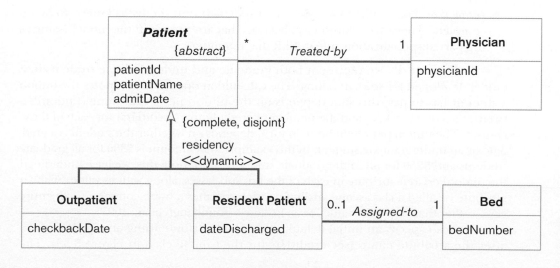

FIGURE 8-32
Example of generalizations, inheritance, and constraints

using exactly the same terminology, the same generalization business rules that we saw with E-R diagramming can be represented.

You can indicate the basis of a generalization by specifying a discriminator next to the path. A discriminator shows which property of an object class is being abstracted by a particular generalization relationship. You can discriminate on only one property at a time. For example, in Figure 8-32, we discriminate the Patient class on the basis of residency.

An instance of a subclass is also an instance of its superclass. For example, in Figure 8-32, an Outpatient instance is also a Patient instance. For that reason, a generalization is also referred to as an Is-a relationship. Also, a subclass inherits all the features from its superclass. For example, in Figure 8-32, in addition to its own special feature—checkbackDate—the Outpatient subclass inherits patientId, patientName, admitDate, and any operations (not shown) from Patient.

Notice that in Figure 8-32, the Patient class is in italics, implying that it is an abstract class. An **abstract class** is a class that has no direct instances but whose descendants may have direct instances (Booch, 1994; Rumbaugh et al., 1991). (*Note:* You can also write the word *abstract* within braces just below the class name. This is especially useful when you generate a class diagram by hand.) A class that can have direct instances (e.g., Outpatient or Resident Patient) is called a **concrete class**. In this example, therefore, Outpatient and Resident Patient can have direct instances, but Patient cannot have any direct instances of its own.

The Patient abstract class participates in a relationship called Treated-by with Physician, implying that all patients—outpatients and resident patients alike—are treated by physicians. In addition to this inherited relationship, the Resident Patient class has its own special relationship called Assigned-to with Bed, implying that only resident patients may be assigned to beds. So, in addition to refining the attributes and operations of a class, a subclass can also specialize the relationships in which it participates.

In Figure 8-32, the words *complete* and *disjoint* have been placed within braces next to the generalization. They indicate semantic constraints among the subclasses (*complete* corresponds to total specialization in the Extended Entity Relationship [EER] notation [see Hoffer et al., 2011], whereas *incomplete* corresponds to partial specialization). Any of the following UML keywords may be used:

- *Overlapping.* A descendant may be descended from more than one of the subclasses (same as the overlapping rule in EER diagramming).
- *Disjoint.* A descendant may not be descended from more than one of the subclasses (same as the disjoint rule in EER diagramming).
- *Complete.* All subclasses have been specified (whether or not shown). No additional subclasses are expected (same as the total specialization rule in EER diagramming).
- *Incomplete.* Some subclasses have been specified, but the list is known to be incomplete. There are additional subclasses that are not yet in the model (same as the partial specialization rule in EER diagramming).

In Figure 8-33, we represent both graduate and undergraduate students in a model developed for student billing. The calc-tuition operation computes the tuition a student has to pay; this sum depends on the tuition per credit hour (tuitionPerCred), the courses taken, and the number of credit hours (creditHrs) for each of those courses. The tuition per credit hour, in turn, depends on whether the student is a graduate or an undergraduate student. In this example, that amount is $300 for all graduate students and $250 for all undergraduate students. To denote this, we have underlined the tuitionPerCred attribute in each of the two subclasses, along with its value. Such an attribute is called a **class-scope attribute**, which specifies a value common to an entire class, rather than a specific value for an instance (Rumbaugh et al., 1991).

You can specify an initial default value of an attribute using an equals sign (=) after the attribute name (see creditHrs for the Course class in Figure 8-33). The

---

Abstract class

A class that has no direct instance, but whose descendents may have direct instances.

Concrete class

A class that can have direct instances.

Class-scope attribute

An attribute of a class that specifies a value common to an entire class, rather than a specific value for an instance.

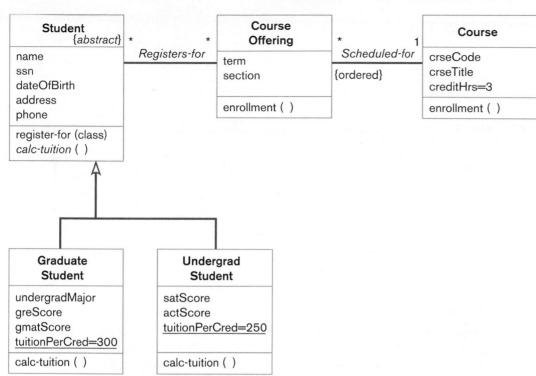

FIGURE 8-33
Polymorphism, abstract operation, class-scope attribute, and ordering

difference between an initial value specification and a class-scope attribute is that the former allows the possibility of different attribute values for the instances of a class, and the latter forces all the instances to share a common value.

In addition to specifying the multiplicity of an association role, you can also specify other properties, for example, if the objects playing the role are ordered or not. In the figure, we placed the keyword constraint "{ordered}" next to the Course Offering end of the Scheduled-for relationship to denote the fact that the offerings for a given course are ordered into a list—say, according to term and section. The default constraint on a role is "{unordered}."

The Graduate Student subclass specializes the abstract Student class by adding four attributes—undergradMajor, greScore, gmatScore, and tuitionPerCred—and by refining the inherited calc-tuition operation. Notice that the operation is shown in italics within the Student class, indicating that it is an abstract operation. An **abstract operation** defines the form or protocol of the operation, but not its implementation. In this example, the Student class defines the protocol of the calc-tuition operation, without providing the corresponding **method** (the actual implementation of the operation). The protocol includes the number and types of the arguments, the result type, and the intended semantics of the operation. The two concrete subclasses, Graduate Student and Undergrad Student, supply their own implementations of the calc-tuition operation. Note that because these classes are concrete, they cannot store abstract operations.

It is important to note that, although the Graduate Student and Undergraduate Student classes share the same calc-tuition operation, they might implement the operation in quite different ways. For example, the method that implements the operation for a graduate student might add a special graduate fee for each course the student takes. The fact that the same operation may apply to two or more classes in different ways is known as **polymorphism**, a key concept in object-oriented systems (Booch, 1994; Rumbaugh et al., 1991). The enrollment operation in Figure 8-33 illustrates another example of polymorphism. Whereas the enrollment operation within Course Offering computes the enrollment for a particular course offering or section, an operation with the same name within Course computes the combined enrollment for all sections of a given course.

Abstract operation
Defines the form or protocol of the operation, but not its implementation.

Method
The implementation of an operation.

Polymorphism
The same operation may apply to two or more classes in different ways.

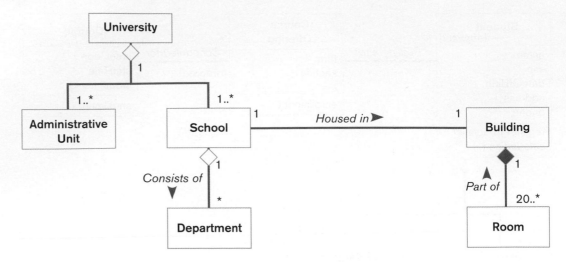

FIGURE 8-34
Aggregation and composition

REPRESENTING AGGREGATION

Aggregation
A part-of relationship between a component object and an aggregate object.

Composition
A part-of relationship in which parts belong to only one whole object, and the parts live and die with the whole object.

An **aggregation** expresses a part-of relationship between a component object and an aggregate object. It is a stronger form of association relationship (with the added "part-of" semantics) and is represented with a hollow diamond at the aggregate end.

Figure 8-34 shows an aggregation structure of a university. Notice that the diamond at one end of the relationship between Building and Room is solid, not hollow. A solid diamond represents a stronger form of aggregation known as **composition**. In composition, a part object belongs to only one whole object; for example, a room is part of only one building. Therefore, the multiplicity on the aggregate end may not exceed one. Parts may be created after the creation of the whole object; for example, rooms may be added to an existing building. However, once part of a composition is created, it lives and dies with the whole; deletion of the aggregate object cascades to its components. If a building is demolished, for example, so are all of its rooms. However, it is possible to delete a part before its aggregate dies, just as it is possible to demolish a room without bringing down a building.

AN EXAMPLE OF CONCEPTUAL DATA MODELING AT HOOSIER BURGER

Chapter 7 structured the process and logic requirements for a new inventory control system for Hoosier Burger. The DFD and decision table (repeated here as Figures 8-35 and 8-36) describe requirements for this new system. The purpose of this system is to monitor and report changes in raw material inventory levels and to issue material orders and payments to suppliers. Thus, the central data entity for this system will be an INVENTORY ITEM, corresponding to data store D1 in Figure 8-22.

Changes in inventory levels are due to two types of transactions: receipt of new items from suppliers and consumption of items from sales of products. Inventory is added upon receipt of new raw materials, for which Hoosier Burger receives a supplier INVOICE (see Process 1.0 in Figure 8-35). Each INVOICE indicates that the supplier has sent a specific quantity of one or more INVOICE_ITEMs, which correspond to Hoosier's INVENTORY ITEMs. Inventory is used when customers order and pay for PRODUCTs. That is, Hoosier makes a SALE for one or more ITEM SALEs, each of which corresponds to a food PRODUCT. Because the real-time customer order processing system is separate from the inventory control system, a source, STOCK ON HAND in Figure 8-35, represents how data flow from the order processing to the inventory control system. Finally, because food PRODUCTs are made up of various INVENTORY ITEMs (and vice versa), Hoosier maintains a

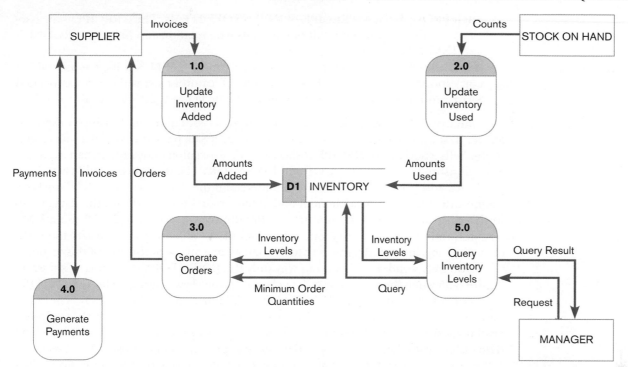

FIGURE 8-35
Level-0 data flow diagram for Hoosier Burger's new logical inventory control system

RECIPE to indicate how much of each INVENTORY ITEM goes into making one PRODUCT. From this discussion, we have identified the data entities required in a data model for the new Hoosier Burger inventory control system: INVENTORY ITEM, INVOICE, INVOICE ITEM, PRODUCT, SALE, ITEM SALE, and RECIPE. To complete the data model, we must determine a necessary relationship between these entities as well as attributes for each entity.

The wording in the previous description tells us much of what we need to know to determine relationships:

• An INVOICE includes one or more INVOICE ITEMs, each of which corresponds to an INVENTORY ITEM. Obviously, an INVOICE ITEM cannot exist without an associated INVOICE, and over time there will be zero-to-many receipts, or INVOICE ITEMs, for an INVENTORY ITEM.

Conditions/ Courses of Action	Rules						
	1	2	3	4	5	6	7
Type of item	P	P	P	P	P	P	N
Time of week	D	W	D	W	D	W	–
Season of year	A	A	S	S	H	H	–
Standing daily order	X		X		X		
Standing weekend order		X		X		X	
Minimum order quantity							X
Holiday reduction					X	X	
Summer reduction			X	X			

FIGURE 8-36
Reduced decision table for Hoosier Burger's inventory reordering

- Each PRODUCT has a RECIPE of INVENTORY ITEMs. Thus, RECIPE is an associative entity supporting a bill-of-materials type relationship between PRODUCT and INVENTORY ITEM.
- A SALE indicates that Hoosier sells one or more ITEM SALEs, each of which corresponds to a PRODUCT. An ITEM SALE cannot exist without an associated SALE, and over time there will be zero-to-many ITEM SALEs for a PRODUCT.

Figure 8-37 shows a class diagram with the classes and relationships described above. In some cases, we include role names (e.g., a Sale plays the role of a transaction in the Sells association). RECIPE is shown as an association class rather than as simply a relationship between PRODUCT and INVENTORY ITEM because it is likely to have attributes and behaviors. Now that we understand the data classes and relationships, we must decide which data element and behaviors are associated with the data classes in this diagram. We have chosen to develop the conceptual data model using UML notation rather than E-R notation, but you should be able to easily translate the UML class diagrams into E-R diagrams (which is left as an exercise at the end of this section).

You may wonder at this point why only the INVENTORY data store is shown in Figure 8-35 when there are seven data classes on the class diagram. The INVENTORY data store corresponds to the INVENTORY ITEM data class in Figure 8-37. The other data classes are hidden inside other processes for which we have not shown lower-level diagrams. In actual requirements structuring steps, you would have to match all data classes with data stores: each data store represents some subset of a class or an E-R diagram, and each data class or entity is included in one or more data stores. Ideally, each data store on a primitive DFD will be an individual data class or entity.

To determine data elements for a data class, we investigate data flows in and out of data stores that correspond to the data class and supplement this with a study of

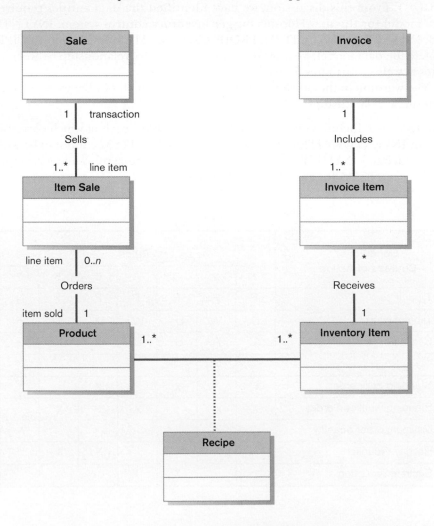

FIGURE 8-37
Preliminary class diagram for Hoosier Burger's inventory control system

decision logic and temporal logic that uses or changes data about the data class. Six data flows are associated with the INVENTORY data store in Figure 8-35. The description of each data flow in the project dictionary or CASE repository would include the data flow's composition, which then tells us what data are flowing in or out of the data store. For example, the Amounts Used data flow coming from Process 2.0 indicates how much to decrement an attribute Quantity_in_Stock due to use of the INVENTORY ITEM to fulfill a customer sale. Thus, the Amounts Used data flow implies that Process 2.0 will first read the relevant INVENTORY ITEM record, then update its Quantity_in_Stock attribute, and finally store the updated value in the record. Structured English for Process 2.0 would depict this logic. Each data flow would be analyzed similarly (space does not permit us to show the analysis for each data flow).

The analysis of data flows for data elements is supplemented by a study of decision logic. For example, consider the decision table in Figure 8-36. One condition used to determine the process of reordering an INVENTORY ITEM involves the Type_of_Item. Thus, Process 3.0 in Figure 8-35 (to which this decision table relates) needs to know this characteristic of each INVENTORY ITEM, so this identifies another attribute of this data class.

An analysis of the DFD and decision table also suggests possible operations for each class. For example, the Inventory Item class will need operations to update quantity on hand, generate replenishment orders, and receive inventory counts.

After considering all data flows in and out of data stores related to data classes, plus all decision and temporal logic related to inventory control, we derive the full class diagram, with attributes and operations, shown in Figure 8-38.

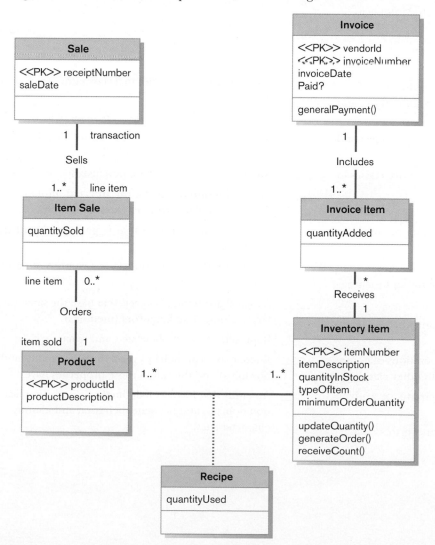

FIGURE 8-38
Final class diagram for Hoosier Burger's inventory control system

SUMMARY

We have presented the process and basic notation used to model the data requirements of an information system using class diagramming. A data model shows the relatively permanent business rules that define the nature of an organization. Rules define characteristics of data such as the unique characteristics of data classes and the relationships between different data classes. A data model shows major categories of data, called *classes* for the UML notation; the associations or relationships between classes; and the attributes (only classes have attributes on a class diagram).

A special type of class called an *association class* is often necessary to represent a many-to-many relationship between classes. Classes are distinct from objects. Each object is distinguished from other instances of the same type by an identifier attribute (or attributes). Relationships are the glue that holds a data model together. The minimum and maximum number of objects that participate in a relationship represent important rules about the nature of the organization, as captured during requirements determination.

KEY TERMS

1. Abstract class
2. Abstract operation
3. Aggregation
4. Association
5. Association role
6. Associative class
7. Behavior
8. Class diagram
9. Class-scope attribute
10. Class-scope operation
11. Composition
12. Concrete class
13. Constructor operation
14. Encapsulation
15. Method
16. Multiplicity
17. Object
18. Object (class)
19. Operation
20. Polymorphism
21. Query operation
22. State
23. Update operation

Match each of the key terms above with the definition that best fits it.

_____ A part object that belongs to only one whole object and that lives and dies with the whole.

_____ A part-of relationship between a component object and an aggregate object.

_____ The same operation may apply to two or more classes in different ways.

_____ The implementation of an operation.

_____ Defines the form or protocol of the operation, but not its implementation.

_____ An attribute of a class that specifies a value common to an entire class, rather than a specific value for an instance.

_____ A class that can have direct instances.

_____ A class that has no direct instances but whose descendants may have direct instances.

_____ An association that has attributes or operations of its own or that participates in relationships with other classes.

_____ Indicates how many objects participate in a given relationship.

_____ The end of an association where it connects to a class.

_____ A relationship among instances of object classes.

_____ An operation that applies to a class rather than an object instance.

_____ An operation that alters the state of an object.

_____ An operation that accesses the state of an object but does not alter the state.

_____ An operation that creates a new instance of a class.

_____ The technique of hiding the internal implementation details of an object from its external view.

_____ A function or a service that is provided by all the instances of a class.

_____ A diagram that shows the static structure of an object-oriented model.

_____ A logical grouping of objects that have the same (or similar) attributes and behaviors (methods).

_____ Represents how an object acts and reacts.

_____ Encompasses an object's properties (attributes and relationships) and the values those properties have.

_____ An entity that has a well-defined role in the application domain, and has state, behavior, and identity characteristics.

REVIEW QUESTIONS

1. Give an example of aggregation. Your example should include at least one aggregate object and three component objects. Specify the multiplicities at each end of all the aggregation relationships.

2. Contrast the following terms:
   a. Object class; object
   b. Abstract class; concrete class

PROBLEMS AND EXERCISES

1. Draw a class diagram, showing the relevant classes, attributes, operations, and relationships for each of the following situations (if you believe that you need to make additional assumptions, clearly state them for each situation):

   a. A company has a number of employees. The attributes of Employee include employeeID (primary key), name, address, and birth date. The company also has several projects. Attributes of Project include projectName and startDate. Each employee may be assigned to one or more projects, or may not be assigned to a project. A project must have at least one employee assigned, and it may have any number of employees assigned. An employee's billing rate may vary by project, and the company wishes to record the applicable billing rate for each employee when assigned to a particular project. At the end of each month, the company mails a check to each employee who has worked on a project during that month. The check amount is based on the billing rate and the hours logged for each project assigned to the employee.

   b. A university has a large number of courses in its catalog. Attributes of Course include courseNumber (primary key), courseName, and units. Each course may have one or more different courses as prerequisites, or a course may have no prerequisites. Similarly, a particular course may be a prerequisite for any number of courses, or it may not be a prerequisite for any other course. The university adds or drops a prerequisite for a course only when the director for the course makes a formal request to that effect.

   c. A laboratory has several chemists who work on one or more projects. Chemists also may use certain kinds of equipment on each project. Attributes of Chemist include name and phoneNo. Attributes of Project include projectName and startDate. Attributes of Equipment include serialNo and cost. The organization wishes to record assignDate—that is, the date when a given equipment item was assigned to a particular chemist working on a specified project—as well as total Hours, that is, the total number of hours the chemist has used the equipment for the project. The organization also wants to track the usage of each type of equipment by a chemist. It does so by computing the average number of hours the chemist has used that equipment

   on all assigned projects. A chemist must be assigned to at least one project and one equipment item. A given equipment item need not be assigned, and a given project need not be assigned either a chemist or an equipment item.

   d. A college course may have one or more scheduled sections, or it may not have a scheduled section. Attributes of Course include courseID, courseName, and units. Attributes of Section include sectionNumber and semester. The value of sectionNumber is an integer (such as 1 or 2) that distinguishes one section from another for the same course, but does not uniquely identify a section. There is an operation called findNumSections that finds the number of sections offered for a given course in a given semester.

   e. A hospital has a large number of registered physicians. Attributes of Physician include physicianID (primary key) and specialty. Patients are admitted to the hospital by physicians. Attributes of Patient include patientID (primary key) and patientName. Any patient who is admitted must have exactly one admitting physician. A physician may optionally admit any number of patients. Once admitted, a given patient must be treated by at least one physician. A particular physician may treat any number of patients, or he or she may not treat any patients. Whenever a patient is treated by a physician, the hospital wishes to record the details of the treatment by including the date, time, and results of the treatment.

2. A student, whose attributes include studentName, Address, phone, and age, may engage in multiple campus-based activities. The university keeps track of the number of years a given student has participated in a specific activity and, at the end of each academic year, mails an activity report to the student showing his or her participation in various activities. Draw a class diagram for this situation.

3. A bank has three types of accounts: checking, savings, and loan. Following are the attributes for each type of account:

   CHECKING: Acct_No, Date_Opened, Balance, Service_Charge

   SAVINGS: Acct_No, Date_Opened, Balance, Interest_Rate

   LOAN: Acct_No, Date_Opened, Balance, Interest_Rate, Payment

Assume that each bank account must be a member of exactly one of these subtypes. At the end of each month, the bank computes the balance in each account and mails a statement to the customer holding that account. The balance computation depends on the type of the account. For example, a checking account balance may reflect a service charge, whereas a savings account balance may include an interest amount. Draw a class diagram to represent the situation. Your diagram should include an abstract class, as well as an abstract operation for computing balance.

4. Convert the class diagram in Figure 8-37 to the equivalent E-R diagram. Compare the two diagrams. Describe what different system specifications are shown in each diagram.

REFERENCES

Booch, G. 1994. *Object-Oriented Analysis and Design with Applications,* 2nd ed. Redwood City, CA: Benjamin Cummings.

Fowler, M. 2000. *UML Distilled: A Brief Guide to the Object Modeling Language,* 2nd ed. Reading, MA: Addison-Wesley.

George, J., D. Batra, J. Valacich, and J. Hoffer. 2007. *Object-Oriented Systems Analysis and Design,* 2nd ed. Upper Saddle River, NJ: Prentice Hall.

Hoffer, J. A., V. Ramesh, and H. Topi. 2011. *Modern Database Management,* 10th ed. Upper Saddle River, NJ: Prentice Hall.

Rumbaugh, J., M. Blaha, W. Premerlani, F. Eddy, and W. Lorensen. 1991. *Object-Oriented Modeling and Design.* Upper Saddle River, NJ: Prentice Hall.

UML Notation Guide. 1997. Document accessed from *www.rational.com/uml.* Accessed February 19. 2009. Copyright held by Rational Software Corporation, Microsoft Corporation, Hewlett-Packard Company, Oracle Corporation, Sterling Software, MCI Systemhouse Corporation, Unisys Corporation, ICON Computing, IntelliCorp, i-Logix, IBM Corporation, ObjecTime Limited, Platinum Technology Incorporated, Ptech Incorporated, Taskon A/S, Reich Technologies, Softeam.

PETRIE ELECTRONICS

Chapter 8: Structuring System Data Requirements

Jim Watanabe, manager of the "No Customer Escapes" project, and assistant director of IT for Petrie Electronics, was sitting in the company cafeteria. He had just finished his house salad and was about to go back to his office when Stephanie Welch sat down at his table. Jim had met Stephanie once, back when he started work at Petrie. He remembered she worked for the database administrator.

"Hi, Jim, remember me?" she asked.

"Sure, Stephanie, how are you? How are things in database land?"

"Can't complain. Sanjay asked me to talk to you about the database needs for your new customer loyalty system." Stephanie's phone binged. She pulled it out of her oversize bag and looked at it. She started to text as she continued to talk to Jim. "How far along are you on your database requirements?"

That's kinda rude, Jim thought. Oh well. "We are still in the early stages. I can send you a very preliminary E-R diagram we have (PE Figure 8-1), along with a description of the major entities."

"OK, that will help. I suspect that you won't have too many new entities to add to what's already in the system," Stephanie responded, still looking at her phone and still texting. She briefly looked up at Jim and smiled slightly before going back to texting. "Just send the E-R to me, and I'll let you know if I have any questions." She stood up, still looking at her phone. "Gotta go," she said, and she walked away.

OK, Jim thought, I need to remember to send Stephanie the preliminary E-R we have. I should probably send her the entity descriptions too (PE Table 8-1), just in case. Jim stood up, carried his tray over to the recycling area of the cafeteria, and went back to his office.

When Jim got back to his office, Sanjay was waiting for him.

"I've got more information on those alternatives we talked about earlier," Sanjay said. "I had one of my employees gather some data on how the alternatives might satisfy our needs." (See the descriptions of the alternatives at the end of Chapter 6.) Sanjay handed Jim a short report. "The matrix shows the requirements and constraints for each alternative and makes it relatively easy to compare them." (See PE Figure 8-2.)

"The matrix favors the XRA CRM system," Jim said, after looking over the report. "It looks like their proposal meets our requirements the best, but the Nova group's proposal does the best job with the constraints."

"Yes, but just barely," Sanjay said. "There is only a five point difference between XRA and Nova, so they are pretty comparable when it comes to constraints. But I think the XRA system has a pretty clear advantage in meeting our requirements."

"XRA seems to be pretty highly rated in your matrix in terms of all of the requirements. You have them ranked better than the other two proposals for implementation, scalability, and vendor support," Jim said. "The '5' you gave them for proven performance is one of the few '5's' you have in your whole matrix."

"That's because they are one of the best companies in the industry to work with," Sanjay responded, "Their reputation is stellar."

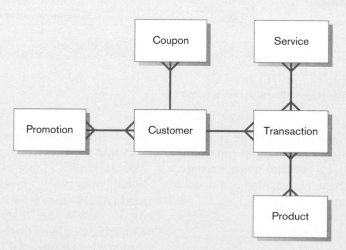

PE FIGURE 8-1
Initial E-R for Petrie's customer loyalty program

PE TABLE 8-1   **Entity Descriptions for the Preliminary E-R Diagram for Petrie's Customer Loyalty System**

Entity	Description
Coupon	A coupon is a special promotion created specifically for an individual customer. A coupon is for a set dollar amount, for example, $10. The customer may use it like cash or like a dollars-off promotion when purchasing products or services. Coupons can only be created for an individual customer based on the points in his or her customer loyalty account. For each dollar value of a coupon, a certain number of points must be redeemed. Coupons must be accounted for when created and when redeemed.
Customer	A customer is someone who buys products and/or services from Petrie Electronics. Customers include both online customers and those who shop in Petrie's brick-and-mortar stores.
Product	An item made available for sale to a Petrie's customer. For example, a product is a 40" Sony LCD HD television. Products can be purchased online or in brick-and-mortar stores.
Promotion	A promotion is a special incentive provided to a customer to entice the customer into buying a specific product or service. For example, a promotion intended to sell BluRay disks may involve 2-for-1 coupons. Promotions are targeted to all customers, or to subsets of customers, not just to individual customers.
Service	A job performed by one of Petrie's associates for a customer. For example, upgrading the memory in a computer by installing new memory cards is a service that Petrie provides for a fee. Services may only be ordered and performed in brick-and-mortar stores, not online.
Transaction	A record that a particular product or service was sold to a specified customer on a particular date. A transaction may involve more than one product or service, and it may involve more than one of a particular kind of product or service. For example, one transaction may involve blank DVDs and pre-recorded DVDs, and the pre-recorded DVDs may all be of the same movie. For members of the loyalty program, each transaction is worth a number of points, depending on the dollar value of the transaction.

Criteria	Weight	Alt A: Rating	SBSI Score	Alt B Rating	XRA Score	Alt C Rating	Nova Score
Requirements							
Effective customer incentives	15	5	75	4	60	4	60
Easy for customers to use	10	3	30	4	40	5	50
Proven performance	10	4	40	5	50	3	30
Easy to implement	5	3	15	4	20	3	15
Scalable	10	3	30	4	40	3	30
Vendor support	10	3	30	4	40	3	30
	60		220		250		215
Constraints							
Cost to buy	15	3	45	4	60	5	75
Cost to operate	10	3	30	4	40	4	40
Time to implement	5	3	15	3	15	3	15
Staff to implement	10	3	30	4	40	3	30
	40		120		155		160
TOTAL	100		340		405		375

PE FIGURE 8-2

Evaluation matrix for customer loyalty proposals

"This looks really promising," Jim said. "Let's see if reality matches what we have here. It's time to put together the formal request for proposal. I'll get that work started today. I hope that all three of these companies decide to bid."

Case Questions

1. Review the data-flow diagrams you developed for questions in the Petrie Electronics case at the end of Chapter 7 (or diagrams given to you by your instructor). Study the data flows and data stored on these diagrams and decided whether you agree with the team's conclusion that the only 6 entity types needed are listed in the case and in PE Figure 8-1. If you disagree, define additional entity types, explain why they are necessary, and modify PE Figure 8-1 accordingly.

2. Again, review the DFDs you developed for the Petrie Electronics case (or those given to you by your instructor). Use these DFDs to identify the attributes of each of the six entities listed in this case plus any additional entities identified in your answer to Question 1. Write an unambiguous definition for each attribute.

Then, redraw PE Figure 8-1 by placing the six (and additional) entities in this case on the diagram along with their associated attributes.

3. Using your answer to Question 2, designate which attribute or attributes form the identifier for each entity type. Explain why you chose each identifier.

4. Using your answer to Question 3, draw the relationships between entity types needed by the system. Remember, a relationship is needed only if the system wants data about associated entity instances. Give a meaningful name to each relationship. Specify cardinalities for each relationship and explain how you decided on each minimum and maximum cardinality on each end of each relationship. State any assumptions you made if the Petrie Electronics cases you have read so far and the answers to questions in these cases do not provide the evidence to justify the cardinalities you choose. Redraw your final E-R diagram in Microsoft Visio.

5. Now that you have developed in your answer to Question 4 a complete E-R diagram for the Petrie Electronics database, what are the consequences of not having an employee entity type in this diagram? Assuming only the attributes you show on the E-R diagram, would any attribute be moved from the entity it is currently associated with to an employee entity type if it were in the diagram? Why or why not?

6. Write project dictionary entries (using standards given to you by your instructor) for all the entities, attributes, and relationships shown in the E-R diagram in your answer to Question 4. How detailed are these entries at this point? What other details still must be filled in? Are any of the entities on the E-R diagram in your answer to Question 4 weak entities? Why? In particular, is the SERVICE entity type a weak entity? If so, why? If not, why not?

7. What date-related attributes did you identify in each of the entity types in your answer to Question 4? Why are each of these needed? Can you make some general observations about why date attributes must be kept in a database based on your analysis of this database?

Design

Chapter 9
Designing Databases

Chapter 10
Designing Forms and Reports

Chapter 11
Designing Interfaces and Dialogues

Chapter 12
Designing Distributed and Internet Systems

PART FOUR

Design

The focus of Part Four is system design, which is often the first phase of the systems development life cycle, the phase in which you and the user develop a concrete understanding of how the system will operate. The activities within design are not necessarily sequential. For example, the design of data, system inputs and outputs, and interfaces interact, allowing you to identify flaws and missing elements. This means that the project dictionary or CASE repository becomes an active and evolving component of systems development management during design. It is only when each design element is consistent with others and each one is satisfactory to the end user that you know that the design phase is complete.

Data are a core system element, and data design and structure are studied in all systems development methodologies. You have seen how data flow diagrams (DFDs) and entity-relationship (E-R) diagrams (as well as use cases and class diagrams from the object-oriented material at the end of prior chapters) are used to depict the data requirements of a system. These diagrams are flexible and allow considerable latitude in how you represent data. For example, you can use one or many data stores with a process in a DFD. E-R diagrams provide more structure, but an entity can still be either very detailed or rather aggregate. When designing databases, you define data in its most fundamental form, called *normalized data. Normalization* is a well-defined method of identifying relationships between each data attribute and representing all the data so that they cannot logically be broken down into more detail. The goal is to rid the data design of unwanted anomalies that would make a database susceptible to errors and inefficiencies. This is the topic of Chapter 9.

In Chapter 10, you will learn the principles and guidelines for usable system inputs and outputs. Your overall goal in formatting the presentation of data to users should be usability: helping users of all types to use the system efficiently, accurately, and with satisfaction. The achievement of these goals can be greatly improved if you follow certain guidelines when presenting data on business forms, visual display screens, printed documents, and other kinds of media. Fortunately, there has been considerable research on how to present data to users, and Chapter 10 summarizes and illustrates the most useful of these guidelines. Chapter 11 is closely related to Chapter 10 and addresses principles you should follow in tying all the system inputs and outputs together into an overall pattern of interaction between users and the system. System interfaces and dialogues form a conversation that provides user access to and navigation between each system function. Chapter 11 focuses on providing specifications for designing effective system interfaces and dialogues and a technique for representing these designs called dialogue diagramming.

For traditional development efforts, before developers can begin the implementation process, questions about multiple users, multiple platforms, and program and data distribution have to be considered. The extent to which the system is Internet-based also has an impact on numerous design issues. The focus of Chapter 12 is on the intricacies of designing distributed and Internet systems.

The deliverables of design include detailed, functional specifications for system inputs, outputs, interfaces, dialogues, and databases. Often these elements are represented in prototypes, or working versions. The project dictionary or CASE repository is updated to include each form, report, interface, dialogue, and relation design. Due to considerable user involvement in reviewing prototypes and specifications during design, and due to the fact that activities within design can be scheduled with considerable overlap in the project baseline plan, a formal review milestone or walkthrough often does not occur after each activity. If prototyping is not done, however, you should conduct a formal walkthrough at the completion of the system design phase.

All of the chapters in Part Four conclude with a Petrie Electronics case. These cases illustrate numerous relevant design activities for an ongoing systems development project within the company.

Designing Databases

Learning Objectives

After studying this chapter, you should be able to:

- Concisely define each of the following key database design terms: *relation, primary key, normalization, functional dependency, foreign key, referential integrity, field, data type, null value, denormalization, file organization, index,* and *secondary key.*

- Explain the role of designing databases in the analysis and design of an information system.

- Transform an entity-relationship (E-R) diagram into an equivalent set of well-structured (normalized) relations.

- Merge normalized relations from separate user views into a consolidated set of well-structured relations.

- Choose storage formats for fields in database tables.

- Translate well-structured relations into efficient database tables.

- Explain when to use different types of file organizations to store computer files.

- Describe the purpose of indexes and the important considerations in selecting attributes to be indexed.

Introduction

In Chapter 8, you learned how to represent an organization's data graphically using an entity-relationship (E-R) or a class diagram. In this chapter, you will learn guidelines for well-structured and efficient database files and about logical and physical database design. It is likely that the human interface and database design steps will happen in parallel, as illustrated in the systems development life cycle (SDLC) in Figure 9-1.

Database design has five purposes:

1. Structure the data in stable structures, called *normalized tables,* that are not likely to change over time and that have minimal redundancy.

FIGURE 9-1
Systems development life cycle with
design phase highlighted

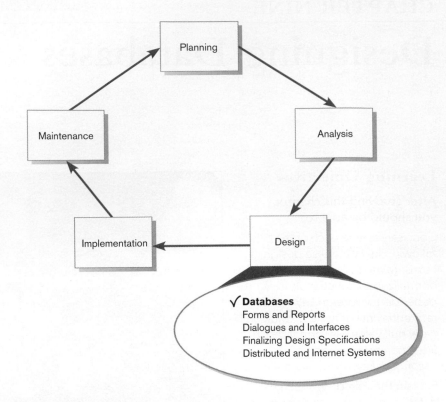

2. Develop a logical database design that reflects the actual data requirements that exist in the forms (hard copy and computer displays) and reports of an information system. This is why database design is often done in parallel with the design of the human interface of an information system.
3. Develop a logical database design from which we can do physical database design. Because most information systems today use relational database management systems, logical database design usually uses a relational database model, which represents data in simple tables with common columns to link related tables.
4. Translate a relational database model into a technical file and database design that balances several performance factors.
5. Choose data storage technologies (such as Read/Write DVD or optical disk) that will efficiently, accurately, and securely process database activities.

The implementation of a database (i.e., creating and loading data into files and databases) is done during the next phase of the systems development life cycle. Because implementation is technology specific, we address implementation issues only at a general level in Chapter 13.

DATABASE DESIGN

File and database design occurs in two steps. You begin by developing a logical database model, which describes data using a notation that corresponds to a data organization used by a database management system. This is the system software responsible for storing, retrieving, and protecting data (such as Microsoft Access, Oracle, or SQL Server). The most common style for a logical database model is the relational database model. Once you develop a clear and precise logical database model, you are ready to prescribe the technical specifications for computer files and databases in which to store the data. A physical database design provides these specifications.

You typically do logical and physical database design in parallel with other systems design steps. Thus, you collect the detailed specifications of data necessary for logical database design as you design system inputs and outputs. Logical database design is driven not only from the previously developed E-R data model for the application or enterprise but also from form and report layouts. You study data elements on these system inputs and outputs and identify interrelationships among the data. As with conceptual data modeling, the work of all systems development team members is coordinated and shared through the project dictionary or repository. The designs for logical databases and system inputs and outputs are then used in physical design activities to specify to computer programmers, database administrators, network managers, and others how to implement the new information system. For this text, we assume that the design of computer programs and distributed information processing and data networks are topics of other courses, so we concentrate on the aspect of physical design most often undertaken by a systems analyst—physical file and database design.

The Process of Database Design

Figure 9-2 shows that database modeling and design activities occur in all phases of the systems development process. In this chapter, we discuss methods that help you finalize logical and physical database designs during the design phase. In logical database design, you use a process called *normalization*, which is a way to build a data model that has the properties of simplicity, nonredundancy, and minimal maintenance.

In most situations, many physical database design decisions are implicit or eliminated when you choose the data management technologies to use with the application. We concentrate on those decisions you will make most frequently and use Oracle to illustrate the range of physical database design parameters you must manage. The interested reader is referred to Hoffer, Ramesh, and Topi (2011) for a more thorough treatment of techniques for logical and physical database design.

There are four key steps in logical database modeling and design:

1. Develop a logical data model for each known user interface (form and report) for the application using normalization principles.
2. Combine normalized data requirements from all user interfaces into one consolidated logical database model; this step is called *view integration*.

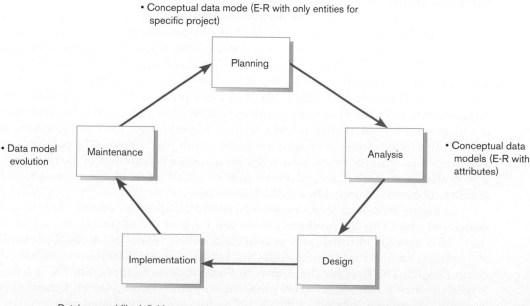

- Enterprise-wide data model (E-R with only entities)
- Conceptual data mode (E-R with only entities for specific project)

- Data model evolution

- Conceptual data models (E-R with attributes)

- Database and file definitions (DBMS specific code)

- Logical data model (relational) and physical file and database design (file organizations)

FIGURE 9-2
Relationship between data modeling and the SDLC

3. Translate the conceptual E-R data model for the application or enterprise, developed without explicit consideration of specific user interfaces, into normalized data requirements.

4. Compare the consolidated logical database design with the translated E-R model and produce, through view integration, one final logical database model for the application.

During physical database design, you use the results of these four key logical database design steps. You also consider definitions of each attribute; descriptions of where and when data are entered, retrieved, deleted, and updated; expectations for response time and data integrity; and descriptions of the file and database technologies to be used. These inputs allow you to make key physical database design decisions, including the following:

- Choosing the storage format (called *data type*) for each attribute from the logical database model; the format is chosen to minimize storage space and to maximize data quality. Data type involves choosing length, coding scheme, number of decimal places, minimum and maximum values, and potentially many other parameters for each attribute.

- Grouping attributes from the logical database model into physical records (in general, this is called selecting a stored record, or data, structure).

- Arranging related records in secondary memory (hard disks and magnetic tapes) so that individual records and groups of records can be stored, retrieved, and updated rapidly (called *file organization*). You should also consider protecting data and recovering data after errors are found.

- Selecting media and structures for storing data to make access more efficient. The choice of media affects the utility of different file organizations. The primary structure used today to make access to data more rapid is key indexes on unique and nonunique keys.

In this chapter, we show how to do each of these logical database design steps and discuss factors to consider in making each physical file and database design decision.

Deliverables and Outcomes

During logical database design, you must account for every data element on a system input or output—form or report—and on the E-R diagram. Each data element (e.g., customer name, product description, or purchase price) must be a piece of raw data kept in the system's database or, in the case of a data element on a system output, the element can be derived from data in the database. Figure 9-3 illustrates the outcomes from the four-step logical database design process listed earlier. Figures 9-3a and 9-3b (step 1) contain two sample system outputs for a customer order processing system at Pine Valley Furniture (PVF). A description of the associated database requirements, in the form of what we call *normalized relations*, is listed below each output diagram. Each relation (think of a relation as a table with rows and columns) is named, and its attributes (columns) are listed within parentheses. The **primary key** attribute—that attribute whose value is unique across all occurrences of the relation—is indicated by an underline, and an attribute of a relation that is the primary key of another relation is indicated by a dashed underline.

In Figure 9-3a, data about customers, products, and the customer orders and associated line items for products are shown. Each of the attributes of each relation either appears in the display or is needed to link related relations. For example, because an order is for some customer, an attribute of ORDER is the associated Customer_ID. The data for the display in Figure 9-3b are more complex. A backlogged product on an order occurs when the amount ordered (Order_Quantity) is less than the amount shipped (Ship_Quantity) for invoices associated with an order. The query refers only to a specified time period, so the Order_Date is needed. The INVOICE Order_Number links invoices with the associated order.

Primary key
An attribute (or combination of attributes) whose value is unique across all occurrences of a relation.

FIGURE 9-3
Simple example of logical data modeling
(a) Highest-volume customer query screen

```
HIGHEST-VOLUME CUSTOMER

ENTER PRODUCT ID.:      M128
START DATE:             11/01/2014
END DATE:               12/31/2014
- - - - - - - - - - - - - - - - - - - -
CUSTOMER ID.:           1256
NAME:                   Commonwealth Builder
VOLUME:                 30
```

This inquiry screen shows the customer with the largest volume of total sales for a specified product during an indicated time period.

Relations:
  CUSTOMER(Customer_ID,Name)
  ORDER(Order_Number,Customer_ID,Order_Date)
  PRODUCT(Product_ID)
  LINE ITEM(Order_Number,Product_ID,Order_Quantity)

(b) Backlog summary report

```
                                        PAGE 1

          BACKLOG SUMMARY REPORT
               11/30/2014

                         BACKLOG
          PRODUCT ID     QUANTITY
             B381           0
             B975           0
             B985           6
             E125          30
              :
             M128           2
              :
```

This report shows the unit volume of each product that has been ordered less that amount shipped through the specified date.

Relations:
  PRODUCT(Product_ID)
  LINE ITEM(Product_ID,Order_Number,Order_Quantity)
  ORDER(Order_Number,Order_Date)
  SHIPMENT(Product_ID,Invoice_Number,Ship_Quantity)
  INVOICE(Invoice_Number,Invoice_Date,Order_Number)

(c) Intergrated set of relations

CUSTOMER(Customer_ID,Name)
PRODUCT(Product_ID)
INVOICE(Invoice_Number,Invoice_Date,Order_Number)
ORDER(Order_Number,Customer_ID,Order_Date)
LINE ITEM(Order_Number,Product_ID,Order_Quantity)
SHIPMENT(Product_ID,Invoice_Number,Ship_Quantity)

FIGURE 9-3 (continued)
(d) Conceptual data model and transformed relations

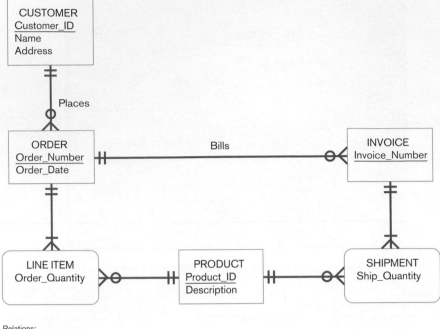

Relations:
CUSTOMER(Customer_ID,Name,Address)
PRODUCT(Product_ID,Description)
ORDER(Order_Number,Customer_ID,Order_Date)
LINE ITEM(Order_Number,Product_ID,Order_Quantity)
INVOICE(Invoice_Number,Order_Number)
SHIPMENT(Invoice_Number,Product_ID,Ship_Quantity)

(e) Final set of normalized relations

CUSTOMER(Customer_ID,Name,Address)
PRODUCT(Product_ID,Description)
ORDER(Order_Number,Customer_ID,Order_Date)
LINE ITEM(Order_Number,Product_ID,Order_Quantity)
INVOICE(Invoice_Number,Order_Number,Invoice_Date)
SHIPMENT(Invoice_Number,Product_ID,Ship_Quantity)

Figure 9-3c (step 2) shows the result of integrating these two separate sets of normalized relations. Figure 9-3d (step 3) shows an E-R diagram for a customer order processing application that might be developed during conceptual data modeling, along with equivalent normalized relations. Finally, Figure 9-3e (step 4) shows a set of normalized relations that would result from reconciling the logical database designs of Figures 9-3c and 9-3d. Normalized relations like those in Figure 9-3e are the primary deliverable from logical database design.

It is important to remember that relations do not correspond to computer files. In physical database design, you translate the relations from logical database design into specifications for computer files. For most information systems, these files will be tables in a relational database. These specifications are sufficient for programmers and database analysts to code the definitions of the database. The coding, done during systems implementation, is written in special database definition and processing languages, such as Structured Query Language (SQL), or by filling in table definition forms, such as with Microsoft Access. Figure 9-4 shows a possible definition for the SHIPMENT relation from Figure 9-3e using Microsoft Access. This display of the SHIPMENT table definition illustrates choices made for several physical database design decisions.

- All three attributes from the SHIPMENT relation, and no attributes from other relations, have been grouped together to form the fields of the SHIPMENT table.
- The Invoice Number field has been given a data type of Text, with a maximum length of 10 characters.

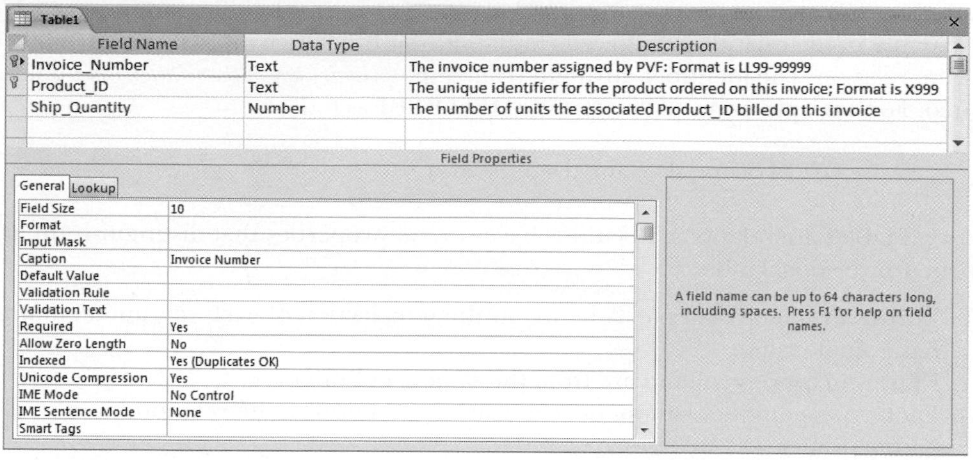

FIGURE 9-4
Definition of shipment table in Microsoft Access

(*Source:* Microsoft Corporation.)

- The Invoice Number field is required because it is part of the primary key for the SHIPMENT table (the value that makes every row of the SHIPMENT table unique is a combination of Invoice Number and Product ID).
- An index is defined for the Invoice Number field, but because there may be several rows in the SHIPMENT table for the same invoice (different products on the same invoice), duplicate index values are allowed (so Invoice Number is what we will call a secondary key).

Many other physical database design decisions were made for the SHIPMENT table, but they are not apparent on the display in Figure 9-4. Further, this table is only one table in the Pine Valley Furniture Company Order Entry database, and other tables and structures for this database are not illustrated in this figure.

RELATIONAL DATABASE MODEL

Many different database models are in use and are the bases for database technologies. Although hierarchical and network models have been popular in the past, these are not used very often today for new information systems. Object-oriented database models are emerging but are still not common. The vast majority of information systems today use the relational database model. The **relational database model** (Codd, 1970) represents data in the form of related tables, or relations. A **relation** is a named, two-dimensional table of data. Each relation (or table) consists of a set of named columns and an arbitrary number of unnamed rows. Each column in a relation corresponds to an attribute of that relation. Each row of a relation corresponds to a record that contains data values for an entity.

Figure 9-5 shows an example of a relation named EMPLOYEE1. This relation contains the following attributes describing employees: Emp_ID, Name, Dept, and Salary. This table has five sample rows, corresponding to five employees.

Relational database model
Data represented as a set of related tables or relations.

Relation
A named, two-dimensional table of data. Each relation consists of a set of named columns and an arbitrary number of unnamed rows.

EMPLOYEE1

Emp_ID	Name	Dept	Salary
100	Margaret Simpson	Marketing	42,000
140	Allen Beeton	Accounting	39,000
110	Chris Lucero	Info Systems	41,500
190	Lorenzo Davis	Finance	38,000
150	Susan Martin	Marketing	38,500

FIGURE 9-5
EMPLOYEE1 relation with sample data

You can express the structure of a relation with a shorthand notation in which the name of the relation is followed (in parentheses) by the names of the attributes in the relation. The identifier attribute (called the *primary key of the relation*) is underlined. For example, you would express EMPLOYEE1 as follows:

EMPLOYEE1(Emp_ID,Name,Dept,Salary)

Not all tables are relations. Relations have several properties that distinguish them from nonrelational tables:

1. Entries in cells are simple. An entry at the intersection of each row and column has a single value.
2. Entries in a given column are from the same set of values.
3. Each row is unique. Uniqueness is guaranteed because the relation has a nonempty primary key value.
4. The sequence of columns can be interchanged without changing the meaning or use of the relation.
5. The rows may be interchanged or stored in any sequences.

Well-Structured Relations

Well-structured relation
A relation that contains a minimum amount of redundancy and that allows users to insert, modify, and delete the rows without error or inconsistencies; also known as a table.

What constitutes a **well-structured relation** (also known as a table)? Intuitively, a well-structured relation contains a minimum amount of redundancy and allows users to insert, modify, and delete the rows in a table without errors or inconsistencies. EMPLOYEE1 (Figure 9-5) is such a relation. Each row of the table contains data describing one employee, and any modification to an employee's data (such as a change in salary) is confined to one row of the table.

In contrast, EMPLOYEE2 (Figure 9-6) contains data about employees and the courses they have completed. Each row in this table is unique for the combination of Emp_ID and Course, which becomes the primary key for the table. This is not a well-structured relation, however. If you examine the sample data in the table, you notice a considerable amount of redundancy. For example, the Emp_ID, Name, Dept, and Salary values appear in two separate rows for employees 100, 110, and 150. Consequently, if the salary for employee 100 changes, we must record this fact in two rows (or more, for some employees).

The problem with this relation is that it contains data about two entities: EMPLOYEE and COURSE. You will learn to use principles of normalization to divide EMPLOYEE2 into two relations. One of the resulting relations is EMPLOYEE1 (Figure 9-5). The other we will call EMP COURSE, which appears with sample data in Figure 9-7. The primary key of this relation is the combination of Emp_ID and Course (we emphasize this by underlining the column names for these attributes).

FIGURE 9-6
Relation with redundancy

EMPLOYEE2

Emp_ID	Name	Dept	Salary	Course	Date_Completed
100	Margaret Simpson	Marketing	42,000	SPSS	6/19/2014
100	Margaret Simpson	Marketing	42,000	Surveys	10/7/2014
140	Alan Beeton	Accounting	39,000	Tax Acc	12/8/2014
110	Chris Lucero	Info Systems	41,500	SPSS	1/22/2014
110	Chris Lucero	Info Systems	41,500	C++	4/22/2014
190	Lorenzo Davis	Finance	38,000	Investments	5/7/2014
150	Susan Martin	Marketing	38,500	SPSS	6/19/2014
150	Susan Martin	Marketing	38,500	TQM	8/12/2014

EMP COURSE

FIGURE 9-7
EMP COURSE relation

Emp_ID	Course	Date_Completed
100	SPSS	6/19/2014
100	Surveys	10/7/2014
140	Tax Acc	12/8/2014
110	SPSS	1/22/2014
110	C++	4/22/2014
190	Investments	5/7/2014
150	SPSS	6/19/2014
150	TQM	8/12/2014

NORMALIZATION

We have presented an intuitive discussion of well-structured relations; however, we need rules and a process for designing them. **Normalization** is a process for converting complex data structures into simple, stable data structures. For example, we used the principles of normalization to convert the EMPLOYEE2 table with its redundancy to EMPLOYEE1 (Figure 9-5) and EMP COURSE (Figure 9-7).

Normalization
The process of converting complex data structures into simple, stable data structures.

Rules of Normalization

Normalization is based on well-accepted principles and rules. There are many normalization rules, more than can be covered in this text (see Hoffer et al. [2011], for a more complete coverage). Besides the five properties of relations outlined previously, there are two other frequently used rules:

1. *Second normal form (2NF).* Each nonprimary key attribute is identified by the whole key (what we call full functional dependency). For example, in Figure 9-7, both Emp_ID and Course identify a value of Date_Completed because the same Emp_ID can be associated with more than one Date_Completed and the same for Course.
2. *Third normal form (3NF).* Nonprimary key attributes do not depend on each other (what we call no transitive dependencies). For example, in Figure 9-5, Name, Dept, and Salary cannot be guaranteed to be unique for one another.

The result of normalization is that every nonprimary key attribute depends upon the whole primary key and nothing but the primary key. We discuss second and third normal form in more detail next.

Functional Dependence and Primary Keys

Normalization is based on the analysis of functional dependence. A **functional dependency** is a particular relationship between two attributes. In a given relation, attribute B is functionally dependent on attribute A if, for every valid value of A, that value of A uniquely determines the value of B (Dutka and Hanson, 1989). The functional dependence of B on A is represented by an arrow, as follows: $A \rightarrow B$ (e.g., Emp_ID $\rightarrow$ Name in the relation of Figure 9-5). Functional dependence does not imply mathematical dependence—that the value of one attribute may be computed from the value of another attribute; rather, functional dependence of B on A means that there can be only one value of B for each value of A. Thus, a given Emp_ID value can have only one Name value associated with it; the value of Name, however, cannot be derived from the value of Emp_ID. Other examples of functional dependencies from Figure 9-3b are in ORDER, Order_Number, Order_Date, and in INVOICE, Invoice_Number, Invoice_Date, and Order_Number.

Functional dependency
A constraint between two attributes in which the value of one attribute is determined by the value of another attribute.

EXAMPLE

A	B	C	D
X	U	X	Y
(Y)	X	Z	X
Z	Y	Y	Y
(Y)	Z	W	Z

FIGURE 9-8
EXAMPLE relation

An attribute may be functionally dependent on two (or more) attributes rather than on a single attribute. For example, consider the relation EMP COURSE (Emp_ID,Course,Date_Completed) shown in Figure 9-7. We represent the functional dependency in this relation as follows:

Emp_ID,Course → Date_Completed (this is sometimes shown as Emp_ID + Course → Date_Completed). In this case, Date_Completed cannot be determined by either Emp_ID or Course alone because Date_Completed is a characteristic of an employee taking a course.

You should be aware that the instances (or sample data) in a relation do not prove that a functional dependency exists. Only knowledge of the problem domain, obtained from a thorough requirements analysis, is a reliable method for identifying a functional dependency. However, you can use sample data to demonstrate that a functional dependency does not exist between two or more attributes. For example, consider the sample data in the relation EXAMPLE(A,B,C,D), shown in Figure 9-8. The sample data in this relation prove that attribute B is not functionally dependent on attribute A because A does not uniquely determine B (two rows with the same value of A have different values of B).

Second Normal Form

Second normal form (2NF)
A relation is in second normal form if every nonprimary key attribute is functionally dependent on the whole primary key.

A relation is in **second normal form (2NF)** if every nonprimary key attribute is functionally dependent on the whole primary key. Thus, no nonprimary key attribute is functionally dependent on part, but not all, of the primary key. Second normal form is satisfied if any one of the following conditions apply:

1. The primary key consists of only one attribute (such as the attribute Emp_ID in relation EMPLOYEE1).
2. No nonprimary key attributes exist in the relation.
3. Every nonprimary key attribute is functionally dependent on the full set of primary key attributes.

EMPLOYEE2 (Figure 9-6) is an example of a relation that is not in second normal form. The shorthand notation for this relation is

EMPLOYEE2(<u>Emp_ID</u>,Name,Dept,Salary,Course,Date_Completed)

The functional dependencies in this relation are the following:

Emp_ID → Name,Dept,Salary
Emp_ID,Course → Date_Completed

The primary key for this relation is the composite key Emp_ID,Course. Therefore, the nonprimary key attributes Name, Dept, and Salary are functionally dependent on only Emp_ID but not on Course. EMPLOYEE2 has redundancy, which results in problems when the table is updated.

To convert a relation to second normal form, you decompose the relation into new relations using the attributes, called *determinants*, that determine other attributes; the determinants are the primary keys of these relations. EMPLOYEE2 is decomposed into the following two relations:

1. EMPLOYEE(<u>Emp_ID</u>,Name,Dept,Salary): This relation satisfies the first second normal form condition (sample data shown in Figure 9-5).
2. EMP COURSE(<u>Emp_ID</u>,Course,Date_Completed): This relation satisfies second normal form condition three (sample data appear in Figure 9-7).

Third normal form (3NF)
A relation is in second normal form and has no functional (transitive) dependencies between two (or more) nonprimary key attributes.

Third Normal Form

A relation is in **third normal form (3NF)** if it is in second normal form and there are no functional dependencies between two (or more) nonprimary key attributes

SALES

Customer_ID	Customer_Name	Salesperson	Region
8023	Anderson	Smith	South
9167	Bancroft	Hicks	West
7924	Hobbs	Smith	South
6837	Tucker	Hernandez	East
8596	Eckersley	Hicks	West
7018	Arnold	Faulb	North

SALES1

Customer_ID	Customer_Name	Salesperson
8023	Anderson	Smith
9167	Bancroft	Hicks
7924	Hobbs	Smith
6837	Tucker	Hernandez
8596	Eckersley	Hicks
7018	Arnold	Faulb

SPERSON

Salesperson	Region
Smith	South
Hicks	West
Hernandez	East
Faulb	North

FIGURE 9-9
Removing transitive dependencies
(a) Relation with transitive dependency

(b) Relation in 3NF

(a functional dependency between nonprimary key attributes is also called a *transitive dependency*). For example, consider the relation SALES (Customer_ID, Customer_Name,Salesperson,Region) (sample data shown in Figure 9-9a).

The following functional dependencies exist in the SALES relation:

1. Customer_ID → Customer_Name,Salesperson,Region (Customer_ID is the primary key.)
2. Salesperson → Region (Each salesperson is assigned to a unique region.)

Notice that SALES is in second normal form because the primary key consists of a single attribute (Customer_ID). However, Region is functionally dependent on Salesperson, and Salesperson is functionally dependent on Customer_ID. As a result, there are data maintenance problems in SALES.

1. A new salesperson (Robinson) assigned to the North region cannot be entered until a customer has been assigned to that salesperson (because a value for Customer_ID must be provided to insert a row in the table).
2. If customer number 6837 is deleted from the table, we lose the information that salesperson Hernandez is assigned to the East region.
3. If salesperson Smith is reassigned to the East region, several rows must be changed to reflect that fact (two rows are shown in Figure 9-9a).

These problems can be avoided by decomposing SALES into the two relations, based on the two determinants, shown in Figure 9-9b. These relations are the following:

SALES1(Customer_ID,Customer_Name,Salesperson)
SPERSON(Salesperson,Region)

Note that Salesperson is the primary key in SPERSON. Salesperson is also a foreign key in SALES1. A **foreign key** is an attribute that appears as a nonprimary key attribute in one relation (such as SALES1) and as a primary key attribute (or part of a primary key) in another relation. You designate a foreign key by using a dashed underline.

A foreign key must satisfy **referential integrity**, which specifies that the value of an attribute in one relation depends on the value of the same attribute in another relation. Thus, in Figure 9-9b, the value of Salesperson in each row of table SALES1

Foreign key

An attribute that appears as a nonprimary key attribute in one relation and as a primary key attribute (or part of a primary key) in another relation.

Referential integrity

A rule that states that either each foreign key value must match a primary key value in another relation or the foreign key value must be null (i.e., have no value).

is limited only to the current values of Salesperson in the SPERSON table. If some sales do not have to have a salesperson, then it is possible for the value of Salesperson to be null (i.e., have no value). Referential integrity is one of the most important principles of the relational model.

TRANSFORMING E-R DIAGRAMS INTO RELATIONS

Normalization produces a set of well-structured relations that contains all of the data mentioned in system inputs and outputs developed in human interface design. Because these specific information requirements may not represent all future information needs, the E-R diagram you developed in conceptual data modeling is another source of insight into possible data requirements for a new application system. To compare the conceptual data model and the normalized relations developed so far, your E-R diagram must be transformed into relational notation, normalized, and then merged with the existing normalized relations.

Transforming an E-R diagram into normalized relations and then merging all the relations into one final, consolidated set of relations can be accomplished in four steps. These steps are summarized briefly here, and then steps 1, 2, and 4 are discussed in detail in the remainder of this chapter.

1. *Represent entities.* Each entity type in the E-R diagram becomes a relation. The identifier of the entity type becomes the primary key of the relation, and other attributes of the entity type become nonprimary key attributes of the relation.
2. *Represent relationships.* Each relationship in an E-R diagram must be represented in the relational database design. How we represent a relationship depends on its nature. For example, in some cases we represent a relationship by making the primary key of one relation a foreign key of another relation. In other cases, we create a separate relation to represent a relationship.
3. *Normalize the relations.* The relations created in steps 1 and 2 may have unnecessary redundancy. So we need to normalize these relations to make them well structured.
4. *Merge the relations.* So far in database design we have created various relations from both a bottom-up normalization of user views and from transforming one or more E-R diagrams into sets of relations. Across these different sets of relations, there may be redundant relations (two or more relations that describe the same entity type) that must be merged and renormalized to remove the redundancy.

Represent Entities

Each regular entity type in an E-R diagram is transformed into a relation. The identifier of the entity type becomes the primary key of the corresponding relation. Each nonkey attribute of the entity type becomes a nonkey attribute of the relation. You should check to make sure that the primary key satisfies the following two properties:

1. The value of the key must uniquely identify every row in the relation.
2. The key should be nonredundant; that is, no attribute in the key can be deleted without destroying its unique identification.

Some entities may have keys that include the primary keys of other entities. For example, an EMPLOYEE DEPENDENT may have a Name for each dependent, but to form the primary key for this entity, you must include the Employee_ID attribute from the associated EMPLOYEE entity. Such an entity whose primary key depends upon the primary key of another entity is called a *weak entity*.

Representation of an entity as a relation is straightforward. Figure 9-10a shows the CUSTOMER entity type for PVF. The corresponding CUSTOMER relation is represented as follows:

CUSTOMER(Customer_ID,Name,Address,City_State_ZIP,Discount)

CUSTOMER
Customer_ID
Name
Address
City_State_Zip
Discount

FIGURE 9-10
Transforming an entity type to a relation
(a) E-R diagram

CUSTOMER

(b) Relations

Customer_ID	Name	Address	City_State_ZIP	Discount
1273	Contemporary Designs	123 Oak St.	Austin, TX 28384	5%
6390	Casual Corner	18 Hoosier Dr.	Bloomington, IN 45821	3%

In this notation, the entity type label is translated into a relation name. The identifier of the entity type is listed first and underlined. All nonkey attributes are listed after the primary key. This relation is shown as a table with sample data in Figure 9-10b.

Represent Relationships

The procedure for representing relationships depends on both the degree of the relationship—unary, binary, ternary—and the cardinalities of the relationship.

Binary 1: N and 1:1 Relationships A binary one-to-many (1:N) relationship in an E-R diagram is represented by adding the primary key attribute (or attributes) of the entity on the one side of the relationship as a foreign key in the relation that is on the many side of the relationship.

Figure 9-11a, an example of this rule, shows the Places relationship (1:N) linking CUSTOMER and ORDER at PVF. Two relations, CUSTOMER and ORDER, were formed from the respective entity types (see Figure 9-11b). Customer_ID, which is the primary key of CUSTOMER (on the one side of the relationship), is added as a foreign key to ORDER (on the many side of the relationship). One special case under this rule was mentioned in the previous section. If the entity on the many side needs the key of the entity on the one side as part of its primary key (this is a so-called weak entity), then this attribute is added, not as a nonkey, but as part of the primary key.

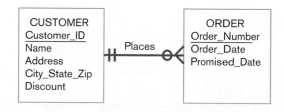

FIGURE 9-11
Representing a 1:N relationship
(a) E-R diagram

CUSTOMER

Customer_ID	Name	Address	City_State_ZIP	Discount
1273	Contemporary Designs	123 Oak St.	Austin, TX 28384	5%
6390	Casual Corner	18 Hoosier Dr.	Bloomington, IN 45821	3%

ORDER

Order_Number	Order_Date	Promised_Date	Customer_ID
57194	3/15/1X	3/28/1X	6390
63725	3/17/1X	4/01/1X	1273
80149	3/14/1X	3/24/1X	6390

(b) Relations

For a binary or unary one-to-one (1:1) relationship between two entities A and B (for a unary relationship, A and B would be the same entity type), the relationship can be represented by any of the following choices:

1. Adding the primary key of A as a foreign key of B
2. Adding the primary key of B as a foreign key of A
3. Both of the above

Binary and Higher-Degree *M:N* Relationships   Suppose that there is a binary many-to-many *(M:N)* relationship (or associative entity) between two entity types A and B. For such a relationship, we create a separate relation C. The primary key of this relation is a composite key consisting of the primary key for each of the two entities in the relationship. Any nonkey attributes associated with the *M:N* relationship are included with relation C.

Figure 9-12a, an example of this rule, shows the Requests relationship *(M:N)* between the entity types ORDER and PRODUCT for PVF. Figure 9-12b shows the three relations (ORDER, ORDER LINE, and PRODUCT) that are formed from the entity types and the Requests relationship. A relation (called ORDER LINE in Figure 9-12b) is created for the Requests relationship. The primary key of ORDER LINE is the combination (Order_Number, Product_ID), which is the respective primary keys of ORDER and PRODUCT. The nonkey attribute Quantity_Ordered also appears in ORDER LINE.

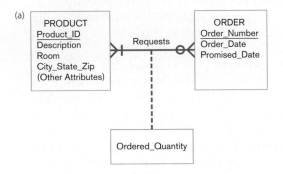

(b) ORDER

Order_Number	Order_Date	Promised_Date
61384	2/17/2014	3/01/2014
62009	2/13/2014	2/27/2014
62807	2/15/2014	3/01/2014

ORDER LINE

Order_Number	Product_ID	Quantity_ Ordered
61384	M128	2
61384	A261	1

PRODUCT

Product_ID	Description	Room	(Other Attributes)
M128	Bookcase	Study	–
A261	Wall unit	Family	–
R149	Cabinet	Study	–

FIGURE 9-12
Representing an *M:N* relationship
(a) E-R diagram
(b) Relations

Occasionally, the relation created from an *M:N* relationship requires a primary key that includes more than just the primary keys from the two related relations. Consider, for example, the following situation:

In this case, Date must be part of the key for the SHIPMENT relation to uniquely distinguish each row of the SHIPMENT table, as follows:

SHIPMENT(Customer_ID,Vendor_ID,Date,Amount)

If each shipment has a separate nonintelligent key, say, a shipment number, then Date becomes a nonkey and Customer_ID and Vendor_ID become foreign keys, as follows:

SHIPMENT(Shipment_Number,Customer_ID,Vendor_ID,Date,Amount)

In some cases, there may be a relationship among three or more entities. In such cases, we create a separate relation that has as a primary key the composite of the primary keys of each of the participating entities (plus any necessary additional key elements). This rule is a simple generalization of the rule for a binary *M:N* relationship.

Unary Relationships   To review, a unary relationship is a relationship between the instances of a single entity type, which are also called *recursive relationships*. Figure 9-13 shows two common examples. Figure 9-13a shows a one-to-many relationship named Manages that associates employees with another employee who is their manager. Figure 9-13b shows a many-to-many relationship that associates certain items with their component items. This relationship is called a *bill-of-materials structure*.

For a unary 1:*N* relationship, the entity type (such as EMPLOYEE) is modeled as a relation. The primary key of that relation is the same as for the entity type. Then a foreign key is added to the relation that references the primary key values. A **recursive foreign key** is a foreign key in a relation that references the primary key values of that same relation. We can represent the relationship in Figure 9-13a as follows:

Recursive foreign key
A foreign key in a relation that references the primary key values of that same relation.

EMPLOYEE(Emp_ID,Name,Birthdate,Manager_ID)

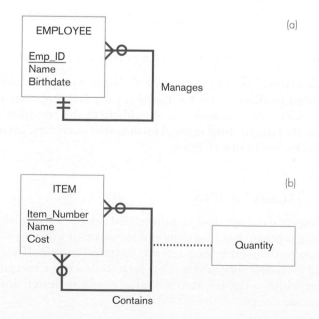

FIGURE 9-13
Two unary relationships
(a) EMPLOYEE with Manages relationship (1:*N*)
(b) Bill-of-materials structure (*M:N*)

TABLE 9-1   E-R Diagrams to Relational Transformation

E-R Structure	Relational Representation
Regular entity	Create a relation with primary key and nonkey attributes.
Weak entity	Create a relation with a composite primary key (which includes the primary key of the entity on which this weak entity depends) and nonkey attributes.
Binary or unary 1:1 relationship	Place the primary key of either entity in the relation for the other entity or do this for both entities.
Binary 1:N relationship	Place the primary key of the entity on the one side of the relationship as a foreign key in the relation for the entity on the many side.
Binary or unary M:N relationship or	Create a relation with a composite primary key using the primary keys of the related entities, plus any nonkey attributes associative entity of the relationship or associative entity.
Binary or unary M:N relationship or associative entity with additional key(s)	Create a relation with a composite primary key using the primary keys of the related entities and additional primary key attributes associated with the relationship or associative entity, plus any nonkey attributes of the relationship or associative entity.
Binary or unary M:N relationship or associative entity with its own key	Create a relation with the primary key associated with the relationship or associative entity, plus any nonkey attributes of the relationship or associative entity and the primary keys of the related entities (as foreign key attributes).
Supertype/subtype	Create a relation for the superclass, which contains the primary relationship key and all nonkey attributes in common with all subclasses, plus create a separate relation for each subclass with the same primary key (with the same or local name) but with only the nonkey attributes related to that subclass.

In this relation, Manager_ID is a recursive foreign key that takes its values from the same set of worker identification numbers as Emp_ID.

For a unary *M:N* relationship, we model the entity type as one relation. Then we create a separate relation to represent the *M:N* relationship. The primary key of this new relation is a composite key that consists of two attributes (which need not have the same name) that both take their values from the same primary key. Any attribute associated with the relationship (such as Quantity in Figure 9-13b) is included as a nonkey attribute in this new relation. We can express the result for Figure 9-13b as follows:

ITEM (Item_Number,Name,Cost)
ITEM-BILL (Item_Number,Component_Number,Quantity)

Summary of Transforming E-R Diagrams to Relations

We have now described how to transform E-R diagrams to relations. Table 9-1 lists the rules discussed in this section for transforming E-R diagrams into equivalent relations. After this transformation, you should check the resulting relations to determine whether they are in third normal form and, if necessary, perform normalization as described earlier in this chapter.

MERGING RELATIONS

As part of the logical database design, normalized relations likely have been created from a number of separate E-R diagrams and various user interfaces. Some of the relations may be redundant—they may refer to the same entities. If so, *you should merge those relations to remove the redundancy*. This section describes merging relations, or view integration, which is the last step in logical database design and prior to physical file and database design.

An Example of Merging Relations

Suppose that modeling a user interface or transforming an E-R diagram results in the following 3NF relation:

EMPLOYEE1(<u>Emp_ID</u>,Name,Address,Phone)

Modeling a second user interface might result in the following relation:

EMPLOYEE2(<u>Emp_ID</u>,Name,Address,Jobcode,Number_of_Years)

Because these two relations have the same primary key (Emp_ID) and describe the same entity, they should be merged into one relation. The result of merging the relations is the following relation:

EMPLOYEE(<u>Emp_ID</u>,Name,Address,Phone,Jobcode,Number_of_Years)

Notice that an attribute that appears in both relations (such as Name in this example) appears only once in the merged relation.

View Integration Problems

When integrating relations, you must understand the meaning of the data and be prepared to resolve any problems that may arise in the process. In this section, we describe and illustrate four problems that arise in view integration: synonyms, homonyms, dependencies between nonkeys, and class/subclass relationships.

Synonyms   In some situations, two or more attributes may have different names but the same meaning, as when they describe the same characteristic of an entity. Such attributes are called **synonyms**. For example, Emp_ID and Employee_Number may be synonyms.

> **Synonym**
> Two different names that are used for the same attribute.

When merging relations that contain synonyms, you should obtain, if possible, agreement from users on a single standardized name for the attribute and eliminate the other synonym. Another alternative is to choose a third name to replace the synonyms. For example, consider the following relations:

STUDENT1(<u>Student_ID</u>,Name)
STUDENT2(<u>Matriculation_Number</u>,Name,Address)

In this case, the analyst recognizes that both the Student_ID and the Matriculation_Number are synonyms for a person's social security number (SSN) and are identical attributes. One possible resolution would be to standardize one of the two attribute names, such as Student_ID. Another option is to use a new attribute name, such as SSN, to replace both synonyms. With the latter approach, merging the two relations would produce the following result:

STUDENT(<u>SSN</u>,Name,Address)

Homonyms   In other situations, a single attribute name, called a **homonym**, may have more than one meaning or describe more than one characteristic. For example, the term *account* might refer to a bank's checking account, savings account, loan account, or other type of account; therefore, *account* refers to different data, depending on how it is used.

> **Homonym**
> A single attribute name that is used for two or more different attributes.

You should be on the lookout for homonyms when merging relations. Consider the following example:

STUDENT1(<u>Student_ID</u>,Name,Address)
STUDENT2(<u>Student_ID</u>,Name,Phone_Number,Address)

In discussions with users, the systems analyst may discover that the attribute Address in STUDENT1 refers to a student's campus address, whereas in STUDENT2 the same attribute refers to a student's home address. To resolve this conflict, we would probably need to create new attribute names, and the merged relation would become

> STUDENT(Student_ID,Name,Phone_Number,Campus_Address, Permanent_ Address)

Dependencies between Nonkeys   When two 3NF relations are merged to form a single relation, dependencies between nonkeys may result. For example, consider the following two relations:

> STUDENT1(Student_ID,Major)
> STUDENT2(Student_ID,Adviser)

Because STUDENT1 and STUDENT2 have the same primary key, the two relations may be merged:

> STUDENT(Student_ID,Major,Adviser)

However, suppose that each major has exactly one adviser. In this case, Adviser is functionally dependent on Major:

> Major → Adviser

If this dependency exists, then STUDENT is in 2NF but not 3NF because it contains a functional dependency between nonkeys. The analyst can create 3NF relations by creating two relations with Major as a foreign key in STUDENT:

> STUDENT(Student_ID,Major)
> MAJOR ADVISER(Major,Adviser)

Class/Subclass   Class/subclass relationships may be hidden in user views or relations. Suppose that we have the following two hospital relations:

> PATIENT1(Patient_ID,Name,Address,Date_Treated)
> PATIENT2(Patient_ID,Room_Number)

Initially, it appears that these two relations can be merged into a single PATIENT relation. However, suppose that there are two different types of patients: inpatients and outpatients. PATIENT1 actually contains attributes common to all patients. PATIENT2 contains an attribute (Room_Number) that is a characteristic only of inpatients. In this situation, you should create *class/subclass* relationships for these entities:

> PATIENT(Patient_ID,Name,Address)
> INPATIENT(Patient_ID,Room_Number)
> OUTPATIENT(Patient_ID,Date_Treated)

LOGICAL DATABASE DESIGN FOR HOOSIER BURGER

Figure 9-14 shows an E-R diagram that has been developed for a new inventory control system at Hoosier Burger. The new system was discussed previously in Chapter 7, where a DFD and decision table (respectively) for the system were created. In this section we show how this E-R model is translated into normalized relations, and how to normalize and then merge the relations for a new report with the relations from the E-R model.

In this E-R model, four entities exist independently of other entities: SALE, PRODUCT, INVOICE, and INVENTORY ITEM. Given the attributes shown in Figure 9-14, we can represent these entities in the following four relations:

> SALE(Receipt_Number,Sale_Date)
> PRODUCT(Product_ID,Product_Description)

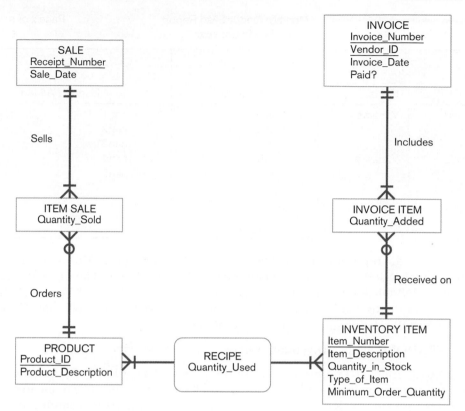

FIGURE 9-14
Final class diagram for Hoosier Burger's inventory control system

INVOICE(Vendor_ID,Invoice_Number,Invoice_Date,Paid?)
INVENTORY ITEM(Item_Number,Item_Description,Quantity_in_Stock,
Minimum_Order_Quantity,Type_of_Item)

The entities ITEM SALE and INVOICE ITEM as well as the associative entity RECIPE each has a composite primary key taken from the entities to which they relate, so we can represent these three entities in the following three relations:

ITEM SALE(Receipt_Number,Product_ID,Quantity_Sold)
INVOICE ITEM (Vendor_ID,Invoice_Number,Item_Number,Quantity_Added)
RECIPE(Product_ID,Item_Number,Quantity_Used)

Because there are no many-to-many, one-to-one, or unary relationships, we have now represented all the entities and relationships from the E-R model. Also, each of the above relations is in 3NF because all attributes are simple, all nonkeys are fully dependent on the whole key, and there are no dependencies between nonkeys in the INVOICE and INVENTORY ITEM relations.

Now suppose that Bob Mellankamp wanted an additional report that was not previously known by the analyst who designed the inventory control system for Hoosier Burger. A rough sketch of this new report, listing volume of purchases from each vendor by type of item in a given month, appears in Figure 9-15. In this report, the same type of item may appear many times if multiple vendors supply the same type of item.

This report contains data about several relations already known to the analyst, including:

- INVOICE(Vendor_ID,Invoice_Number,Invoice_Date): Primary keys and the date are needed to select invoices in the specified month of the report.
- INVENTORY ITEM(Item_Number,Type_of_Item): Primary key and a nonkey in the report.
- INVOICE ITEM (Vendor_ID,Invoice_Number,Item_Number,Quantity_Added): Primary keys and the raw quantity of items invoiced that are subtotaled by vendor and type of item in the report.

FIGURE 9-15
Hoosier Burger Monthly Vendor Load
Report

	Monthly Vendor Load Report for Month: xxxxx		Page x of n

Vendor		Type of Item	Total Quantity Added
ID	Name		
V1	V1name	aaa	nnn1
		bbb	nnn2
		ccc	nnn3
V2	V2name	bbb	nnn4
		mmm	nnn5
x			
x			
x			

In addition, the report includes a new attribute—Vendor_Name. After some investigation, an analyst determines that Vendor_ID → Vendor_Name. The whole primary key of the INVOICE relation is Vendor_ID and Invoice_Number, so if Vendor_Name were part of the INVOICE relation, this relation would violate the 3NF rule. Thus, a new VENDOR relation must be created as follows:

VENDOR(Vendor_ID,Vendor_Name)

Now, Vendor_ID not only is part of the primary key of INVOICE but also is a foreign key referencing the VENDOR relation. Hence, there must be a one-to-many relationship from VENDOR to INVOICE. The systems analyst determines that an invoice must come from a vendor, and there is no need to keep data about a vendor unless the vendor invoices Hoosier Burger. An updated E-R diagram, reflecting these enhancements for new data needed in the monthly vendor load report, appears in Figure 9-16. The normalized relations for this database are as follows:

SALE(Receipt_Number,Sale_Date)
PRODUCT(Product_ID,Product_Description)

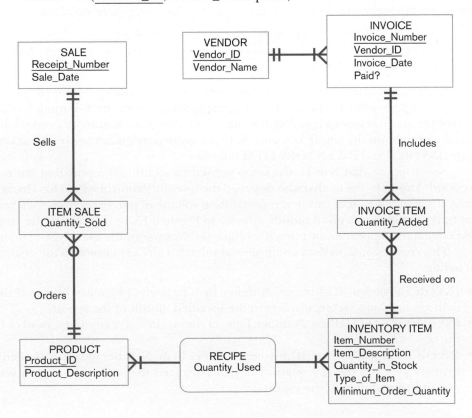

FIGURE 9-16
Class diagram corresponding to normalized relations of Hoosier Burger's inventory control system

INVOICE(Vendor_ID,Invoice_Number,Invoice_Date,Paid?)
INVENTORY ITEM(Item_Number,Item_Description,Quantity_in_Stock,
Minimum_Order_Quantity,Type_of_Item)
ITEM SALE(Receipt_Number,Product_ID,Quantity_Sold)
INVOICE ITEM(Vendor_ID,Invoice_Number,Item_Number,Quantity_Added)
RECIPE(Product_ID,Item_Number,Quantity_Used)
VENDOR(Vendor_ID,Vendor_Name)

PHYSICAL FILE AND DATABASE DESIGN

Designing physical files and databases requires certain information that should have been collected and produced during prior SDLC phases. This information includes the following:

- Normalized relations, including volume estimates
- Definitions of each attribute
- Descriptions of where and when data are used: entered, retrieved, deleted, and updated (including frequencies)
- Expectations or requirements for response time and data integrity
- Descriptions of the technologies used for implementing the files and database so that the range of required decisions and choices for each is known

Normalized relations are, of course, the result of logical database design. Statistics on the number of rows in each table as well as the other information listed above may have been collected during requirements determination in systems analysis. If not, these items need to be discovered to proceed with database design.

We take a bottom-up approach to reviewing physical file and database design. Thus, we begin the physical design phase by addressing the design of physical fields for each attribute in a logical data model.

DESIGNING FIELDS

A **field** is the smallest unit of application data recognized by system software, such as a programming language or database management system. An attribute from a logical database model may be represented by several fields. For example, a student name attribute in a normalized student relation might be represented as three fields: last name, first name, and middle initial. In general, you will represent each attribute from each normalized relation as one or more fields. The basic decisions you must make in specifying each field concern the type of data (or storage type) used to represent the field and data integrity controls for the field.

Field
The smallest unit of named application data recognized by system software.

Choosing Data Types

A **data type** is a coding scheme recognized by system software for representing organizational data. The bit pattern of the coding scheme is usually immaterial to you, but the space to store data and the speed required to access data are of consequence in the physical file and database design. The specific file or database management software you use with your system will dictate which choices are available to you. For example, Table 9-2 lists the most commonly used data types available in Oracle 10g.

Selecting a data type balances four objectives that will vary in degree of importance, depending on the application:

1. Minimize storage space
2. Represent all possible values of the field
3. Improve data integrity for the field
4. Support all data manipulations desired on the field

Data type
A coding scheme recognized by system software for representing organizational data.

TABLE 9-2 **Commonly Used Data Types in Oracle 10i**

Data Type	Description
VARCHAR2	Variable-length character data with a maximum length of 4000 characters; you must enter a maximum field length (e.g., VARCHAR2(30) for a field with a maximum length of 30 characters). A value less than 30 characters will consume only the required space.
CHAR	Fixed-length character data with a maximum length of 255 characters; default length is 1 character (e.g., CHAR(5) for a field with a fixed length of five characters, capable of holding a value from 0 to 5 characters long).
LONG	Capable of storing up to two gigabytes of one variable-length character data field (e.g., to hold a medical instruction or a customer comment).
NUMBER	Positive and negative numbers in the range 10^{-130} to 10^{126}; can specify the precision (total number of digits to the left and right of the decimal point) and the scale (the number of digits to the right of the decimal point) (e.g., NUMBER(5) specifies an integer field with a maximum of 5 digits and NUMBER(5, 2) specifies a field with no more than five digits and exactly two digits to the right of the decimal point).
DATE	Any date from January 1, 4712 B.C. to December 31, 4712 A.D.; date stores the century, year, month, day, hour, minute, and second.
BLOB	Binary large object, capable of storing up to four gigabytes of binary data (e.g., a photograph or sound clip).

You want to choose a data type for a field that minimizes space, represents every possible legitimate value for the associated attribute, and allows the data to be manipulated as needed. For example, suppose a quantity sold field can be represented by a Number data type. You would select a length for this field that would handle the maximum value, plus some room for growth of the business. Further, the Number data type will restrict users from entering inappropriate values (text), but it does allow negative numbers (if this is a problem, application code or form design may be required to restrict the values to positive ones).

Be careful—the data type must be suitable for the life of the application; otherwise, maintenance will be required. Choose data types for future needs by anticipating growth. Also, be careful that date arithmetic can be done so that dates can be subtracted or time periods can be added to or subtracted from a date.

Several other capabilities of data types may be available with some database technologies. We discuss a few of the most common of these features next: calculated fields and coding and compression techniques.

Calculated Fields   It is common for an attribute to be mathematically related to other data. For example, an invoice may include a total due field, which represents the sum of the amount due on each item on the invoice. A field that can be derived from other database fields is called a **calculated field** (or a computed field or a derived field). Recall that a functional dependency between attributes does not imply a calculated field. Some database technologies allow you to explicitly define calculated fields along with other raw data fields. If you specify a field as calculated, you would then usually be prompted to enter the formula for the calculation; the formula can involve other fields from the same record and possibly fields from records in related files. The database technology will either store the calculated value or compute it when requested.

Coding and Compression Techniques   Some attributes have very few values from a large range of possible values. For example, suppose that each product from PVF has a finish attribute, with possible values of Birch, Walnut, Oak, and so forth. To store this attribute as Text might require 12, 15, or even 20 bytes to represent the longest finish value. Suppose that even a liberal estimate is that PVF will never have more than 25 finishes. Thus, a single alphabetic or alphanumeric character would be more than sufficient. We not only reduce storage space but also increase integrity

---

Calculated field
A field that can be derived from other database fields. Also known as a computed field or a derived field.

(by restricting input to only a few values), which helps to achieve two of the physical file and database design goals. Codes also have disadvantages. If used in system inputs and outputs, they can be more difficult for users to remember, and programs must be written to decode fields if codes will not be displayed.

Controlling Data Integrity

Accurate data are essential for compliance with new national and international regulations, such as Sarbanes-Oxley (SOX) and Basel II. COBIT (Control Objectives for Information and Related Technologies) and ITIL (IT Infrastructure Library), provide standards, guidelines, and rules for corporate governance, risk assessment, security, and controls of data. These preventive controls are best and consistently applied if designed into the database and enforced by the database management system (DBMS). Data integrity controls can be viewed very positively during audits for compliance with regulations. These controls are only as good as the underlying field data controls.

We have already explained that data typing helps control data integrity by limiting the possible range of values for a field. There are additional physical file and database design options you might use to ensure higher-quality data. Although these controls can be imposed within application programs, it is better to include these as part of the file and database definitions so that the controls are guaranteed to be applied all the time as well as uniformly for all programs. There are four popular data integrity control methods: default value, range control, referential integrity, and null value control.

- *Default value.* A **default value** is the value a field will assume unless an explicit value is entered for the field. For example, the city and state of most customers for a particular retail store will likely be the same as the store's city and state. Assigning a default value to a field can reduce data entry time (the field can simply be skipped during data entry) and data entry errors, such as typing *IM* instead of *IN* for *Indiana.*

- *Range control.* Both numeric and alphabetic data may have a limited set of permissible values. For example, a field for the number of product units sold may have a lower bound of zero, and a field that represents the month of a product sale may be limited to the values JAN, FEB, and so forth.

- *Referential integrity.* As noted earlier in this chapter, the most common example of referential integrity is cross-referencing between relations. For example, consider the pair of relations in Figure 9-17a. In this case, the values for the foreign key Customer_ID field within a customer order must be limited to the set of Customer_ID values from the CUSTOMER relation; we would not want to accept an order for a nonexisting or unknown customer. Referential integrity may be useful in other instances. Consider the employee relation example in Figure 9-17b. In this example, the EMPLOYEE relation has a field of Supervisor_ID. This field refers to the Employee_ID of the employee's supervisor and should have referential integrity on the Employee_ID field within the same relation. Note in this case that, because some employees do not have supervisors, this is a weak referential integrity constraint because the value of a Supervisor_ID field may be empty.

- *Null value control.* A **null value** is a special field value, distinct from a zero, blank, or any other value, that indicates that the value for the field is missing or otherwise unknown. It is not uncommon that when it is time to enter data—for example, a new customer—you might not know the customer's phone number. The question is whether a customer, to be valid, must have a value for this field. The answer for this field is probably no, initially, because most data processing can continue without knowing the customer's phone number. Later, a null value may not be allowed when you are ready to ship a product to the customer. On the other hand, you must always know a value for the Customer_ID field. Due to

Default value
A value a field will assume unless an explicit value is entered for that field.

Null value
A special field value, distinct from zero, blank, or any other value, that indicates that the value for the field is missing or otherwise unknown.

FIGURE 9-17
Examples of referential intergrity field controls
(a) Referential integrity between relations

CUSTOMER(**Customer_ID**,Cust_Name,Cust_Address, . . .)

CUST_ORDER(Order_ID,**Customer_ID**,Order_Date, . . .)
 and Customer_ID may not be null because every order must be for
 some existing customer

(b) Referential integrity within a relation

EMPLOYEE(**Employee_ID,Supervisor_ID**,Empl_Name, . . .)
 and Superviosr_ID may be null because not all employees have supervisors

referential integrity, you cannot enter any customer orders for this new customer without knowing an existing Customer_ID value, and customer name is essential for visual verification of correct data entry. Besides using a special null value when a field is missing its value, you can also estimate the value, produce a report indicating rows of tables with critical missing values, or determine whether the missing value matters when computing needed information.

DESIGNING PHYSICAL TABLES

Physical table
A named set of rows and columns that specifies the fields in each row of the table.

A relational database is a set of related tables (tables are related by foreign keys referencing primary keys). In logical database design, you grouped into a relation those attributes that concern some unifying, normalized business concept, such as a customer, product, or employee. In contrast, a **physical table** is a named set of rows and columns that specifies the fields in each row of the table. A physical table may or may not correspond to one relation. Whereas normalized relations possess properties of well-structured relations, the design of a physical table has two goals different from those of normalization: efficient use of secondary storage and data processing speed.

The efficient use of secondary storage (disk space) relates to how data are loaded on disks. Disks are physically divided into units (called pages) that can be read or written in one machine operation. Space is used efficiently when the physical length of a table row divides close to evenly into the length of the storage unit. For many information systems, this even division is very difficult to achieve because it depends on factors, such as operating system parameters, outside the control of each database. Consequently, we do not discuss this factor of physical table design in this text.

Denormalization
The process of splitting or combining normalized relations into physical tables based on affinity of use of rows and fields.

A second and often more important consideration when selecting a physical table design is efficient data processing. Data are most efficiently processed when they are stored close to one another in secondary memory, thus minimizing the number of input/output (I/O) operations that must be performed. Typically, the data in one physical table (all the rows and fields in those rows) are stored close together on disk. **Denormalization** is the process of splitting or combining normalized relations into physical tables based on affinity of use of rows and fields. In Figure 9-18a, a normalized product relation is split into separate physical tables, each containing only engineering, accounting, or marketing product data; the primary key must be included in each table. Note that the Description and Color attributes are repeated in both the engineering and marketing tables because these attributes relate to both kinds of data. In Figure 9-18b, a customer relation is denormalized by putting rows from different geographic regions into separate tables. In both cases, the goal is to create tables that contain only the data used together in programs. By placing data used together close to one another on disk, the number of disk I/O operations needed to retrieve all the data needed by a program is minimized.

Normalized Product Relation
    Product(Product_ID,Description,Drawing_Number,Weight,Color,Unit_Cost,
        Burden_Rate,Price,Product_Manager)

Denormalized Functional Area Product Relations for Tables
    Engineering: E_Product(Product_ID,Description,Drawing_Number,Weight,Color)
    Accounting: A_Product(Product_ID,Unit_Cost,Burden_Rate)
    Marketing: M_Product(Product_ID,Description,Color,Price,Product_Manager)

FIGURE 9-18
Examples of denormalization
(a) Denormalization by columns

Normalized Customer Table

CUSTOMER

Customer_ID	Name	Region	Annual_Sales
1256	Rogers	Atlantic	10,000
1323	Temple	Pacific	20,000
1455	Gates	South	15,000
1626	Hope	Pacific	22,000
2433	Bates	South	14,000
2566	Bailey	Atlantic	12,000

(b) Denormalization by rows

Denormalized Regional Customer Tables

A_CUSTOMER

Customer_ID	Name	Region	Annual_Sales
1256	Rogers	Atlantic	10,000
2566	Bailey	Atlantic	12,000

P_CUSTOMER

Customer_ID	Name	Region	Annual_Sales
1323	Temple	Pacific	20,000
1626	Hope	Pacific	22,000

S_CUSTOMER

Customer_ID	Name	Region	Annual_Sales
1455	Gates	South	15,000
2433	Bates	South	14,000

The capability to split a table into separate sections, often called *partitioning*, is possible with most relational database products. With Oracle, there are three types of table partitioning:

1. *Range partitioning.* Partitions are defined by nonoverlapping ranges of values for a specified attribute (so separate tables are formed of the rows whose specified attribute values fall in indicated ranges).

2. *Hash partitioning.* A table row is assigned to a partition by an algorithm and then maps the specified attribute value to a partition.
3. *Composite partitioning.* Combines range and hash partitioning by first segregating data by ranges on the designated attribute, and then within each of these partitions, it further partitions by hashing on the designated attribute.

Each partition is stored in a separate contiguous section of disk space, which Oracle calls a *tablespace.*

Denormalization can increase the chance of errors and inconsistencies that normalization avoided. Further, denormalization optimizes certain data processing activities at the expense of others, so if the frequencies of different processing activities change, the benefits of denormalization may no longer exist (Finkelstein, 1988).

Various forms of denormalization, which involves combining data from several normalized tables, can be done, but there are no hard-and-fast rules for deciding when to denormalize data. Here are three common situations (Rodgers, 1989) in which denormalization across tables often makes accessing related data faster (see Figure 9-19 for illustrations):

1. *Two entities with a one-to-one relationship.* Figure 9-19a shows student data with optional data from a standard scholarship application that a student may complete. In this case, one record could be formed with four fields from the STUDENT and SCHOLARSHIP APPLICATION FORM normalized relations. (Note: In this case, fields from the optional entity must have null values allowed.)
2. *A many-to-many relationship (associative entity) with nonkey attributes.* Figure 9-19b shows price quotes for different items from different vendors. In this case, fields from ITEM and PRICE QUOTE relations might be combined into one physical table to avoid having to combine all three tables together. (*Note:* This may create considerable duplication of data—in the example, the ITEM fields, such as Description, would repeat for each price quote—and excessive updating if duplicated data change.)
3. *Reference data.* Figure 9-19c shows that several ITEMs have the same STORAGE INSTRUCTIONS and that STORAGE INSTRUCTIONS relate only to ITEMs. In this case, the storage instruction data could be stored in the ITEM table, thus reducing the number of tables to access but also creating redundancy and the potential for extra data maintenance.

Arranging Table Rows

Physical file
A named set of table rows stored in a contiguous section of secondary memory.

The result of denormalization is the definition of one or more physical files. A computer operating system stores data in a **physical file**, which is a named set of table rows stored in a contiguous section of secondary memory. A file contains rows and columns from one or more tables, as produced from denormalization. To the operating system (e.g., Windows, Linux, or UNIX), each table may be one file or the whole database may be in one file, depending on how the database technology and database designer organize data. The way the operating system arranges table rows in a file is called a **file organization**. With some database technologies, the systems designer can choose from among several organizations for a file.

File organization
A technique for physically arranging the records of a file.

If the database designer has a choice, he or she chooses a file organization for a specific file that will provide the following:

1. Fast data retrieval
2. High throughput for processing transactions
3. Efficient use of storage space
4. Protection from failures or data loss
5. Minimal need for reorganization
6. Accommodation of growth
7. Security from unauthorized use

FIGURE 9-19
Possible denormalization situations
(a) Two entities with a one-to-one relationship

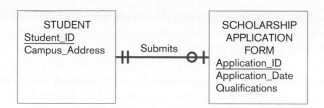

Normalized relations:
  STUDENT(Student_ID,Campus_Address,Application_ID)
  APPLICATION(Application_ID,Application_Date,Qualifications,Student_ID)

Denormalized relation:
  STUDENT(Student_ID,Campus_Address,Application_Date,Qualifications)
  and Application_Date and Qualifications may be null

(**Note:** We assume Application_ID is not necessary when all fields are stored in one record, but this field can be included if it is required application data.)

(b) A many-to-many relationship with nonkey attributes

Normalized relations:
  VENDOR(Vendor_ID,Address,Contact_Name)
  ITEM(Item_ID,Description)
  PRICE QUOTE(Vendor_ID,Item_ID,Price)

Denormalized relations:
  VENDOR(Vendor_ID,Address,Contact_Name)
  ITEM-QUOTE(Vendor_ID,Item_ID,Description,Price)

(c) Reference data

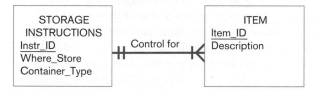

Normalized relations:
  STORAGE(Instr_ID,Where_Store,Container_Type)
  ITEM(Item_ID,Description,Instr_ID)

Denormalized relation
  ITEM(Item_ID,Description,Where_Store,Container_Type)

Often these objectives conflict, and you must select an organization for each file that provides a reasonable balance among the criteria within the resources available.

To achieve these objectives, many file organizations use a pointer. A **pointer** is a field of data that can be used to locate a related field or row of data. In most cases, a pointer contains the address of the associated data, which has no business meaning. Pointers are used in file organizations when it is not possible to store related data next to each other. Because this is often the case, pointers are common. In most cases, fortunately, pointers are hidden from a programmer. Because a database designer may need to decide if and how to use pointers, however, we introduce the concept here.

Literally hundreds of different file organizations and variations have been created, but we outline the basics of three families of file organizations used in most

Pointer
A field of data that can be used to locate a related field or row of data.

file management environments: sequential, indexed, and hashed, as illustrated in Figure 9-20. You need to understand the particular variations of each method available in the environment for which you are designing files.

Sequential file organization
A file organization in which rows in a file are stored in sequence according to a primary key value.

Sequential File Organizations   In a **sequential file organization**, the rows in the file are stored in sequence according to a primary key value (see Figure 9-20a). To locate a particular row, a program must normally scan the file from the beginning until the desired row is located. A common example of a sequential file is the alphabetic list of persons in the white pages of a phone directory (ignoring any index that may be included with the directory). Sequential files are very fast if you want to process rows sequentially, but they are impractical for random row retrievals. Deleting rows can cause wasted space or the need to compress the file. Adding rows requires rewriting the file, at least from the point of insertion. Updating a row may also require rewriting the file, unless the file organization supports rewriting over the updated row only. Only one sequence can be maintained without duplicating the rows.

Indexed file organization
A file organization in which rows are stored either sequentially or nonsequentially, and an index is created that allows software to locate individual rows.

Index
A table used to determine the location of rows in a file that satisfy some condition.

Secondary key
One or a combination of fields for which more than one row may have the same combination of values.

Indexed File Organizations   In an **indexed file organization**, the rows are stored either sequentially or nonsequentially, and an index is created that allows the application software to locate individual rows (see Figure 9-20b). Like a card catalog in a library, an index is a structure that is used to determine the rows in a file that satisfy some condition. Each entry matches a key value with one or more rows. An **index** can point to unique rows (a primary key index, such as on the Product_ID field of a product table) or to potentially more than one row. An index that allows each entry to point to more than one record is called a **secondary key** index. Secondary key indexes are important for supporting many reporting requirements and for providing rapid ad hoc data retrieval. An example would be an index on the Finish field of a product table.

One of the most powerful capabilities of indexed file organizations is the ability to create multiple indexes, similar to the title, author, and subject indexes in a library. Search results from the multiple indexes can be combined very quickly to find those records with precisely the combination of values sought. The example in Figure 9-20b, typical of many index structures, illustrates that indexes can be built on top of indexes, creating a hierarchical set of indexes, and the data are stored sequentially in many contiguous segments. For example, to find the record with key "Hoosiers," the file organization would start at the top index and take the pointer after the entry P, which points to another index for all keys that begin with the letters G through P in the alphabet. Then the software would follow the pointer after the H in this index, which represents all those records with keys that begin with the letters G through H. Eventually, the search through the indexes either locates the desired record or indicates that no such record exists. The reason for storing the data in many contiguous segments is to allow room for some new data to be inserted in sequence without rearranging all the data.

The main disadvantages to indexed file organizations are the extra space required to store the indexes and the extra time necessary to access and maintain indexes. Usually these disadvantages are more than offset by the advantages. Because the index is kept in sequential order, both random processing and sequential processing are practical. Also, because the index is separate from the data, you can build multiple index structures on the same data file (just as in the library, where there are multiple indexes on author, title, subject, and so forth). With multiple indexes, software may rapidly find records that have compound conditions, such as records of books by Tom Clancy on espionage.

The decision of which indexes to create is probably the most important physical database design task for relational database technology, such as Microsoft Access, Oracle, DB2, and similar systems. Indexes can be created for both primary and secondary keys. When using indexes, there is a trade-off between improved performance for retrievals and degrading performance for inserting, deleting, and updating the rows

FIGURE 9-20
Comparison of file organizations
(a) Sequential

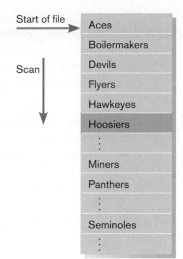

(b) Indexed

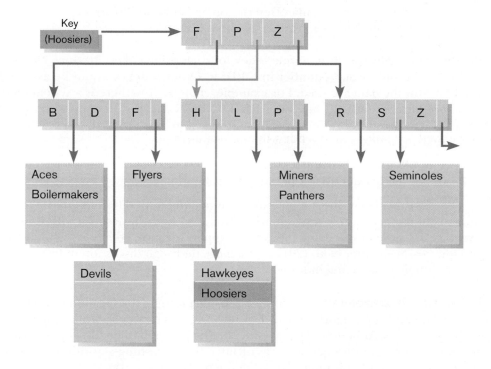

(c) Hashed

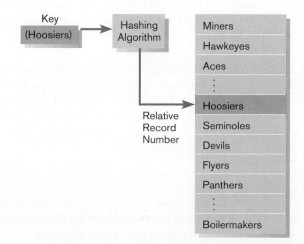

in a file. Thus, indexes should be used generously for databases intended primarily to support data retrievals, such as for decision support applications. Because they impose additional overhead, indexes should be used judiciously for databases that support transaction processing and other applications with heavy updating requirements.

Here are some guidelines for choosing indexes for relational databases (Gibson, Hughes, and Remington, 1989):

1. Specify a unique index for the primary key of each table (file). This selection ensures the uniqueness of primary key values and speeds retrieval based on those values. Random retrieval based on primary key value is common for answering multitable queries and for simple data maintenance tasks.
2. Specify an index for foreign keys. As in the first guideline, this speeds processing of multitable queries.
3. Specify an index for nonkey fields that are referenced in qualification and sorting commands for the purpose of retrieving data.

To illustrate the use of these rules, consider the following relations for PVF:

PRODUCT(Product_Number,Description,Finish,Room,Price)
ORDER(Order_Number,Product_Number,Quantity)

You would normally specify a unique index for each primary key: Product_Number in PRODUCT and Order_Number in ORDER. Other indexes would be assigned based on how the data are used. For example, suppose that there is a system module that requires PRODUCT and PRODUCT_ORDER data for products with a price below $500, ordered by Product_Number. To speed up this retrieval, you could consider specifying indexes on the following nonkey attributes:

1. Price in PRODUCT because it satisfies rule 3
2. Product_Number in ORDER because it satisfies rule 2

Because users may direct a potentially large number of different queries to the database, and especially for a system with a lot of ad hoc queries, you will probably have to be selective in specifying indexes to support the most common or frequently used queries. See Hoffer et al. (2011) for a more thorough discussion of factors and rules of thumb for selecting indexes.

Hashed file organization
A file organization in which the address of each row is determined using an algorithm.

Hashed File Organizations   In a **hashed file organization**, the location of each row is determined using an algorithm (see Figure 9-20c) that converts a primary key value into a row address. Although there are several variations of hashed files, in most cases the rows are located nonsequentially as dictated by the hashing algorithm. Thus, sequential data processing is impractical. On the other hand, retrieval of random rows is very fast. There are issues in the design of hashing file organizations, such as how to handle two primary keys that translate into the same address, but again, these issues are beyond our scope (see Hoffer et al. [2011] for a thorough discussion).

Summary of File Organizations   The three families of file organizations—sequential, indexed, and hashed—cover most of the file organizations you will have at your disposal as you design physical files and databases. Table 9-3 summarizes the comparative features of these file organizations. You can use this table to help choose a file organization by matching the file characteristics and file processing requirements with the features of the file organization.

Designing Controls for Files

Two of the goals of physical table design mentioned earlier are protection from failures or data loss and security from unauthorized use. These goals are achieved primarily by

TABLE 9-3 Comparative Features of Sequential, Indexed, and Hashed File Organizations

Factor	File Organization		
	Sequential	Indexed	Hashed
Storage space	No wasted space	No wasted space for data, but extra space for index	Extra space may be needed to allow for addition and deletion of records
Sequential retrieval on primary key	Very fast	Moderately fast	Impractical
Random retrieval on primary key	Impractical	Moderately fast	Very fast
Multiple key retrieval	Possible, but requires scanning whole file	Very fast with multiple indexes	Not possible
Deleting rows	Can create wasted space or require reorganizing	If space can be dynamically allocated, this is easy, but requires maintenance of indexes	Very easy
Adding rows	Requires rewriting file	If space can be dynamically allocated, this is easy, but requires maintenance of indexes	Very easy, except multiple keys with same address require extra work
Updating rows	Usually requires rewriting file	Easy, but requires maintenance of indexes	Very easy

implementing controls on each file. Data integrity controls, a primary type of control, were mentioned earlier in this chapter. Two other important types of controls address file backup and security.

It is almost inevitable that a file will be damaged or lost, due to either software or human errors. When a file is damaged, it must be restored to an accurate and reasonably current condition. A file and database designer has several techniques for file restoration, including:

- Periodically making a backup copy of a file
- Storing a copy of each change to a file in a transaction log or audit trail
- Storing a copy of each row before or after it is changed

For example, a backup copy of a file and a log of rows after they were changed can be used to reconstruct a file from a previous state (the backup copy) to its current values. This process would be necessary if the current file were so damaged that it could not be used. If the current file is operational but inaccurate, then a log of before images of rows can be used in reverse order to restore a file to an accurate but previous condition. Then a log of the transactions can be reapplied to the restored file to bring it up to current values. It is important that the information system designer make provisions for backup, audit trail, and row image files so that data files can be rebuilt when errors and damage occur.

An information system designer can build data security into a file by several means, including the following:

- Coding, or encrypting, the data in the file so that they cannot be read unless the reader knows how to decrypt the stored values.
- Requiring data file users to identify themselves by entering user names and passwords, and then possibly allowing only certain file activities (read, add, delete, change) for selected users to selected data in the file.
- Prohibiting users from directly manipulating any data in the file, but rather force programs and users to work with a copy (real or virtual) of the data they need; the copy contains only the data that users or programs are allowed to manipulate, and the original version of the data will change only after changes to the copy are thoroughly checked for validity.

Security procedures such as these all add overhead to an information system, so only necessary controls should be included.

PHYSICAL DATABASE DESIGN
FOR HOOSIER BURGER

A set of normalized relations and an associated E-R diagram for Hoosier Burger (Figure 9-16) were presented in the section Logical Database Design for Hoosier Burger earlier in this chapter. The display of a complete design of this database would require more documentation than space permits in this text, so we illustrate in this section only a few key decisions from the complete physical database.

As outlined in this chapter, to translate a logical database design into a physical database design, you need to make the following decisions:

- Create one or more fields for each attribute and determine a data type for each field.
- For each field, decide if it is calculated; needs to be coded or compressed; must have a default value or picture; or must have range, referential integrity, or null value controls.
- For each relation, decide if it should be denormalized to achieve desired processing efficiencies.
- Choose a file organization for each physical file.
- Select suitable controls for each file and the database.

Remember, the specifications for these decisions are made in physical database design, and then the specifications are coded in the implementation phase using the capabilities of the chosen database technology. These database technology capabilities determine what physical database design decisions you need to make. For example, for Oracle, which we assume is the implementation environment for this illustration, the only choice for file organization is indexed, so the file organization decision becomes how to decide which primary and secondary key attributes on which to build indexes.

We illustrate these physical database design decisions only for the INVOICE table. The first decision most likely would be whether to denormalize this table. Based on the suggestions for possible denormalization presented in this chapter, the only possible denormalization of this table would be to combine it with the VENDOR table. Because each invoice must have a vendor, and the only additional data about vendors not in the INVOICE table is the Vendor_Name attribute, this is a good candidate for denormalization. Because Vendor_Name is not very volatile, repeating Vendor_Name in each invoice for the same vendor will not cause excessive update maintenance. If Vendor_ Name is often used with other invoice data when invoice data are displayed, then this would be a good candidate for denormalization. So the denormalized relation to be transformed into a physical table is:

INVOICE(Vendor_ID,Invoice_Number,Invoice_Date,Paid?,Vendor_Name)

The next decision can be what indexes to create. The guidelines presented in this chapter suggest creating an index for the primary key, all foreign keys, and secondary keys used for sorting and qualifications in queries. So we create a primary key index on the combined fields Vendor_ID and Invoice_Number. INVOICE has no foreign keys. To determine what fields are used as secondary keys in query sorting and qualification clauses, we would need to know the content of queries. Also, it would be helpful to know query frequency because indexes do not provide much performance efficiency for infrequently run queries. For simplicity, suppose there were only two frequently run queries that reference the INVOICE table, as follows:

1. Display all the data about all unpaid invoices due this week.
2. Display all invoices ordered by vendor, show all unpaid invoices first, then all paid invoices, and order the invoices of each category in reverse sequence by invoice date.

In the first query, both the Paid? and Invoice_Date fields are used for quali-fication. Paid?, however, may not be a good candidate for an index because there are only two values for this field. The systems analyst would need to discover what percentage of invoices on file are unpaid. If this value is more than 10 percent, then an index on Paid? would not likely be helpful. Invoice_Date is a more discriminating field, so an index on this field would be helpful.

In the second query, Vendor_ID, Paid?, and Invoice_Date are used for sort-ing. Vendor_ID and Invoice_Date are discriminating fields (most values occur in less than 10 percent of the rows), so indexes on these fields will be helpful. Assuming less than 10 percent of the invoices on file are unpaid, then it would make sense to create the following indexes to make these two queries run as efficiently as possible:

1. *Primary key index:* Vendor_ID and Invoice_Number
2. *Secondary key indices:* Vendor_ID,Invoice_Date, and Paid?

We do not illustrate security and other types of controls because these deci-sions are very dependent on unique capabilities of the technology and a complex analysis of what data which users have the right to read, modify, add, or delete.

ELECTRONIC COMMERCE APPLICATION: DESIGNING DATABASES

Like many other analysis and design activities, designing the database for an Internet-based electronic commerce application is no different than the process fol-lowed when designing the database for other types of applications. In the last chapter, you read how Jim Woo and the PVF development team designed the human interface for the WebStore. In this section, we examine the processes Jim followed when trans-forming the conceptual data model for the WebStore into a set of normalized relations.

PINE
VALLEY
FURNITURE

Designing Databases for Pine Valley Furniture's WebStore

The first step Jim took when designing the database for the WebStore was to review the conceptual data model—the E-R diagram—developed during the analysis phase of the SDLC (see Figure 8-22 for a review). Given that there were no associative entities—many-to-many relationships—in the diagram, he began by identifying four distinct entity types, which he named:

> CUSTOMER
> ORDER
> INVENTORY
> SHOPPING_CART

Once reacquainted with the conceptual data model, he examined the lists of at-tributes for each entity. He noted that three types of customers were identified dur-ing conceptual data modeling, namely, corporate customers, home office customers, and student customers. Yet all were referred to simply as a "customer." Nonetheless, because each type of customer had some unique information (attributes) that other types of customers did not, Jim created three additional entity types, or subtypes, of customers:

> CORPORATE
> HOME_OFFICE
> STUDENT

Table 9-4 lists the common and unique information about each customer type. As Table 9-4 implies, four separate relations are needed to keep track of customer information without having anomalies. The CUSTOMER relation is used to capture common attributes, whereas the additional relations are used to capture information unique to each distinct customer type. To identify the type of customer within the

TABLE 9-4 Common and Unique Information about Each Customer Type

Common Information About ALL Customer Types		
Corporate Customer	**Home Office Customer**	**Student Customer**
Customer ID	Customer ID	Customer ID
Address	Address	Address
Phone Number	Phone Number	Phone Number
E-Mail Address	E-Mail Address	E-Mail Address

Unique Information About EACH Customer Type		
Corporate Customer	**Home Office Customer**	**Student Customer**
Corporate Name	Customer Name	Customer Name
Shipping Method	Corporate Name	School
Buyer Name	Fax Number	
Fax Number		

CUSTOMER relation easily, a Customer_Type attribute is added to the CUSTOMER relation. Thus, the CUSTOMER relation consists of:

CUSTOMER(Customer_ID,Address,Phone,E-mail,Customer_Type)

To link the CUSTOMER relation to each of the separate customer types—CORPORATE, HOME_OFFICE, and STUDENT—all share the same primary key, Customer_ID, in addition to the attributes unique to each. This results in the following relations:

CORPORATE(Customer_ID,Corporate_Name,Shipping_Method,Buyer_Name,Fax)
HOME_OFFICE(Customer_ID,Customer_Name,Corporate_Name,Fax)
STUDENT(Customer_ID,Customer_Name,School)

In addition to identifying all the attributes for customers, Jim also identified the attributes for the other entity types. The results of this investigation are summarized in Table 9-5. As described in Chapter 8, much of the order-related information is captured and tracked within PVF's Purchasing Fulfillment System. This means that the ORDER relation does not need to track all the details of the order because the Purchasing Fulfillment System produces a detailed invoice that contains all order details such as the list of ordered products, materials used, colors, quantities, and other such information. To access this invoice information, a foreign key, Invoice_ID, is included in the ORDER relation. To identify easily which orders belong to a specific customer, the Customer_ID attribute is also included in ORDER. Two additional attributes, Return_Code and Order_Status, are also included in ORDER.

TABLE 9-5 Attributes for Order, Inventory, and Shopping Cart Entities

Order	Inventory	Shopping_Cart
Order_ID (primary key)	Inventory_ID (primary key)	Cart_ID (primary key)
Invoice_ID (foreign key)	Name	Customer_ID (foreign key)
Customer_ID (foreign key)	Description	Inventory_ID (foreign key)
Return_Code	Size	Material
Order_Status	Weight	Color
	Materials	Quantity
	Colors	
	Price	
	Lead_Time	

The Return_Code is used to track the return of an order more easily—or a product within an order—whereas Order_Status is a code used to represent the state of an order as it moves through the purchasing fulfillment process. This results in the following ORDER relation:

> ORDER(Order_ID,Invoice_ID,Customer_ID,Return_Code,Order_Status)

In the INVENTORY entity, two attributes—Materials and Colors—could take on multiple values but were represented as single attributes. For example, Materials represents the range of materials that a particular inventory item could be constructed from. Likewise, Colors is used to represent the range of possible product colors. PVF has a long-established set of codes for representing materials and colors; each of these complex attributes is represented as a single attribute. For example, the value "A" in the Colors field represents walnut, dark oak, light oak, and natural pine, whereas the value "B" represents cherry and walnut. Using this coding scheme, PVF can use a single character code to represent numerous combinations of colors. This results in the following INVENTORY relation:

> INVENTORY(Inventory_ID,Name,Description,Size,Weight,Materials,Colors, Price,Lead_Time)

Finally, in addition to Cart_ID, each shopping cart contains the Customer_ID and Inventory_ID attributes so that each item in a cart can be linked to a particular inventory item and to a specific customer. In other words, both the Customer_ID and Inventory_ID attributes are foreign keys in the SHOPPING_CART relation. Recall that the SHOPPING_CART is temporary and is kept only while a customer is shopping. When a customer actually places the order, the ORDER relation is created and the line items for the order—the items in the shopping cart—are moved to the Purchase Fulfillment System and stored as part of an invoice. Because we also need to know the selected material, color, and quantity of each item in the SHOPPING_CART, these attributes are included in this relation. This results in the following:

> SHOPPING_CART(Cart_ID,Customer_ID,Inventory_ID,Material,Color, Quantity)

Now that Jim has completed the database design for the WebStore, he has shared all the design information with his project team so that the design can be turned into a working database during implementation. We read more about the WebStore's implementation in the next chapter.

SUMMARY

Databases are defined during the design phase of the systems development life cycle. They are designed usually in parallel with the design of system interfaces. To design a database, a systems analyst must understand the conceptual database design for the application, usually specified by an E-R diagram, and the data requirements of each system interface (report, form, screen, etc.). Thus, database design is a combination of top-down (driven by an E-R diagram) and bottom-up (driven by specific information requirements in system interfaces) processes. Besides data requirements, systems analysts must also know physical data characteristics (e.g., length and format), frequency of use of the system interfaces, and the capabilities of database technologies.

An E-R diagram is transformed into normalized relations by following well-defined principles, which are summarized in Table 9-1. For example, each entity becomes a relation and each many-to-many relationship or associative entity also becomes a relation. These principles also specify how to add foreign keys to relations to represent one-to-many relationships.

Separate sets of normalized relations are merged (a process called *view integration*) to create a consolidated logical database design. The different sets of relations come from the conceptual E-R diagram for the application, known human system interfaces (reports, screens, forms, etc.), and known or anticipated queries for data that meet certain qualifications. The result of merging is a comprehensive, normalized set of relations for the application. Merging is not simply a mechanical process. A systems analyst must address issues of synonyms, homonyms, and functional dependencies between nonkeys during view integration.

Fields in the physical database design represent the attributes (columns) of relations in the logical database design. Each field must have a data type as well as potentially other characteristics such as a coding scheme to simplify the storage of business data, a default value, picture (or template) control, range control, referential integrity control, or null value control. A storage format is chosen to balance four objectives: (1) minimize storage space, (2) represent all possible values of the field, (3) improve data integrity for the field, and (4) support all data manipulations desired on the field.

Whereas normalized relations possess properties of well-structured relations, the design of a physical table attempts to achieve two goals different from those of normalization: efficient use of secondary storage and data processing speed. Efficient use of storage means that the amount of extra (or overhead) information is minimized. Therefore, sequential file organizations are efficient in the use of storage because little or no extra information, besides the meaningful business data, is kept. Data processing speed is achieved by storing data close together that are used together and by building extra information in the database, which allows data to be quickly found based on primary or secondary key values or by sequence.

Table 9-3 summarizes the performance characteristics of different types of file organizations. The systems analyst must decide which performance factors are most important for each application and the associated database. These factors are storage space; sequential retrieval speed; random row retrieval speed; speed of retrieving data based on multiple key qualifications; and the speed to perform data maintenance activities of row deletion, addition, and updating.

An index is information about the primary or secondary keys of a file. Each index entry contains the key value and a pointer to the row that contains that key value. An index facilitates rapid retrieval to rows for queries that involve AND, OR, and NOT qualifications of keys (e.g., all products with a maple finish and unit cost greater than $500 or all products in the office furniture product line). When using indices, there is a trade-off between improved performance for retrievals and degrading performance for inserting, deleting, and updating the rows in a file. Thus, indices should be used generously for databases intended primarily to support data retrievals, such as for decision support applications. Because they impose additional overhead, indices should be used judiciously for databases that support transaction processing and other applications with heavy updating requirements. Typically, you create indices on a file for its primary key, foreign keys, and other attributes used in qualification and sorting clauses in queries, forms, reports, and other system interfaces.

KEY TERMS

1. Calculated field
2. Data type
3. Default value
4. Denormalization
5. Field
6. File organization
7. Foreign key
8. Functional dependency
9. Hashed file organization
10. Homonym
11. Index
12. Indexed file organization
13. Normalization
14. Null value
15. Physical file
16. Physical table
17. Pointer
18. Primary key
19. Recursive foreign key
20. Referential integrity
21. Relation
22. Relational database model
23. Second normal form (2NF)
24. Secondary key
25. Sequential file organization
26. Synonym
27. Third normal form (3NF)
28. Well-structured relation

Match each of the key terms above to the definition that best fits it.

_____ A named, two-dimensional table of data. Each relation consists of a set of named columns and an arbitrary number of unnamed rows.

_____ A relation that contains a minimum amount of redundancy and allows users to insert, modify, and delete the rows without errors or inconsistencies.

_____ The process of converting complex data structures into simple, stable data structures.

_____ A particular relationship between two attributes.

_____ A relation for which every nonprimary key attribute is functionally dependent on the whole primary key.

_____ A relation that is in second normal form and that has no functional (transitive) dependencies between two (or more) nonprimary key attributes.

_____ An attribute that appears as a nonprimary key attribute in one relation and as a primary key attribute (or part of a primary key) in another relation.

_____ An integrity constraint specifying that the value (or existence) of an attribute in one relation depends on the value (or existence) of the same attribute in another relation.

_____ A foreign key in a relation that references the primary key values of that same relation.

_____ Two different names that are used for the same attribute.

_____ A single attribute name that is used for two or more different attributes.

_____ The smallest unit of named application data recognized by system software.

_____ A coding scheme recognized by system software for representing organizational data.

_____ A field that can be derived from other database fields.

_____ A value a field will assume unless an explicit value is entered for that field.

_____ A special field value, distinct from a zero, blank, or any other value, that indicates that the value for the field is missing or otherwise unknown.

_____ A named set of rows and columns that specifies the fields in each row of the table.

_____ The process of splitting or combining normalized relations into physical tables based on affinity of use of rows and fields.

_____ A named set of table rows stored in a contiguous section of secondary memory.

_____ A technique for physically arranging the records of a file.

_____ A field of data that can be used to locate a related field or row of data.

_____ The rows in the file are stored in sequence according to a primary key value.

_____ The rows are stored either sequentially or nonsequentially, and an index is created that allows software to locate individual rows.

_____ A table used to determine the location of rows in a file that satisfy some condition.

_____ One or a combination of fields for which more than one row may have the same combination of values.

_____ The address for each row is determined using an algorithm.

_____ An attribute whose value is unique across all occurrences of a relation.

_____ Data represented as a set of related tables or relations.

REVIEW QUESTIONS

1. What is the purpose of normalization?
2. List five properties of relations.
3. What problems can arise when merging relations (view integration)?
4. How are relationships between entities represented in the relational data model?
5. What is the relationship between the primary key of a relation and the functional dependencies among all attributes within that relation?
6. How is a foreign key represented in relational notation?
7. Can instances of a relation (sample data) prove the existence of a functional dependency? Why or why not?
8. In what way does the choice of a data type for a field help to control the integrity of that field?
9. What is the difference between how a range control statement and a referential integrity control statement are handled by a file management system?
10. What is the purpose of denormalization? Why might you not want to create one physical table or file for each relation in a logical data model?
11. What factors influence the decision to create an index on a field?
12. Explain the purpose of data compression techniques.
13. What are the goals of designing physical tables?
14. What are the seven factors that should be considered in selecting a file organization?

PROBLEMS AND EXERCISES

1. Assume that, at PVF, products are composed of components, products are assigned to salespersons, and components are produced by vendors. Also assume that, in the relation PRODUCT(Prodname, Salesperson, Compname, Vendor), Vendor is functionally dependent on Compname and Compname is functionally dependent on Prodname. Eliminate the transitive dependency in this relation and form 3NF relations.

2. Transform the E-R diagram of Figure 8-23 into a set of 3NF relations. Make up a primary key where needed and one or more nonkey attributes for each entity.

3. Consider the E-R diagram of Figure 9-21.
   a. Transform this E-R diagram into a set of 3NF relations.
   b. State and justify all referential integrity rules for the 3NF relations you created in Problem and Exercise 3a.

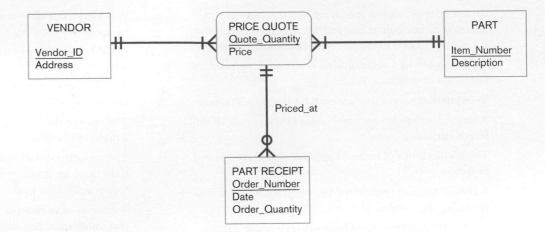

FIGURE 9-21
E-R diagram for Problem and
Exercise 3

4. Consider the list of individual 3NF relations below. These relations were developed from several separate normalization activities.

   PATIENT(Patient_ID,Room_Number,Admit_Date, Address)

   ROOM(Room_Number,Phone,Daily_Rate)

   PATIENT(Patient_Number,Treatment_Description, Address)

   TREATMENT(Treatment_ID,Description,Cost)

   PHYSICIAN(Physician_ID,Name,Department)

   PHYSICIAN(Physician_ID,Name,Supervisor_ID)

   a. Merge these relations into a consolidated set of 3NF relations. State whatever assumptions you consider necessary (including but not limited to foreign keys) to resolve any potential problems you identify in the merging process.

   b. Draw an E-R diagram for your answer to Problem and Exercise 4a.

5. Consider the following 3NF relations about a sorority or fraternity:

   MEMBER(Member_ID,Name,Address,Dues_Owed)

   OFFICE(Office_Name,Officer_ID,Term_Start_Date, Budget)

   EXPENSE(Ledger_Number,Office_Name,Expense_Date, Amt_Owed)

   PAYMENT(Check_Number,Expense_Ledger_Number, Amt_Paid)

   RECEIPT(Member_ID,Receipt_Date,Dues_Received)

   COMMITTEE(Committee_ID,Officer_in_Charge)

   WORKERS(Committee_ID,Member_ID)

   a. Foreign keys are not indicated in these relations. Decide which attributes are foreign keys and justify your decisions.

   b. Draw an E-R diagram for these relations, using your answer to Problem and Exercise 5a.

   c. Explain the assumptions you made about cardinalities in your answer to Problem and Exercise 5b. Explain why it is said that the E-R data model is more expressive or more semantically rich than the relational data model.

6. Consider the following functional dependencies:

   Applicant_ID → Applicant_Name

   Applicant_ID → Applicant_Address

   Position_ID → Position_Title

   Position_ID → Date_Position_Opens

   Position_ID → Department

   Applicant_ID + Position_ID → Date_Applied

   Applicant_ID + Position_ID + Date_ Interviewed →

   a. Represent these attributes with 3NF relations. Provide meaningful relation names.

   b. Represent these attributes using an E-R diagram. Provide meaningful entity and relationship names.

7. Suppose you were designing a file of student records for your university's placement office. One of the fields that would likely be in this file is the student's major. Develop a coding scheme for this field that achieves the objectives outlined in this chapter for field coding.

8. In Problem and Exercise 3, you developed integrated normalized relations. Choose primary keys for the files that would hold the data for these relations. Did you use attributes from the relations for primary keys or did you design new fields? Why or why not?

9. Suppose you created a file for each relation in your answer to Problem and Exercise 3. If the following queries represented the complete set of accesses to this database, suggest and justify what primary and secondary key indices you would build.

   a. For each PART in Item_Number order list in Vendor_ID, sequence all the vendors and their associated prices for that part.

   b. List all PART RECEIPTs, including related PART fields for all the parts received on a particular day.

   c. For a particular VENDOR, list all the PARTs and their associated prices that VENDOR can supply.

10. Suppose you were designing a default value for the marital status field in a student record at your university. What possible values would you consider and why? How would the default value change depending on other factors, such as type of student (undergraduate, graduate, professional)?

11. Consider Figure 9-19b. Explain a query that would likely be processed more quickly using the denormalized relations rather than the normalized relations.

12. Model a set of typical family relationships—spouse, father, and mother—in a single 3NF relation. Also include nonkey attributes name and birth date. Assume that each person has only one spouse, one father, and one mother. Show foreign keys with dashed underlining.

FIELD EXERCISES

1. Locate library books or articles that discuss additional normal forms other than second and third normal forms. Describe each of these additional normal forms and give examples of each. How are these additional normal forms different from those presented in this chapter? What additional benefit does their use provide?

2. Find a systems analyst or database administrator within a company that uses a database management system to organize the company's corporate data. Ask this person to describe how he or she uses normalization and which level of normal form is used for each database table. Are all tables in third normal form? Why do they denormalize, if they do?

3. Find a systems analyst or database administrator within a company that uses a database management system to organize the company's corporate data. Ask this person to describe what "additional information" should be collected during requirements analysis that is needed for file and database design but that is not very useful for earlier phases of systems development.

4. Find out what database management systems are available at your university for student use. Investigate which data types these DBMSs support. Compare these DBMSs based on the data types supported and suggest which types of applications each DBMS is best suited for based on this comparison. Also investigate the capabilities these DBMSs have for creating indexes. What limitations are imposed on index creation? These might include constraints such as the maximum number of indexes per table, what fields or combinations of fields can be indexed, and how indexes are used in query processing.

5. Find out what database management systems are available at your university for student use. Investigate what physical file and database design decisions need to be made. Compare this list of decisions with those discussed in this chapter. For physical database and design decisions (or options) not discussed in this chapter, investigate what choices you have and how you should choose among them. Submit a report to your instructor with your findings.

REFERENCES

Codd, E. F. 1970. "A Relational Model of Data for Large Relational Databases." *Communications of the ACM* 13(6): 77–87.

Dutka, A. F., and H. H. Hanson. 1989. *Fundamentals of Data Normalization.* Reading, MA: Addison-Wesley.

Finkelstein, R. 1988. "Breaking the Rules Has a Price." *Database Programming & Design* 1 (June): 11–14.

Gibson, M., C. Hughes, and W. Remington. 1989. "Tracking the Trade-Offs with Inverted Lists." *Database Programming & Design* 2 (January): 28–34.

Hoffer, J. A., V. Ramesh, and H. Topi. 2011. *Modern Database Management,* 10th ed. Upper Saddle River, NJ: Prentice Hall.

Rodgers, U. 1989. "Denormalization: Why, What, and How?" *Database Programming & Design* 2(12): 46–53.

PETRIE ELECTRONICS

Chapter 9: Designing Databases

Jim Watanabe, assistant director of IT for Petrie Electronics, and the manager of the "No Customer Escapes" customer loyalty system project, was walking down the hall from his office to the cafeteria. It was 4 P.M., but Jim was nowhere close to going home yet. The deadlines he had imposed for the project were fast approaching. His team was running behind, and he had a lot of work to do over the next week to try to get things back on track. He needed to get some coffee for what was going to be a late night.

As Jim approached the cafeteria, he saw Sanjay Agarwal and Sam Waterston walking toward him. Sanjay was in charge of systems integration for Petrie, and Sam was one of the company's top interface designers. They were both on the customer loyalty program team. They were having an intense conversation as Jim approached.

"Hi guys," Jim said.

"Oh, hi, Jim," Sanjay replied. "Glad I ran into you—we are moving ahead on the preliminary database designs. We're translating the earlier conceptual designs into physical designs."

"Who's working on that, Stephanie?" Jim asked. Stephanie Welch worked for Petrie's database administrator.

"Yes," Sanjay replied. "But she is supervising a couple of interns who have been assigned to her for this task."

"So how is that going? Has she approved their work?"

"Yeah, I guess so. It all seems to be under control."

"I don't want to second-guess Stephanie, but I'm curious about what they've done."

"Do you really have time to review interns' work?" Sanjay asked. "OK, let me send you the memo Stephanie sent me (PE Figure 9-1)."

MEMO

To: Stephanie Welch
From: Xin Zhu & Anton Washington
Re: Preliminary physical database design for "No Customer Escapes"
Date: June 1, 2013

We were charged with converting the conceptual database designs for the customer loyalty system to physical database designs. We started with one of the initial ERDs (see PE Figure 8-1), designed at a very high level. The ERD identified six entities: Customer, Product, Service, Promotion, Transaction, and Coupon. We discovered that all of these entities are already defined in Petrie's existing systems. The only entity not already defined is Coupon. Product and Service are defined as part of the product database. Promotion is defined as part of the marketing database. Customer and Transaction are defined as part of the core database.

However, after considerable consideration, we are not sure if some of these already identified and defined entities are the same as those identified in the preliminary ERD we were given. Specifically, we have questions about Customer, Transaction and Promotion.

Customer: The Customer entity is more complex than it appears. There are several ways to think about the instances of this entity. For example, we can divide Customers into those who shop online and those who shop in the brick-and-mortar stores. And there is of course some overlap. The biggest distinction between these two groups is that we know the names of (and other information about) the Customers who shop online, but we may have very little identifying information about those who shop only in the stores. For example, if an individual shops only at a store and pays only with cash, that individual meets the definition of Customer (see PE Table 8-1), but we collect no data on that individual at all. We raise these issues to call attention to the relationship between Customers and members of the customer loyalty program: All members are Customers, but not all Customers are members. We suggest that the entity called Customer in the preliminary ERD be renamed 'Member,' as we think that is a better name for this entity. We are prepared to map out the table design when this change is approved.

Transaction: Petrie already has a relational table called Transaction, but that applies to all transactions in all stores and online. The customer loyalty program focuses on the transactions of its Members, so the program involves only a subset of Transactions. We suggest that the ERD be redesigned to take this fact into account, and that what is now called Transaction be renamed 'Member Transaction.' The relational tables should then be designed accordingly.

Promotion: Petrie already has a relational table called Promotion. Again, the customer loyalty program, while having some interest in general promotions, focuses primarily on promotions created specifically for Members of the program. What is called Promotion in the ERD is really a subset of all of Petrie's promotions. We recommend a name change to 'Member Promotion' with the associated relational table design.

Finally, for the Coupon entity, which is new, we note from the ERD that Coupon only has one relationship, and that is with Customer. As it is a one-to-many relationship, the PK from Customer will be an FK in Coupon. We recommend the following table design: COUPON(Coupon ID,Customer ID, Creation Date, Expiration Date, Value)

PE FIGURE 9-1
Memo on issues related to physical database design for Petrie Electronic's customer loyalty program

"You're right, I don't have time," Jim said, "But I'm curious. It won't take long to read the memo, right?"

"OK, I'll send it as soon as I get back to my desk."

"OK, thanks." Jim walked on to the cafeteria, and he poured himself a big cup of coffee.

Case Questions

1. In the questions associated with the Petrie Electronics case at the end of Chapter 8, you were asked to modify the E-R diagram given in PE Figure 8-1 to include any other entities and the attributes you identified from the Petrie's cases. Review your answers to these questions, and add any additional needed relations to the document in PE Figure 9-1.

2. Study you answer to Question 1. Verify that the relations you say represent the Petrie Electronics database are in third normal form. If they are, explain why. If they are not, change them so that they are.

3. The E-R diagram you developed in questions in the Petrie Electronics case at the end of Chapter 8 should have shown minimum cardinalities on both ends of each relationship. Are minimum cardinalities represented in some way in the relations in your answer to Question 2? If not, how are minimum cardinalities enforced in the database?

4. Using your answer to Question 2, select data types, formats, and lengths for each attribute of each relation. Use the data types and formats supported by Microsoft Access. What data type should be used for nonintelligent primary keys?

5. Complete all table and field definitions for the Petrie Electronics case database using Microsoft Access. Besides the decisions you have made in answers to the preceding questions, fill in all other field definition parameters for each field of each table.

6. The one decision for a relational database that usually influences efficiency the most is index definition. What indexes do you recommend for this database? Justify your selection of each index.

7. Using Microsoft Visio, develop an E-R diagram with all the supporting database properties for decisions you made in Questions 1–6. Can all the database design decisions you made be documented in Visio? Finally, use Visio to generate Microsoft Access table definitions. Did the table generation create the table definitions you would create manually?

CHAPTER TEN

Designing Forms and Reports

Learning Objectives

After studying this chapter, you should be able to:

- Explain the process of designing forms and reports and the deliverables for their creation.

- Apply the general guidelines for formatting forms and reports.

- Use color and know when color improves the usability of information.

- Format text, tables, and lists effectively.

- Explain how to assess usability and describe how variations in users, tasks, technology, and environmental characteristics influence the usability of forms and reports.

- Discuss guidelines for the design of forms and reports for Internet-based electronic commerce systems.

Introduction

In this chapter, you will learn what guidelines to follow when designing forms and reports. In general, forms are used to present or collect information on a single item, such as a customer, product, or event. Forms can be used for both input and output. Reports, on the other hand, are used to convey information on a collection of items. Form and report design is a key ingredient for successful systems. Because users often equate the quality of a system with the quality of its input and output methods, you can see that the design process for forms and reports is an especially important activity. And because information can be collected and formatted in many ways, gaining an understanding of design do's and don'ts and the trade-offs between various formatting options is useful for all systems analysts.

In the next section, the process of designing forms and reports is briefly described, and we also provide guidance on the deliverables produced during this process. Guidelines for formatting information are then provided that serve as the building blocks for designing all forms and reports. We then describe methods for assessing the usability of form and report designs. The chapter concludes by examining how to design forms and reports for Internet-based electronic commerce applications.

DESIGNING FORMS AND REPORTS

This is the second chapter that focuses on system design within the systems development life cycle (see Figure 10-1). In this chapter, we describe issues related to the design of system inputs and outputs—forms and reports. In Chapter 11, we focus on the design of dialogues and interfaces, which are how users interact with systems. Due to the highly related topics and guidelines in these two chapters, they form one conceptual body of guidelines and illustrations that jointly guide the design of all aspects of system inputs and outputs. In each of these chapters, your objective is to gain an understanding of how you can transform information gathered during analysis into a coherent design. Although all system design issues are related, topics discussed in this chapter on designing forms and reports are especially relative to those in the following chapter—the design of dialogues and interfaces.

System inputs and outputs—forms and reports—were identified during requirements structuring. The kinds of forms and reports the system will handle were established as part of the design strategy formed at the end of the analysis phase of the systems development process. During analysis, however, you may not have been concerned with the precise appearance of forms and reports; your concerns likely focused on which forms or reports need to exist and their contents. You may have distributed prototypes of forms and reports that emerged during analysis as a way to confirm requirements with users. Forms and reports are integrally related to various diagrams developed during requirements structuring. For example, every input form will be associated with a data flow entering a process on a data flow diagram (DFD), and every output form or report will be a data flow produced by a process on a DFD. This means that the contents of a form or report correspond to the data elements contained in the associated data flow. Further, the data on all forms and reports must consist of data elements in data stores and on the E-R data model for the application, or must be computed from these data elements. (In rare instances, data simply go from system input to system output without being stored within the system.) It is common that, as you design forms and reports, you will discover flaws in DFDs and E-R diagrams; these diagrams should be updated as designs evolve.

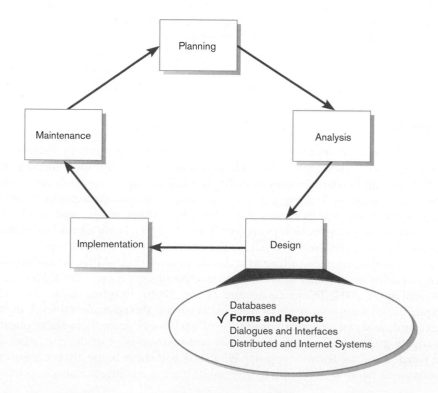

FIGURE 10-1
Systems development life cycle with logical design phase highlighted

TABLE 10-1   **Common Types of Business Reports**

Report Name	Description
Scheduled Reports	Reports produced at predefined intervals—daily, weekly, or monthly—to support the routine informational needs of an organization.
Key-Indicator Reports	Reports that provide a summary of critical information on a recurring basis.
Exception Reports	Reports that highlight data that are out of the normal operating range.
Drill-Down Reports	Reports that provide details behind the summary values on a key-indicator or exception report.
Ad-hoc Reports	Unplanned information requests in which information is gathered to support a nonroutine decision.

Form

A business document that contains some predefined data and may include some areas where additional data are to be filled in. An instance of a form is typically based on one database record.

Report

A business document that contains only predefined data; it is a passive document used solely for reading or viewing. A report typically contains data from many unrelated records or transactions.

If you are unfamiliar with computer-based information systems, it will be helpful to clarify exactly what we mean by a form or report. A **form** is a business document that contains some predefined data and often includes some areas where additional data are to be filled in. Most forms have a stylized format and are usually not in a simple row and column format. Examples of business forms are product order forms, employment applications, and class registration sheets. Traditionally, forms have been displayed on a paper medium, but today video display technology allows us to duplicate the layout of almost any printed form, including an organizational logo or any graphic, on a video display terminal. Forms displayed on a video display may be used for data display or data entry. Additional examples of forms are an electronic spreadsheet, a computer sign-on or menu, and an ATM transaction layout. On the Internet, form interaction is the standard method of gathering and displaying information when consumers order products, request product information, or query account status.

A **report** is a business document that contains only predefined data; it is a passive document used solely for reading or viewing. Examples of reports include invoices, weekly sales summaries by region and salesperson, or a pie chart of population by age categories (see Table 10-1). We usually think of a report as printed on paper, but it may be printed to a computer file, a visual display screen, or some other medium such as microfilm. Often a report has rows and columns of data, but a report may be of any format—for example, mailing labels. Frequently, the differences between a form and a report are subtle. A report is only for reading and often contains data about multiple unrelated records in a computer file. In contrast, a form typically contains data from only one record or is based on one record, such as data about one customer, one order, or one student. The guidelines for the design of forms and reports are very similar.

The Process of Designing Forms and Reports

Designing forms and reports is a user-focused activity that typically follows a prototyping approach (see Figure 6-7). First, you must gain an understanding of the intended user and task objectives by collecting initial requirements during requirements determination. During this process, several questions must be answered. These questions attempt to answer the "who, what, when, where, and how" related to the creation of all forms or reports (see Table 10-2). Gaining an understanding of these questions is a required first step in the creation of any form or report.

For example, understanding who the users are—their skills and abilities—will greatly enhance your ability to create an effective design (Lazar, 2004; McCracken, Wolfe, and Spoll, 2004; Te'eni, Carey, and Zhang, 2006). In other words, are your users experienced computer users or novices? What are the educational level, business background, and task-relevant knowledge of each user? Answers to these questions will provide guidance for both the format and content of your designs. Also, what is the purpose of the form or report? What task will users be performing and what information is needed to complete this task? Other questions are also important to

TABLE 10-2 **Fundamental Questions When Designing Forms and Reports**

1. Who will use the form or report?
2. What is the purpose of the form or report?
3. When is the form or report needed and used?
4. Where does the form or report need to be delivered and used?
5. How many people need to use or view the form or report?

consider. Where will the users be when performing this task? Will users have access to online systems or will they be in the field? Also, how many people will need to use this form or report? If, for example, a report is being produced for a single user, the design requirements and usability assessment will be relatively simple. A design for a larger audience, however, may need to go through a more extensive requirement collection and usability assessment process.

After collecting the initial requirements, you structure and refine this information into an initial prototype. Structuring and refining the requirements are completed independently of the users, although you may need to occasionally contact users in order to clarify some issue overlooked during analysis. Finally, you ask users to review and evaluate the prototype. After reviewing the prototype, users may accept the design or request that changes be made. If changes are needed, you will repeat the construction–evaluate–refinement cycle until the design is accepted. Usually, several iterations of this cycle occur during the design of a single form or report. As with any prototyping process, you should make sure that these iterations occur rapidly in order to gain the greatest benefits from this design approach.

The initial prototype may be constructed in numerous environments, including Windows, Linux, Macintosh, or HTML. The obvious choice is to employ standard development tools used within your organization. Often, initial prototypes are simply mock screens that are not working modules or systems. Mock screens can be produced from a word processor, computer graphics design package, electronic spreadsheet, or even on paper (Snyder, 2003). It is important to remember that the focus of this activity is on the *design*—content and layout—of forms and reports; of course, you must also consider how specific forms and reports will be implemented. It is fortunate that tools for designing forms and reports are rapidly evolving, making development faster and easier. In the past, inputs and outputs of all types were typically designed by hand on a coding or layout sheet. For example, Figure 10-2 shows the layout of a data input form using a coding sheet.

Although coding sheets are still used, their importance has diminished due to significant changes in system operating environments and the evolution of automated design tools. Prior to the creation of graphical operating environments, for example, analysts designed many inputs and outputs that were 80 columns (characters) by 25 rows, the standard dimensions for most video displays. These limits in screen dimensions are radically different in graphical operating environments such as Microsoft's Windows or the Web, where font sizes and screen dimensions can change from user to user. Consequently, the creation of new tools and development environments was needed to help analysts and programmers develop these graphical and flexible designs. Figure 10-3 shows an example of the same data input form as designed in Microsoft's Visual Basic .NET. Note the variety of fonts, sizes, and highlighting that was used. Given the need for rapid, iterative development when designing forms and reports, tools that seamlessly move prototype designs to functional systems are becoming standard in most professional development organizations.

Deliverables and Outcomes

Each systems development life cycle (SDLC) phase helps you to construct a system. In order to move from phase to phase, each activity produces a type of deliverable

FIGURE 10-2
The layout of a data input form
using a coding sheet

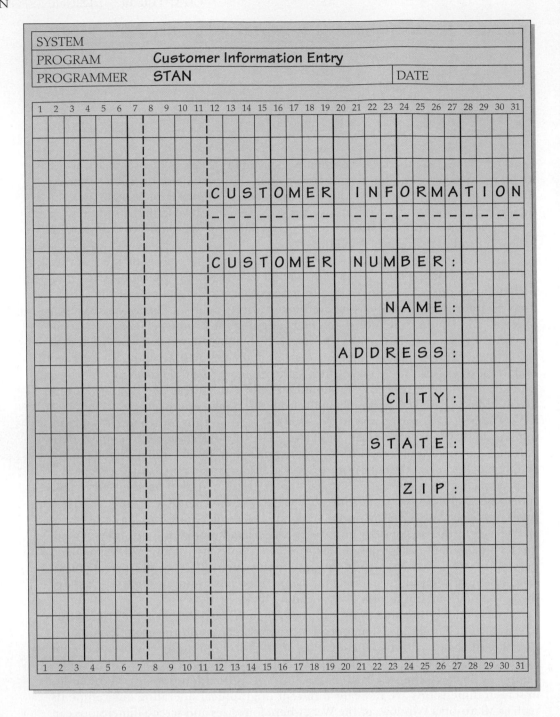

that is used in a later phase or activity. For example, within the project initiation and planning phase of the SDLC, the Baseline Project Plan serves as input to many subsequent SDLC activities. In the case of designing forms and reports, design specifications are the major deliverables and are inputs to the system implementation phase. Design specifications have three sections:

1. Narrative overview
2. Sample design
3. Testing and usability assessment

The first section of a design specification contains a general overview of the characteristics of the target users, tasks, system, and environmental factors in which the form or report will be used. The purpose is to explain to those who will actually develop the

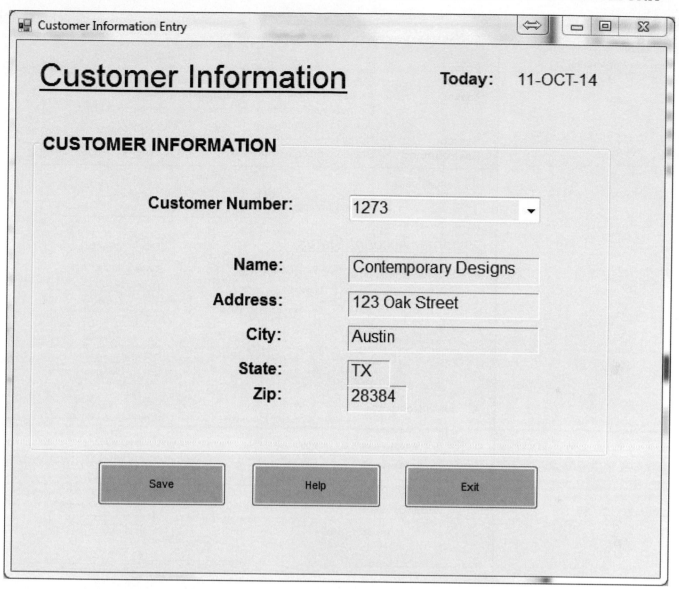

FIGURE 10-3
A data input screen designed in Microsoft's Visual Basic .NET

(*Source:* Microsoft Corporation.)

final form why this form exists and how it will be used so that they can make the appropriate implementation decisions. In this section, you list general information and the assumptions that helped shape the design. For example, Figure 10-4 shows an excerpt of a design specification for a Customer Account Status form for Pine Valley Furniture (PVF). The first section of the specification, Figure 10-4a, provides a narrative overview containing the relevant information to developing and using the form within PVF. The overview explains the tasks supported by the form, where and when the form is used, characteristics of the people using the form, the technology delivering the form, and other pertinent information. For example, if the form is delivered on a visual display terminal, this section would describe the capabilities of this device, such as whether it has a touch screen and whether color and a mouse are available.

In the second section of the specification, Figure 10-4b, a sample design of the form is shown. This design may be hand drawn using a coding sheet, although in most instances, it is developed using standard development tools. Using actual development tools allows the design to be more thoroughly tested and assessed. The final section of the specification, Figure 10-4c, provides all testing and usability assessment information. Procedures for assessing designs are described later in this chapter. Some specification information may be irrelevant when designing some forms and reports. For example, the design of a simple Yes/No selection form may

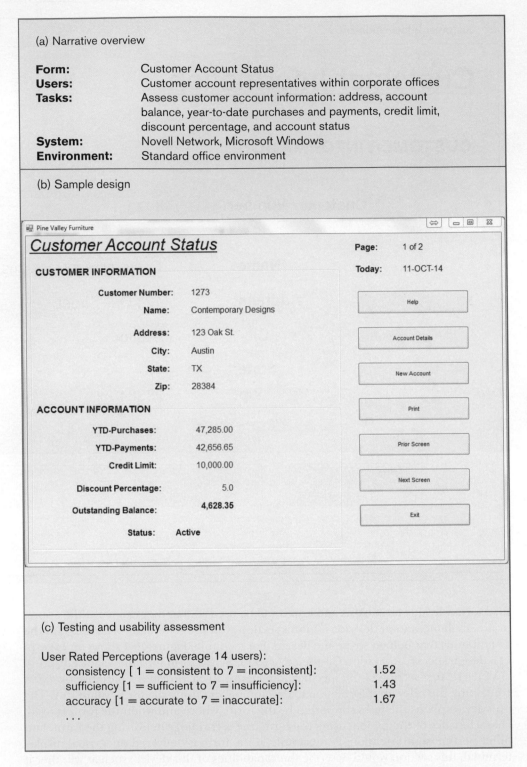

(a) Narrative overview

Form:	Customer Account Status
Users:	Customer account representatives within corporate offices
Tasks:	Assess customer account information: address, account balance, year-to-date purchases and payments, credit limit, discount percentage, and account status
System:	Novell Network, Microsoft Windows
Environment:	Standard office environment

(b) Sample design

Pine Valley Furniture

Customer Account Status

Page: 1 of 2

Today: 11-OCT-14

CUSTOMER INFORMATION

Customer Number:	1273
Name:	Contemporary Designs
Address:	123 Oak St.
City:	Austin
State:	TX
Zip:	28384

ACCOUNT INFORMATION

YTD-Purchases:	47,285.00
YTD-Payments:	42,656.65
Credit Limit:	10,000.00
Discount Percentage:	5.0
Outstanding Balance:	4,628.35
Status:	Active

Help

Account Details

New Account

Print

Prior Screen

Next Screen

Exit

(c) Testing and usability assessment

User Rated Perceptions (average 14 users):

consistency [1 = consistent to 7 = inconsistent]:	1.52
sufficiency [1 = sufficient to 7 = insufficiency]:	1.43
accuracy [1 = accurate to 7 = inaccurate]:	1.67
. . .	

be so straightforward that no usability assessment is needed. Also, much of the narrative overview may be unnecessary unless intended to highlight some exception that must be considered during implementation.

FORMATTING FORMS AND REPORTS

A wide variety of information can be provided to users of information systems, ranging from text to video to audio. As technology continues to evolve, a greater variety of data types will be used. Unfortunately, a definitive set of rules for delivering every type of information to users has yet to be defined, and these rules are continuously evolving along with the rapid changes in technology. Nonetheless, a large body of human–computer interaction research has provided numerous general guidelines for formatting information. Many of these guidelines will undoubtedly apply to the formatting of information on yet-to-be-determined devices. Keep in mind that the mainstay of designing usable forms and reports requires your active interaction with users. If this single and fundamental activity occurs, it is likely that you will create effective designs.

For example, one of the greatest challenges for designing mobile applications that run on devices like the iPhone is the human-computer interface. In particular, the small video display of these devices presents significant challenges for application designers. Nevertheless, as these and other computing devices evolve and gain popularity, standard guidelines will emerge to make the process of designing interfaces for these devices much less challenging.

General Formatting Guidelines

Over the past several years, industry and academic researchers have investigated how the format of information influences individual task performance and perceptions of usability. Through this work, several guidelines for formatting information have emerged (see Table 10-3). These guidelines reflect some of the general truths that apply to the formatting of most types of information (for more information, the interested reader should see the books by Flanders and Peters, 2002; Johnson, 2007; Krug, 2006; Nielson, 1999; Nielson and Loranger, 2006; and Shneiderman, Plaisant, Cohen, and Jacobs, 2009). The differences between a well-designed form or report and one that is poorly designed will often be obvious. For example, Figure 10-5a shows a poorly designed form for viewing the current account balance for a PVF customer. Figure 10-5b (page 2 of 2) is a better design that incorporates several general guidelines from Table 10-3.

The first major difference between the two forms has to do with the title. The title on Figure 10-5a is ambiguous, whereas the title on Figure 10-5b clearly and specifically describes the contents of the form. The form in Figure 10-5b also includes the date on which the form was generated so that, if printed, it will be clear to the reader when this occurred. Figure 10-5a displays information that is extraneous

---

TABLE 10-3 General Guidelines for the Design of Forms and Reports

Meaningful Titles:

Clear and specific titles describing content and use of form or report

Revision date or code to distinguish a form or report from prior versions

Current date, which identifies when the form or report was generated

Valid date, which identifies on what date (or time) the data in the form or report were accurate

Meaningful Information:

Only needed information should be displayed

Information should be provided in a manner that is usable without modification

Balance the Layout:

Information should be balanced on the screen or page

Adequate spacing and margins should be used

All data and entry fields should be clearly labeled

Design an Easy Navigation System:

Clearly show how to move forward and backward

Clearly show where you are (e.g., page 1 of 3)

Notify user when on the last page of a multipaged sequence

FIGURE 10-5
Contrasting customer information forms
(Pine Valley Furniture)

(*Source:* Microsoft Corporation.)
(a) Poorly designed form

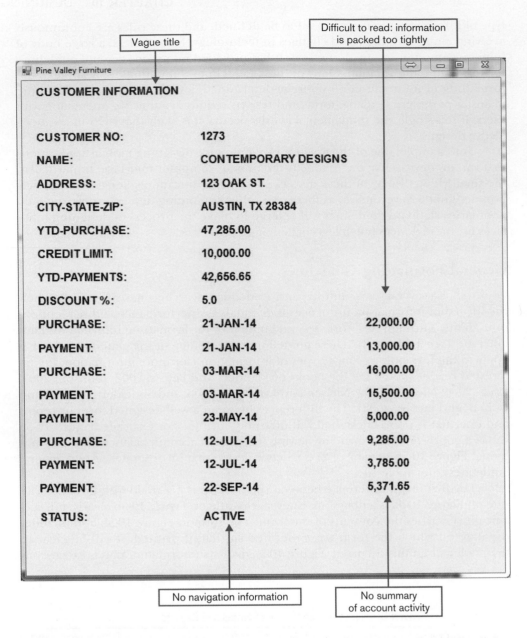

**TABLE 10-4 Methods of
Highlighting**

Blinking and audible tones
Color differences
Intensity differences
Size differences
Font differences
Reverse video
Boxing
Underlining
All capital letters
Offsetting the position of
nonstandard information

to the intent of the form—viewing the current account balance—and provides information that is not in the most useful format for the user. For example, Figure 10-5a provides all customer data as well as account transactions and a summary of year-to-date purchases and payments. The form does not, however, provide the current outstanding balance of the account; a user who desires this information must make a manual calculation. The layout of information between the two forms also varies in balance and information density. Gaining an understanding of the skills of the intended system users and the tasks they will be performing is invaluable when constructing a form or report. By following these general guidelines, your chances of creating effective forms and reports will be enhanced. In the next sections, we will discuss specific guidelines for highlighting information, using color, displaying text, and presenting numeric tables and lists.

Highlighting Information

As display technologies continue to improve, a greater variety of methods will be available to you for highlighting information. Table 10-4 provides a list of the most

FIGURE 10-5 (continued)
(b) Improved design for form

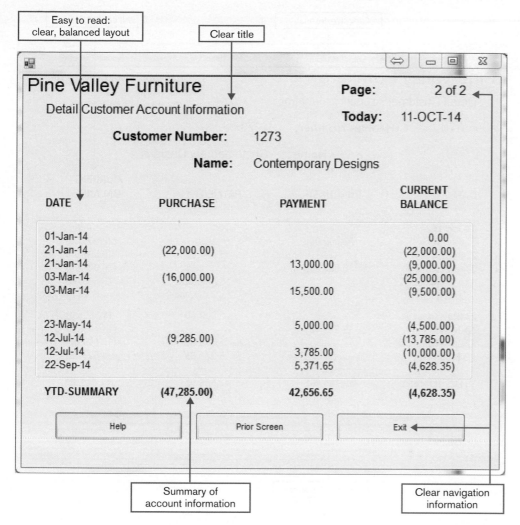

commonly used methods for highlighting information. Given this vast array of options, it is more important than ever to consider how highlighting can be used to enhance an output and not prove a distraction. In general, highlighting should be used sparingly to draw the user to or away from certain information and to group together related information. There are several situations when highlighting can be a valuable technique for conveying special information:

- Notifying users of errors in data entry or processing
- Providing warnings to users regarding possible problems such as unusual data values or an unavailable device
- Drawing attention to keywords, commands, high-priority messages, and data that have changed or gone outside normal operating ranges

Additionally, many highlighting techniques can be used singularly or in tandem, depending upon the level of emphasis desired by the designer. Figure 10-6 illustrates a form where several types of highlighting are used. In this example, boxes clarify different categories of data, capital letters and different fonts distinguish labels from actual data, and bold is used to draw attention to important data.

Much research has focused on the effects of varying highlighting techniques on task performance and user perceptions. A general guideline resulting from this research is that highlighting should be used conservatively. For example, blinking and audible tones should be used only to highlight critical information requiring an immediate response from the user. Once a response is made, these highlights should be turned off. Additionally, highlighting methods should be consistently used

FIGURE 10-6
Customer account status display using various highlighting techniques (Pine Valley Furniture)

(*Source:* Microsoft Corporation.)

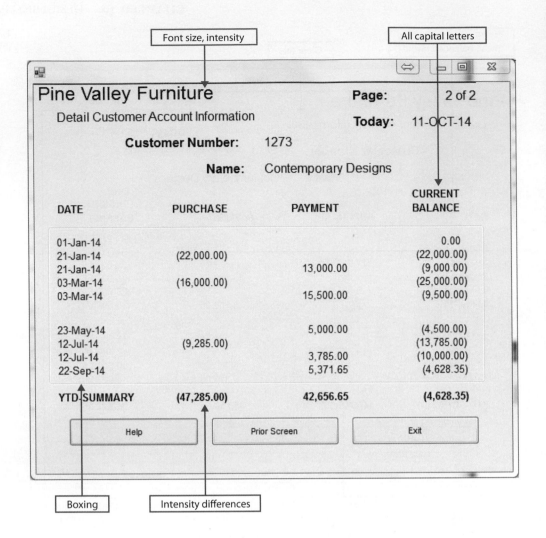

and selected based upon the level of importance of the emphasized information. It is also important to examine how a particular highlighting method appears on all possible output devices that could be used with the system. For example, some color combinations may convey appropriate information on one display configuration but wash out and reduce legibility on another.

The continued evolution of graphical operating environments such as Windows, Macintosh, and the Web has provided designers with some standard highlighting guidelines. However, these guidelines are often quite vague and are continuously evolving, leaving a great deal of control in the hands of the systems developer. Therefore, in order for organizations to realize the benefits of using standard graphical operating environments—such as reduced user training time and interoperability among systems—you must be disciplined in how you use highlighting.

Color versus No Color

Color is a powerful tool for the designer in influencing the usability of a system. When applied appropriately, color provides many potential benefits to forms and reports, which are summarized in Table 10-5. As the use of color displays became widely available during the 1980s, a substantial amount of color versus no color research was conducted. The objective of this research was to gain a better understanding of the effects of color on human task performance (e.g., see Benbasat, Dexter, and Todd, 1986).

The general findings from this research were that the use of color had positive effects on user task performance and perceptions when the user was under time

TABLE 10-5   Benefits and Problems from Using Color

Benefits from Using Color:
Soothes or strikes the eye.
Accents an uninteresting display.
Facilitates subtle discriminations in complex displays.
Emphasizes the logical organization of information.
Draws attention to warnings.
Evokes more emotional reactions.
Problems from Using Color:
Color pairings may wash out or cause problems for some users (e.g., color blindness).
Resolution may degrade with different displays.
Color fidelity may degrade on different displays.
Printing or conversion to other media may not easily translate.

(*Source:* Based on Shneiderman, et al., 2009; Benbasat, Dexter, and Todd, 1986.)

constraints for the completion of a task. Color was also beneficial for gaining greater understanding from a display or chart. An important conclusion from this research was that color was not universally better than no color. *The benefits of color only seem to apply if the information is first provided to the user in the most appropriate presentation format.* That is, if information is most effectively displayed in a bar chart, color can be used to enhance or supplement the display. If information is displayed in an inappropriate format, color has little or no effect on improving understanding or task performance.

Several problems are associated with using color, also summarized in Table 10-5. Most of these dangers are related more to the technical capabilities of the display and hard-copy devices than misuse. However, color blindness is a particular user issue that is often overlooked in the design of systems; approximately 8 percent of the males in the European and North American communities have some form of color blindness (Shneiderman et al., 2009). It is recommended that you first design video displays for monochrome and allow color (or better yet, a flexible palette of colors) to be a user-activated option. Shneiderman et al. (2009) also suggest that you limit the number of colors and where they are applied, using color primarily as a tool to assist in the highlighting and formatting of information.

Displaying Text

In business-related systems, textual output is becoming increasingly important as text-based applications such as electronic mail, bulletin boards, and information services (e.g., Dow Jones) are more widely used. The display and formatting of system help screens, which often contain lengthy textual descriptions and examples, is one example of textual data that can benefit from following a few simple guidelines that have emerged from past research. These guidelines appear in Table 10-6. The first

TABLE 10-6   Guidelines for Displaying Text

Case	Display text in mixed uppercase and lowercase and use conventional punctuation.
Spacing	Use double spacing if space permits. If not, place a blank line between paragraphs.
Justification	Left-justify text and leave a ragged-right margin.
Hyphenation	Do not hyphenate words between lines.
Abbreviations	Use abbreviations and acronyms only when they are widely understood by users and are significantly shorter than the full text.

FIGURE 10-7
Contrasting the display of textual help
information

(*Source:* Microsoft Corporation.)
(a) Poorly designed help screen with many
violations of the general guidelines for
displaying text

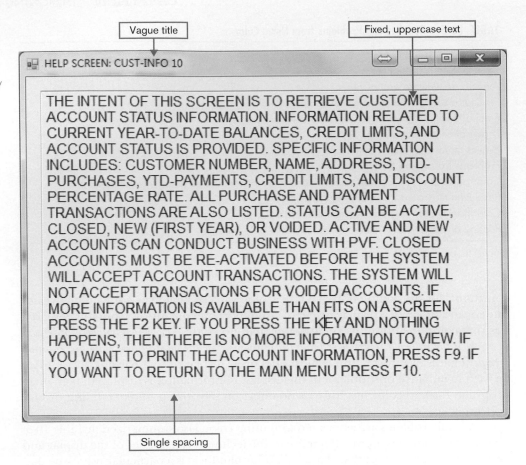

guideline is simple: You should display text using common writing conventions such as mixed uppercase and lowercase letters and appropriate punctuation. For large blocks of text, if space permits, text should be double-spaced. However, if the text is short, or rarely used, it may be appropriate to use single spacing and place a blank line between each paragraph. You should also left-justify text and use a ragged-right margin—research shows that a ragged-right margin makes it easier to find the next line of text when reading than when text is both left and right justified.

When displaying textual information, you should also be careful not to hyphenate words between lines or use obscure abbreviations and acronyms. Users may not know whether the hyphen is a significant character if it is used to continue words across lines. Information and terminology that are not widely understood by the intended users may significantly influence the usability of the system. Thus, you should use abbreviations and acronyms only if they are significantly shorter than the full text and are commonly known by the intended system users. Figure 10-7 shows two versions of a help screen from an application system at PVF. Figure 10-7a shows many violations of the general guidelines for displaying text, whereas Figure 10-7b shows the same information but follows the general guidelines for displaying text. Formatting guidelines for the entry of text and alphanumeric data are also a very important topic. These guidelines are presented in Chapter 11, "Designing Interfaces and Dialogues," where we focus on issues of human–computer interaction.

Designing Tables and Lists

Unlike textual information, where context and meaning are derived through reading, the context and meaning of tables and lists are derived from the format of the information. Consequently, the usability of information displayed in tables and alphanumeric lists is likely to be much more heavily influenced by effective layout than

FIGURE 10-7 (continued)
(b) An improved design for a help screen

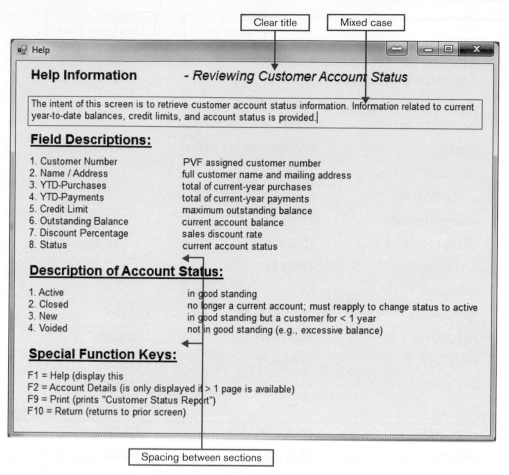

most other types of information display. As with the display of textual information, tables and lists can also be greatly enhanced by following a few simple guidelines. These are summarized in Table 10-7. You should review these guidelines and carefully apply them to ensure that your tables and lists are highly usable.

TABLE 10-7 *General Guidelines for Displaying Tables and Lists*

Use Meaningful Labels:

All columns and rows should have meaningful labels.

Labels should be separated from other information by using highlighting.

Redisplay labels when the data extend beyond a single screen or page.

Formatting Columns, Rows, and Text:

Sort in a meaningful order (e.g., ascending, descending, or alphabetic).

Place a blank line between every five rows in long columns.

Similar information displayed in multiple columns should be sorted vertically (ie., read from top to bottom, not left to right).

Columns should have at least two spaces between them.

Allow white space on printed reports for user to write notes.

Use a single typeface, except for emphasis.

Use same family of typefaces within and across displays and reports.

Avoid overly fancy fonts.

Formatting Numeric, Textual, and Alphanumeric Data:

Right-justify *numeric data* and align columns by decimal points or other delimiter.

Left-justify *textual data*. Use short line length, usually 30–40 characters per line (this is what newspapers use, and it is easier to speed-read).

Break long sequences of *alphanumeric data* into small groups of three to four characters each.

FIGURE 10-8
Contrasting the display of tables and lists
(Pine Valley Furniture)

(*Source:* Microsoft Corporation.)
(a) Poorly designed form

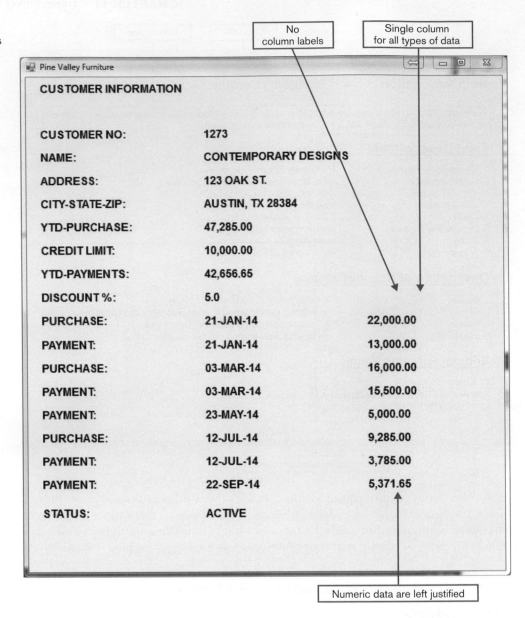

Figure 10-8 displays two versions of a form design from a PVF application system that displays customer year-to-date transaction information in a table format. Figure 10-8a displays the information without consideration of the guidelines presented in Table 10-7, and Figure 10-8b (only page 2 of 2 is shown) displays this information after consideration of these guidelines.

One key distinction between these two display forms relates to labeling. The information reported in Figure 10-8b has meaningful labels that more clearly stand out as labels compared with the display in Figure 10-8a. Transactions are sorted by date, and numeric data are right justified and aligned by decimal point in Figure 10-8b, which helps to facilitate scanning. Adequate space is left between columns, and blank lines are inserted after every five rows in Figure 10-8b to help ease the finding and reading of information. Such spacing also provides room for users to annotate data that catch their attention. Use of the guidelines presented in Table 10-7 helped the analyst to create an easy-to-read layout of the information for the user.

Most of the guidelines in Table 10-7 are rather obvious, but this and other tables serve as a quick reference to validate that your form and report designs will

FIGURE 10-8 (continued)
(b) Improved design for form

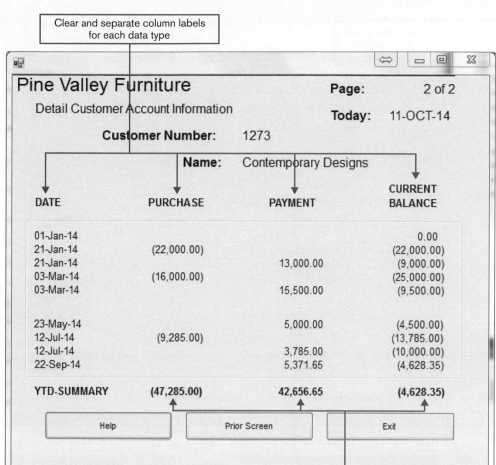

be usable. It is beyond our scope here to discuss each of these guidelines, but you should read each carefully and think about why each is appropriate. For example, why are labels repeated on subsequent screens and pages (the third guideline in Table 10-7)? One explanation is that pages may be separated or copied and the original labels will no longer be readily accessible to the reader of the data. Why should long alphanumeric data (see the last guideline) be broken into small groups? (If you have a credit card or bank check, look at how your account number is displayed.) One reason is that the characters will be easier to remember as you read and type them. Another reason is that there will be a natural and consistent place to pause when you speak them over the phone; for example, when you are placing a phone order for products in a catalog.

When you design the display of numeric information, you must determine whether a table or a graph should be used. A considerable amount of research focusing on this topic has been conducted (e.g., see Jarvenpaa and Dickson [1988] for very specific guidelines on the use of tables and graphs). In general, this research has found that tables are best when the user's task is related to finding an individual data value from a larger data set, whereas line and bar graphs are more appropriate for gaining an understanding of data changes over time (see Table 10-8). For example, if the marketing manager for PVF needed to review the actual sales of a particular salesperson for a particular quarter, a tabular report like the one shown in Figure 10-9 would be most useful. This report has been annotated to emphasize good report-design practices. The report has both a printed date as well as a clear indication, as part of the report title, of the period over which the data apply. There is also sufficient

TABLE 10-8   Guidelines for Selecting Tables versus Graphs

Use Tables For:

Reading individual data values

Use Graphs For:

Providing a quick summary of data

Detecting trends over time

Comparing points and patterns of different variables

Forecasting activities

Reporting vast amounts of information when relatively simple impressions are to be drawn

(*Source:* Based on Jarvenpaa and Dickson, 1988.)

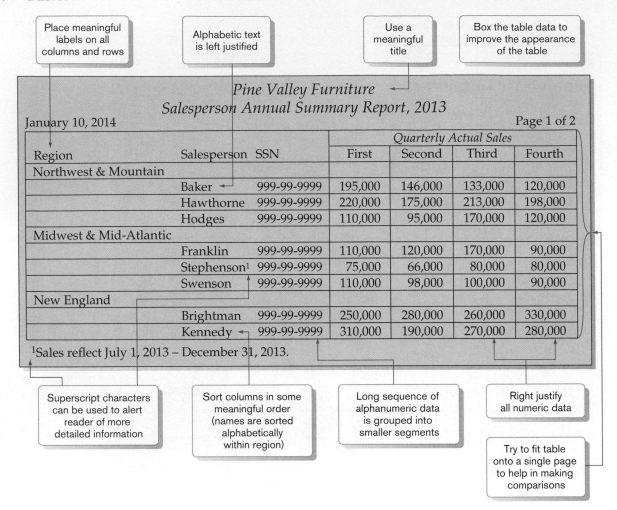

FIGURE 10-9
Tabular report illustrating numerous
design guidelines (Pine Valley Furniture)

white space to provide some room for users to add personal comments and observations. Often, to provide such white space, a report must be printed in landscape, rather than portrait, orientation. Alternatively, if the marketing manager wished to compare the overall sales performance of each sales region, a line or bar graph would be more appropriate (see Figure 10-10). As with other formatting considerations, the key determination as to when you should select a table or a graph is the task being performed by the user.

Paper versus Electronic Reports

When a report is produced on paper rather than on a computer display, there are some additional things that you need to consider. For example, laser printers (especially color laser printers) and ink jet printers allow you to produce a report that looks exactly as it does on the display screen. Thus, when using these types of printers, you can follow our general design guidelines to create a report with high usability. However, other types of printers are not able to closely reproduce the display screen image onto paper. For example, many business reports are produced using high-speed impact printers that produce characters and a limited range of graphics by printing a fine pattern of dots. The advantages of impact printers are that they are very fast, very reliable, and relatively inexpensive. Their drawbacks are that they have a limited ability to produce graphics and have a somewhat lower print quality. In other words, they are good at rapidly producing reports that contain primarily alphanumeric information, but they cannot exactly replicate a screen report

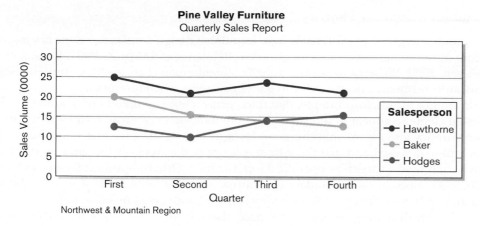

FIGURE 10-10
Graphs for comparison
(a) Line graph

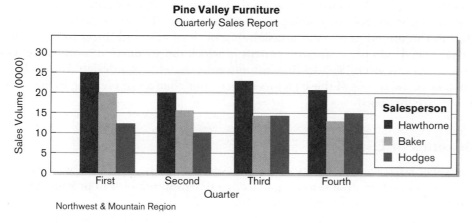

(b) Bar graph

onto paper. Because of this, impact printers are mostly used for producing large batches of reports, such as a batch of phone bills for a telephone company, on a wide range of paper widths and types. When designing reports for impact printers, you use a coding sheet like that displayed in Figure 10-2, although coding sheets for designing printer reports typically can have up to 132 columns. Like the process for designing all forms and reports, you follow a prototyping process and carefully control the spacing of characters in order to produce a high-quality report. However, unlike other form and report designs, you may be limited in the range of formatting, text types, and highlighting options. Nonetheless, you can easily produce a highly usable report of any type if you carefully and creatively use the formatting options that are available.

ASSESSING USABILITY

There are many factors to consider when you design forms and reports. The objective for designing forms, reports, and all human–computer interactions is usability. Usability typically refers to the following three characteristics:

1. *Speed.* Can you complete a task efficiently?
2. *Accuracy.* Does the output provide what you expect?
3. *Satisfaction.* Do you like using the output?

In other words, usability means that your designs should assist, not hinder, user performance. Thus, **usability** refers to an overall evaluation of how a system performs in supporting a particular user for a particular task. In the remainder of this section, we describe numerous factors that influence usability and several techniques for assessing the usability of a design.

Usability
An overall evaluation of how a system performs in supporting a particular user for a particular task.

Usability Success Factors

Research and practical experience have found that design consistency is the key ingredient in designing usable systems (Cooper, Reimann, and Cronin, 2007; Krug, 2006; Nielsen, 2000; Nielsen and Loranger, 2006; Shneiderman et al., 2009). Consistency significantly influences users' ability to gain proficiency when interacting with a system. Consistency means, for example, that titles, error messages, menu options, and other design elements appear in the same place and look the same on all forms and reports. Consistency also means that the same form of highlighting has the same meaning each time it is used and that the system will respond in roughly the same amount of time each time a particular operation is performed. Other important factors include efficiency, ease (or understandability), format, and flexibility. Each of these usability factors, with associated guidelines, is described in more detail in Table 10-9.

When designing outputs, you must also consider the context in which the screens, forms, and reports will be used. As mentioned, numerous characteristics play an important role in shaping a system's usability. These characteristics are related to the intended users and task being performed in addition to the technological, social, and physical environment in which the system and outputs are used. Table 10-10 lists several factors that influence the usability of a design. Your role is to gain a keen awareness of these factors so that your chances of creating highly usable designs are increased.

Measures of Usability

User-friendliness is a term often used, and misused, to describe system usability. Although the term is widely used, it is too vague from a design standpoint to provide adequate information because it means different things to different people. Consequently, most development groups use several methods for assessing usability, including the following considerations (Shneiderman and Plaisant, 2004; Te'eni et al., 2006):

- Time to learn
- Speed of performance
- Rate of errors
- Retention over time
- Subjective satisfaction

TABLE 10-9   General Design Guidelines for Usability of Forms and Reports

Usability Factor	Guidelines for Achievement of Usability
Consistency	Consistent use of terminology, abbreviations, formatting, titles, and navigation within and across outputs. Consistent response time each time a function is performed.
Efficiency	Formatting should be designed with an understanding of the task being performed and the intended user. Text and data should be aligned and sorted for efficient navigation and entry. Entry of data should be avoided where possible (e.g., computing rather than entering totals).
Ease	Outputs should be self-explanatory and not require users to remember information from prior outputs in order to complete tasks. Labels should be extensively used, and all scales and units of measure should be clearly indicated.
Format	Information format should be consistent between entry and display. Format should distinguish each piece of data and highlight, not bury, important data. Special symbols, such as decimal places, dollar signs, and ±signs, should be used as appropriate.
Flexibility	Information should be viewed and retrieved in a manner most convenient to the user. For example, users should be given options for the sequence in which to enter or view data and for use of shortcut keystrokes, and the system should remember where the user stopped during the last use of the system.

TABLE 10-10   **Characteristics for Consideration When Designing Forms and Reports**

Characteristic	Consideration for Form and Report Design
User	Issues related to experience, skills, motivation, education, and personality should be considered.
Task	Tasks differ in amount of information that must be obtained from or provided to the user. Task demands such as time pressure, cost of errors, and work duration (fatigue) will influence usability.
System	The platform on which the system is constructed will influence interaction styles and devices.
Environment	Social issues such as the users' status and role should be considered in addition to environmental concerns such as lighting, sound, task interruptions, temperature, and humidity. The creation of usable forms and reports may necessitate changes in the users' physical work facilities.

(*Source:* Based on Norman, 1991.)

In assessing usability, you can collect information by observation, interviews, keystroke capturing, and questionnaires. Time to learn simply reflects how long it takes the average system user to become proficient using the system. Equally important is the extent to which users remember how to use inputs and outputs over time. The manner in which the processing steps are sequenced and the selection of one set of keystrokes over others can greatly influence learning time, the user's task performance, and error rates. For example, the most commonly used functions should be quickly accessed with the fewest number of steps possible (e.g., pressing one key to save your work). Additionally, the layout of information should be consistent, both *within and across* applications, whether the information is delivered on a screen display or on a hard-copy report.

ELECTRONIC COMMERCE APPLICATIONS: DESIGNING FORMS AND REPORTS FOR PINE VALLEY FURNITURE'S WEBSTORE

Designing the forms and reports for an Internet-based electronic commerce application is a central and critical design activity. Because this is where a customer will interact with a company, much care must be put into its design. Like the process followed when designing the forms and reports for other types of systems, a prototyping design process is most appropriate. Although the techniques and technology for building Internet sites are rapidly evolving, several general design guidelines have emerged. In this section, we examine some of these as they apply to the design of PVF's WebStore.

General Guidelines

The rapid deployment of Internet websites has resulted in having countless people design sites who, arguably, have limited ability to do so. To put this into perspective, consider the following classic quote from Web design guru Jakob Nielsen (1999), which is still relevant today:

> If the [Web's] growth rate does not slow down, the Web will reach 200 million sites sometime during 2003. . . . The world has about 20,000 user interface professionals. If all sites were to be professionally designed by a single UI professional, we can conclude that every UI professional in the world would need to design one Web site every working hour from now on to meet demand. This is obviously not going to happen. (pp. 65–66)

There are three possible solutions to the problem:

1. Make it possible to design reasonably usable sites without having UI expertise.
2. Train more people in good Web design.
3. Live with poorly designed sites that are hard to use.

When designing forms and reports, there are several errors that are specific to website design. It is unfortunately beyond the scope of this book to critically examine all possible design problems within contemporary websites. Here, we will simply summarize those errors that commonly occur and that are particularly detrimental to the user's experience (see Table 10-11). Fortunately, there are numerous excellent sources for learning more about designing useful websites (Ash, 2008; Cooper et al., 2007; Flanders and Peters, 2002; Johnson, 2007; Krug, 2006; Nielson, 1999, 2000; Nielsen and Loranger, 2006; Shneiderman et al., 2009; *www.useit.com*; *www.webpagesthatsuck.com*).

Designing Forms and Reports at Pine Valley Furniture

When Jim Woo and the PVF development team focused on designing the forms and reports (i.e., the "pages") for the WebStore, they first reviewed many popular electronic commerce websites. From this review, they established the following design guidelines:

- Use lightweight graphics.
- Establish forms and data integrity rules.
- Use template-based HTML.

TABLE 10-11 **Common Errors When Designing the Layout of Web Pages**

Error	Recommendation
Nonstandard Use of GUI Widgets	Make sure that when using standard design items, they behave in accordance with major interface design standards. For example, the rules for radio buttons state that they are used to select one item among a set of items, that is, not confirmed until "OK'ed" by a user. In many websites selecting radio buttons is used as both *selection* and *action*.
Anything That Looks Like Advertising	Because research on Web traffic has shown that many users have learned to stop paying attention to Web advertisements, make sure that you avoid designing any legitimate information in a manner that resembles advertising (e.g., banners, animations, pop-ups).
Bleeding-Edge Technology	Make sure that users don't need the latest browsers or plug-ins to view your site.
Scrolling Test and Looping Animations	Avoid scrolling text and animations because they are both hard to read and users often equate such content with advertising.
Nonstandard Link Colors	Avoid using nonstandard colors to show links and for showing links that users have already used; nonstandard colors will confuse the user and reduce ease of use.
Outdated Information	Make sure your site is continuously updated so that users "feel" that the site is regularly maintained and updated. Outdated content is a sure way to lose credibility.
Slow Download Times	Avoid using large images, lots of images, unnecessary animations, or other time-consuming content that will slow the downloading time of a page.
Fixed-Formatted Text	Avoid fixed-formatted text that requires users to scroll horizontally to view content or links
Displaying Long Lists as Long Pages	Avoid requiring users to scroll down a page to view information, especially navigational controls. Manage information by showing only *N* items at a time, using multiple pages, or by using a scrolling container within the window.

In order to ensure that all team members understood what was meant by each guideline, Jim organized a design briefing to explain how each guideline would be incorporated into the WebStore interface design.

Lightweight Graphics

In addition to easy menu and page navigation, the PVF development team wants a system where Web pages load quickly. A technique that can assist in making pages load quickly is the use of lightweight graphics. **Lightweight graphics** is the use of small, simple images that allow a page to load as quickly as possible. "Using lightweight graphics allows pages to load quickly and helps users to reach their final location in the site—hopefully the point of purchase area—as quickly as possible. Large color images will only be used for displaying detailed product pictures that customers explicitly request to view," explained Jim. Experienced Web designers have found that customers are not willing to wait at each hop of navigation for a page to load, just so they have to click and wait again. The quick feedback that a website with lightweight graphics can provide will help to keep customers at the WebStore longer.

Lightweight graphics
The use of small, simple images to allow a Web page to be displayed more quickly.

Forms and Data Integrity Rules

Because the goal of the WebStore is to have users place orders for products, all forms that request information should be clearly labeled and provide adequate room for input. If a specific field requires a specific input format such as a date of birth or phone number, it must provide a clear example for the user so that data errors can be reduced. Additionally, the site must clearly designate which fields are optional, which are required, and which have a range of values.

Jim emphasized, "All of this seems to be overkill, but it makes processing the data much simpler. Our site will check all data before submitting it to the server for processing. This will allow us to provide quicker feedback to the user on any data entry error and eliminate the possibility of writing erroneous data into the permanent database. Additionally, we want to provide a disclaimer to reassure our customers that the data will be used only for processing orders, that it will never be sold to marketers, and that it will be kept strictly confidential."

Template-Based HTML

When Jim talked with the consultants about the WebStore during the analysis phase, they emphasized the advantages of using **template-based HTML**. He was told that when displaying individual products, it would be very advantageous to try to have a few "templates" that could be used to display the entire product line. In other words, not every product needs its own page—the development time for that would be far too great. Jim explained,

Template-Based HTML
Templates to display and process common attributes of higher-level, more abstract items.

We need to look for ways to write a module once and reuse it. This way, a change requires modifying one page, not 700. Using HTML templates will help us create an interface that is very easy to maintain. For example, a desk and a filing cabinet are two completely different products. Yet both have an array of finishes to choose from. Logically, each item requires the same function—namely, "display all finishes." If designed correctly, this function can be applied to all products in the store. On the other hand, if we write a separate module for each product, it would require us to change each and every module every time we make a product change, like adding a new finish. But a function such as "display all finishes," written once and associated with all appropriate products, will require the modification of one generic or "abstract" function, not hundreds.

SUMMARY

This chapter focused on a primary product of information systems: forms and reports. As organizations move into more complex and competitive business environments with greater diversity in the workforce, the quality of the business processes will determine success. One key to designing quality business processes is the delivery of the right information to the right people, in the right format, at the right time. The design of forms and reports concentrates on this goal. A major difficulty of this process comes from the great variety of information-formatting options available to designers.

Specific guidelines should be followed when designing forms and reports. These guidelines, proven over years of experience with human–computer interaction, help you to create professional, usable systems. This chapter presented a variety of guidelines covering the use of titles, layout of fields, navigation between pages or screens, highlighting of data, use of color, format of text and numeric data, appropriate use and layout of tables and graphs, avoidance of bias in information display, and achievement of usable forms and reports.

Form and report designs are created through a prototyping process. Once created, designs may be stand-alone or integrated into actual working systems. The purpose, however, is to show users what a form or report will look like when the system is implemented. The outcome of this activity is the creation of a specification document where characteristics of the users, tasks, system, and environment are outlined along with each form and report design. Performance testing and usability assessments may also be included in the design specification.

The goal of form and report design is usability. Usability means that users can use a form or report quickly, accurately, and with a high level of satisfaction. To be usable, designs must be consistent, efficient, self-explanatory, well formatted, and flexible. These objectives are achieved by applying a wide variety of guidelines concerning aspects such as navigation; the use of highlighting and color; and the display of text, tables, and lists.

KEY TERMS

1. Form
2. Lightweight graphics
3. Report
4. Template-Based HTML
5. Usability

Match each of the key terms above with the definition that best fits it.

_____ The use of templates to display and process common attributes of high-level, more abstract items.

_____ An overall evaluation of how a system performs in supporting a particular user for a particular task.

_____ A business document that contains only predefined data; it is a passive document used only for reading or viewing.

It typically contains data from many unrelated records or transactions.

_____ A business document that contains some predefined data and may include some areas where additional data are to be filled in. An instance on such a document is typically based on one database record.

_____ The use of small, simple images to allow a Web page to be displayed more quickly.

REVIEW QUESTIONS

1. Describe the prototyping process of designing forms and reports. What deliverables are produced from this process? Are these deliverables the same for all types of system projects? Why or why not?

2. What initial questions must be answered for an analyst to build an initial prototype of a system output?

3. When can highlighting be used to convey special information to users?

4. Discuss the benefits, problems, and general design process for the use of color when designing system output.

5. How should textual information be formatted on a help screen?

6. What type of labeling can you use in a table or list to improve its usability?

7. What column, row, and text formatting issues are important when designing tables and lists?

8. Describe how numeric, textual, and alphanumeric data should be formatted in a table or list.

9. What is meant by usability and what characteristics of an interface are used to assess a system's usability?

10. What measures do many development groups use to assess a system's usability?

11. List and describe common website design errors.

12. Provide some examples where variations in users, tasks, systems, and environmental characteristics might affect the design of system forms and reports.

PROBLEMS AND EXERCISES

1. Imagine that you are to design a budget report for a colleague at work using a spreadsheet package. Following the prototyping discussed in the chapter (see also Figure 6-7), describe the steps you would take to design a prototype of this report.

2. Consider a system that produces budget reports for your department at work. Alternatively, consider a registration system that produces enrollment reports for a department at a university. For whichever system you choose, answer the following design questions: Who will use the output? What is the purpose of the output? When is the output needed and when is the information that will be used within the output available? Where does the output need to be delivered? How many people need to view the output?

3. Imagine the worst possible reports from a system. What is wrong with them? List as many problems as you can. What are the consequences of such reports? What could go wrong as a result? How does the prototyping process help guard against each problem?

4. Imagine an output display form for a hotel registration system. Using a software package for drawing such as Microsoft Visio, follow the design suggestions in this chapter and design this form entirely in black and white. Save the file and then, following the color design suggestions in this chapter, redesign the form using color. Based on this exercise, discuss the relative strengths and weaknesses of each output form.

5. Consider reports you might receive at work (e.g., budgets or inventory reports) or at a university (e.g., grade reports or transcripts). Evaluate the usability of these reports in terms of speed, accuracy, and satisfaction. What could be done to improve the usability of these outputs?

6. List the PC-based software packages you like to use. Describe each package in terms of the following usability characteristics: time to learn, speed of performance, rate of errors by users, retention over time, and subjective satisfaction. Which of these characteristics has made you want to continue to use this package?

7. Given the guidelines presented in this chapter, identify flaws in the design of the Report of Customers that follows. What assumptions about users and tasks did you make in order to assess this design? Redesign this report to correct these flaws.

Report of Customers 26-Oct-14

Cust-ID	Organization
AC-4	A.C. Nielson Co.
ADTRA-20799	Adran
ALEXA-15812	Alexander & Alexander, Inc.
AMERI-1277	American Family Insurance
AMERI-28157	American Residential Mortgage
ANTAL-28215	Antalys
ATT-234	AT&T Residential Services
ATT-534	AT&T Consumer Services
. . .	
DOLE-89453	Dole United, Inc.
DOME-5621	Dome Caps, Inc.
DO-67	Doodle Dandies
. . .	
ZNDS-22267	Zenith Data System

8. Review the guidelines for attaining usability of forms and reports in Table 10-9. Consider an online form you might use to register a guest at a hotel. For each usability factor, list two examples of how this form could be designed to achieve that dimension of usability. Use examples other than those mentioned in Table 10-9.

9. How can differences in user, task, system, or the environment influence the design of a form or report? Provide an example that contrasts characteristics for each difference.

10. Go to the Internet and find commercial websites that demonstrate each of the common errors listed in Table 10-11.

FIELD EXERCISES

1. Find your last grade report. Given the guidelines presented in this chapter, identify flaws in the design of this grade report. Redesign this report to correct these flaws.

2. As stated in this chapter, most forms and reports are designed for contemporary information systems by using software to prototype output. Packages such as Microsoft Visual Studio .NET have very sophisticated output design modules. Gain access to such a tool at your university or where you work and study all the features the software provides for the design of printed output. Write a report that lists and explains all the features for layout, highlighting, summarizing data, etc.

3. Investigate the displays used in another field (e.g., aviation). What types of forms and reports are used in this field? What standards, if any, are used to govern the use of these outputs?

4. Interview a variety of people you know about the different types of forms and reports they use in their jobs. Ask to examine a few of these documents and answer the following questions for each one:

   a. What types of tasks does each support and how is it used?

   b. What types of technologies and devices are used to deliver each one?

   c. Assess the usability of each form or report. Is each usable? Why or why not? How could each be improved?

5. Scan the annual reports of a dozen or so companies for the past year. These reports can usually be obtained from a university library. Describe the types of information and the ways that information has been presented in these reports. How have color and graphics been used to improve the usability of information? Describe any instances where formatting has been used to hide or enhance the understanding of information.

6. Choose a PC-based software package you like to use and choose one that you don't like to use. Interview other users to determine their evaluations of these two packages. Ask each individual to evaluate each package in terms of speed, accuracy, and satisfaction as described in this chapter. Is there a consensus among these evaluations or do the respondents' evaluations differ from each other or from your own evaluations? Why?

REFERENCES

Ash, T. Landing Page Optimization: *The Definitive Guide to Testing and Tuning for Conversion*. New York: Sybex, 2008.

Benbasat, I., A. S. Dexter, and P. Todd. 1986. "The Influence of Color and Graphical Information Presentation in a Managerial Decision Simulation." *Human—Computer Interaction* 2: 65–92.

Cooper, A., R. Reimann, and D. Cronin. 2007. *About Face 3: The Essentials of Interaction Design*. New York: Wiley and Sons.

Flanders, V. and D. Peters. 2002. *Son of Web Pages That Suck: Learn Good Design by Looking at Bad Design*. Alameda, CA: Sybex Publishing.

Jarvenpaa, S. L., and G. W. Dickson. 1988. "Graphics and Managerial Decision Making: Research Based Guidelines." *Communications of the ACM* 31(6): 764–74.

Johnson, J. 2007. *GUI Bloopers 2.0: Common User Interface Design Don'ts and Dos*, 2nd ed. New York: Morgan Kaufmann.

Krug, S. 2006. *Don't Make Me Think: A Common Sense Approach to Web Usability*, 2nd ed. Upper Saddle River, NJ: Prentice Hall.

Lazar, J. 2004. *User-Centered Web Development: Theory into Practice*. Sudbury, MA: Jones & Bartlett.

McCracken, D. D., R. J. Wolfe, and J. M. Spoll. 2004. *User-Centered Web Site Development: A Human–Computer Interaction Approach*. Upper Saddle River, NJ: Prentice Hall.

Nielsen, J. 1999. "User Interface Directions for the Web." *Communications of the ACM* 42(1): 65–71.

Nielsen, J. 2000. *Designing Web Usability: The Practice of Simplicity*. Indianapolis, IN: New Riders Publishing.

Nielsen, J., and H. Loranger. 2006. *Prioritizing Web Usability*. Upper Saddle River, NJ: Prentice Hall.

Norman, K. L. 1991. *The Psychology of Menu Selection*. Norwood, NJ: Ablex.

Shneiderman, B., C. Plaisant, M. Cohen, and S. Jacobs. 2009. *Designing the User Interface: Strategies for Effective Human-Computer Interaction*, 5th ed. Reading, MA. Addison-Wiley.

Snyder, C. 2003. *Paper Prototyping: The Fast and Easy Way to Design and Refine User Interfaces*. San Francisco: Morgan Kaufmann Publishers.

Sun Microsystems. 2001. *Java Look and Feel Guidelines*. Palo Alto, CA: Sun Microsystems.

Te'eni, D., J. Carey, and P. Zhang. 2006. *Human–Computer Interaction: Developing Effective Organizational Information Systems*. New York: John Wiley & Sons.

PETRIE ELECTRONICS

Chapter 10: Designing Forms and Reports

It was late. Sally Fukuyama, assistant director of marketing, knocked on the slightly open door of Jim Watanabe's office. Jim was the project director for the "No Customer Escapes" customer loyalty system for Petrie Electronics

"Yeah, come in," Jim called.

"Hi Jim," Sally said, pushing the door open further. "Are you getting ready to leave?"

"Well, I was thinking about it, but something tells me that I'm probably not leaving any time soon. What's up?"

"I just got an e-mail from John [John Smith, the head of marketing at Petrie]. He has a whole bunch of reports he wants this system to generate," Sally replied. She took the stuffed manila folder in her hand and dropped it on Jim's desk.

"What is all this?" he moaned.

"John says all of these reports are absolutely essential. He says you should be able to generate all of the necessary data from the new customer loyalty system."

"It will take forever to work out the specific designs on all of these reports," Jim said. "I'm going to need a lot of help on this." Jim dropped the folder on his desk.

"Sorry, Jim," Sally said. "I'll help you tomorrow, but I really need to go."

"OK, bye," Jim said, as Sally left his office.

He opened the folder and started to look at what was there. Some of the report requirements were more complete than others. One of the reports near the top of the heap focused on listing the best customers, based on how much they had spent in a particular month. "I'll start with this one," Jim thought. "I think I'll do a quick design in Excel."

Jim worked on the report design for 15 minutes. His first cut is featured in PE Figures 10-1 and 10-2. PE Figure 10-1 shows the high-level summary report, which lists only the names of the customers, where they are from, and the total they spent during a given month. PE Figure 10-2 shows the details of what each customer bought.

"Well," Jim thought, "These are certainly practical designs for these reports. They show what John says he wants, but they sure are ugly. I wonder how I can make them look better. No time for that now. I have to start work on all of these other report designs. How many are there? A hundred? Sure seems like it. Maybe I can get the interns to work on some of this. It would be good for them."

Jim looked over the next suggestion for a report from John's stack of requests.

Case Questions

1. How would you make the reports in PE Figures 10-1 and 10-2 "look better"? After you improve the design of the reports, explain why you make the changes you did.

2. What other reports do you think John would ask for, based on the data that would be available from Petrie's customer loyalty system? Make a list. Then take the first two reports on your list and design how they would look.

3. Using the text as one source and what you can find on the Internet as another source, make a list of the 10 most important things to consider when designing reports.

	A	B	C	D	E	F
1		*Petrie's Best Customers by Monthly Purchases*				
2						
3			*March 2014*			
4						
5	Customer				Grand	
6	Customer Name	Customer ID	Home City	State	Total	
7	Francesca Jones	43218765-991	New Orleans	LA	3327.65	
8	Ahmad Walgreens	12345678-990	Yuba City	CA	2134.35	
9	Wilma Sanchez	45645699-990	Lamoni	IA	2038.75	
10	Sylvia Pollock		Los Angeles	CA	1988.94	
11	William Peace		Tampa	FL	1645.87	
12	Jose Gonzalez		Atlanta	GA	1543.34	
13	D'Andre Martinez		New York	NY	1109.15	
14	John Smith		Las Vegas	NV	1065.34	

PE FIGURE 10-1
Initial design for Best Customers Monthly Summary Report

PE FIGURE 10-2
Initial design for Best Customers Monthly
Detail Report

	A	B	C	D	E	F	G	H	I	J
1						*Petrie's Best Customers by Monthly Purchases*				
2										
3						*March 2014*				
4										
5	Customer				Purchases					Grand
6	Customer Name	Customer ID	Home City	State	Quantity	SKU	Description	Amount	Total	Total
7	Francesca Jones	43218765-991	New Orleans	LA	2	67890	50" Panasonic 3D TV	1398.95	2797.90	3327.65
8					4	98000	8' HDVI cables	69.95	279.80	
9					1	44441	Flat screen TV stand	249.95	249.95	
10	Ahmad Walgreens	12345678-990	Yuba City	CA	1	34567	19" computer monitor	99.99	99.99	2134.35
11					1	34447	Dell desktop	345.56	345.56	
12					1	34889	HP laser printer P1102w	149.95	149.95	
13					1	67890	50" Panasonic 3D TV	1398.95	1398.95	
14					2	98000	8' HDVI cables	69.95	139.90	
15	Wilma Sanchez	45645699-990	Lamoni	IA	1	67890	50" Panasonic 3D TV	1398.95	1398.95	2038.75
16					1	44441	Flat screen TV stand	249.95	249.95	
17					1	67888	Petri's 7.1 surround sound set	249.95	249.95	
18					2	98000	8' HDVI cables	69.95	139.90	

4. Do you belong to any customer loyalty programs? Such as an airline's frequent flyer program or a program at a national retailer? If not, maybe your parents or other relatives do. Take the monthly report that a loyalty program sends to customers. Identify all of the data elements needed to create the report and use that information to create an ER diagram.

CHAPTER ELEVEN

Designing Interfaces and Dialogues

Learning Objectives

After studying this chapter, you should be able to:

- Explain the process of designing interfaces and dialogues and the deliverables for their creation.

- Contrast and apply several methods for interacting with a system.

- List and describe various input devices and discuss usability issues for each in relation to performing different tasks.

- Describe and apply the general guidelines for designing interfaces and specific guidelines for layout design, structuring data entry fields, providing feedback, and system help.

- Design human–computer dialogues and understand how dialogue diagramming can be used to design dialogues.

- Design graphical user interfaces.

- Discuss guidelines for the design of interfaces and dialogues for Internet-based electronic commerce systems.

Introduction

In this chapter, you will learn about system interface and dialogue design. Interface design focuses on how information is provided to and captured from users; dialogue design focuses on the sequencing of interface displays. Dialogues are analogous to a conversation between two people. The grammatical rules followed by each person during a conversation are analogous to the interface. Thus, the design of interfaces and dialogues is the process of defining the manner in which humans and computers exchange information. A good human–computer interface provides a uniform structure for finding, viewing, and invoking the different components of a system. This chapter complements

Chapter 10, which addressed design guidelines for the content of forms and reports. Here, you will learn about navigation between forms, alternative ways for users to cause forms and reports to appear, and how to supplement the content of forms and reports with user help and error messages, among other topics.

We then describe the process of designing interfaces and dialogues and the deliverables produced during this activity. This is followed by a section that describes interaction methods and devices. Next, interface design is described. This discussion focuses on layout design, data entry, providing feedback, and designing help. We then examine techniques for designing human–computer dialogues. Finally, we examine the design of interfaces and dialogues within electronic commerce applications.

DESIGNING INTERFACES AND DIALOGUES

This is the third chapter that focuses on design within the systems development life cycle (see Figure 11-1). In Chapter 10, you learned about the design of forms and reports. As you will see, the guidelines for designing forms and reports also apply to the design of human–computer interfaces.

The Process of Designing Interfaces and Dialogues

Similar to designing forms and reports, the process of designing interfaces and dialogues is a user-focused activity. This means that you follow a prototyping methodology of iteratively collecting information, constructing a prototype, assessing usability, and making refinements. To design usable interfaces and dialogues, you must answer the same who, what, when, where, and how questions used to guide the design of forms and reports (see Table 10-2). Thus, this process parallels that of designing forms and reports (see Lazar, 2004; McCracken et al., 2004).

Deliverables and Outcomes

The deliverable and outcome from system interface and dialogue design is the creation of a design specification. This specification is also similar to the specification produced for form and report designs—with one exception. Recall that the design specification document discussed in Chapter 10 had three sections (see Figure 10-4):

1. Narrative overview
2. Sample design
3. Testing and usability assessment

   For interface and dialogue designs, one additional subsection is included: a section outlining the dialogue sequence—the ways a user can move from one display

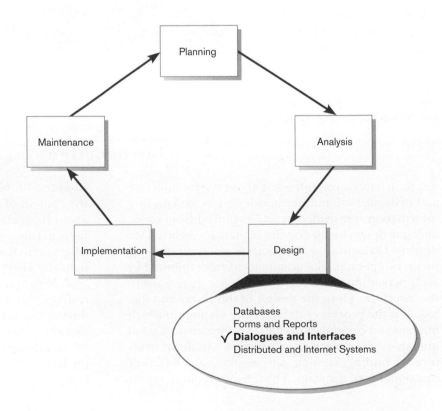

FIGURE 11-1
Systems development life cycle

Design Specification

1. Narrative Overview
   a. Interface/Dialogue Name
   b. User Characteristics
   c. Task Characteristics
   d. System Characteristics
   e. Environmental Characteristics
 	
2. Interface/Dialogue Designs
   a. Form/Report Designs
   b. Dialogue Sequence Diagram(s) and Narrative Description
 	
3. Testing and Usability Assessment
   a. Testing Objectives
   b. Testing Procedures
   c. Testing Results
     i) Time to Learn
     ii) Speed of Performance
     iii) Rate of Errors
     iv) Retention over Time
     v) User Satisfaction and Other Perceptions

FIGURE 11-2
Specification outline for the design of interfaces and dialogues

to another. Later in this chapter, you will learn how to design a dialogue sequence by using dialogue diagramming. An outline for a design specification for interfaces and dialogues is shown in Figure 11-2.

INTERACTION METHODS AND DEVICES

The human–computer **interface** defines the ways in which users interact with an information system. All human–computer interfaces must have an interaction style and use some hardware device(s) for supporting this interaction. In this section, we describe various interaction methods and guidelines for designing usable interfaces.

Interface
A method by which users interact with an information system.

Methods of Interacting

When designing the user interface, the most fundamental decision you make relates to the methods used to interact with the system. Given that there are numerous approaches for designing the interaction, we briefly provide a review of those most commonly used. (Readers interested in learning more about interaction methods are encouraged to see the books by Johnson [2007], Seffah and Javahery [2003], Shneiderman et al. [2009], and Te'eni et al. [2006].) Our review will examine the basics of five widely used styles: command language, menu, form, object, and natural language. We will also describe several devices for interacting, focusing primarily on their usability for various interaction activities.

Command Language Interaction In **command language interaction**, the user enters explicit statements to invoke operations within a system. This type of interaction requires users to remember command syntax and semantics. For example, to copy a file named PAPER.DOC from one storage location (C:) to another (A:) using Microsoft's disk operating system (DOS), a user would type:

Command language interaction
A human–computer interaction method whereby users enter explicit statements into a system to invoke operations.

   COPY C:PAPER.DOC A:PAPER.DOC

   Command language interaction places a substantial burden on the user to remember names, syntax, and operations. Most newer or large-scale systems no longer rely entirely on a command language interface. Yet command languages are good for experienced users, for systems with a limited command set, and for rapid interaction with the system.

A relatively simple application such as a word processor may have hundreds of commands for operations such as saving a file, deleting words, canceling the current action, finding a specific piece of data, or switching between windows. Some of the burden of assigning keys to actions has been taken off users' shoulders through the development of user interface standards such as those for the Macintosh, Microsoft Windows, or Java (Apple Computer, 1993; McKay, 1999; Sun Microsystems, 2001). For example, Figure 11-3a shows a help screen from Microsoft Word describing keyboard shortcuts, and Figure 11-3b shows the same screen for Microsoft PowerPoint. Note how many of the same keys have been assigned the same function. Also note that designers still have great flexibility in how they interpret and implement these standards. This means that you still need to pay attention to usability factors and conduct formal assessments of designs.

Menu Interaction   A significant amount of interface design research has stressed the importance of a system's ease of use and understandability. **Menu interaction** is a means by which many designers have accomplished this goal. A menu is simply a list of options; when an option is selected by the user, a specific command is invoked or another menu is activated. Menus have become the most widely used interface method because the user only needs to understand simple signposts and route options to effectively navigate through a system.

Menus can differ significantly in their design and complexity. The variation of their design is most often related to the capabilities of the development environment,

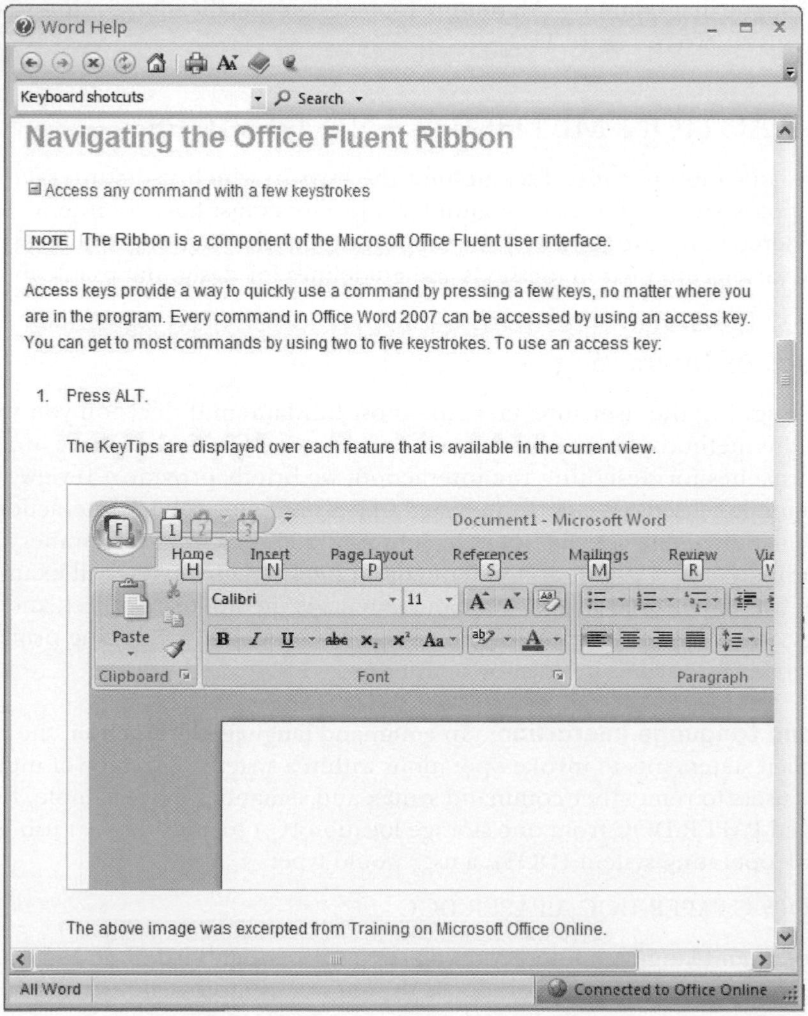

Menu interaction
A human–computer interaction method in which a list of system options is provided and a specific command is invoked by user selection of a menu option.

FIGURE 11-3
Function key assignments in Microsoft Office 2007

(*Source:* Microsoft Corporation.)
(a) Help screen from Microsoft Word describing keyboard commands

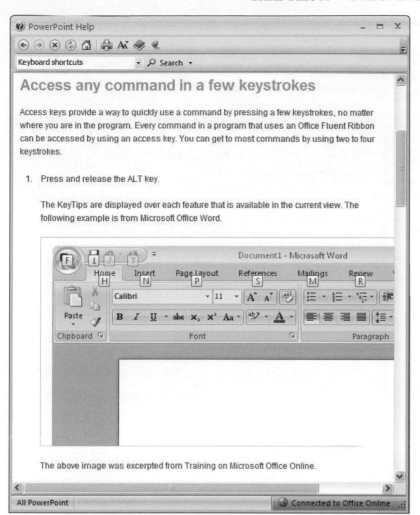

FIGURE 11-3 (continued)
(b) Help screen from Microsoft PowerPoint describing keyboard commands

the skills of the developer, and the size and complexity of the system. For smaller and less complex systems with limited system options, you may use a single menu or a linear sequence of menus. A single menu has obvious advantages over a command language but may provide little guidance beyond invoking the command. An example of a single menu can be found on the "Tabs" settings under the "Options" menu in the popular Firefox Web browser, as shown in Figure 11-4.

For large and more complex systems, you can use menu hierarchies to provide navigation between menus. These hierarchies can be simple tree structures or variations wherein children menus have multiple parent menus. Some of these hierarchies may allow multilevel traversal. Variations as to how menus are arranged can greatly influence the usability of a system. For instance, Microsoft has recently deployed its "ribbon menu" system in recent Office products. Figure 11-5 shows a variety of ways in which menus can be structured and traversed. An arc on this diagram signifies the ability to move from one menu to another. Although more complex menu structures provide greater user flexibility, they may also confuse users about exactly where they are in the system. Structures with multiple parent menus also require the application to remember which path has been followed so that users can correctly backtrack.

There are two common methods for positioning menus. With a **pop-up menu** (also called a *dialogue box*), menus are displayed near the current cursor position so users don't have to move the position or their eyes to view system options (Figure 11-6a). A pop-up menu has a variety of potential uses. One is to show a list of commands relevant to the current cursor position (e.g., delete, clear, copy, or validate current field). Another is to provide a list of possible values (from a look-up

Pop-up menu

A menu-positioning method that places a menu near the current cursor position.

FIGURE 11-4
Single-level menu on the "Tabs" settings
funder the "Options" menu in the Firefox
Web browser

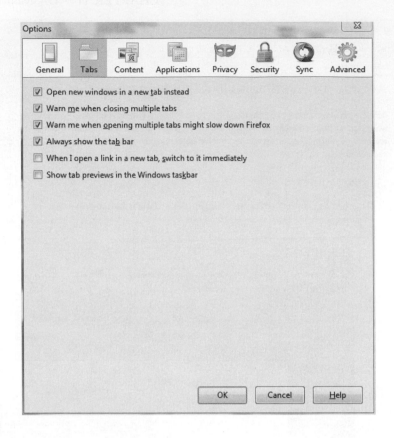

FIGURE 11-5
Various types of menu configurations
(*Source:* Based on Shneiderman et al., 2009.)

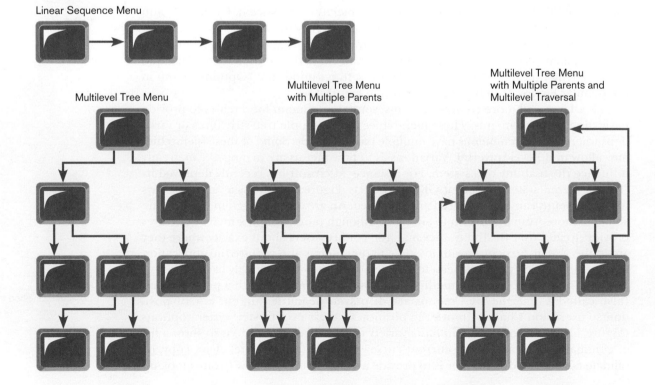

Single Menu

Linear Sequence Menu

Multilevel Tree Menu

Multilevel Tree Menu
with Multiple Parents

Multilevel Tree Menu
with Multiple Parents and
Multilevel Traversal

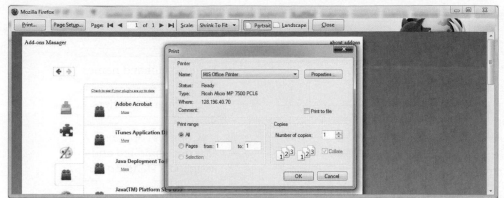

FIGURE 11-6
Menus from Microsoft Word 2003
(*Source:* Microsoft Corporation.)
(a) Pop-up menu

(b) Drop-down menu

table) to fill in for the current field. For example, in a customer order form, a list of current customers could pop up next to the customer number field so the user can select the correct customer without having to know the customer's identifier. With a **drop-down menu**, menus drop down from the top line of the display (Figure 11-6b). Drop-down menus have become very popular in recent years because they provide consistency in menu location and operation among applications and efficiently use display space. Most advanced operating environments, such as Microsoft Windows or the Apple Macintosh, provide a combination of both pop-up and drop-down menus.

When designing menus, several general rules should be followed, and these are summarized in Table 11-1. For example, each menu should have a meaningful title and be presented in a meaningful manner to users. A menu option of Quit, for instance, is

Drop-down menu

A menu-positioning method that places the access point of the menu near the top line of the display; when accessed, menus open by dropping down onto the display.

TABLE 11-1   Guidelines for Menu Design

Wording	• Each menu should have a meaningful title
	• Command verbs should clearly and specifically describe operations
	• Menu items should be displayed in mixed uppercase and lowercase letters and have a clear, unambiguous interpretation
Organization	• A consistent organizing principle should be used that relates to the tasks the intended users perform; for example, related options should be grouped together, and the same option should have the same wording and codes each time it appears
Length	• The number of menu choices should not exceed the length of the screen
	• Submenus should be used to break up exceedingly long menus
Selection	• Selection and entry methods should be consistent and reflect the size of the application and sophistication of the users
	• How the user is to select each option and the consequences of each option should be clear (e.g., whether another menu will appear)
Highlighting	• Highlighting should be minimized and used only to convey selected options (e.g., a check mark) or unavailable options (e.g., dimmed text)

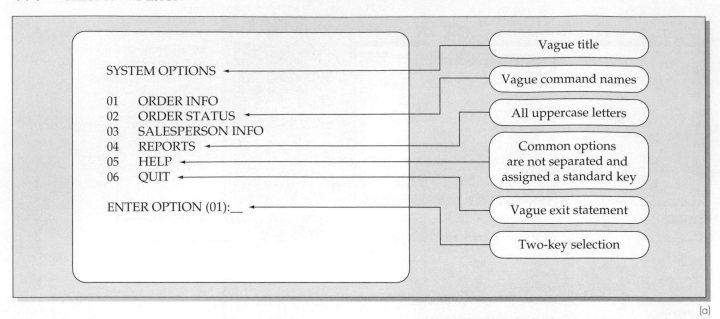

(a)

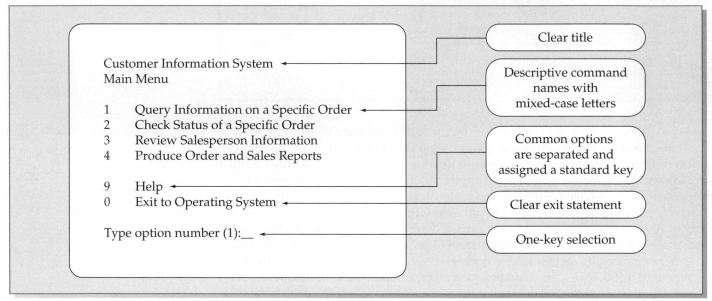

(b)

FIGURE 11-7
Contrasting menu designs
(a) Poor menu design
(b) Improved menu design

ambiguous—does it mean return to the previous screen or exit the program? To more easily see how to apply these guidelines, Figure 11-7 contrasts a poorly designed menu with a menu that follows the menu design guidelines. Annotations on the two parts of this figure highlight poor and improved menu interface design features.

Many advanced programming environments provide powerful tools for designing menus. For example, Microsoft's Visual Basic .NET allows you to quickly design a menu structure for a system. For example, Figure 11-8 shows a design form in which a menu structure is being defined; menu items are added by selecting the "Type Here" tags and typing the words that represent each item on the menu. With the use of a few easily invoked options, you can also assign shortcut keys to menu items, connect help screens to individual menu items, define submenus, and set usage properties (see the Properties window within Figure 11-8). Usage properties,

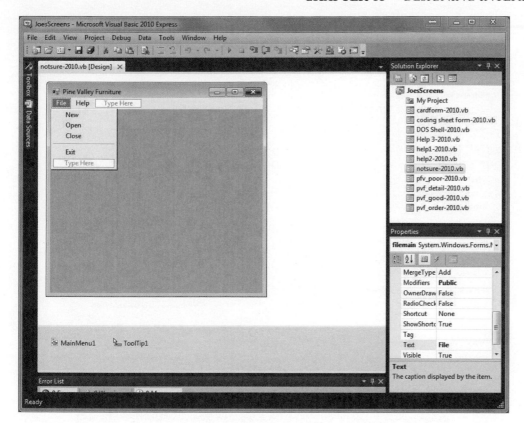

FIGURE 11-8
Menu building with Microsoft Visual
Basic .NET
(*Source:* Microsoft Corporation.)

for example, include the ability to dim the color of a menu item while a program is running, indicating that a function is currently unavailable. Menu-building tools allow a designer to quickly and easily prototype a design that will look exactly as it will in the final system.

Form Interaction The premise of **form interaction** is to allow users to fill in the blanks when working with a system. Form interaction is effective for both the input and presentation of information. An effectively designed form includes a self-explanatory title and field headings, has fields organized into logical groupings with distinctive boundaries, provides default values when practical, displays data in appropriate field lengths, and minimizes the need to scroll windows (Shneiderman and Plaisant, 2004). You saw many other design guidelines for forms in Chapter 10. Form interaction is the most commonly used method for data entry and retrieval in business-based systems. Figure 11-9 shows a form from the Google Advanced Search Engine. Using interactive forms, organizations can easily provide all types of information to Web surfers.

Object-Based Interaction The most common method for implementing **object-based interaction** is through the use of icons. **Icons** are graphic symbols that look like the processing option they are meant to represent. Users select operations by pointing to the appropriate icon with some type of pointing device. The primary advantages to icons are that they take up little screen space and can be quickly understood by most users. An icon may also look like a button that, when selected or depressed, causes the system to take an action relevant to that form, such as cancel, save, edit a record, or ask for help. For example, Figure 11-10 illustrates an icon-based interface when setting Options within the Firefox Web browser.

Natural Language Interaction One branch of artificial intelligence research studies techniques for allowing systems to accept inputs and produce outputs in

Form interaction
A highly intuitive human–computer interaction method whereby data fields are formatted in a manner similar to paper-based forms.

Object-based interaction
A human–computer interaction method in which symbols are used to represent commands or functions.

Icon
Graphical picture that represents specific functions within a system.

FIGURE 11-9
Example of form interaction from the
Google Advanced Search Engine
(*Source:* Google.)

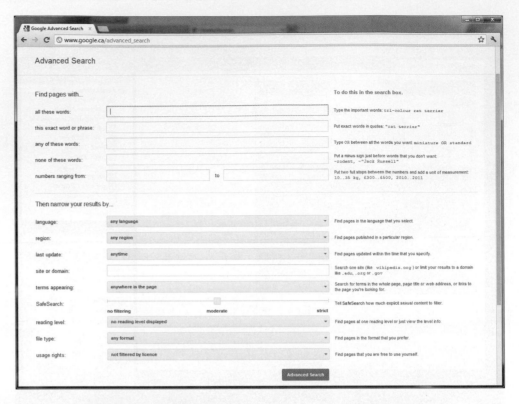

FIGURE 11-10
Object-based (icon) interface from the
Option menu in the Firefox Web browser

Natural language interaction
A human–computer interaction method
whereby inputs to and outputs from a
computer-based application are in a
conventional spoken language such as
English.

a conventional language such as English. This method of interaction is referred
to as **natural language interaction**. Presently, natural language interaction is not
as viable an interaction style as the other methods presented. Current implementations can be tedious, frustrating, and time consuming for the user and
are often built to accept input in narrowly constrained domains (e.g., database
queries). Natural language interaction is being applied within both keyboard
and voice entry systems.

Hardware Options for System Interaction

In addition to the variety of methods used for interacting with a system, there is also a growing number of hardware devices employed to support this interaction (see Table 11-2 for a list of interaction devices along with brief descriptions of the typical usage of each). The most fundamental and widely used device is the keyboard, which is the mainstay of most computer-based applications for the entry of alphanumeric information. Keyboards vary, from the typewriter kind of keyboards used with personal computers to special-function keyboards on point-of-sale or shop-floor devices. The growth in graphical user environments, however, has spurred the broader use of pointing devices such as mice, joysticks, trackballs, and graphics tablets. The creation of notebook and pen-based computers with trackballs, joysticks, or pens attached directly to the computer has also brought renewed interest to the usability of these various devices.

Research has found that each device has its strengths and weaknesses. These strengths and weaknesses must guide your selection of the appropriate devices to aid users in their interaction with an application. The selection of an interaction device must be made during logical design because different interfaces require different devices. Table 11-3 summarizes much of the usability assessment research by relating each device to various types of human–computer interaction problems. For example, for many applications, keyboards do not give users a precise feel for cursor movement, do not provide direct feedback on each operation, and can be a slow way to enter data (depending on the typing skill of the user). Another means to gain an understanding of device usability is to highlight which devices have been found most useful for completing specific tasks. The results of this research are summarized in Table 11-4. The rows of this table list common user–computer interaction tasks, and the columns show three criteria for evaluating the usability of the different devices. After reviewing these three tables, it should be evident that no device is perfect and that some are more appropriate for performing some tasks than others. To design

TABLE 11-2   Common Devices for Interacting with an Information System

Device	Description and Primary Characteristics or Usage
Keyboard	Users push an array of small buttons that represent symbols that are then translated into words and commands. Keyboards are widely understood and provide considerable flexibility for interaction.
Mouse	A small plastic box that users push across a flat surface and whose movements are translated into cursor movement on a computer display. Buttons on the mouse tell the system when an item is selected. A mouse works well on flat desks but may not be practical in dirty or busy environments, such as a shop floor or check-out area in a retail store. Newer pen-based mice provide the user with more of the feel of a writing implement.
Joystick	A small vertical lever mounted on a base that steers the cursor on a computer display. Provides similar functionality to a mouse.
Trackball	A sphere mounted on a fixed base that steers the cursor on a computer display. A suitable replacement for a mouse when work space for a mouse is not available.
Touch Screen	Selections are made by touching a computer display. This works well in dirty environments or for users with limited dexterity or expertise.
Light Pen	Selections are made by pressing a pen-like device against the screen. A light pen works well when the user needs to have a more direct interaction with the contents of the screen.
Graphics Tablet	Moving a pen-like device across a flat tablet steers the cursor on a computer display. Selections are made by pressing a button or by pressing the pen against the tablet. This device works well for drawing and graphical applications.
Voice	Spoken words are captured and translated by the computer into text and commands. This is most appropriate for users with physical challenges or when hands need to be free to do other tasks while interacting with the application.

TABLE 11-3   Summary of Interaction Device Usability Problems

Device	Problem						
	Visual Blocking	User Fatigue	Movement Scaling	Durability	Adequate Feedback	Speed	Pointing Accuracy
Keyboard	□	□	■	□	■	■	□
Mouse	□	□	■	□	■	□	□
Joystick	□	□	■	□	■	□	■
Trackball	□	□	■	■	■	□	□
Touch Screen	■	■	□	■	□	□	■
Light Pen	■	■	□	□	□	□	■
Graphics Tablet	□	□	■	□	■	□	□
Voice	□	□	■	□	■	□	■

Key:

□ = little or no usability problems

■ = potentially high usability problems for some applications

Visual Blocking = extent to which device blocks display when using

User Fatigue = potential for fatigue over long use

Movement Scaling = extent to which device movement translates to equivalent screen movement

Durability = lack of durability or need for maintenance (e.g., cleaning) over extended use

Adequate Feedback = extent to which device provides adequate feedback for each operation

Speed = cursor movement speed

Pointing Accuracy = ability to precisely direct cursor

(*Source:* Based on Blattner and Schultz, 1988.)

TABLE 11-4   Summary of General Conclusions from Experimental Comparisons of Input Devices in Relation to Specific Task Activities

Task	Most Accurate	Shortest Positioning	Most Preferred
Target Selection	trackball, graphics tablet, mouse, joystick	touch screen, light pen, mouse, graphics tablet, trackball	touch screen, light pen
Text Selection	mouse	mouse	—
Data Entry	light pen	light pen	—
Cursor Positioning	—	light pen	—
Text Correction	light pen, cursor keys	light pen	light pen
Menu Selection	touch screen	—	keyboard, touch screen

Key:

Target Selection = moving the cursor to select a figure or item

Text Selection = moving the cursor to select a block of text

Data Entry = entering information of any type into a system

Cursor Positioning = moving the cursor to a specific position

Text Correction = moving the cursor to a location to make a text correction

Menu Selection = activating a menu item

— = no clear conclusion from the research

(*Source:* Based on Blattner and Schultz, 1988.)

the most effective interfaces for a given application, you should understand the capabilities of various interaction methods and devices.

DESIGNING INTERFACES

Building on the information provided in Chapter 10 on the design of content for forms and reports, here we discuss issues related to the design of interface layouts. This discussion provides guidelines for structuring and controlling data entry fields, providing feedback, and designing online help. Effective interface design requires that you gain a thorough understanding of each of these concepts.

Designing Layouts

To ease user training and data recording, you should use standard formats for computer-based forms and reports similar to those used on paper-based forms and reports for recording or reporting information. A typical paper-based form for reporting customer sales activity is shown in Figure 11-11. This form has several general areas common to most forms:

- Header information
- Sequence and time-related information
- Instruction or formatting information
- Body or data details
- Totals or data summary

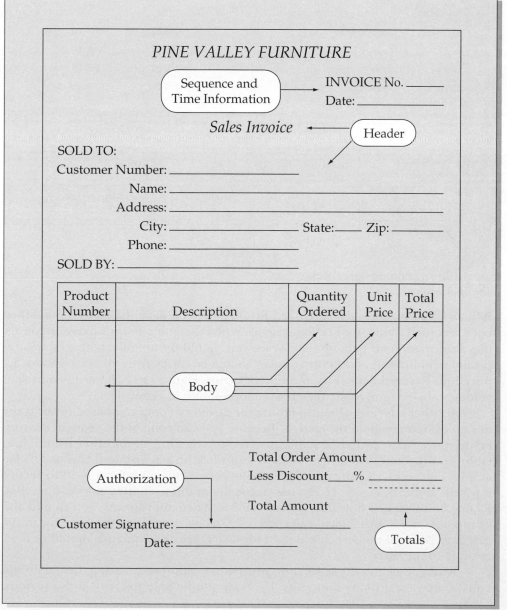

FIGURE 11-11
Paper-based form for reporting customer sales activity (Pine Valley Furniture)

FIGURE 11-12
Computer-based form reporting customer sales activity (Pine Valley Furniture)

(*Source:* Microsoft Corporation.)

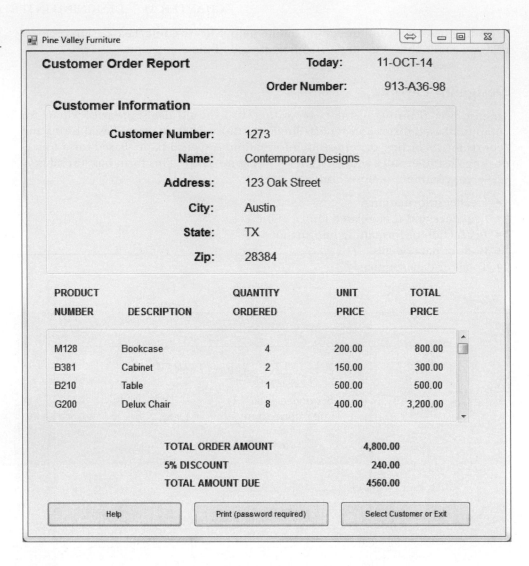

- Authorization or signatures
- Comments

In many organizations, data are often first recorded on paper-based forms and then later recorded within application systems. When designing layouts to record or display information on paper-based forms, you should try to make both as similar as possible. Additionally, data entry displays should be consistently formatted across applications to speed data entry and reduce errors. Figure 11-12 shows an equivalent computer-based form to the paper-based form shown in Figure 11-11.

Another concern when designing the layout of computer-based forms is the design of between-field navigation. Because you can control the sequence for users to move between fields, standard screen navigation should flow from left to right and top to bottom just as when you work on paper-based forms. For example, Figure 11-13 contrasts the flow between fields on a form used to record business contacts. Figure 11-13a uses a consistent left-to-right, top-to-bottom flow. Figure 11-13b uses a flow that is nonintuitive. When appropriate, you should also group data fields into logical categories with labels describing the contents of the category. Areas of the screen not used for data entry or commands should be inaccessible to the user.

When designing the navigation procedures within your system, flexibility and consistency are primary concerns. Users should be able to freely move forward

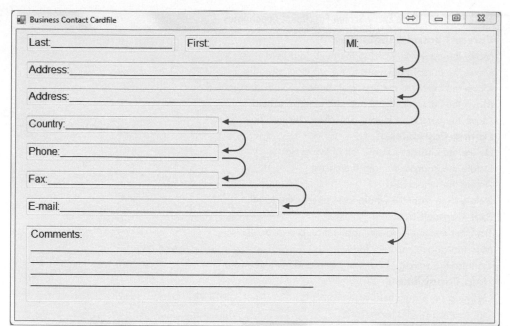

FIGURE 11-13
Contrasting the navigation flow within
a data entry form
(*Source:* Microsoft Corporation.)
(a) Proper flow between data entry fields

(b) Poor flow between data entry fields

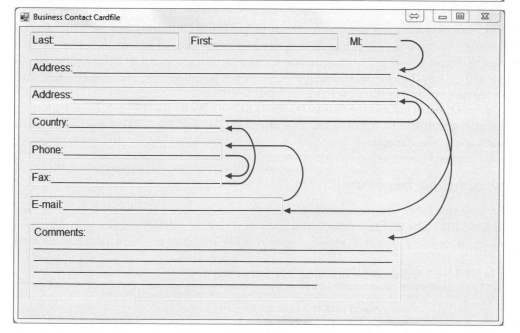

and backward or to any desired data entry fields. Users should be able to navigate each form in the same way or in as similar a manner as possible. Additionally, data should not usually be permanently saved by the system until the user makes an explicit request to do so. This allows the user to abandon a data entry screen, back up, or move forward without adversely affecting the contents of the permanent data.

Consistency extends to the selection of keys and commands. Each key or command should have only one function, and this function should be consistent throughout the entire system and across systems, if possible. Depending upon the application, various types of functional capabilities will be required to provide smooth navigation and data entry. Table 11-5 provides a list of the functional requirements for providing smooth and easy navigation within a form. For example, a functional and consistent interface will provide common ways for users to move the cursor to different places on the form, edit characters and fields,

TABLE 11-5   Data Entry Screen Functional Capabilities

Cursor Control Capabilities:
Move the cursor forward to the next data field
Move the cursor backward to the previous data field
Move the cursor to the first, last, or some other designated data field
Move the cursor forward one character in a field
Move the cursor backward one character in a field

Editing Capabilities:
Delete the character to the left of the cursor
Delete the character under the cursor
Delete the whole field
Delete data from the whole form (empty the form)

Exit Capabilities:
Transmit the screen to the application program
Move to another screen/form
Confirm the saving of edits or go to another screen/form

Help Capabilities:
Get help on a data field
Get help on a full screen/form

move among form displays, and obtain help. These functions may be provided by keystrokes, mouse or other pointing device operations, or menu selection or button activation. It is possible that, for a single application, all of the functional capabilities listed in Table 11-5 may not be needed in order to create a flexible and consistent user interface. Yet the capabilities that are used should be consistently applied to provide an optimal user environment. As with other tables in Chapters 10 and 11, Table 11-5 can serve as a checklist for you to validate the usability of user interface designs.

Structuring Data Entry

Several rules should be considered when structuring data entry fields on a form (see Table 11-6). The first is simple, but it is often violated by designers. To minimize data entry errors and user frustration, never require the user to enter information that

TABLE 11-6   Guidelines for Structuring Data Entry Fields

Entry	Never require data that are already online or that can be computed; for example, do not enter customer data on an order form if those data can be retrieved from the database, and do not enter extended prices that can be computed from quantity sold and unit prices.
Defaults	Always provide default values when appropriate; for example, assume today's date for a new sales invoice, or use the standard product price unless overridden.
Units	Make clear the type of data units requested for entry; for example, indicate quantity in tons, dozens, pounds, etc.
Replacement	Use character replacement when appropriate; for example, allow the user to look up the value in a table or automatically fill in the value once the user enters enough significant characters.
Captioning	Always place a caption adjacent to fields; see Table 11-7 for caption options.
Format	Provide formatting examples when appropriate; for example, automatically show standard embedded symbols, decimal points, credit symbol, or dollar sign.
Justify	Automatically justify data entries; numbers should be right justified and aligned on decimal points, and text should be left justified.
Help	Provide context-sensitive help when appropriate; for example, provide a hot key, such as the F1 key, that opens the help system on an entry that is most closely related to where the cursor is on the display.

TABLE 11-7   Options for Entering Text

Options	Example
Line Caption	Phone Number ( ) - _____
Drop Caption	( ) - _____ Phone Number
Boxed Caption	Phone Number
Delimited Characters	(| | | |) | | |- | | | | Phone Number
Check-Off Boxes	Method of payment (check one) ❏ Check ❏ Cash ❏ Credit card: Type

is already available within the system or information that can be easily computed by the system. For example, never require the user to enter the current date and time because each of these values can be easily retrieved from the computer system's internal calendar and clock. By allowing the system to do this, the user simply confirms that the calendar and clock are working properly.

Other rules are equally important. For example, suppose that a bank customer is repaying a loan on a fixed schedule with equal monthly payments. Each month when a payment is sent to the bank, a clerk needs to record into a loan processing system that the payment has been received. Within such a system, default values for fields should be provided whenever appropriate. This means that only in the instances where the customer pays *more or less* than the scheduled amount should the clerk have to enter data into the system. In all other cases, the clerk would simply verify that the check is for the default amount provided by the system and press a single key to confirm the receipt of payment.

When entering data, the user should also not be required to specify the dimensional units of a particular value. For example, a user should not be required to specify that an amount is in dollars or that a weight is in tons. Field formatting and the data entry prompt should make clear the type of data being requested. In other words, a caption describing the data to be entered should be adjacent to each data field. Within this caption, it should be clear to the user what type of data is being requested. As with the display of information, all data entered onto a form should automatically justify in a standard format (e.g., date, time, money). Table 11-7 illustrates a few options appropriate for printed forms. For data entry on video display terminals, you should highlight the area in which text is entered so that the exact number of characters per line and number of lines are clearly shown. You can also use check boxes or radio buttons to allow users to choose standard textual responses. And you can use data entry controls to ensure that the proper type of data (alphabetic or numeric, as required) is entered. Data entry controls are discussed next.

Controlling Data Input

One objective of interface design is to reduce data entry errors. As data are entered into an information system, steps must be taken to ensure that the input is valid. As a systems analyst, you must anticipate the types of errors users may make and design features into the system's interfaces to avoid, detect, and correct data entry mistakes. Several types of data errors are summarized in Table 11-8. In essence, data errors can occur from appending extra data onto a field, truncating characters off a field, transcripting the wrong characters into a field, or transposing one or more characters within a field. Systems designers have developed numerous tests and techniques for catching invalid data before saving or transmission, thus improving the likelihood that data will be valid (see Table 11-9 for a summary of these techniques). These tests

TABLE 11-8   Sources of Data Errors

Data Error	Description
Appending	Adding additional characters to a field
Truncating	Losing characters from a field
Transcripting	Entering invalid data into a field
Transposing	Reversing the sequence of one or more characters in a field

TABLE 11-9   Validation Tests and Techniques to Enhance the Validity of Data Input

Validation Test	Description
Class or Composition	Test to ensure that data are of proper type (e.g., all numeric, all alphabetic, all alphanumeric)
Combinations	Test to see if the value combinations of two or more data fields are appropriate or make sense (e.g., does the quantity sold make sense given the type of product?)
Expected Values	Test to see if data are what is expected (e.g., match with existing customer names, payment amount, etc.)
Missing Data	Test for existence of data items in all fields of a record (e.g., is there a quantity field on each line item of a customer order?)
Pictures/Templates	Test to ensure that data conform to a standard format (e.g., are hyphens in the right places for a student ID number?)
Range	Test to ensure data are within proper range of values (e.g., is a student's grade point average between 0 and 4.0?)
Reasonableness	Test to ensure data are reasonable for situation (e.g., pay rate for a specific type of employee)
Self-Checking Digits	Test where an extra digit is added to a numeric field in which its value is derived using a standard formula (see Figure 11-14)
Size	Test for too few or too many characters (e.g., is social security number exactly nine digits?)
Values	Test to make sure values come from set of standard values (e.g., two-letter state codes)

and techniques are often incorporated into both data entry screens and intercomputer data transfer programs.

Practical experience has also found that it is much easier to correct erroneous data before they are permanently stored in a system. Online systems can notify a user of input problems as data are being entered. When data are processed online as events occur, it is much less likely that data validity errors will occur and not be caught. In an online system, most problems can be easily identified and resolved before permanently saving data to a storage device, using many of the techniques described in Table 11-9. However, in systems where inputs are stored and entered (or transferred) in batch, the identification and notification of errors is more difficult. Batch processing systems can, however, reject invalid inputs and store them in a log file for later resolution.

Most of the tests and techniques shown in Table 11-9 are widely used and straightforward. Some of these tests can be handled by data management technologies, such as a database management system (DBMS), to ensure that they are applied for all data maintenance operations. If a DBMS cannot perform these tests, then you must design the tests into program modules. An example of one item that is a bit sophisticated, self-checking digits, is shown in Figure 11-14. The figure provides a description and an outline of how to apply the technique as well as a short example. The example shows how a check digit is added to a field before data entry or transfer. Once entered or transferred, the check digit algorithm is again applied to the field to "check" whether the check digit received obeys the calculation. If it does, it is likely (but not guaranteed because two different values could yield the same check digit) that no data transmission or entry error occurred. If the transferred value does not equal the calculated value, then some type of error occurred.

In addition to validating the data values entered into a system, controls must be established to verify that all input records are correctly entered and that they are processed only once. A common method used to enhance the validity of entering batches of data records is to create an audit trail of the entire sequence of data entry, processing, and storage. In such an audit trail, the actual sequence, count, time, source location, human operator, and so on are recorded into a separate transaction log in the event of a data input or processing error. If an error occurs, corrections

FIGURE 11-14
Using check digits to verify data correctness

Description	Techniques where extra digits are added to a field to assist in verifying its accuracy
Method	1. Multiply each digit of a numeric field by a weighting factor (e.g., 1, 2, 1, 2, _). 2. Sum the results of weighted digits. 3. Divide sum by modulus number (e.g., 10). 4. Subtract remainder of division from modulus number to determine check digit. 5. Append check digits to field.
Example	Assume a numeric part number of: 12473 1-2. Multiply each digit of part number by weighting factor from right to left and sum the results of weighted digits:

$$
\begin{array}{ccccc}
1 & 2 & 4 & 7 & 3 \\
\times 1 & \times 2 & \times 1 & \times 2 & \times 1 \\
\hline
1 \;+ & 4 \;+ & 4 \;+ & 14 \;+ & 3 \;=\; 26
\end{array}
$$

3. Divide sum by modulus number.

$$26/10 = 2 \text{ remainder } 6$$

4. Subtract remainder from modulus number to determine check digit.

$$\text{check digit} = 10 - 6 = 4$$

5. Append check digits to field.

Field value with appended check digit $= 124734$

can be made by reviewing the contents of the log. Detailed logs of data inputs are not only useful for resolving batch data entry errors and system audits, but they also serve as a powerful method for performing backup and recovery operations in the case of a catastrophic system failure. These types of file and database controls are discussed further in Hoffer et al. (2011).

Providing Feedback

When talking with a friend, you would be concerned if he or she did not provide you with feedback by nodding and replying to your questions and comments. Without feedback, you would be concerned that he or she was not listening, likely resulting in a less-than-satisfactory experience. Similarly, when designing system interfaces, providing appropriate feedback is an easy method for making a user's interaction more enjoyable; not providing feedback is a sure way to frustrate and confuse. There are three types of system feedback:

1. Status information
2. Prompting cues
3. Error or warning messages

Status Information Providing status information is a simple technique for keeping users informed of what is going on within a system. For example, relevant status information such as displaying the current customer name or time, placing appropriate titles on a menu or screen, or identifying the number of screens following the current one (e.g., Screen 1 of 3) all provide needed feedback to the user. Providing status information during processing operations is especially important if the operation takes longer than a second or two. For example, when opening a file you might display "Please wait while I open the file" or, when performing a large calculation, flash the message "Working…" to the user. Further, it is important to tell the user that besides working, the system has accepted the user's input and that the input was in the correct form. Sometimes it is important to give the user a chance to obtain more feedback. For example, a function key could toggle between showing a "Working…" message and giving more specific information as each intermediate step is accomplished. Providing status information will reassure users that nothing is wrong and will make them feel in command of the system, not vice versa.

Prompting Cues   A second feedback method is to display prompting cues. When prompting the user for information or action, it is useful to be specific in your request. For example, suppose a system prompted users with the following request:

READY FOR INPUT: _____

With such a prompt, the designer assumes that the user knows exactly what to enter. A better design would be specific in its request, possibly providing an example, default values, or formatting information. An improved prompting request might be as follows:

Enter the customer account number (123–456–7): ____-____-__

Errors and Warning Messages   A final method available to you for providing system feedback is using error and warning messages. Practical experience has found that a few simple guidelines can greatly improve their usefulness. First, messages should be specific and free of error codes and jargon. Additionally, messages should never scold the user and should attempt to guide the user toward a resolution. For example, a message might say, "No customer record found for that Customer ID. Please verify that digits were not transposed." Messages should be in user, not computer, terms. Hence, such terms as *end of file*, *disk I/O error*, or *write protected* may be too technical and not helpful for many users. Multiple messages can be useful so that a user can get more detailed explanations if wanted or needed. Also, error messages should appear in roughly the same format and placement each time so that they are recognized as error messages and not as some other information. Examples of good and bad messages are provided in Table 11-10. Using these guidelines, you will be able to provide useful feedback in your designs. A special type of feedback is answering help requests from users. This important topic is described next.

Providing Help

Designing how to provide help is one of the most important interface design issues you will face. When designing help, you need to put yourself in the user's place. When accessing help, the user likely does not know what to do next, does not understand what is being requested, or does not know how the requested information needs to be formatted. A user requesting help is much like a ship in distress, sending

TABLE 11-10   Examples of Poor and Improved Error Messages

Poor Error Messages	Improved Error Messages
ERROR 56 OPENING FILE	The file name you typed was not found. Press F2 to list valid file names.
WRONG CHOICE	Please enter an option from the menu.
DATA ENTRY ERROR	The prior entry contains a value outside the range of acceptable values. Press F9 for list of acceptable values.
FILE CREATION ERROR	The file name you entered already exists. Press F10 if you want to overwrite it. Press F2 if you want to save it to a new name.

TABLE 11-11   Guidelines for Designing Usable Help

Guideline	Explanation
Simplicity	Use short, simple wording, common spelling, and complete sentences. Give users only what they need to know, with ability to find additional information.
Organize	Use lists to break information into manageable pieces.
Show	Provide examples of proper use and the outcomes of such use.

an SOS. In Table 11-11, we provide our SOS guidelines for the design of system help: simplicity, organize, and show. Our first guideline, *simplicity,* suggests that help messages should be short and to the point, and use words that enable understanding. This leads to our second guideline, *organize,* which means that help messages should be written so that information can be easily absorbed by users. Practical experience has found that long paragraphs of text are often difficult for people to understand. A better design organizes lengthy information in a manner that is easier for users to digest through the use of bulleted and ordered lists. Finally, it is often useful to explicitly *show* users how to perform an operation and the outcome of procedural steps. Figures 11-15a and 11-15b show the contrasts between two help screen designs, one that employs our guidelines and one that does not.

Many commercially available systems provide extensive system help. For example, Table 11-12 lists the range of help available in a popular electronic spreadsheet. Many systems are also designed so that users can vary the level of detail provided. Help may be provided at the system level, screen or form level, and individual field level. The ability to provide field level help is often referred to as "context-sensitive" help. For some applications, providing context-sensitive help for all system options is a tremendous undertaking that is virtually a project in itself. If you do decide to design an extensive help system with many levels of detail, you must be sure that you know exactly what the user needs help with, or your efforts may confuse users more

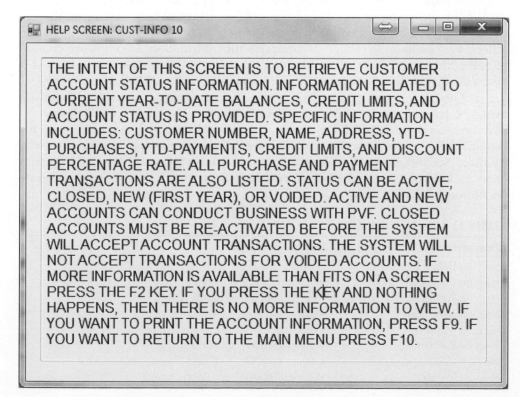

THE INTENT OF THIS SCREEN IS TO RETRIEVE CUSTOMER ACCOUNT STATUS INFORMATION. INFORMATION RELATED TO CURRENT YEAR-TO-DATE BALANCES, CREDIT LIMITS, AND ACCOUNT STATUS IS PROVIDED. SPECIFIC INFORMATION INCLUDES: CUSTOMER NUMBER, NAME, ADDRESS, YTD-PURCHASES, YTD-PAYMENTS, CREDIT LIMITS, AND DISCOUNT PERCENTAGE RATE. ALL PURCHASE AND PAYMENT TRANSACTIONS ARE ALSO LISTED. STATUS CAN BE ACTIVE, CLOSED, NEW (FIRST YEAR), OR VOIDED. ACTIVE AND NEW ACCOUNTS CAN CONDUCT BUSINESS WITH PVF. CLOSED ACCOUNTS MUST BE RE-ACTIVATED BEFORE THE SYSTEM WILL ACCEPT ACCOUNT TRANSACTIONS. THE SYSTEM WILL NOT ACCEPT TRANSACTIONS FOR VOIDED ACCOUNTS. IF MORE INFORMATION IS AVAILABLE THAN FITS ON A SCREEN PRESS THE F2 KEY. IF YOU PRESS THE KEY AND NOTHING HAPPENS, THEN THERE IS NO MORE INFORMATION TO VIEW. IF YOU WANT TO PRINT THE ACCOUNT INFORMATION, PRESS F9. IF YOU WANT TO RETURN TO THE MAIN MENU PRESS F10.

FIGURE 11-15
Contrasting help screens
(*Source:* Microsoft Corporation.)
(a) Poorly designed help display

FIGURE 11-15 (continued)
(b) Improved design for help display

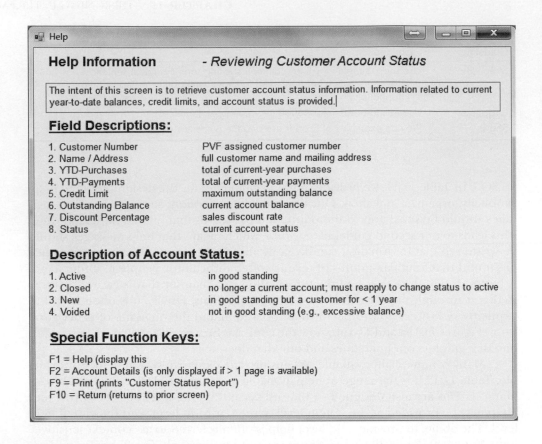

than help them. After leaving a help screen, users should always return to where they were prior to requesting help. If you follow these simple guidelines, you will likely design a highly usable help system.

As with the construction of menus, many programming environments provide powerful tools for designing system help. For example, Microsoft's HTML Help environment allows you to quickly construct hypertext-based help systems. In this environment, you use a text editor to construct help pages that can be easily linked to other pages containing related or more specific information. Linkages are created by embedding special characters into the text document that make words hypertext buttons—that is, direct linkages—to additional information. HTML Help transforms the text document into a hypertext document. For example, Figure 11-16 shows a hypertext-based help screen from the Firefox browser. Hypertext-based help systems have become the standard for most commercial applications. This has occurred for two primary reasons. First, standardizing system help across applications eases user training. Second, hypertext allows users to selectively access the level of help they need, making it easier to provide effective help for both novice and experienced users within the same system.

TABLE 11-12   Types of Help

Type of Help	Example of Question
Help on Help	How do I get help?
Help on Concepts	What is a customer record?
Help on Procedures	How do I update a record?
Help on Messages	What does "Invalid File Name" mean?
Help on Menus	What does "Graphics" mean?
Help on Function Keys	What does each Function key do?
Help on Commands	How do I use the "Cut" and "Paste" commands?
Help on Words	What do "Merge" and "Sort" mean?

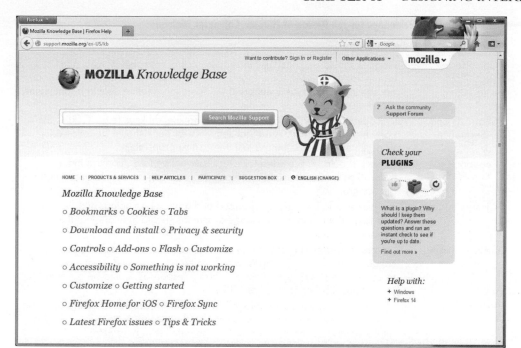

FIGURE 11-16
Hypertext-based help system from Firefox
(*Source:* Mozilla Firefox.)

DESIGNING DIALOGUES

The process of designing the overall sequences that users follow to interact with an information system is called *dialogue design*. A **dialogue** is the sequence in which information is displayed to and obtained from a user. As the designer, your role is to select the most appropriate interaction methods and devices (described earlier) and to define the conditions under which information is displayed to and obtained from users. The dialogue design process consists of three major steps:

1. Designing the dialogue sequence
2. Building a prototype
3. Assessing usability

> **Dialogue**
> The sequence of interaction between a user and a system.

A few general rules that should be followed when designing a dialogue are summarized in Table 11-13. For a dialogue to have high usability, it must be consistent in form, function, and style. All other rules regarding dialogue design are mitigated by the consistency guideline. For example, the effectiveness of how well errors are handled or feedback is provided will be significantly influenced by consistency in design. If the system does not consistently handle errors, the user will often be at a loss as to why certain things happen.

One example of these guidelines concerns removing data from a database or file (see the Reversal entry in Table 11-13). It is good practice to display the information that will be deleted before making a permanent change to the file. For example, if the customer service representative wanted to remove a customer from the database, the system should ask only for the customer ID in order to retrieve the correct customer account. Once found, and before allowing the confirmation of the deletion, the system should display the account information. For actions making permanent changes to system data files and when the action is not commonly performed, many system designers use the *double-confirmation* technique. With this technique, users must confirm their intention twice before being allowed to proceed.

Designing the Dialogue Sequence

Your first step in dialogue design is to define the sequence. In other words, you must first gain an understanding of how users might interact with the system. This means

TABLE 11-13 Guidelines for the Design of Human–Computer Dialogues

Guideline	Explanation
Consistency	Dialogues should be consistent in sequence of actions, keystrokes, and terminology (e.g., the same labels should be used for the same operations on all screens, and the location of the same information should be the same on all displays).
Shortcuts and Sequence	Allow advanced users to take shortcuts using special keys (e.g., CTRL-C to copy highlighted text). A natural sequence of steps should be followed (e.g., enter first name before last name, if appropriate).
Feedback	Feedback should be provided for every user action (e.g., confirm that a record has been added, rather than simply putting another blank form on the screen).
Closure	Dialogues should be logically grouped and have a beginning, middle, and end (e.g., the last in the sequence of screens should indicate that there are no more screens).
Error Handling	All errors should be detected and reported; suggestions on how to proceed should be made (e.g., suggest why such errors occur and what user can do to correct the error). Synonyms for certain responses should be accepted (e.g., accept either "t," "T," or "TRUE").
Reversal	Dialogues should, when possible, allow the user to reverse actions (e.g., undo a deletion); data should not be deleted without confirmation (e.g., display all the data for a record the user has indicated is to be deleted).
Control	Dialogues should make the user (especially an experienced user) feel in control of the system (e.g., provide a consistent response time at a pace acceptable to the user).
Ease	It should be a simple process for users to enter information and navigate between screens (e.g., provide means to move forward, backward, and to specific screens, such as first and last).

(*Source:* Based on Shneiderman et al., 2009.)

that you must have a clear understanding of user, task, technological, and environmental characteristics when designing dialogues. Suppose that the marketing manager at Pine Valley Furniture (PVF) wants sales and marketing personnel to be able to review the year-to-date transaction activity for any PVF customer. After talking with the manager, you both agree that a typical dialogue between a user and the Customer Information System for obtaining this information might proceed as follows:

1. Request to view individual customer information
2. Specify the customer of interest
3. Select the year-to-date transaction summary display
4. Review customer information
5. Leave system

As a designer, once you understand how a user wishes to use a system, you can then transform these activities into a formal dialogue specification.

A formal method for designing and representing dialogues is **dialogue diagramming**. Dialogue diagrams have only one symbol, a box with three sections; each box represents one display (which might be a full screen or a specific form or window) within a dialogue (see Figure 11-17). The three sections of the box are used as follows:

1. *Top:* Contains a unique display reference number used by other displays for referencing it.
2. *Middle:* Contains the name or description of the display.
3. *Bottom:* Contains display reference numbers that can be accessed from the current display.

All lines connecting the boxes within dialogue diagrams are assumed to be bidirectional and thus do not need arrowheads to indicate direction. This means that users are allowed to move forward and backward between adjacent displays. If

Dialogue diagramming
A formal method for designing and representing human–computer dialogues using box and line diagrams.

FIGURE 11-17
Sections of a dialogue diagramming box

Unique Reference
Number of Display — Top

Name or Description
of Display — Middle

Reference Numbers
of Return Displays — Bottom

you desire only unidirectional flows within a dialogue, arrowheads should be placed on one end of the line. Within a dialogue diagram, you can easily represent the sequencing of displays, the selection of one display over another, or the repeated use of a single display (e.g., a data entry display). These three concepts—sequence, selection, and iteration—are illustrated in Figure 11-18.

Continuing with our PVF example, Figure 11-19 shows a partial dialogue diagram for processing the marketing manager's request. In this diagram, the analyst placed the request to view year-to-date customer information within the context of the overall Customer Information System. The user must first gain access to the system through a log-on procedure (item 0). If log-on is successful, a main menu is displayed that has four items (item 1). Once the user selects the Individual Customer Information (item 2), control is transferred to the Select Customer display (item 2.1). After a customer is selected, the user is presented with an option to view customer information four different ways (item 2.1.1). Once the user views the customer's year-to-date transaction activity (item 2.1.1.2), the system will allow the user to back up to select a different customer (2.1), return to the main menu (1), or exit the system (see bottom of item 2.1.1.2).

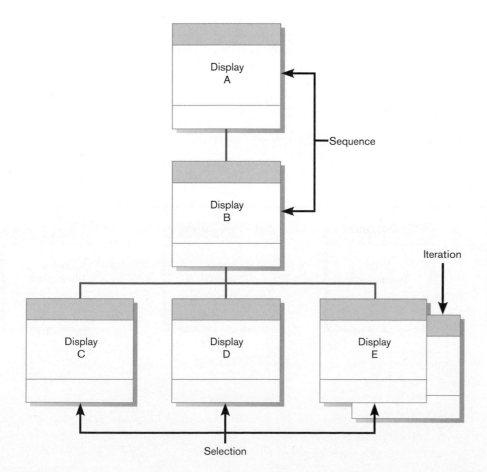

FIGURE 11-18
Dialogue diagram illustrating sequence, selection, and iteration

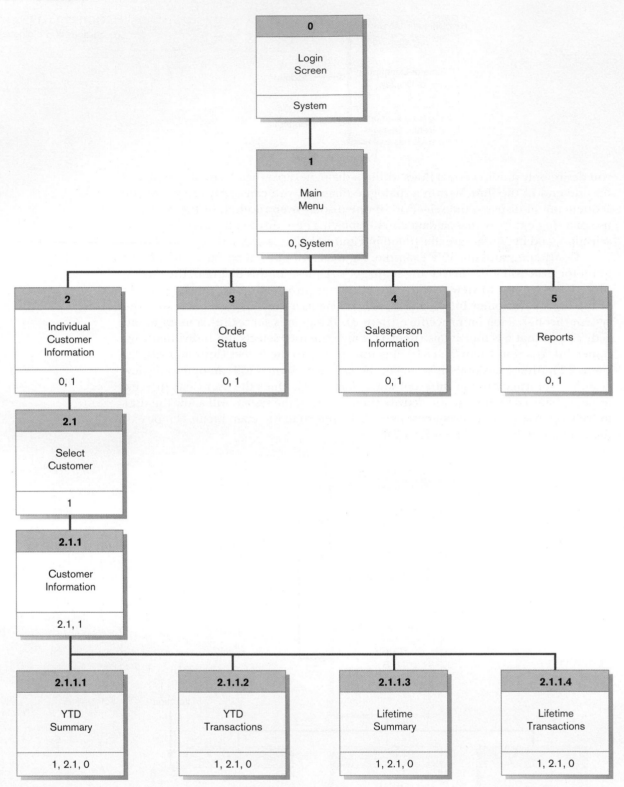

FIGURE 11-19
Dialogue diagram for the Customer
Information System (Pine Valley Furniture)

Building Prototypes and Assessing Usability

Building dialogue prototypes and assessing usability are often optional activities. Some systems may be very simple and straightforward; others may be more complex but are extensions to existing systems where dialogue and display standards have already been established. In either case, you may not be required to build prototypes and do a formal assessment. However, for many other systems, it is critical that you build prototype displays and then assess the dialogue; this can pay numerous dividends later in the systems development life cycle (e.g., it may be easier to implement a system or train users on a system they have already seen and used).

Building prototype displays is often a relatively easy activity if you use graphical development environments such as Microsoft's Visual Studio .NET. Some systems development environments include easy-to-use input and output (form, report, or window) design utilities. Tools called "prototypers" or "demo builders" allow you to quickly design displays and show how an interface will work within a full system. These demo systems allow users to enter data and move through displays as if using the actual system. Such activities are not only useful for you to show how an interface will look and feel, they are also useful for assessing usability and for performing user training long before actual systems are completed. In the next section, we extend our discussion of interface and dialogue design to consider issues specific to graphical user interface environments.

DESIGNING INTERFACES AND DIALOGUES IN GRAPHICAL ENVIRONMENTS

Graphical user interface (GUI) environments have become the de facto standard for human–computer interaction. Although all of the interface and dialogue design guidelines presented previously apply to designing GUIs, additional issues that are unique to these environments must be considered. Here, we briefly discuss some of these issues.

Graphical Interface Design Issues

When designing GUIs for an operating environment such as Microsoft Windows or the Apple Macintosh, numerous factors must be considered. Some factors are common to all GUI environments, whereas others are specific to a single environment. We will not, however, discuss the subtleties and details of any single environment. Instead, our discussion will focus on a few general truths that experienced designers mention as critical to the design of usable GUIs (Cooper et al., 2007; Krug, 2006; Nielsen and Loranger, 2006; Shneiderman et al., 2009). In most discussions of GUI programming, two rules repeatedly emerge as comprising the first step to becoming an effective GUI designer:

1. Become an expert user of the GUI environment.
2. Understand the available resources and how they can be used.

The first step should be an obvious one. The greatest strength of designing within a standard operating environment is that *standards* for the behavior of most system operations have already been defined. For example, how you cut and paste, set up your default printer, design menus, or assign commands to functions have been standardized both within and across applications. This allows experienced users of one GUI-based application to easily learn a new application. Thus, in order to design effective interfaces in such environments, you must first understand how other applications have been designed so that you will adopt the established standards for "look and feel." Failure to adopt the standard conventions in a given environment will result in a system that will likely frustrate and confuse users.

FIGURE 11-20
Highlighting GUI design standards
(*Source:* University of Arizona.)

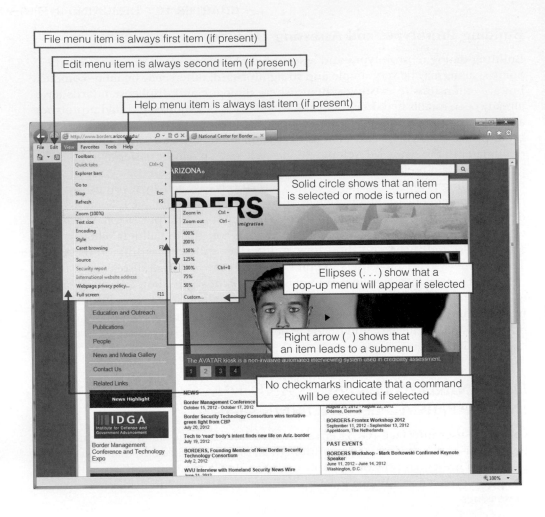

The second rule—gaining an understanding of the available resources and how they can be used—is a much larger undertaking. For example, within Windows you can use menus, forms, and boxes in many ways. In fact, the flexibility with which these resources *can be used* versus the established standards for how most designers *actually use* these resources makes design especially challenging. For example, you have the ability to design menus using all uppercase text, putting multiple words on the top line of the menu, and other nonstandard conventions. Yet the standards for menu design require that top-level menu items consist of one word and follow a specific ordering. Numerous other standards for menu design have also been established (see Figure 11-20 for illustrations of many of these standards). Failure to follow standard design conventions will likely prove very confusing to users.

In GUIs, information is requested by placing a window (or form) on the visual display screen. Like menu design, forms can also have numerous properties that can be mixed and matched (see Table 11-14). For example, properties about a form determine whether a form is resizable or movable after being opened. Because properties define how users can actually work with a form, the effective application of properties is fundamental to gaining usability. This means that, in addition to designing the layout of a form, you must also define the "personality" of the form with its characteristic properties. Fortunately, numerous GUI design tools have been developed that allow you to "visually" design forms and interactively engage properties. Interactive GUI design tools have greatly facilitated the design and construction process.

In addition to the issues related to interface design, the sequencing of displays turns out to be a bit more challenging in graphical environments. This topic is discussed next.

TABLE 11-14   Common Properties of Windows and Forms in a GUI Environment That Can Be Active or Inactive

Property	Explanation
Modality	Requires users to resolve the request for information before proceeding (e.g., need to cancel or save before closing a window)
Resizable	Allows users to resize a window or form (e.g., to make room to see other windows that are also on the screen)
Movable	Allows users to move a window or form (e.g., to allow another window to be seen)
Maximize	Allows users to expand a window or form to a full-size screen (e.g., to avoid distraction from other active windows or forms)
Minimize	Allows users to shrink a window or form to an icon (e.g., to get the window out of the way while working on other active windows)
System Menu	Allows a window or form to also have a system menu to directly access system-level functions (e.g., to save or copy data)

(*Source:* Based on Wagner, 1994.)

Dialogue Design Issues in a Graphical Environment

When designing a dialogue, your goal is to establish the sequence of displays (full screens or windows) that users will encounter when working with the system. Within many GUI environments, this process can be a bit more challenging due to the GUI's ability to suspend activities (without resolving a request for information or exiting the application altogether) and switch to another application or task. For example, within Microsoft Word, the spell-checker executes independently from the general word processor. This means that you can easily jump between the spell-checker and word processor without exiting either one. Conversely, when selecting the print operation, you must either initiate printing or abort the request before returning to the word processor. This is an example of the concept of "modality" described in Table 11-14. Thus, Windows-type environments allow you to create forms that either *require* the user to resolve a request before proceeding (print example) or *selectively choose* to resolve a request before proceeding (the spell-checker). Creating dialogues that allow the user to jump from application to application or from module to module within a given application requires that you carefully think through the design of dialogues.

One easy way to deal with the complexity of designing advanced GUIs is to require users to *always* resolve all requests for information before proceeding. For such designs, the dialogue diagramming technique is an adequate design tool. This, however, would make the system operate in a manner similar to a traditional non-GUI environment where the sequencing of displays is tightly controlled. The drawback to such an approach would be the failure to capitalize on the task-switching capabilities of these environments. Consequently, designing dialogues in environments where the sequence between displays cannot be predetermined offers significant challenges to the designer. Using tools such as dialogue diagramming helps analysts to better manage the complexity of designing graphical interfaces.

ELECTRONIC COMMERCE APPLICATION: DESIGNING INTERFACES AND DIALOGUES FOR PINE VALLEY FURNITURE'S WEBSTORE

Designing the human interface for an Internet-based electronic commerce application is a central and critical design activity. Because this is where a customer will interact with a company, much care must be put into its design. Like the process followed

PINE VALLEY FURNITURE

when designing the interface for other types of systems, a prototyping design process is most appropriate when designing the human interface for an Internet electronic commerce system. Although the techniques and technology for building the human interface for Internet sites is rapidly evolving, several general design guidelines have emerged. In this section, we examine some of these as they apply to the design of PVF's WebStore.

General Guidelines

Over the years, interaction standards have emerged for virtually all of the commonly used desktop computing environments such as Windows or Macintosh. However, some interface design experts believe that the growth of the Web has resulted in a big step backward for interface design. One problem, as discussed in Chapter 10, is that countless nonprofessional developers are designing commercial Web applications. In addition to this, there are four other important contributing factors (Johnson, 2007):

1. The Web's single "click-to-act" method of loading static hypertext documents (i.e., most buttons on the Web do not provide click feedback)
2. Limited capabilities of most Web browsers to support finely grained user interactivity
3. Limited agreed-upon standards for encoding Web content and control mechanisms
4. Lack of maturity of Web scripting and programming languages as well as limitations in commonly used Web GUI component libraries

In addition to these contributing factors, designers of Web interfaces and dialogues are often guilty of many design errors. Although not inclusive of all possible errors, Table 11-15 summarizes those errors that are particularly troublesome. Fortunately, there are numerous excellent sources on how to avoid these and other interface and design errors (Cooper et al., 2007; Flanders and Peters, 2002; Johnson, 2007; Nielsen, 1999, 2000; Seffah and Javahery, 2003; *www.useit.com; www.webpagesthatsuck.com*).

Designing Interfaces and Dialogues at Pine Valley Furniture

To establish design guidelines for the human–computer interface, Jim Woo and the PVF development team again reviewed many popular electronic commerce websites. The key feature they wanted to incorporate into the design was an interface with "menu-driven navigation with cookie crumbs." In order to ensure that all team members understood what was meant by this guideline, Jim organized a design briefing to explain how this feature would be incorporated into the WebStore interface.

Menu–Driven Navigation with Cookie Crumbs

After reviewing several sites, the team concluded that menus should stay in the exact same place throughout the entire site. They concluded that placing a menu in the same location on every page will help customers to more quickly become familiar with the site and therefore more rapidly navigate through it. Experienced Web developers know that the quicker customers can reach a specific destination at a site, the quicker they can purchase the product they are looking for or get the information they seek. Jim emphasized this point by stating, "These details may seem silly, but the second a user finds themselves 'lost' in our site, they're gone. One mouse click and they're no longer shopping at Pine Valley, but at one of our competitor's sites."

A second design feature, and one that is being used on many electronic commerce sites, is cookie crumbs. **Cookie crumbs** are "tabs" on a Web page that show a user where he or she is on a site and where he or she has been. These tabs are hypertext links that the user can use to quickly move backward in the site. For example,

Cookie crumbs
The technique of placing "tabs" on a Web page that show a user where he or she is on a site and where he or she has been.

TABLE 11-15 **Common Errors When Designing the Interface and Dialogues of Websites**

Error	Description
Opening New Browser Window	Avoid opening a new browser window when a user clicks on a link unless it is clearly marked that a new window will be opened; users may not see that a new window has been opened, which will complicate navigation, especially moving backward.
Breaking or Slowing Down the Back Button	Make sure users can use the back button to return to prior pages. Avoid opening new browser windows, using an immediate redirect where, when users click the back button, they are pushed forward to an undesired location, or prevent caching such that each click of the back button requires a new trip to the server.
Complex URLs	Avoid overly long and complex URLs because it makes it more difficult for users to understand where they are and can cause problems if users want to e-mail page locations to colleagues.
Orphan Pages	Avoid having pages with no "parent" that can be reached by using a back button; requires users to "hack" the end of the URL to get back to some other prior page.
Scrolling Navigation Pages	Avoid placing navigational links below where a page opens because many users may miss these important options that are below the opening window.
Lack of Navigation Support	Make sure your pages conform to users' expectations by providing commonly used icon links such as a site logo at the top or other major elements. Also place these elements on pages in a consistent manner.
Hidden Links	Make sure you leave a border around images that are links, don't change link colors from normal defaults, and avoid embedding links within long blocks of text.
Links That Don't Provide Enough Information	Avoid not turning off link-marking borders so that links clearly show which links users have clicked and which they have not. Make sure users know which links are internal anchor points versus external links, and indicate if a link brings up a separate browser window from those that do not. Finally, make sure link images and text provide enough information to users so that they understand the meaning of the link.
Buttons That Provide No Click Feedback	Avoid using image buttons that don't clearly change when being clicked; use Web GUI toolkit buttons, HTML form-submit buttons, or simple textual links.

suppose that a site is four levels deep, with the top level called "Entrance," the second called "Products," the third called "Options," and the fourth called "Order." As the user moves deeper into the site, a tab is displayed across the top of the page showing the user where he or she is, giving the user the ability to quickly jump backward one or more levels. In other words, when first entering the store, a tab will be displayed at the top (or some other standard place) of the screen with the word "Entrance." After moving down a level, two tabs will be displayed, "Entrance" and "Products." After selecting a product on the second level, a third level is displayed where a user can choose product options. When this level is displayed, a third tab is produced with the label "Options." Finally, if the customer decides to place an order and selects this option, a fourth-level screen is displayed and a fourth tab is displayed with the label "Order." In summary:

1. Level 1: Entrance
2. Level 2: Entrance → Products
3. Level 3: Entrance → Products → Options
4. Level 4: Entrance → Products → Options → Order

By using cookie crumbs, users know exactly how far they have wandered from "home." If each tab is a link, users can quickly jump back to a broader part of the store should they not find exactly what they are looking for. Cookie crumbs serve two important purposes. First, they allow users to navigate to a point previously visited and will ensure that they are not lost. Second, they clearly show users where they have been and how far they have gone from home.

SUMMARY

In this chapter, our focus was to acquaint you with the process of designing human–computer interfaces and dialogues. It is imperative that you understand the characteristics of various interaction methods (command language, menu, form, object, natural language) and devices (keyboard, mouse, joystick, trackball, touch screen, light pen, graphics tablet, voice). No single interaction style or device is the most appropriate in all instances: Each has its strengths and weaknesses. You must consider the characteristics of the intended users, the tasks being performed, and various technical and environmental factors when making design decisions.

The chapter also reviewed design guidelines for computer-based forms. You learned that most forms have a header, sequence or time-related information, instructions, a body, summary data, authorization, and comments. Users must be able to move the cursor position, edit data, exit with different consequences, and obtain help. Techniques for structuring and controlling data entry were presented along with guidelines for providing feedback, prompts, and error messages. A simple, well-organized help function that shows examples of proper use of the system should be provided. A variety of help types were reviewed.

Next, guidelines for designing human–computer dialogues were presented. These guidelines are consistency, allowing for shortcuts, providing feedback and closure on tasks, handling errors, allowing for operations reversal, giving the user a sense of control, and ease of navigation. We also discussed dialogue diagramming as a design tool. Assessing the usability of dialogues and procedures was also reviewed. Several interface and dialogue design issues were described within the context of designing GUIs. These included the need to follow standards to provide the capabilities of modality, resizing, moving, and maximizing and minimizing windows, and to offer a system menu choice. This discussion highlighted how concepts presented earlier in this chapter can be applied or augmented in these emerging environments. Finally, interface and dialogue design issues for Internet-based applications were discussed, and several common design errors were highlighted.

Our goal was to provide you with a foundation for building highly usable human–computer interfaces. As more and more development environments provide rapid prototyping tools for the design of interfaces and dialogues, many complying with common interface standards, the difficulty of designing usable interfaces will be reduced. However, you still need a solid understanding of the concepts presented in this chapter in order to succeed. Learning to use a computer system is like learning to use a parachute—if a person fails on the first try, odds are he or she won't try again (Blattner and Schultz, 1988). If this analogy is true, it is important that a user's first experience with a system be a positive one. By following the design guidelines outlined in this chapter, your chances of providing a positive first experience to users will be greatly enhanced.

KEY TERMS

1. Command language interaction
2. Cookie crumbs
3. Dialogue
4. Dialogue diagramming
5. Drop-down menu
6. Form interaction
7. Icon
8. Interface
9. Menu interaction
10. Natural language interaction
11. Object-based interaction
12. Pop-up menu

Match each of the key terms above with the definition that best fits it.

_____ A method by which users interact with information systems.

_____ A human–computer interaction method whereby users enter explicit statements into a system to invoke operations.

_____ A formal method for designing and representing human–computer dialogues using box and line diagrams.

_____ A menu-positioning method that places a menu near the current cursor position.

_____ A human–computer interaction method whereby a list of system options is provided and a specific command is invoked by user selection of a menu option.

_____ The technique of placing "tabs" on a Web page that show a user where he or she is on a site and where he or she has been.

_____ A menu-positioning method that places the access point of the menu near the top line of the display; when accessed, menus open by dropping down onto the display.

_____ A highly intuitive human–computer interaction method whereby data fields are formatted in a manner similar to paper-based forms.

_____ A human–computer interaction method whereby symbols are used to represent commands or functions.

_____ Graphical pictures that represent specific functions within a system.

_____ A human–computer interaction method whereby inputs to and outputs from a computer-based application are in a conventional speaking language such as English.

_____ The sequence of interaction between a user and a system.

REVIEW QUESTIONS

1. Contrast the following terms:
   a. Dialogue, interface
   b. Command language interaction, form interaction, menu interaction, natural language interaction, object-based interaction
   c. Drop-down menu, pop-up menu

2. Describe the process of designing interfaces and dialogues. What deliverables are produced from this process? Are these deliverables the same for all types of system projects? Why or why not?

3. Describe five methods of interacting with a system. Is one method better than all others? Why or why not?

4. Describe several input devices for interacting with a system. Is one device better than all others? Why or why not?

5. Describe the general guidelines for the design of menus. Can you think of any instances when it would be appropriate to violate these guidelines?

6. List and describe the general sections of a typical business form. Do computer-based and paper-based forms have the same components? Why or why not?

7. List and describe the functional capabilities needed in an interface for effective entry and navigation. Which capabilities are most important? Why? Will this be the same for all systems? Why or why not?

8. Describe the general guidelines for structuring data entry fields. Can you think of any instances when it would be appropriate to violate these guidelines?

9. Describe four types of data errors.

10. Describe the methods used to enhance the validity of data input.

11. Describe the types of system feedback. Is any form of feedback more important than the others? Why or why not?

12. Describe the general guidelines for designing usable help. Can you think of any instances when it would be appropriate to violate these guidelines?

13. What steps do you need to follow when designing a dialogue? Of the guidelines for designing a dialogue, which is most important? Why?

14. Describe the properties of windows and forms in a GUI environment. Which property do you feel is most important? Why?

15. List and describe the common interface and dialogue design errors found on Web sites.

PROBLEMS AND EXERCISES

1. Consider software applications that you regularly use that have menu interfaces, whether they be PC- or mainframe-based applications. Evaluate these applications in terms of the menu design guidelines outlined in Table 11-1.

2. Consider the design of a registration system for a hotel. Following the design specification items in Figure 11-2, briefly describe the relevant users, tasks, and displays involved in such a system.

3. Imagine the design of a system used to register students at a university. Discuss the user, task, system, and environmental characteristics (see Table 10-10) that should be considered when designing the interface for such a system.

4. For the three common methods of system interaction—command language, menus, and objects—recall a software package that you have used recently and list what you liked and disliked about each package with regard to the interface. What were the strengths and weaknesses of each interaction method for this particular program? Which type of interaction do you prefer for which circumstances? Which type do you believe will become most prevalent? Why?

5. Briefly describe several different business tasks that are good candidates for form-based interaction within an information system.

6. List the physical input devices described in this chapter that you have seen or used. For each device, briefly describe your experience and provide your personal evaluation. Do your personal evaluations parallel the evaluations provided in Tables 11-3 and 11-4?

7. Propose some specific settings where natural language interaction would be particularly useful and explain why.

8. Examine the help systems for some software applications that you use. Evaluate each using the general guidelines provided in Table 11-11.

9. Design one sample data entry screen for a hotel registration system using the data entry guidelines provided in this chapter (see Table 11-6). Support your design with arguments for each of the design choices you made.

10. Describe some typical dialogue scenarios between users and a hotel registration system. For hints, reread the section in this chapter that provides sample dialogue between users and the Customer Information System at PVF.

11. Represent the dialogues from the previous question through the use of dialogue diagrams.

12. List four contributing factors that have acted to impede the design of high-quality interfaces and dialogues on Internet-based applications.

13. Go to the Internet and find commercial websites that demonstrate each of the common errors listed in Table 11-15.

FIELD EXERCISES

1. Research the topic "natural language" at your library. Determine the status of applications available with natural language interaction. Forecast how long it will be before natural language capabilities are prevalent in information systems use.

2. Examine two PC-based GUIs (e.g., Microsoft's Windows and Macintosh). If you do not own these interfaces, you are likely to find them at your university or workplace, or at a computer retail store. You may want to supplement your hands-on evaluation with recent formal evaluations published on the Web. In what ways are these two interfaces similar and different? Are these interfaces intuitive? Why or why not? Is one more intuitive than the other? Why or why not? Which interface seems easier to learn? Why? What types of system requirements does each interface have? What are the costs of each interface? Which do you prefer? Why?

3. Interview a variety of people about the various ways they interact, in terms of inputs, with systems at their workplaces. What types of technologies and devices are used to deliver these inputs? Are the input methods and devices easy to use, and do they help these people complete their tasks effectively and efficiently? Why or why not? How could these input methods and devices be improved?

4. Interview systems analysts and programmers in an organization where GUIs are used. Describe the ways that these interfaces are developed and used. How does the use of such interfaces enhance or complicate the design of interfaces and dialogues?

REFERENCES

Apple Computer. 1993. *Macintosh Human Interface Guidelines.* Reading, MA: Addison-Wesley.

Blattner, M., and E. Schultz. 1988. "User Interface Tutorial." Presented at the 1988 Hawaii International Conference on System Sciences, Kona, Hawaii, January.

Cooper, A., R. Reimann, and D. Cronin, D. 2007. *About Face 3: The Essentials of Interaction Design.* New York: Wiley and Sons.

Flanders, V., and D. Peters. 2002. *Son of Web Pages That Suck: Learn Good Design by Looking at Bad Design.* Alameda, CA: Sybex Publishing.

Hoffer, J. A., V. Ramesh, and H. Topi. 2011. *Modern Database Management,* 10th ed. Upper Saddle River, NJ: Prentice Hall.

Johnson, J. 2007. *GUI Bloopers 2.0: Common User Interface Design Don'ts and Dos,* 2nd ed. New York: Morgan Kaufmann.

Krug, S. 2006. *Don't Make Me Think: A Common Sense Approach to Web Usability,* 2nd ed. Upper Saddle River, NJ: Prentice Hall.

Lazar, J. 2004. *User-Centered Web Development: Theory into Practice.* Sudbury, MA: Jones & Bartlett.

McCracken, D. D., R. J. Wolfe, and J. M. Spoll. 2004. *User-Centered Web Site Development: A Human–Computer Interaction Approach.* Upper Saddle River, NJ: Prentice Hall.

McKay, E. N. 1999. *Developing User Interfaces for Microsoft Windows.* Redmond, WA: Microsoft Press.

Nielsen, J. 1999. "User Interface Directions for the Web." Communications of the ACM 42 (1): 65–71.

Nielsen, J. 2000. Designing Web Usability: The Practice of Simplicity. Indianapolis, IN: New Riders Publishing.

Nielsen, J., and H. Loranger. 2006. *Prioritizing Web Usability.* Upper Saddle River, NJ: Prentice Hall.

Seffah, A., and H. Javahery. 2003. *Multiple User Interfaces: Cross-Platform Applications and Context-Aware Interfaces.* New York: John Wiley & Sons.

Shneiderman, B., C. Plaisant, M. Cohen, and S. Jacobs. 2009. *Designing the User Interface: Strategies for Effective Human-Computer Interaction,* 5th ed. Reading, MA. Addison-Wiley.

Sun Microsystems. 2001. *Java Look and Feel Design Guidelines.* Reading, MA: Addison-Wesley.

Te'eni, D., J. Carey, and P. Zhang. 2006. *Human–Computer Interaction: Developing Effective Organizational Information Systems.* New York: John Wiley & Sons.

Wagner, R. 1994. "A GUI Design Manifesto." *Paradox Informant* 5(6): 36–42.

PETRIE ELECTRONICS

Chapter 11: Designing Interfaces and Dialogues

Jim Watanabe, project director for the "No Customer Escapes" customer loyalty system for Petrie Electronics, walked into the conference room. Sally Fukuyama, from marketing, and Sanjay Agarwal, from IT, were already there. Also at the meeting was Sam Waterston, one of Petrie's key interface designers.

"Good morning," Jim said. "I'm glad everyone could be here today. I know you are all busy, but we need to make some real progress on the customer account area for 'No Customer Escapes.' We have just awarded the development of the system to XRA, and once all the documents are signed, they will be coming over to brief us on the implementation process and our role in it."

"I'm sorry," Sally said, "I don't understand. If we are licensing their system, what's left for us to do? Don't we just install the system and we're done?" Sally took a big gulp of coffee from her cup.

"I wish it was that easy," Jim said. "While it is true that we are licensing their system, there are many parts of it that we need to customize for our own particular needs. One obvious area where we need to customize is all of the human interfaces. We don't want the system to look generic to our loyal customers—we need to make it unique to Petrie."

"And we have to integrate the XRA system with our own operations," added Sanjay. "For example, we have to integrate our existing marketing and product databases with the XRA CRM (see PE Figure 7-2). That's just one piece of all the technical work we have to do."

"We've already done some preliminary work on system functionality and the conceptual database," Jim said. "I want to start working on interface issues now. That's why Sam is here. What we want to do today is start work on how the customer account area should look and operate. And Sally, the customer loyalty site is a great opportunity for marketing. We can advertise specials and other promotions to our best customers on this site. Maybe we could use it to show offers that are only good for members of our loyalty program."

"Oh yeah," Sally replied, "That's a great idea. How would that look?"

"I have ideas," said Sam. Using a drawing program on a tablet PC, he started to draw different zones that would be part of the interface. "Here at the top we would have a simple banner that says 'Petrie's' and the name of the program."

"It's not really going to be called 'No Customer Escapes,' is it?" asked Sally.

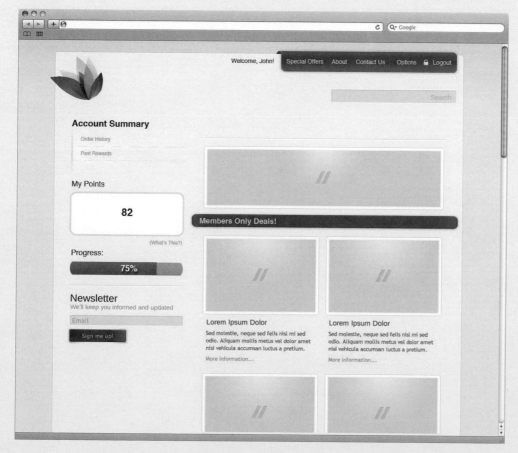

PE FIGURE 11-1
Preliminary design for the customer account area

(*Source*: Microsoft Corporation.)

"No, that's an internal name," replied Jim, "but I don't know what the real name will be yet."

"OK, so the real name of the program will go in the banner, after 'Petrie's.' Then on the left side, we'll have a sidebar that has overview information about the customer account, things like name and points balance," said Sam, drawing in a sidebar on the left of the screen. "There will also be links to more detailed information about the account, so the customer can see more details on past transactions and on his or her profile."

"So the rest of the screen is open. That would be a perfect place for marketing information," suggested Sally. "Would we want just one big window for marketing? Maybe we could divide it up into additional windows, so we could use one to focus on general promotions and one to advertise 'member only' promotions?"

"Yeah, we can do that," said Sam.

Just then Jim's phone beeped. Jim looked at it. Uh-oh, it was an urgent message from his boss, the director of IT. "Sorry, I need to take care of this immediately," he told the group. "Can you guys work on this some more and then send me some of the screen designs you come up with?"

Later that afternoon, after the crisis was over, Jim sat back down at his desk for the time in what seemed like a very long time. He glanced over his e-mail and noticed there was a message from Sam. Attached was a preliminary design for the customer account area. Jim opened it and looked it over (PE Figure 11-1). Hmmm, not bad, he thought. This is a good place for us to start.

Case Questions

1. Using the guidelines from this chapter and other sources, evaluate the usability of the page design depicted in PE Figure 11-1.
2. Chapter 11 encourages the design of a help system early in the design of the human interface. How would you incorporate help into the interface shown in PE Figure 11-1?
3. Describe how cookie crumbs could be used in this system. Are cookie crumbs a desirable navigation aid for this system? Why or why not?
4. The page design depicted in PE Figure 11-1 links to an Order History page. Sketch a similar layout for the Order History page, following guidelines from Chapter 11.
5. Describe how the use of template-based HTML might be leveraged in the design of the "No Customer Escapes" system.

Designing Distributed and Internet Systems

Learning Objectives

After studying this chapter, you should be able to:

- Define the key terms *client/server architecture, local area network (LAN), distributed database,* and *middleware.*

- Distinguish between file server and client/server environments and contrast how each is used in a LAN.

- Describe alternative designs for distributed systems and their trade-offs.

- Describe how standards shape the design of Internet-based systems.

- Describe options for ensuring Internet design consistency.

- Describe how site management issues can influence customer loyalty and trustworthiness as well as system security.

- Discuss issues related to managing online data, including context development, online transaction processing (OLTP), online analytical processing (OLAP), and data warehousing.

Introduction

Advances in computing technology and the rapid evolution of graphical user interfaces, networking, and the Internet are changing the way today's computing systems are being used to meet ever more demanding business needs. In many organizations, existing stand-alone personal computers are being linked together to form networks that support workgroup computing (this process is sometimes called *upsizing*). At the same time, other organizations (or even the same organization) are migrating mainframe applications to personal computers, workstations, and networks (this process is sometimes called downsizing) to take advantage of the greater cost-effectiveness of these environments. Organizations are also using the Internet and mobile services for delivering applications to internal and external customers. This Internet-based computing model focused on cloud-based, service-oriented architectures (see Chapter 2).

A variety of new opportunities and competitive pressures are driving the trend toward these technologies. Corporate restructuring—mergers, acquisitions, and consolidations—requires the connection or replacement of existing stand-alone applications. Similarly, corporate downsizing has caused individual managers to have a broader span of control, thus requiring access to a wider range of data, applications, and people.

443

Applications are being downsized from expensive mainframes to both public and private cloud-based architectures that are much more cost effective, scalable, and manageable. The explosion of electronic and mobile commerce is today's biggest driver for developing new types of systems. How distributed and Internet systems are designed can significantly influence system performance, usability, and maintenance.

In this chapter, we describe several technologies that are being used to upsize, downsize, and distribute information systems and data. These technologies are local area network (LAN)–based database management systems (DBMSs), client/server DBMSs, and the Internet. The capabilities and issues surrounding these technologies are the foundation for understanding how to migrate single-processor applications and designs into multiprocessor, distributed computing environments.

DESIGNING DISTRIBUTED AND INTERNET SYSTEMS

In this section, we briefly discuss the process and deliverables in designing distributed and Internet systems. Given the direction of organizational change and technological evolution, it is likely that most future systems development efforts will need to consider the issues surrounding the design of distributed systems.

The Process of Designing Distributed and Internet Systems

This is the last chapter in the text that deals with system design within the systems development life cycle (see Figure 12-1). In the previous chapters on system design, specific techniques for representing and refining data, screens, interfaces, and design specifications were presented. In this chapter, however, no specific techniques will be presented on how to represent the design of distributed and Internet systems because no generally accepted techniques exist. Alternatively, we will focus on increasing your awareness of common environments for deploying these systems and the issues you will confront surrounding their design and implementation. To distinguish between distributed and Internet-focused system design, we will use "distributed" to refer to LAN-based file server and client/server architectures.

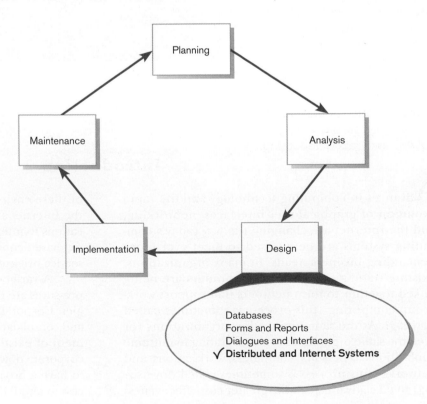

FIGURE 12-1
Systems development life cycle

Designing distributed and Internet systems is much like designing single-location systems. The primary difference is that, because such a system will be deployed over two or more locations, many more design issues must be considered that will influence the reliability, availability, and survivability of the system once it is implemented. Because distributed and Internet systems have more components than a single-location system—that is, more processors, networks, locations, data, and so on—there are more potential places for a failure to occur. Consequently, various strategies can be used when designing and implementing these systems to minimize points of failure.

Thus, when designing distributed and Internet systems, you will need to consider numerous trade-offs. To create effective designs, you need to understand the characteristics of the architectures commonly used to support these systems.

Deliverables and Outcomes

When designing distributed and Internet systems, the deliverable is a document that will consolidate the information that must be considered when implementing a system design. Figure 12-2 lists the types of information that must be considered when implementing such a system. In general, the information that must be considered is the site, processing needs, and data information for each location (or processor) in the distributed environment. Specifically, information related to physical distances between locations, counts and usage patterns by users, building and location infrastructure issues, personnel capabilities, data usage (use, create, update, or destroy), and local organizational processes must be described. Additionally, the pros and cons of various implementation solutions for each location should be reviewed. The collection of this information, in conjunction with the physical design information already developed, will provide the basis for implementing the information system in the distributed environment. Note, however, that our discussion assumes that any required network infrastructure is already in place. In other words, we focus only on those issues in which you will likely have a choice.

1. Description of Site (for each site)
   a. geographical information
   b. physical location
   c. infrastructure information
   d. personnel characteristics (education, technical skills, etc.)
   e. . . .

2. Description of Data Usage (for each site)
   a. data elements used
   b. data elements created
   c. data elements updated
   d. data elements deleted

3. Description of Business Process (for each site)
   a. list of processes
   b. description of processes

4. Contrasts of Alternative IS Architectures for Site, Data, and Process Needs (for each site)
   a. pros and cons of no technological support
   b. pros and cons of non-networked, local system
   c. pros and cons of various distributed configurations
   d. . . .

FIGURE 12-2
Outcomes and deliverables from designing distributed systems

DESIGNING DISTRIBUTED SYSTEMS

In this section, we focus on issues related to the design of distributed systems that use LAN-based file server or client/server architectures. The section begins by providing a high-level description of both architectures. This is followed by a brief description of various configurations for designing client/server systems.

Designing Systems for LANs

Personal computers and workstations can be used as stand-alone systems to support local applications. However, organizations have discovered that if data are valuable to one employee, they are probably also valuable to other employees in the same workgroup or in other workgroups. By interconnecting their computers, workers can exchange information electronically and can also share devices such as laser printers that may be too expensive to be utilized by only a single user.

A **local area network (LAN)** supports a network of personal computers, each with its own storage; each computer is able to share common devices and software attached to the LAN. Each PC and workstation on a LAN is typically within 100 feet of another, with a total network cable length of less than 1 mile. At least one computer (a microcomputer or larger) is designated as a file server, on which shared databases and applications are stored. The LAN modules of a DBMS, for example, add concurrent access controls, possibly extra security features, and query or transaction queuing management to support concurrent access from multiple users of a shared database.

File Servers   In a basic LAN environment (see Figure 12-3), all data manipulation occurs at the workstations from which data are requested. One or more file servers are attached to the LAN. A **file server** is a device that manages file operations and is shared by each client PC that is attached to the LAN. In a file server configuration, each file server acts as an additional hard disk for each client PC. For example, your PC might recognize a logical F: drive, which is actually a disk volume stored on a file

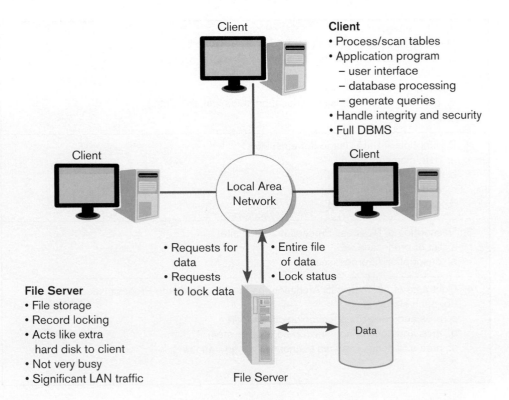

FIGURE 12-3
File server model

Client
Client
• Process/scan tables
• Application program
   – user interface
   – database processing
   – generate queries
• Handle integrity and security
• Full DBMS

Client

Local Area Network

Client

File Server
• File storage
• Record locking
• Acts like extra hard disk to client
• Not very busy
• Significant LAN traffic

• Requests for data
• Requests to lock data

• Entire file of data
• Lock status

File Server

Data

server on the LAN. Programs on your PC refer to files on this drive by the typical path specification, using this drive and any directories, as well as the file name.

When using a DBMS in a file server environment, each client PC is authorized to use the DBMS application program on that PC. Thus, there is one database on the file server and many concurrently running copies of the DBMS on each active PC client. The primary characteristic of a client-based LAN is that all data manipulation is performed at the client PC, not at the file server. The file server acts simply as a shared data storage device and is an extension of a typical PC. It also provides additional resources (e.g., disk drives, shared printing) and collaborative applications (e.g., e-mail) in addition to the shared data. Software at the file server only queues access requests; it is up to the application program at each client PC, working with the copy of the DBMS on that PC, to handle all data management functions. This means that, in an application that wants to view a single customer account record in a database stored on the server, the file containing all customer account records will be sent over the network to the PC. Once at the PC, the file will be searched to find the desired record. Additionally, data security checks and file and record locking are done at the client PCs in this environment, making multiple-user application development a relatively complex process.

Limitations of File Servers   There are three limitations when using file servers on LANs:

1. Excessive data movement
2. The need for a powerful client workstation
3. Decentralized data control

First, when using a file server architecture, considerable data movement is generated across the network. For example, when an application program running on a client PC in Pine Valley Furniture (PVF) wants to access the birch products, the whole product table is transferred to the client PC; then the table is scanned at the client to find the few desired records. Thus, the server does very little work, the client is busy with extensive data manipulation, and the network is transferring large blocks of data (see Figure 12-4). Consequently, a client-based LAN places a considerable burden on the client PC to carry out functions that have to be performed on all clients and creates a high network traffic load.

Second, because each client workstation must devote memory to a full version of the DBMS, there is less room on the client PC to rapidly manipulate data in high-speed random access memory (RAM). Often, data must be swapped between RAM and a relatively slower hard disk when processing a particularly large database. Further, because the client workstation does most of the work, each client must be rather powerful to provide a suitable response time. File server architectures also benefit from having a very fast hard disk and cache memory in both clients and the server to enhance their ability to transfer files to and from the network, RAM, and hard disk.

File Server Architecture

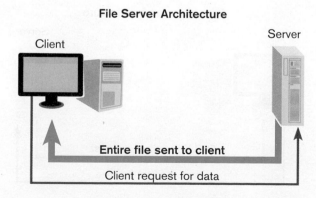

FIGURE 12-4
File servers transfer files when data are requested from a client

Third, and possibly most important, the DBMS copy in each workstation must manage the shared database integrity. In addition, each application program must recognize, for example, locks on data and take care to initiate the proper locks. A lock is necessary to stop users from accessing data that are in the process of being updated. Thus, application programmers must be rather sophisticated to understand the various subtle conditions that can arise in a multiple-user database environment. Programming in such an environment is complex because you have to program each application with the proper concurrency, recovery, and security controls.

Designing Systems for a Client/Server Architecture

Client/server architecture
A LAN-based computing environment in which a central database server or engine performs all database commands sent to it from client workstations, and application programs on each client concentrate on user interface functions.

An improvement in LAN-based systems is the **client/server architecture** in which application processing is divided (not necessarily evenly) between client and server. The client workstation is most often responsible for managing the user interface, including presenting data, and the database server is responsible for database storage and access, such as query processing. The typical client/server architecture is illustrated in Figure 12-5.

In the typical client/server architecture, all database recovery, security, and concurrent access management are centralized at the server; this is the responsibility of each user workstation in a simple LAN. These central DBMS functions are often called a **database engine** in a client/server environment (Hoffer et al., 2011). Some people refer to the central DBMS functions as the *back-end functions* and the client-based delivery of applications to users using PCs and workstations as the *front-end applications*. Further, in the client/server architecture, the server executes all requests for data so that only data that match the requested criteria are passed across the network to client stations. This is a significant advantage of client/server over simple file server designs. Because the server provides all shared database services, this leaves the **client** software to concentrate on user interface and data manipulation functions. The trade-off is that the server must be more powerful than the server in a file server environment.

Database engine
The (back-end) portion of the client/server database system running on the server that provides database processing and shared access functions.

Client
The (front-end) portion of the client/server database system that provides the user interface and data manipulation functions.

An application built using the client/server architecture is also different from a centralized database system on a mainframe. The primary difference is that each client is an *intelligent* part of the application processing system. In other words, the application program executed by a user is running on the client, not on the server. The application program handles all interactions with the user and local devices (printer, keyboard, screen, etc.). Thus, there is a division of duties between the server (database engine) and the client: The database engine handles all database access and control functions, and the client handles all user interaction and data manipulation functions. The client PC sends database commands to the database engine for processing. Alternatively, in a mainframe environment, all parts of the information system are managed and executed by the central computer.

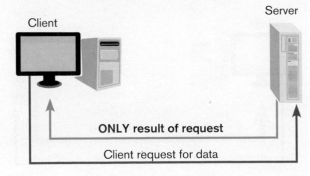

Client/Server Architecture

Client

Server

ONLY result of request

Client request for data

FIGURE 12-5
The required data after a request from a client

Another advantage of client/server architectures is the ability to decouple the client environment from the server environment. Clients can consist of multiple types (e.g., different computers, operating systems, and application programs), which means that the client can be running any application system that can generate the proper commands (often SQL) to request data from the server. For example, the application program might be written in Visual Basic, a report writer, a sophisticated screen painter, or any fourth-generation language that has an **application program interface (API)** for the database engine. The database engine might be DB2 on an IBM mainframe or midrange computer or MySQL, Sybase, or Oracle running on a variety of platforms. An API calls library routines that transparently route SQL commands from the front-end client application to the database server. An API might work with existing front-end software, such as a third-generation language or custom report generator, and it might include its own facilities for building applications. When APIs exist for several program development tools, then you have considerable independence to develop client applications in the most convenient front-end programming environment, yet still draw from a common server database. With some APIs, it is possible to access data from both the client and server in one database operation, as if the data were in one location managed by one DBMS (see Hoffer et al., 2011; Kroenke, 2010).

Application program interface (API)
Software building blocks that are used to ensure that common system capabilities, such as user interfaces and printing, as well as modules are standardized to facilitate data exchange between clients and servers.

Client/Server Advantages and Cautions   Several significant benefits can be realized by adopting a client/server architecture:

1. It allows companies to leverage the benefits of microcomputer technology. Today's workstations deliver impressive computing power at a fraction of the cost of a mainframe.
2. It allows most processing to be performed close to the source of processed data, thereby improving response times and reducing network traffic.
3. It facilitates the use of graphical user interfaces and visual presentation techniques commonly available for workstations.
4. It allows for and encourages the acceptance of open systems.

Many vendors of relational DBMSs and other LAN-based technologies have migrated or are attempting to migrate their products into the client/server environment. However, products that were not designed from the beginning under a client/server architecture may have problems adapting to this environment (see Levinson [2003] for a discussion of such issues). This is because new issues and new spins on old issues arise in this new environment. These issues and areas include compatibility of data types, query optimization, distributed databases, data administration of distributed data, CASE tool code generators, cross-operating system integration, and more. In general, the client/server environment has few tools for systems design and performance monitoring. As versions of different front- and back-end tools change, problems may arise with compatibility, until the API evolves, and these problems must be handled directly by the programmer, not by development tools.

Now that you have an understanding of the general differences between file server and client/server architectures, we will next discuss how data can be managed within a distributed environment. After discussing data management options, we will present several design alternatives for distributed systems. All LAN-based distributed system designs are implemented using some configuration of the general file server or client/server architectures and data management options.

Alternative Designs for Distributed Systems

A clear trend in systems design is to move away from central mainframe systems and stand-alone PC applications to some form of system that distributes data and processing across multiple computers. In this section, we briefly review the major differences between file servers and database servers. In the following section, we discuss the trade-offs among various ways to separate processing between clients and servers.

TABLE 12-1 Several Differences between File Server and Client/Server Architectures

Characteristic	File Server	Client/Server
Processing	Client only	Both client and server
Concurrent Data Access	Low—managed by each client	High—managed by server
Network Usage	Large file and data transfers	Efficient data transfers
Database Security and Integrity	Low—managed by each client	High—managed by server
Software Maintenance	Low—software changes just on server	Mixed—some new parts must be delivered to each client
Hardware and System Software Flexibility	Client and server decoupled and can be mixed	Need for greater coordination between client and server

Choosing between File Server and Client/Server Architectures Both file server and client/server architectures use personal computers and workstations and are interconnected using a LAN. Yet, a file server architecture is very different from a client/server architecture. A file server architecture supports only the distribution of data, whereas the client/server architecture supports both the distribution of data and the distribution of processing. This is an important distinction that has ramifications for systems design.

Table 12-1 summarizes some of the key differences between file server and client/server architectures. Specifically, a file server architecture is the simplest method for interconnecting PCs and workstations. In this architecture, the file server simply acts as a shared storage device for all clients on the network. Entire programs and databases must be transferred to each client when accessed. This means that a file server architecture is most appropriate for applications that are relatively small in size with little or no concurrent data access by multiple users.

Alternatively, a client/server architecture overcomes many of the limitations of the file server architecture because both the client and server share the processing workload of a task and transfer only needed information. Because of this, many organizations have migrated very large applications with extensive data sharing requirements to client/server environments. In fact, client/server computing has become the workhorse architecture for many organizations where multiple clients are likely to be working concurrently with the same data. Also, if the systems and databases are relatively large in size, the client/server architecture is preferred because of the client and server's ability to distribute the work and to transfer only needed information (e.g., only a single record if that is all that is needed rather than an entire database, as in a file server environment).

Advanced Forms of Client/Server Architectures Client/server architectures represent the way different application system functions can be distributed between client and server computers. These variations are based on the concept that there are three general components to any information system:

1. *Data management.* These functions manage all interaction between software and files and databases, including data retrieval/querying, updating, security, concurrency control, and recovery.
2. *Data presentation.* These functions manage just the interface between system users and the software, including the display and printing of forms and reports and possibly validating system inputs.
3. *Data analysis.* These functions transform inputs into outputs, including simple summarization to complex mathematical modeling such as regression analysis.

Different client/server architectures distribute, or partition, each of these functions to one or both of the client or server computers, and increasingly, into a third computer, referred to as the **application server**. In fact, it is becoming commonplace

Application server
A computing server where data analysis functions primarily reside.

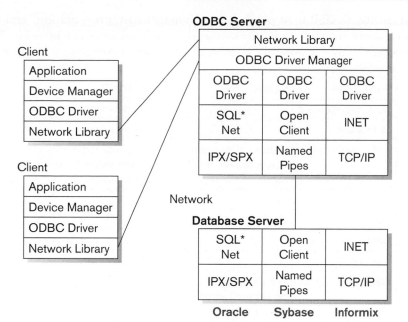

FIGURE 12-6
ODBC middleware environment

to use three or more distinct computers in many advanced client/server architectures (see Bass et al., 2003; Rosenfeld and Morville, 2002). This evolution in client/server computing has resulted in two new terms, **three-tiered client/server** and **middleware**, to represent this evolution. Three-tiered client/server architectures combine three logical and distinct applications—data management, presentation, and analysis—into a single information system application. Middleware brings together distinct hardware, software, and communication technologies in order to create a three-tiered client/server environment.

A typical use of middleware is shown in Figure 12-6. This figure shows how client applications can access databases on database servers. Open Database Connectivity (ODBC) is a Microsoft standard for database middleware. ODBC drivers, residing on both client and server computers, allow, for example, an Access query to retrieve data stored in an Oracle database as if they were in an Access database. By referring to a different ODBC driver, the Access database could reference data in an Informix database using the same query.

There are three primary reasons for creating three-tiered client/server architectures (Bass et al., 2003; Hoffer et al., 2011; Kroenke, 2010). First, applications can be partitioned in a way that best fits organizational computing needs. For example, in a traditional two-tiered client/server system, the application (data analysis) resides on the client, which would access information such as customer data from a database server. In a three-tiered architecture, data analysis can reside on a powerful application server, resulting in substantially faster response times for users. In addition, a three-tiered architecture provides greater flexibility by allowing the partitioning of applications in different ways for different users, thus optimizing performance.

A second advantage of the three-tiered architecture is that because most or all of the data analysis is contained in the application server, making global changes or customizing processes for individual users is relatively easy. This allows developers to easily create custom versions of large-scale systems without creating a completely separate system. Finally, because the data analysis is separate from the user interface, it is a lot easier to change one or both without having a major maintenance effort. By separating the data analysis from the data presentation (the user interface), either can be changed independently without affecting the other, greatly simplifying system maintenance. The combinations of these three benefits—application partitioning, easier customization, and easier maintenance—are driving many organizations to adopt this powerful alternative to standard client/server computing.

Three-tiered client/server
Advanced client/server architectures in which there are three logical and distinct applications—data management, presentation, and analysis—that are combined to create a single information system.

Middleware
A combination of hardware, software, and communication technologies that brings data management, presentation, and analysis together into a three-tiered client/server environment.

Given the flexibility of placing data management, presentation, and analysis on two or more separate machines, countless architectures are possible. In practice, however, only six possible configurations have emerged (see Figure 12-7).

FIGURE 12-7
Types of client/server architectures
(a) Distributed presentation

FUNCTION	CLIENT	SERVER
Data management		All data management
Data analysis		All data analysis
Data presentation	Data for presentation on server are reformatted for presentation to user	Data delivered to client using server presentation technologies

(b) Remote presentation

FUNCTION	CLIENT	SERVER
Data management		All data management
Data analysis		All data analysis
Data presentation	Data from analysis on server are formatted for presentation to user	

(c) Remote data management

FUNCTION	CLIENT	SERVER
Data management		All data management
Data analysis	Raw data from server are retrieved and analyzed	
Data presentation	All data presentation	

(d) Distributed function

FUNCTION	CLIENT	SERVER
Data management		All data management
Data analysis	Selective data from server retrieved and analyzed	Selective data from server retrieved and analyzed, then transmitted to client
Data presentation	All data presentation, from analyses on both server and client	

(e) Distributed database

FUNCTION	CLIENT	SERVER
Data management	Local data management	Shared management of data on server
Data analysis	Data retrieved from both client and server for analysis	
Data presentation	All data presentation	

(f) Distributed processing

FUNCTION	CLIENT	SERVER
Data management	Local data management	Shared management of data on server
Data analysis	Data retrieved from both client and server for analysis	Data retrieved from server for analysis, then sent to client for further analysis and presentation
Data presentation	All data presentation	

TABLE 12-2   Approaches to Designing Client/Server Architectures

Approach	Architecture Description
Distributed Presentation	This architecture is used to freshen up the delivery of existing server-based applications to distributed clients. Often the server is a mainframe, and the existing mainframe code is not changed. Here, technologies called "screen scrappers" work on the client machines to simply reformat mainframe screen data in a more appealing and easier-to-use interface.
Remote Presentation	This architecture places all data presentation functions on the client machine so that the client has total responsibility for formatting data. This architecture gives you greater flexibility compared to the distributed presentation style because the presentation on the client will not be constrained by having to be compatible with applications on the server.
Remote Data Management	This architecture places all software on the client except for the data management functions. This form is the closest to what we have called client/server earlier in the chapter. A mixed client environment may be more difficult to support than in the previous architectures because you must learn multiple analysis programming environments, not just those for presentation tools.
Distributed Function	This architecture splits analysis functions between the client and server, leaving all presentation on the client and all data management on the server. This is a very difficult environment in which to develop, test, and maintain software due to the potential for considerable coordination between analysis functions on both client and server.
Distributed Database	This architecture places all functionality on the client, except data storage and management that is divided between client and server. Although possible today, this is a very unstable architecture because it requires considerable compatibility and communication between software on the client and server, which may never have been meant to be compatible.
Distributed Processing	This architecture combines the best features of distributed function and distributed database by splitting both of these across client and server, with presentation functions under the exclusive responsibility of the client machine. This permits even greater flexibility because analysis functions and data both can be located wherever it makes the most sense.

Technology is available that will allow you to develop an application using any of these six architectures; however, automated development tools do not yet have equal code-generation capabilities for each. A brief description of each architecture is provided in Table 12-2.

As the designer of information systems, you have more choices available to you today than ever before. You must weigh the factors discussed previously and outlined in Table 12-2 to determine a distributed system design that will be most beneficial to the organization. As with other physical design decisions, organizational standards may limit your choices, and you will make such application design decisions in cooperation with other system professionals. You, however, are in the best position to understand user requirements and to estimate the ramifications of distributed system design decisions on response time and other factors for the user.

DESIGNING INTERNET SYSTEMS

The vast majority of new systems development in organizations focuses on Internet-based applications. The Internet can be used for delivering internal organizational systems, business-to-business systems, or business-to-consumer systems. The rapid migration to Internet-based systems should not be a surprise; it is motivated by the desire to take advantage of the global computing infrastructure of the Internet and the comprehensive set of tools and standards that has been developed. However, as with any other type of system, there are numerous choices that have to be made when designing an Internet application. The design choices you make can greatly influence the ease of development and the future maintainability of any system. In this section, we focus on several fundamental issues that must be considered when designing Internet-based systems.

Internet Design Fundamentals

Standards play a major role when designing Internet-based systems (Zeldman, 2006). In this section, we examine many fundamental and emerging building blocks of the Internet and how each of these pieces influences system design.

Standards Drive the Internet   Designing Internet-based systems is much simpler than designing traditional client/server systems because of the use of standards. For example, information is located throughout the Internet via the use of the standard **domain naming system (BIND)** (the "B" in BIND refers to Berkeley, California, where the standard was first developed at the University of California [BIND stands for Berkeley Internet Name Domain]; for more information see *www.isc.org/products/BIND/bind-history.html*). BIND provides the ability to locate information using common domain names that are translated into corresponding Internet Protocol (IP) addresses. For example, the domain name *www.wsu.edu* translates to 134.121.1.61.

Universal user access on a broad variety of clients is achieved through a standardized communication protocol: the **Hypertext Transfer Protocol (HTTP)**. HTTP is the agreed-upon format for exchanging information on the World Wide Web (see *www .w3.org/Protocols/* for more information). The HTTP protocol defines how messages are formatted and transmitted as well as how Web servers and browsers respond to commands. For example, when you enter a URL into your browser, an HTTP command is sent to the appropriate Web server requesting the desired Web page.

Beyond the naming standards of BIND and the transfer mechanism of HTTP, an Internet-based system has another advantage over other types of systems: the **Hypertext Markup Language (HTML)**. HTML is the standard language for representing content on the Web through the use of hundreds of command tags. Examples of command tags include those to bold text (. . .), to create tables (<table>. . .</table>), or to insert links onto a Web page (<A *href=http://www.wsu .edu/>* Washington State University Home Page,).

Having standardized naming (BIND), translating (HTTP), and formatting (HTML) enables designers to quickly craft systems because much of the complexity of the design and implementation is removed. These standards also free the designer from much of the worry of delivering applications over a broad range of computing devices and platforms. Together BIND, HTTP, and HTML provide a standard for designers when developing Internet-based applications. In fact, without these standards, the Internet as we know it would not be possible.

Separating Content and Display   As a method to build first-generation electronic commerce applications, HTML has been a tremendous success. It is a very easy language to learn, and there are countless tools to assist in authoring Web pages. In addition to its ease of use, it is also extremely tolerant of variations in usage, such as the use of both uppercase and lowercase letters for representing the same command or even the leniency to allow some commands to not *require* closing tags. However, HTML's simplicity also greatly limits its power (Castro, 2001). For example, most of HTML's tags are formatting oriented, making it difficult to distinguish data from formatting information. Additionally, because formatting information is inherently embedded into HTML documents, the migration of electronic commerce applications to emerging types of computing devices—such as wireless handheld computers—is much more difficult. Some new devices, for example, wireless Internet phones,

---

Domain naming system (BIND)

A method for translating Internet domain names into Internet Protocol (IP) addresses. BIND stands for Berkeley Internet Name Domain.

Hypertext Markup Language (HTML)

The standard language for representing content on the Web through the use of hundreds of command tags.

Hypertext Transfer Protocol (HTTP)

A communication protocol for exchanging information on the Internet.

cannot display HTML due to limited screen space and other limitations. To address this problem, new languages are being developed to separate content (data) from its display. For example, the Microsoft Visual Studio .NET development environment seamlessly builds applications that are optimized for a device's display and networking characteristics (Esposito, 2002; Wigley and Roxburgh, 2002).

A language specifically designed to separate content from display is **eXtensible Markup Language (XML)** (see Castro, 2001; *www.w3.org/XML/*). XML is a lot like HTML, with tags, attributes, and values, but it also allows designers to create customized tags, enabling the definition, transmission, validation, and interpretation of data between applications. For electronic commerce applications, XML is rapidly growing in popularity. Whereas HTML has a fixed set of tags, designers can create custom languages—called *vocabularies*—for any type of application in XML. This ability to create customized languages is at the root of the power of XML; however, this power comes at a price. Whereas HTML is very forgiving on the formatting of tags, XML is very strict. Additionally, as mentioned earlier, XML documents do not contain any formatting information. XML tags simply define what the data mean. For this reason, many believe that HTML will remain a popular tool for developing personal Web pages, but that XML will become the tool of choice for commercial Internet applications.

Future Evolution The infrastructure currently supporting HTML-based data exchange is the same infrastructure that will support the widespread use of XML and other emerging standards. As we move beyond desktop computers and standard Web browsers, the greatest driver of change and evolution of Internet standards will be the desire to support wireless mobile computing devices. Wireless mobile computing devices are often referred to as *thin-client technologies*. Thin clients such as network PCs, handheld computers, and wireless phones are being designed to operate as clients in Internet-based environments (see Figure 12-8). **Thin clients** are most appropriate for doing a minimal amount of client-side processing, essentially displaying information sent to the client from the server. Alternatively, a workstation that can provide significant amounts of client-side storage and processing is referred to as a fat client. Current PC workstations connected to the Internet can be thought of as fat clients. For desktop PC workstations, Internet browsers render content marked up in HTML documents. However, as thin clients gain in popularity, designing applications in XML will enable content to be displayed more effectively on any client device, regardless of the screen size or resolution (see Figure 12-9).

Regardless of whether the device is a smartphone, tablet, or a desktop PC, the use of standards will drive Internet-based system design. A well-designed system will isolate the content presentation from the business logic and data, allowing any Internet-capable device to become part of the overall distributed system. Techniques to ensure the consistency of the site's appearance for any type of device are discussed next.

eXtensible Markup Language (XML)
An Internet-authoring language that allows designers to create customized tags, enabling the definition, transmission, validation, and interpretation of data between applications.

Thin client
A client device designed so that most processing and data storage occur on the server.

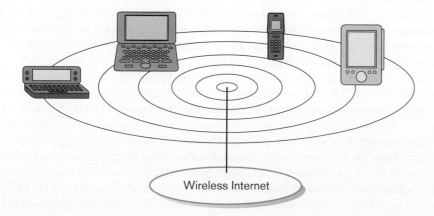

FIGURE 12-8
Thin clients used to access the Internet

Site Consistency

A consistent "look and feel" is fundamental to conveying the image that a site is
professionally designed. A site with high consistency is also much easier for users to
navigate, and it is much easier for users to anticipate the meaning of links. From a
practical standpoint, it is a poor design decision to not enforce a standard look and
feel to an entire site. Development and maintenance can become a nightmare when
implementing changes to colors, fonts, or other elements across thousands of Web
pages within a site. In this section, we discuss ways to help you enforce design consistency across an entire site and to simplify page maintenance.

Cascading Style Sheets   One of the biggest difficulties in developing a large-scale
website is maintaining consistency throughout the site with regard to color, background, fonts, and other page elements. Experienced website designers have found
that the use of **Cascading Style Sheets (CSSs)** can greatly simplify site maintenance and
also ensure that pages are consistent. CSSs are simply a set of style rules that tell a Web
browser how to present a document. Although there are various ways of linking these
style rules to HTML documents (see *www.w3.org/Style/CSS/*), the simplest method is
to use HTML's STYLE element; that is, to embed the style elements within each page.
To do this, style elements can be placed in the document HEAD element, which is
generally not displayed and contains information such as the page's title, keywords
that may be useful to search engines, and the style rules for the page. This method,
however, is not the best method for implementing CSSs because each page will have to
be changed if a single change is made to a site's style. The best way to implement CSSs
is through the use of linked style sheets. Using this method, through HTML's LINK
element, only a single file needs to be updated when changing style elements across an

Cascading Style Sheets (CSSs)
A set of style rules that tells a Web browser
how to present a document.

Sample Command:	
LINK HREF="style5.css" REL=StyleSheet TYPE="text/css" TITLE="Common Background Style" MEDIA="screen, print">	
Command Parameters:	
HREF="filename or URL"	Indicate the location of the linked object or document.
REL="relationship"	Specify the type of relationship between the document and linked object or document.
TITLE="object or document title"	Declare the title of the linked object or document.
TYPE="object to document type"	Declare the type of linked object or document.
MEDIA="type of media"	Declare the type of medium or media to which the style sheet will be applied (e.g., screen, print, projection, aural, braille, tty, tv, all).

FIGURE 12-10
Using HTML's link command for Cascading Style Sheets

entire site. The LINK element indicates some sort of a relationship between an HTML document and some other object or file (see Figure 12-10). CSSs are the most basic way to implement a standard style design within a website.

eXtensible Style Language A second and more sophisticated method for implementing standard page styles throughout a site is via the **eXtensible Style Language (XSL)**. XSL is a specification for separating style from content when generating HTML documents (see *www.w3.org/TR/xsl/* for more information). XSL allows designers to apply single style templates to multiple pages in a manner similar to that of Cascading Style Sheets. XSL allows designers to dictate how Web pages are displayed and whether the client device is a Web browser, handheld device, speech output, or some other media. In other words, XSL provides designers with specifications that allow XML content to be seamlessly displayed on various client devices. This method of separating style from content is a significant departure from how normal HTML content is displayed.

eXtensible Style Language (XSL)
A specification for separating style from content when generating HTML documents.

In practical terms, XSL allows designers to separate presentation logic from site content. This separation standardizes a site's "look and feel" without having to customize to the capabilities of individual devices. Given the rapid evolution of devices (e.g., desktop computers, mobile computing devices, and televisions), XSL is a powerful method to ensure that information is displayed in a consistent manner and uses the capabilities of the client device. XSL-based formatting consists of two parts:

1. *Methods for transforming* XML documents into a generic comprehensive form
2. *Methods for formatting* the generic comprehensive form into a device-specific form

In other words, XML content, queried from a remote data source, is formatted based on rules within an associated XSL style sheet (see Figure 12-11). This content is then translated to a device-specific format and displayed to the user. For example, if the user has made the request from a Web browser, the presentation layer will produce an HTML document. If the request has been made from a wireless mobile phone, the content will be delivered as a Wireless Markup Language (WML) document.

Other Site Consistency Issues In addition to using style sheets to enforce consistency in the design of a website, it is also important that designers adopt standards for defining page and link titles. Every Web page should have a title that helps users better navigate through the site (see Nielsen, 1996; Nielsen and Loranger, 2006). Page titles are used as the default description in bookmark lists, in history lists, and within listings retrieved from search engines. Given this variety of use, page titles need to be clear

FIGURE 12-11
Combining XML data with XSL style sheet to format content

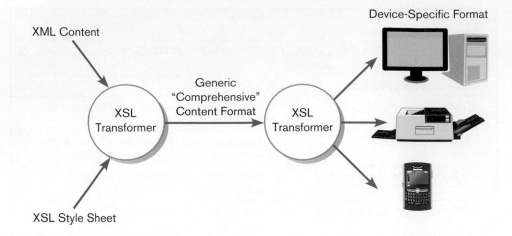

and meaningful. However, care should be taken so that overly long titles—those of more than 10 words—are not used. The selection of the actual words for the title is also extremely important. When selecting a title, two key issues should be considered:

1. *Use unique titles.* Give each page a unique identity that represents its purpose and assists user navigation. If all pages have the same title, a user will have difficulty returning to a prior page from the history list or distinguishing pages from the results of a search engine.
2. *Choose words carefully.* Given that titles are used for summarizing page content, choose words that assist users. Bookmark lists, history archives, and search engine results may be listed alphabetically; eliminate the use of articles such as *an*, *a,* or *the* at the beginning of the title. Likewise, don't use a title such as "Welcome to My Company," but instead use "My Company—Home Page." The latter title will not only be much more meaningful to users, it will also provide a standard model for defining the titles of subsequent titles within your site (e.g., "My Company—Feedback" or "My Company—Products").

A major problem on the Internet is that many users do not know where they are going when they follow a hyperlink. In addition to taking great care when choosing link names, most browsers support the ability to pop up a brief explanation of the link before a user selects it (see Figure 12-12). Using link titles, users will be less likely to follow the wrong link and have more success when navigating your site. Some guidelines for defining link titles are summarized in Table 12-3.

This section highlighted issues that focused on the need for design consistency within an Internet website. Experienced designers have learned that consistency not only makes the site easier for users, it also greatly simplifies site implementation and maintenance. It should be clear that careful attention to issues of design consistency will yield tremendous benefits.

Design Issues Related to Site Management

Maintenance is part of the ongoing management of a system. Many design issues will significantly influence the long-term successful operation of a system. Therefore, in this section we will discuss those issues that are particularly important when designing an Internet-based system.

Customer Loyalty and Trustworthiness In order for your website to become the preferred method for your customers to interact with you, they must feel that the site—and their data—are secure. There are many ways that the design of the site can convey trustworthiness to your users. Customers build trust from positive experiences gained while interacting with a site (McKnight et al., 2002). According to Web

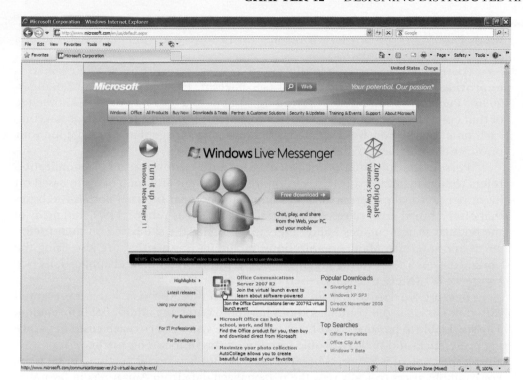

FIGURE 12-12
Using link titles to explain hyperlink

design guru Jakob Nielsen (1999), designers can convey trustworthiness in a website in the following ways:

1. *Design quality.* A professional appearance and clear navigation convey respect for customers and an implied promise of good service.
2. *Up-front disclosure.* Immediately inform users of all aspects of the customer relationship (e.g., shipping charges, data privacy policy); this conveys an open and honest relationship.
3. *Comprehensive, correct, and current content.* Up-to-date content conveys a commitment to provide users with the most up-to-date information.
4. *Connected to the rest of the Web.* Linking to outside sites is a sign of confidence and lends credibility; an isolated site feels like it may have something to hide.

In addition to these methods, protecting your customers' data will also be a significant factor for conveying trustworthiness. For example, many users are reluctant to disclose their e-mail address for fear of getting frequent unsolicited messages (spam). As a result, many users have learned to provide a secondary e-mail address—using services such as Hotmail or Yahoo! mail—when trust has not yet been established. Consequently, if you need to gather a customer's e-mail address or other information, you should disclose why this is being done and how this information

TABLE 12-3 Guidelines for Link Titles

Guideline	Description
Appropriate Information to Include	• Name of site (or subsite) link will lead to if different from current site • Details about the type of information found on the destination page • Warnings about the selection of the link (e.g., "password required")
Length	Usually less than 80 characters—shorter is better
Limit Usage	Only add titles to links that are not obvious

(*Source:* Based on Nielsen, 1998a.)

will be used in the future (e.g., information will be used only for order confirmation). Failure to consider how you convey trust to your customers may result in a system design that is not a success.

Another way to increase loyalty and to convey trustworthiness to customers is to provide useful, *personalized* content (see Nielsen, 1998b; Nielsen and Loranger, 2006). **Personalization** refers to providing content to a user based upon knowledge of that customer. For example, once you register and place orders on Amazon.com, each time you visit you are presented with a customized page that is based upon your prior purchase behavior.

Personalization should not be confused with customization. **Customization** refers to sites that allow a user to customize the content and look of a site based on his or her personal preferences. For example, the popular Internet portals—websites that offer a broad array of resources and services, such as Yahoo!, MSN, and many of the popular search engines—allow users to customize the site based on their preferences and interests. Many organizations, including universities, are also using the portal concept for delivering organization-specific information and applications (see Figure 12-13) (Nielsen, 2003; Nielsen and Loranger, 2006).

Because a personalized site *knows* you, each time you visit you are presented with new personalized content without having to enter any additional information. The site is able to personalize content because the system learns each customer's buying preferences and builds a profile based upon this history. This method for personalizing site content is a success because users do not have to do anything to set it up. For example, users typically view the personalized data from Amazon.com favorably. To personalize each customer's content, Amazon compares a user's prior purchases with the purchasing behavior of millions of other customers to reliably make purchase recommendations that may never have occurred to a customer. Amazon does a nice job of not making personalization recommendations too obtrusive so that if the system makes a bad guess at what the user might be interested in, the user isn't annoyed by having the site trying to be smarter than it actually is. For example, many users visit Amazon.com to purchase books as gifts for friends; using these data to personalize the site may impede the user's experience when shopping for personal items.

Web Pages Must Live Forever   For commercial Internet sites, your pages must live forever. There are four primary reasons why professional developers have come to this conclusion (Nielsen, 1998c):

Personalization
Providing Internet content to a user based upon knowledge of that customer.

Customization
Internet sites that allow users to customize the content and look of the site based on their personal preferences.

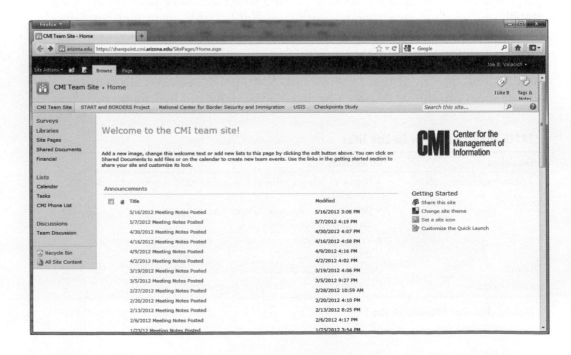

FIGURE 12-13
Organizations are providing customized portals for users

1. *Customer bookmarks.*   Because customers may bookmark any page on your site, you cannot ever remove a page without running the risk of losing customers who would not find a working link if they encountered a dead link.
2. *Links from other sites.*   Like your customers who bookmark pages, other sites may link directly to pages within your site; removing a page may result in losing customer referrals.
3. *Search engine referrals.*   Because search engines are often slow to update their databases, this is another source for old and dead pages.
4. *Old content adds value.*   In addition to these practical issues, many users may actually find value from old content. Old content can remain useful to users because of historic interest, old product support, or background information for recent events. Additionally, the cost of keeping old content is relatively small. However, it is important to maintain old content so that links do not die and that obsolete or misleading information is corrected or removed. Finally, make sure that you explicitly date old content, provide disclaimers that point out what no longer applies or is accurate, and provide forward-pointing links to current pages.

You should not conclude from this discussion that Web content cannot change and evolve. However, you should now understand that the links themselves cannot die. In other words, when users bookmark a page and return to your site, this link should return something useful to the user; otherwise, you run the risk of losing the customer. With a small amount of maintenance on your site's old content, you will provide a valuable resource to your customers. It should be obvious that customers who visit your site infrequently should easily be able find what they are looking for; otherwise, they will become frustrated, leave, and not come back.

System Security   A paradox lies in the fact that, within a distributed system, security and ease of use are in conflict with each other. A secure system is often much less "user-friendly," whereas an easy-to-use system is often less secure. When designing an Internet-based system, successful sites strike an appropriate balance between security and ease of use. For example, many sites that require a password for site entry provide the functionality of "remember my password." This feature will make a user's experience at a site much more convenient and smooth, but it also results in a less secure environment. By remembering the password, anyone utilizing the user's computer potentially has access to the initial user's account and personal information.

In addition, if you must require customers to register to use your site and gain access via passwords, experienced designers have learned that it is best to delay customer registration and not to require registration to gain access to the top levels of the site. If you ask for registration too early, before you have demonstrated value to a new customer, you run the risk of turning away the customer (Nielsen, 1997; Nielsen and Loranger, 2006). Once a customer chooses to register on your site, make sure that the process is as simple as possible. Also, if possible, store user information in client- or server-side cookies rather than requiring users to reenter information each time they visit your site. Of course, if your site requires high security (e.g., a stock trading site), you may want to require users to enter an explicit password at each visit. Security is clearly a double-edged sword. Too much and you might turn customers away; too little and you run the risk of losing customers because they do not trust the security of the site. Careful system design is needed to achieve the right balance between security and ease of use.

Managing Online Data

Modern organizations are said to be drowning in data but starving for information. Despite the mixed metaphor, this statement seems to portray quite accurately the situation in many organizations. The advent of Internet-based electronic commerce has resulted in the collection of an enormous amount of customer and transactional data. How these data are collected, stored, and manipulated is a significant factor influencing the success of a commercial Internet website. In this section, we discuss system design issues for managing online data.

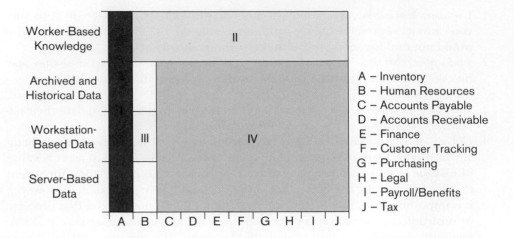

Context development
A method that helps analysts to better understand how a system fits within existing business activities and data.

Integration depth
A measurement of how far into the existing technology infrastructure a system penetrates.

Organizational breadth
A measurement that tracks the core business functions affected by a system.

Context Development   Gaining an understanding of how a new system fits within the context of an organization's existing application portfolio is a fundamental part of effectively designing and managing system data. This understanding is necessary to ensure that data can be effectively collected, stored, and managed. **Context development** is a method that helps analysts gain an understanding of how a system fits within existing business activities and data. Two metrics—integration depth and organizational breadth—can be used to define a system's context. **Integration depth** reflects how far into the existing technology infrastructure a system penetrates. A "deep" integration both retrieves data from and sends data directly into existing systems. A "shallow" system will have minimal real-time coexistence with existing data sources. **Organizational breadth** tracks the core business functions that are affected by a system. A "wide" breadth reflects a situation in which many distinct organizational areas have some type of interaction with the system; a "narrow" breadth reflects a situation in which very few departments use or access the system.

Simply put, the context of an Internet-based system is an assessment of the integration depth and the organizational breadth of a system and can be represented using a graph with an x-axis and a y-axis. The x-axis measures the business functions affected by the system, and the y-axis measures how far a system penetrates into the existing technology infrastructure. For example, Figure 12-14 is a context graph for an organization, like PVF, that has a broad range of organizational functions and a variety of information systems. Table 12-4 provides a relative comparison of the systems shown in Figure 12-14, each with varying depth and breadth. With this understanding, analysts are able to better understand how a new system will fit within an organization's existing application portfolio.

Online Transaction Processing   Figure 12-15 shows the level-0 data flow diagram (DVD) from PVF's WebStore application. Each of the processes defined in the DFD can be viewed as an autonomous transaction. For example, Process 5.0, Add/Modify Account Profile, shows one input from the Customer (Customer Information) and two outputs: one output to the Customer (Customer Information/ID) and

TABLE 12-4   Comparison of Relative System Context

Breadth	Context
Narrow	Inventory control
Wide	Knowledge management across all business functions
Narrow	System that gathers information residing on employees' workstations in HR department
Wide	Enterprise Resource Planning (ERP) system

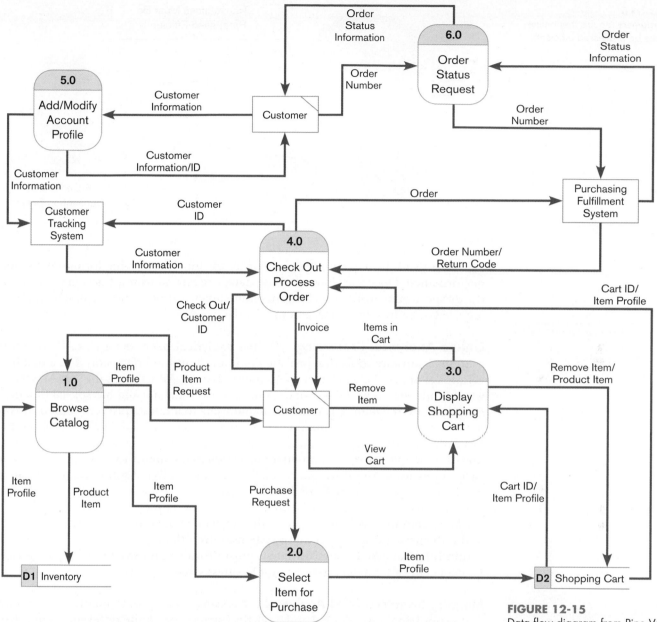

FIGURE 12-15
Data flow diagram from Pine Valley Furniture's WebStore

one output to the Customer Tracking System (Customer Information). All of these operations are transactions. **Online transaction processing (OLTP)** refers to immediate automated responses to the requests of users (Hoffer et al., 2011). OLTP systems are designed to specifically handle multiple concurrent transactions. Typically, these transactions have a fixed number of inputs and a specified output such as those represented in the DFD in Figure 12-15. Common transactions include receiving user information, processing orders, and generating sales receipts.

Consequently, OLTP is a big part of interactive electronic commerce applications on the Internet. Because customers can be located virtually anywhere in the world, it is critical that transactions be processed efficiently (see Figure 12-16). The speed at which DBMSs can process transactions is therefore an important design decision when building Internet systems. In addition to the technology chosen to process the transactions, how the data are organized is also a major factor in determining system performance. Although the database operations behind most transactions are relatively simple, designers often spend considerable time making

Online transaction processing (OLTP)
The immediate automated responses to the requests of users.

FIGURE 12-16
Global customers require that online transactions be processed efficiently

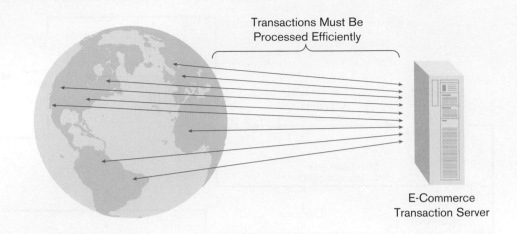

Transactions Must Be
Processed Efficiently

E-Commerce
Transaction Server

adjustments to the database design in order to "tune" processing for optimal system performance. Once they have all these data, organizations must design ways to gain the greatest value from its collection; online analytical processing is one method being used to analyze these vast amounts of data.

Online analytical processing (OLAP)
The use of graphical software tools that provide complex analysis of data stored in a database.

Online Analytical Processing   **Online analytical processing (OLAP)** refers to graphical software tools that provide complex analysis of data stored in a database (Hoffer et al., 2011). The chief component of an OLAP system is the "OLAP server," which understands how data are organized in the database and has special functions for analyzing the data. OLAP tools enable users to analyze different dimensions of data, beyond the data summaries and aggregations of normal database queries. For example, OLAP can provide time series and trend analysis views of data, data drill-downs to deeper levels of consolidation, as well as the ability to answer "what if" and "why" questions. An OLAP query for PVF might be: "What would be the effect on wholesale furniture costs if wood prices increased by 10 percent and transportation costs decreased by 5 percent?" Managers use the complex query capabilities of an OLAP system to answer questions within executive information, decision support, and enterprise resource planning systems. Given the high volume of transactions within Internet-based systems, analysts must provide extensive OLAP capabilities to managers in order to gain the greatest business value.

Merging Transaction and Analytical Processing   The requirements for designing and supporting transactional and analytical systems are quite different. In a distributed online environment, performing real-time analytical processing will diminish the performance of transaction processing. For example, complex analytical queries from an OLAP system require the locking of data resources for extended periods of execution time, whereas transactional events—data insertions and simple queries—are fast and can often occur simultaneously. Thus, a well-tuned and responsive transaction system may have uneven performance for customers while analytical processing occurs. As a result, many organizations replicate all transactions on a second server, so that analytical processing does not slow customer transaction processing performance. This replication typically occurs in batches during off-peak hours when site traffic volumes are at a minimum.

Operational systems
Systems that are used to interact with customers and run a business in real time.

Informational systems
Systems designed to support decision making based on stable point-in-time or historical data.

The systems that are used to interact with customers and run a business in real time are called the **operational systems**. Examples of operational systems are sales order processing and reservation systems. The systems designed to support decision making based on stable point-in-time or historical data are called **informational systems**. The key differences between operational and informational systems are shown in Table 12-5. Increasingly, data from informational systems are being consolidated with other organizational data into a comprehensive

TABLE 12-5   Comparison of Operational and Informational Systems

Characteristic	Operational System	Informational System
Primary Purpose	Run the business on a current basis	Support managerial decision making
Type of Data	Current representation of state of the business	Historical or point-in-time (snapshot)
Primary Users	Online customers, clerks, salespersons, administrators	Managers, business analysts, customers (checking status, history)
Scope of Usage	Narrow vs. simple updates and queries	Broad vs. complex queries and analysis
Design Goal	Performance	Ease of access and use

data warehouse from which OLAP tools can be used to extract the greatest and broadest understanding from the data.

Data Warehousing

A **data warehouse** is a subject-oriented, integrated, time-variant, nonvolatile collection of data used in support of management decision making (Hoffer et al., 2011; Inmon and Hackathorn, 1994; Kroenke, 2010). The meaning of each key term in this definition is presented in the following list:

> **Data warehouse**
> A subject-oriented, integrated, time-variant, nonvolatile collection of data used in support of management decision making.

1. *Subject-oriented.* A data warehouse is organized around the key subjects (or high-level entities) of the enterprise, such as customers, patients, students, or products.
2. *Integrated.* Data housed in the data warehouse are defined using consistent naming conventions, formats, encoding structures, and related characteristics collected from many operational systems within the organization and from external data sources.
3. *Time-variant.* Data in the data warehouse have a time dimension so that they may be used as historical records pertaining to the business.
4. *Nonvolatile.* Data in the data warehouse are loaded and refreshed from operational systems, but the data cannot be updated by end users.

In other words, data warehouses contain a broad range of data that, if analyzed appropriately, can provide a broad and coherent picture of business conditions at a single point in time. The basic architectures used for data warehouses are either a generic, two-level architecture or a more sophisticated, three-level architecture. Of course, there are more complex architectures beyond these two basic models; however, this topic is beyond the scope of our discussion.

The generic two-level architecture is shown in Figure 12-17. Building a data warehouse using this architecture requires four basic steps:

1. Data are extracted from the various source systems' files and databases. A large organization may have dozens or hundreds of such files and databases.
2. The data from the various source systems are transformed and integrated before being loaded into the data warehouse.
3. The data warehouse is a read-only database organized for decision support. It contains both detailed and summary data.
4. Users access the data warehouse by means of a variety of query languages and analytical tools.

The two-level architecture represents the earliest model, but it is still widely used today. The two-level architecture works well in small- to medium-sized companies with a limited number of hardware and software platforms and a relatively homogeneous computing environment (Galemmo et al., 2003). However, for larger companies with a large number of data sources and a heterogeneous computing environment, this approach leads to problems in maintaining data quality and managing the data extraction processes (Devlin, 1997; Hoffer et al., 2011; Kroenke,

Operational
Environment

Decision Support
Environment

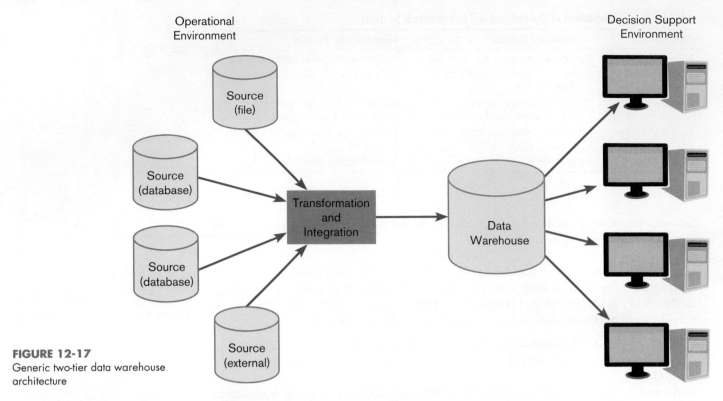

FIGURE 12-17
Generic two-tier data warehouse
architecture

2010). These problems, together with the trend toward distributed computing, have led to the expanded, three-level architecture shown in Figure 12-18. The three-level architecture has the following components:

1. Operational systems and data
2. An enterprise data warehouse
3. Data marts

The first major difference between the two-level and the three-level architectures is the enterprise data warehouse. An **enterprise data warehouse (EDW)** is a centralized, integrated data repository that serves as the control point of all data made available to end users. This single data source drives decision support applications throughout the entire organization. The EDW's purpose is twofold:

1. A centralized control point ensures the quality and integrity of data before they are made available to end users.
2. The single data source provides an accurate, consolidated historical record of business for time-sensitive data.

Although the EDW is the single source of all data for decision support, it is not typically accessed directly by end users. For most large organizations, the EDW is simply too large and too complex for users to navigate for most decision support applications. This leads to the second major difference between the two- and three-level architectures. In a two-level architecture, users access data directly from the data warehouse via decision support tools. However, in a three-level architecture, users access data that have been derived from the EDW that are stored in data marts.

A **data mart** is a data warehouse that is limited in scope. A data mart may be a physically separate subset of data extracted from the EDW, or it may be a customized, logical view of the data in the EDW of relevance to a class of users. Data marts contain selected information from the EDW such that each data mart is customized for the decision support applications of a particular end-user group. For example, an organization may have several data marts that are customized for a particular type of user, such as a marketing data mart or a finance data mart.

Enterprise data warehouse (EDW)
A centralized, integrated data warehouse that is the control point and single source of all data made available to end users for decision support applications throughout the entire organization.

Data mart
A data warehouse that is limited in scope; its data are obtained by selecting and (where appropriate) summarizing data from the enterprise data warehouse.

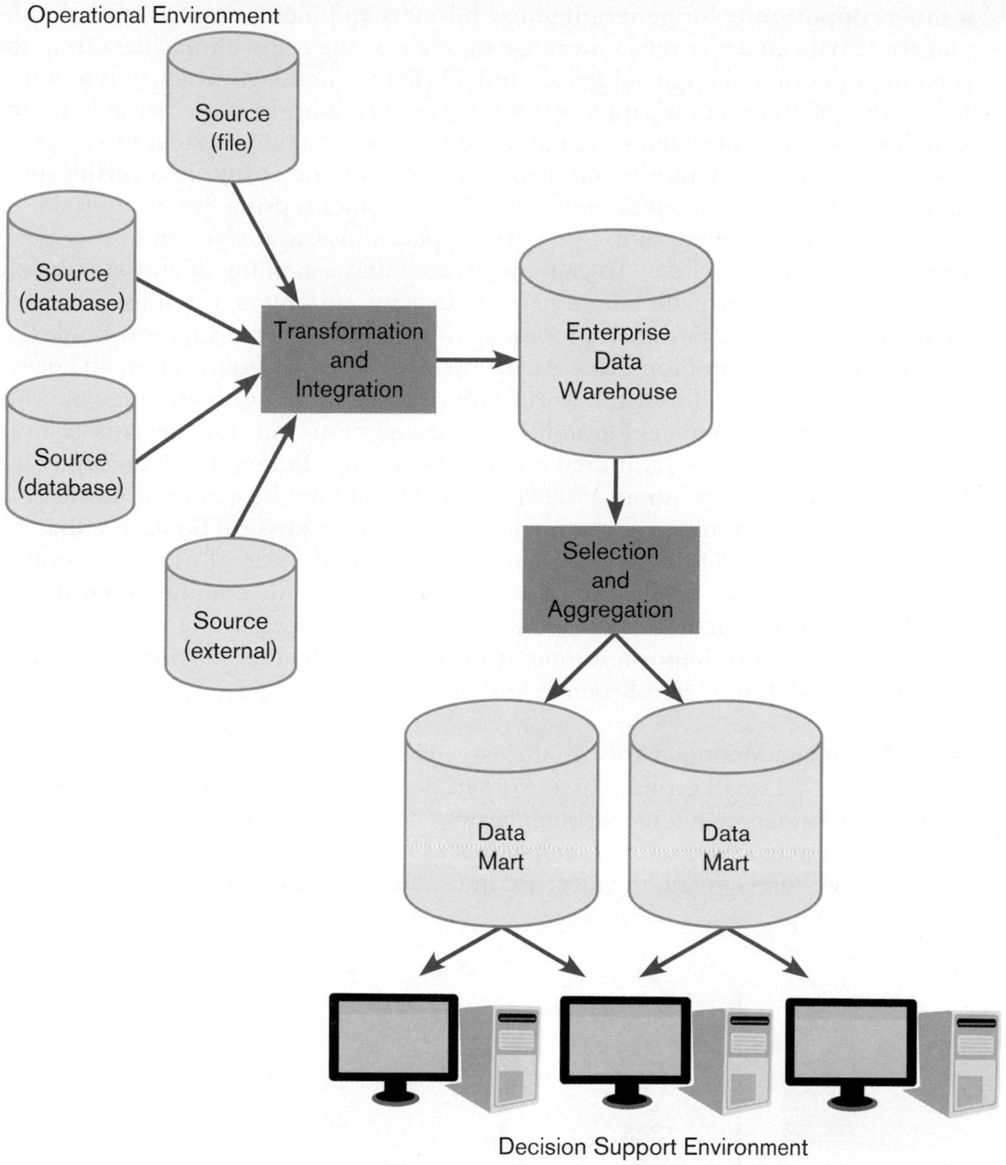

Operational Environment

FIGURE 12-18
Three-tier warehouse architecture

Decision Support Environment

A data warehouse for an Internet-based business can become huge. The data coming from Internet activities often include records of clickstream user actions, such as which links were clicked and in which sequence. Analysis of data warehouse clickstream data can then be used to customize and personalize a marketing message to a customer during a visit to the website. For example, on a travel website, one customer might first look at airplane flight itineraries and then look at books about related travel destinations; this would suggest the display of certain ads on that customer's Web page. Another customer who typically looks at rental cars when checking flight information would receive different ads.

The use of clickstream and other event-based data stored in the EDW (e.g., purchase transactions, help desk inquiries, sales staff contacts) allows an organization to create an active data warehouse. For example, in a banking environment, suppose that a customer receives a large electronic direct deposit of funds into an account from, say, a Treasury note interest payment. Within minutes of this transaction, the same customer may independently log on to the bank's Internet banking site to pay utility bills. Typically, the bank will be using separate operational applications to manage electronic direct funds transfers and Internet banking. Without an EDW, the bank will have no way to link these transactions in real time and may miss a timely opportunity for generating new business and increasing customer loyalty and trust. With an active data enterprise warehouse, the transactional data from the separate operational applications are quickly fed to the EDW, which acts as a hub for sending messages to separate operational applications. In this environment, the bank can develop rules that would allow the Internet banking system to recognize the opportunity to attempt to automatically cross-sell the customer a certificate of deposit or other investment account while the customer is using the Internet site.

Some Internet electronic commerce applications can receive and process millions of transactions per day. To gain the greatest understanding of customer behavior and to ensure adequate system performance for customers, you must effectively manage online data. For example, Amazon.com is the world's largest bookstore with more than five million titles. Amazon.com is open 24 hours a day, 365 days a year, with customers all over the world ordering books and a broad range of other products. Amazon's servers log millions of transactions per day. Amazon is a vast departure from a more traditional physical bookstore. In fact, the largest physical bookstore carries "only" about 170,000 titles. It would not be economically feasible to build a physical bookstore the size of Amazon.com; a physical bookstore that carried Amazon.com's 5 million titles would need to be the size of nearly 50 football fields! The key to effectively designing an online electronic commerce business is clearly the effective management of online data. In this section, we provided a very brief overview of this important topic (to learn more about managing online data, see Hoffer et al. [2011] and Kroenke [2010]).

Website Content Management   In the early days of the Internet, websites were often maintained by a small group of overworked developers; sites were often filled with outdated information and inconsistent layouts. To gain consistency in website appearance, organizations have utilized templates and stylesheets as described previously. To make sure websites contain the most accurate and up-to-date information, often from

FIGURE 12-19
A content management system allows content from multiple sources to be stored separately from its formatting to ease web site management

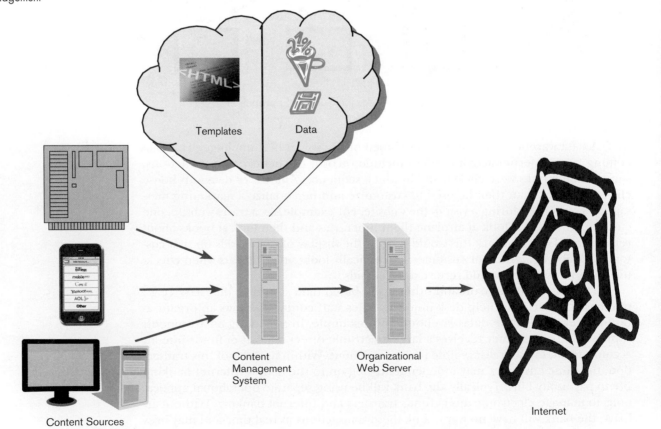

Content Sources

Templates   Data

Content Management System

Organizational Web Server

Internet

multiple sources, many organizations have turned to using a **content management system (CMS)**. A CMS is a special type of software application for collecting, organizing, and publishing website content from multiple organizational data sources, such as data warehouses, personnel databases, inventories, and so on. This content is stored in a single repository along with templates for formatting any type of Web page within the organization's website. Because content and formatting is separated by the CMS, the same underlying content can be presented differently to different audiences—customers, employees, or suppliers—as well as for different devices (see Figure 12-19).

Additionally, a CMS allows numerous content developers and sources to provide updated information for a website, without having to know anything about HTML. For example, a personnel manager could author a new job description and post it at the CMS server using a standard word-processing program like Microsoft Word. Once stored at the CMS server, the job posting text can be merged with a standard template that automatically formats it into a standard Web page. Once formatted, the Webmaster can review and approve the job posting before it is published onto the publicly (intranet, Internet, or extranet) viewed website. In this way, organizations can facilitate timely updates to their Web sites from throughout the organization, without having to wait for Web developers to author pages. This separation of content, appearance, and publishing greatly improves organizational workflow and site management. It is only through a CMS that organizations can deploy sophisticated websites, containing thousands of pages with rapidly changing content (e.g., visit a popular website with constantly changing content like cnn.com and imagine how to keep this site up to date without a CMS).

Content management system (CMS)

A special type of software application for collecting, organizing, and publishing website content.

ELECTRONIC COMMERCE APPLICATION: DESIGNING A DISTRIBUTED ADVERTISEMENT SERVER FOR PINE VALLEY FURNITURE'S WEBSTORE

In this chapter, we have examined numerous issues to consider when designing Internet-based systems. As we saw in the prior two chapters, prototyping can be useful in conceptualizing the look and feel of a website. The look and feel of a site is a function of the data presentation layer within an Internet-based application. Prototyping also provides a view of the transactions and processes within the system. Transactions and processes are managed by the middle layer, data analysis, of a three-tiered architecture. In this section, we will see how a distributed system, the advertisement rotation system, is integrated into PVF's WebStore.

In the prior two chapters, you read how Jim Wood defined specifications for the forms and reports as well as the interface and dialogues for PVF's WebStore. In this design work, he and his development team concluded that they wanted the human–computer interface of their site to have four key features:

1. Menu-driven navigation with cookie crumbs
2. Lightweight graphics
3. Form and data integrity rules
4. Template-based HTML

To demonstrate these features to the team, Jim built a prototype (see Figure 12-20).

Advertising on Pine Valley Furniture's WebStore

Having reviewed Jim Woo's throwaway prototype of the WebStore, Jackie Judson wanted to assess the feasibility of adding advertisements to the site. She came up with the following list of potential benefits for including advertising:

- Potential to increase revenue generated from the WebStore
- Potential to create cross-promotions and alliances with other online commerce systems
- Potential to provide customers with improved service when looking for additional products that accessorize PVF's product line

FIGURE 12-20
Initial prototype of the WebStore

Jim agreed with the principles of advertising on the site and researched advertising examples on an array of different Internet sites. He compiled the following list of potential concerns that need to be addressed in the system design in order to implement a successful advertisement rotation system within the WebStore:

- Advertisements must be served quickly so that site performance is not affected.
- Advertisements must be uniform in size and resolution, so they do not *disrupt* the site layout.
- Advertisement links must not redirect the user's browser away from the WebStore.

Designing the Advertising Component

To begin the process, Jim modified the style sheets of the initial prototype to include a space where the advertisement would appear. Because all advertisements would be approved by the marketing department before being included in the rotation, Jim could rely on the fact that they would be uniform in size and resolution. If an advertisement is clicked, a new, smaller window is opened and directed to the advertiser's site. The link is not direct, though. It is first directed to the advertising server within the WebStore system, the same server the advertisement came from. This "click-thru" transaction is logged, and the user is sent to the appropriate destination.

Jim identified two distinct sets of data that would be generated by the advertisement rotation system: the number of advertisements served and the number of "click-thru's." The data being generated must be stored quickly and function transparently within the overall operation of the system. The transactional requirements of the advertisement system are the following:

1. Determine which advertisements apply, based on where the user is in the WebStore.
2. Personalize the advertisement if the identity of the user has been established and his or her preferences are known.
3. Check for any seasonal or promotional advertisements.
4. Log the transaction.

FIGURE 12-21
Adding advertising to the WebStore
prototype

These requirements are part of the business rules that govern the rotation system. Jim and Jackie want these parameters to be flexible and scalable so that future systems can incorporate these rules. To demonstrate how an advertisement might be placed on the WebStore, Jim modified the prototype to include a small banner ad (see Figure 12-21).

Designing the Management Reporting Component

Once the transactional requirements of the system were established, Jackie turned her attention to what reports she and other upper-level managers would like to see generated. Jim immediately began to write down all of the demographic information stored in the customer-tracking system and cross-referenced it to the information stored when an advertisement was clicked. This led Jim and Jackie to identify numerous potential analytical queries that tied information from the customer-tracking system with the transactional data in the advertisement rotation system. A few of the queries they came up with included the following:

- "How many women, when shopping for desks, clicked on an advertisement for lamps?"
- "How many advertisements were served to shoppers looking at filing cabinets?"
- "How many people clicked on the first advertisement they saw?"
- "How many people clicked on an advertisement and then purchased something from the WebStore?"

Being able to analyze these and other results will provide critical feedback from targeted marketing campaigns, seasonal promotions, and product tie-ins. Using a distributed, transaction-based advertisement system in the WebStore will keep maintenance costs low and should increase the revenue generated from the site. Information derived from the analytical queries of advertisement transaction data increases the site's value even further.

Jackie and Jim reviewed the advertising model with the entire marketing staff. Many of the client account reps expressed interest in seeking a partnership with frequent customers to do advertising on the site. Junior sales staff members were eager to sell advertising space with the knowledge that they could provide purchasers with feedback on "click-thru" rates and overall advertisement views. One of the graphic designers even produced an advertisement on the spot for an upcoming product release. Everyone seemed to agree that the advertisement rotation system would increase the value of the WebStore to PVF.

SUMMARY

This chapter covered various issues and technologies involved in the sharing of systems and data by multiple people across space and time in distributed and Internet systems. You learned about the client/server architecture, which is being used to network personal computers and workstations (upsizing) and to replace older mainframe applications (downsizing). Components of the client/server architecture, including LANs, database servers, application programming interfaces, and application development tools, were described.

Two common types of LAN-based architectures—file server and client/server—were compared. It was shown that the newer client/server technologies have significant advantages over the older file servers. We also outlined in this chapter the evolution of distributed systems and three-tiered client/server technologies that are giving analysts more options for distributed system design.

When designing Internet-based systems, standardized location naming, content translating, and document formatting enable designers to quickly craft systems because much of the complexity of the design and implementation is removed. These standards also free the designer from much of the worry of delivering applications over a broad range of computing devices and platforms. Many commercial Internet applications are vast and can contain thousands of distinct pages. A consistent "look and feel" is fundamental to conveying the image that the site is professionally designed. A site with high consistency is also much easier for users to navigate and is much more intuitive so users can anticipate the meaning of links. Two techniques for enforcing consistency when designing large-scale Web applications are the use of CSSs and the XSL. Given the desire to deliver the Internet to a broader array of client devices, there is

a trend to separate Web content from its delivery. Electronic commerce applications are particularly embracing this trend and are adopting standards such as XML to author Web data and XSL to manage content formatting. In addition to using style sheets to enforce consistency in the design of a website across client devices, it is also important that designers adopt standards for defining page and link titles. Finally, a successful design will make users feel that the site—and their data—are secure. Customers build trust from positive experiences interacting with a site. Taking steps to convey trustworthiness will help to attract and retain customers.

Gaining an understanding of how a new system fits within the context of an organization's existing application portfolio is a fundamental part of designing and managing system data. The major source of data within an Internet-based application is through the accumulation of customer transactions. OLTP refers to the collection and immediate response to the requests of users interacting with a Web application. To improve decision making, organizations use OLAP to analyze the vast amounts of transaction data. OLAP refers to graphical software tools that provide complex analysis of data stored in a database or data warehouse. The purpose of a data warehouse is to consolidate and integrate data from a variety of sources and to format those data in a context that is appropriate for making accurate business decisions.

We did not have the space in this chapter to address several additional issues concerning distributed and Internet systems. Many of these issues are handled by other systems professionals, such as database administrators, telecommunications experts, and computer security specialists. Systems analysts must work closely with other professionals to build sound distributed systems.

KEY TERMS

1. Application program interface (API)
2. Application server
3. Cascading Style Sheets (CSSs)
4. Client
5. Client/server architecture
6. Content management system (CMS)
7. Context development
8. Customization
9. Data mart
10. Data warehouse
11. Database engine
12. Domain naming system (BIND)
13. Enterprise data warehouse (EDW)
14. File server
15. Hypertext Markup Language (HTML)
16. Hypertext Transfer Protocol (HTTP)
17. Informational systems
18. Integration depth
19. Local area network (LAN)
20. Middleware
21. Online analytical processing (OLAP)
22. Online transaction processing (OLTP)
23. Operational systems
24. Organizational breadth
25. Personalization
26. Thin client
27. Three-tiered client/server
28. eXtensible Markup Language (XML)
29. eXtensible Style Language (XSL)

Match each of the key terms above with the definition that best fits it.

_____ The cabling, hardware, and software used to connect workstations, computers, and file servers located in a confined geographical area (typically within one building or campus).

_____ A device that manages file operations and is shared by each client PC attached to a LAN.

_____ A LAN-based computing environment in which a central database server or engine performs all database commands sent to it from client workstations, and application programs on each client concentrate on user interface functions.

_____ The (back-end) portion of the client/sever database system running on the server that provides database processing and shared access functions.

_____ The (front-end) portion of the client/server database system that provides the user interface and data manipulation functions.

_____ Software building blocks that are used to ensure that common system capabilities, such as user interfaces and printing, and modules are standardized to facilitate the data exchange between clients and servers.

_____ A computing server where data analysis functions primarily reside.

_____ Advanced client/server architectures in which there are three logical and distinct applications—data management, presentation, and analysis—that are combined to create a single information system.

_____ A combination of hardware, software, and communication technologies that brings together data management, presentation, and analysis into a three-tiered client/server environment.

_____ A method for translating Internet domain names into Internet Protocol (IP) addresses.

_____ A communications protocol for exchanging information on the Internet.

_____ The standard language for representing content on the Web through the use of hundreds of command tags.

_____ An Internet authoring language that allows designers to create customized tags, enabling the definition, transmission, validation, and interpretation of data between applications.

_____ A client device designed so that most processing and data storage occur on the server.

_____ A set of style rules that tell a Web browser how to present a document.

_____ A specification for separating style from content when generating HTML documents.

_____ Providing Internet content to users based upon knowledge of that customer.

_____ Internet sites that allow users to customize information to their personal preferences.

_____ A method that helps analysts to better understand how a system fits within existing business activities and data.

_____ A measurement of how far into the existing technology infrastructure a system penetrates.

_____ A measurement that tracks the core business functions affected by a system.

_____ The immediate automated responses to the requests of users.

_____ The use of graphical software tools that provide complex analysis of data stored in a database.

_____ Systems that are used to interact with customers and run a business in real time.

_____ Systems designed to support decision making based on stable point-in-time or historical data.

_____ A subject-oriented, integrated, time-variant, nonvolatile collection of data used in support of management decision making.

_____ A centralized, integrated data warehouse that is the control point and single source of all data made available to end users for decision support applications throughout the entire organization.

_____ A data warehouse that is limited in scope; its data are obtained by selecting and (where appropriate) summarizing data from the enterprise data warehouse.

_____ A special type of software application for collecting, organizing, and publishing website content.

REVIEW QUESTIONS

1. Contrast the following terms.
   a. File server, client/server architecture, local area network (LAN)
   b. Hypertext Markup Language (HTML), Hypertext Transfer Protocol (HTTP), domain naming system (BIND)
   c. Cascading Style Sheets (CSSs), eXtensible Style Language (XSL)
   d. Personalization, customization
   e. Operational system, informational system
   f. Integration depth, organizational depth
   g. Online transaction processing (OLTP), online analytical processing (OLAP)
   h. Data warehouse, enterprise data warehouse, data mart

2. Describe the limitations of a file server architecture.

3. Describe the advantages of a client/server architecture.

4. Summarize the six possible architectures for client/server systems.

5. Summarize the reasons for using a three-tiered client/server architecture.

6. Explain the role of middleware in client/server computing.

7. In what ways do Internet standards such as BIND, HTTP, and HTML assist designers in building Internet-based systems?

8. Why is it important to separate content from display when designing an Internet-based electronic commerce system?

9. How can CSSs and XSL help to ensure design consistency when designing an Internet-based electronic commerce system?

10. Discuss how you can instill customer loyalty and trust-worthiness when designing an Internet-based electronic commerce system.

11. Why is it important that "Web pages live forever" when designing an Internet-based electronic commerce system?

12. Why do many commercial websites have both operational and informational systems?

13. Briefly describe and contrast the components of a two-tier versus a three-tier data warehouse.

14. What is a data mart and why do some organizations use it to support organizational decision making?

PROBLEMS AND EXERCISES

1. Under what circumstances would you recommend that a file server approach, as opposed to a client/server approach, be used for a distributed information system application? What warnings would you give the prospective user of this file server approach? What factors would have to change for you to recommend the move to a client/server approach?

2. Develop a table that summarizes the comparative capabilities of the six client/server architectures. You might start with Table 12-1 for some ideas.

3. Suppose you are responsible for the design of a new order entry and sales analysis system for a national chain of auto part stores. Each store has a PC that supports office functions. The company also has regional managers who travel from store to store working with the local managers to promote sales. There are four national offices for the regional managers, who each spend about 1 day a week in their office and 4 on the road. Stores place orders to replenish stock on a daily basis, based on sales history and inventory levels. The company uses high-speed dial-in lines and modems to connect store PCs into the company's main computer. Each regional manager has a laptop computer with a modem and a network connection for times when the manager is in the office. Would you recommend a client/server distributed system for this company and, if so, which architecture would you recommend? Why?

4. The Internet is a network of networks. Using the terminology of this chapter, what type of distributed network architecture is used on the Internet?

5. Suppose you were designing applications for a standard file server environment. One issue discussed in this chapter for this distributed processing environment is that the application software on each client PC must share in the responsibilities for data management. One data management problem that can arise is that applications running concurrently on two clients may want to update the same data at the same time. What could you do to manage this potential conflict? Is there any way this conflict might result in both PCs making no progress (in other words, going into an infinite loop)? How might you avoid such problems?

6. An extension of the three-tiered client/server architecture is the *n*-tiered architecture, in which there are many specialized application servers. Extend the reasons for the three-tiered architecture to an *n*-tiered architecture.

7. You read in this chapter about the advantages of client/server architectures. What operational and management problems can be created by client/server architectures? Considering both the advantages and disadvantages of the client/server model, suggest the characteristics of an application that could be implemented in a client/server architecture.

8. Obtain access to a typical PC DBMS, such as Microsoft Access. What steps do you have to follow to link an Access database to a database on a server? Do any of these steps change depending on the DBMS on the server?

9. There is a movement toward wireless mobile computing using thin-client technology. Go to the Web and visit some of the major computer vendors that are producing thin-client products such as handheld computers, smartphones, and PDAs. Investigate the features of each category of device and prepare a report that contrasts each type of device on at least the following criteria: screen size and color, networking options and speed, permanent memory, and embedded applications.

10. Building on the research conducted in Problem and Exercise 9, what challenges does each device present for designers when delivering an electronic commerce application? Are some devices more suitable for supporting some applications than others?

11. Design consistency within an Internet site is an important way to build customer loyalty and trustworthiness. Visit one of your favorite websites and analyze this site for design consistency. Your analysis should consider general layout, colors and fonts, labeling, links, and other such items.

12. Go to the Web and find a site that provides personalized content and a site that allows you to customize the site's content to your preferences. Prepare a report that compares and contrasts personalization and customization. Is one method better than the other? Why or why not?

13. Data warehousing is an important part of most large-scale commercial electronic commerce sites. Assume you are an executive with a leading company like Amazon.com; develop a list of questions that you would like to be answered by analyzing information within your company's data warehouse.

FIELD EXERCISES

1. Visit an organization that has installed a LAN. Explore the following questions.

   a. Inventory all application programs that are delivered to client PCs using a file server architecture. How many users use each application? What are their professional and technical skills? What business processes are supported by the application? What data are created, read, updated, or destroyed in each application? Could the same business processes be performed without using technology? If so, how? If not, why not?

   b. Inventory all application programs that are delivered to client PCs using a client/server architecture. How many users use each application? What are their professional and technical skills? What business processes are supported by the application? What data are created, read, updated, or destroyed in each application? Could the same business processes be performed without using technology? If so, how? If not, why not?

2. Scan the literature and determine the various LAN operating systems available. Describe the relative strengths and weaknesses of these systems. Do these systems seem to be adequate for distributed information system needs in organizations? Why or why not? Determine the current sales volume and approximate market shares for these systems. Why are they selling so well?

3. In this chapter, file servers were described as one way of providing information to users of a distributed information system. What file servers are available, and what are their relative strengths, weaknesses, and costs? What other types of servers are available and/or for what other uses are file servers employed (e.g., print servers)?

4. Examine the capabilities of a client/server API environment. List and describe the types of client-based operations that you can perform with the API. List and describe the types of server-based operations that you can perform with the API. How are these operations the same/different?

5. The references in this chapter point to a number of sources that provide website design guidelines (see additional references in the References list). Visit these sites and summarize, in a report, guidelines not addressed in this chapter. Did you find inconsistencies or contradictions across the sites you studied? Why do these differences exist?

REFERENCES

Bass, L., P. Clements, and R. Kazman. 2003. *Software Architecture in Practice*, 2nd ed. Boston: Addison-Wesley.

Castro, E. 2001. *XML for the World Wide Web*. Berkeley, CA: Peachpit Press.

Devlin, B. 1997. *Data Warehouse: From Architecture to Implementation*. Reading, MA: Addison-Wesley Longman.

Esposito, D. 2002. *Applied XML Programming for Microsoft .NET*. Redmond, WA: Microsoft Press.

Galemmo, N., C. Imhoff, and J. Geiger. 2003. *Mastering Data Warehouse Design: Relational and Dimensional Techniques*. New York: John Wiley & Sons.

Hoffer, J. A., V. Ramesh, and H. Topi. 2011. *Modern Database Management*, 10th ed. Upper Saddle River, NJ: Prentice Hall.

Inmon, W. H., and R. D. Hackathorn. 1994. *Using the Data Warehouse*. New York: John Wiley & Sons.

Kroenke, D. M. 2010. *Database Processing*, 11th ed. Upper Saddle River, NJ: Prentice Hall.

Levinson, J. 2003. *Building Client/Sever Applications Under VB .NET: An Example-Driven Approach*. Berkeley, CA: APress.

McKnight, D. H., V. Choudhury, and C. Kacmar. 2002. "Developing and Validating Trust Measures for E-Commerce: An Integrative Typology." *Information Systems Research* 13(3): 334–59.

Nielsen, J. 1996. "Marginalia of Web Design." November. Available at *www.useit.com/alertbox/9611.html*. Accessed February 14, 2009.

Nielsen, J. 1997. "Loyalty on the Web." August 1. Available at *www.useit.com/alertbox/9708a.html*. Accessed February 14, 2009.

Nielsen, J. 1998a. "Using Link Titles to Help Users Predict Where They Are Going." January 11. *www.useit.com/alertbox/980111.html*. Accessed February 14, 2009.

Nielsen, J. 1998b. "Personalization Is Over-Rated." October 4. Available at *www.useit.com/alertbox/981004.html*. Accessed February 14, 2009.

Nielsen, J. 1998c. "Web Pages Must Live Forever." November 29. Available at *www.useit.com/alertbox/981129.html*. Accessed February 14, 2009.

Nielsen, J. 1999. "Trust or Bust: Communicating Trustworthiness in Web Design." March 7. Available at *www.useit.com/alertbox/990307.html*. Accessed February 14, 2009.

Nielsen, J. 2003. "Intranet Portals: A Tool Metaphor for Corporate Information." March 31. Available at *www.useit.com/alertbox/20030331.html*. Accessed February 14, 2009.

Nielsen, J., and H. Loranger. 2006. *Prioritizing Web Usability*. Upper Saddle River, NJ: Prentice Hall.

Rosenfeld, L., and P. Morville. 2002. *Information Architecture for the World Wide Web: Designing Large-Scale Web Sites*. Sebastopol, CA: O'Reilly & Associates.

Wigley, A., and P. Roxburgh. 2002. *Building .NET Applications for Mobile Devices*. Redmond, WA: Microsoft Press.

Zeldman, J. 2006. *Designing with Web Standards*. Indianapolis, IN: Peach Pit Press.

PETRIE ELECTRONICS

Chapter 12: Designing Distributed and Internet Systems

Stephanie Welsh worked for Petrie's database administrator. She had been overseeing two interns who were helping her translate conceptual database designs into physical designs. As they were finishing up the task she had assigned to them, she realized that they would soon need something else to do.

She called Sanjay Agarwal, one of the most talented interface designers in Petrie's IT shop.

"Hi Sanjay, this is Stephanie. Got a minute?"

"For you, I can make the time," Sanjay replied.

"Well, this is not about me, this is about my two interns. They are about done with the database work I assigned them. They need something else to do, and I thought of you. Aren't you starting work on some of the customized web designs for 'No Customer Escapes?'"

"Yep," Sanjay said. "That's next on my list of two thousand things I have to do this week."

"So I can send them over? That will be great. They are both good workers and very bright, so I think you will get a lot out of them."

"How much do they know about Web interface design?" Sanjay asked.

"Not much, I don't think."

"Well, that's not the answer I wanted. OK, I know what I'll do. I'll have them derive a list of guiding principles for good Web interface design. They can start by looking at the website design principles listed on Jakob Nielsen's site [*www.useit.com*]. His site is extensive, with many short articles of helpful hints for making websites usable."

After visiting Nielsen's site, the interns came up with the list of guidelines featured in PE Figure 12-1.

Case Questions

1. Visit the Nielsen website and update PE Figure 12-1 based on guidelines and articles posted since this list was compiled. Add only elements you believe are essential and relevant to the design of "No Customer Escapes."

2. Review Chapters 10 and 11. Combine into your answer to Case Question 1 guidelines from these chapters. How unique do you consider the human interface design guidelines for a website to be from general application design guidelines? Justify your answer.

3. Search for other Web-based resources, besides the Nielsen website, for website design. (Hint: Look at the references at the end of this and prior chapters.) In what ways do the design guidelines you find contradict your answer to Case Question 2? Explain the differences.

4. Chapter 12 introduced the concepts of loyalty and trustworthiness as necessary for customers to interact with a website. What elements could be added to a customer loyalty site such as "No Customer Escapes" to improve the levels of loyalty and trustworthiness of Petrie's customers?

PE FIGURE 12-1

Guidelines for design of Petrie's "No Customer Escapes"

(*Source:* Adapted from the following sources: Jakob Nielsen website *www.useit.com*, specifically pages: *www.useit.com/alertbox/20001112.html, www.useit.com/alertbox/9605.html, www.useit.com/papers/heuristic/heuristic_list.html, www.useit.com/alertbox/20000416.html,* and *www.useit.com/alertbox/990502.html.*)

Feature	Guideline
Interacting menus– avoid	An interacting menu changes when users select something in another menu on the same page. Users get very confused when options come and go, and it is often hard to make a desired option visible when it depends on a selection in a different widget.
Very long menus– avoid	Very long menus that require scrolling make it impossible for users to see all their choices in one glance. It's often better to present such long lists of options as a regular HTML list of traditional hypertext links.
Menus of abbreviations– avoid	It is usually faster for users to simply type the abbreviation (e.g., a two-character state code) than to select it from a drop-down menu. Free-form input requires validation by a code on the Web page or on the server.
Menus of well-known data– avoid	Selecting well-known data, such as month, city, or country, often breaks the flow of typing for users and creates other data entry problems.
Frames– use sparingly	Frames can be confusing when a user tries to print a page or when trying to link to another site. Frames can prevent a user from e-mailing a URL to other users and can be more clumsy for inexperienced users.

PE FIGURE 12-1 (continued)

Feature	Guideline
Moving page elements– use sparingly	Moving images have an overpowering effect on the human peripheral vision and can distract a user from productive use of other page content. Moving text may be difficult to read.
Scrolling– minimize	Some users will not scroll beyond the information that is visible on the screen. Thus, critical content and navigation elements should be obvious (on the top of the page, possibly in a frame on the top of the page so that these elements never leave the page).
Context– emphasize	Don't assume that users know as much about your site as you do. Users have difficulty finding information, so they need support in the form of a strong sense of structure and place. Start your design with a good understanding of the structure of the information from the user's perspective and communicate this structure explicitly to the user.
System status– make visible	The system should always keep users informed about what is going on, through appropriate feedback within a reasonable amount of time.
Language– use user's terms	The system should consistently speak the users' language, with words, phrases, and concepts familiar to the user, rather than system-oriented terms. Follow real-world conventions, making information appear in a natural and logical order.
Fixing mistakes– make it easy	Users often choose system functions by mistake and will need a clearly marked "emergency exit" to leave the unwanted state without having to go through an extended dialogue. Support undo and redo and default values. Even better than good error messages is a careful design that prevents a problem from occurring in the first place.
Actions– make them obvious	Make objects, actions, and options visible. The user should not have to remember information from one part of the dialogue to another. Instructions for use of the system should be visible or easily retrievable whenever appropriate.
Customize– for flexibility and efficiency	Design the system for both novice and experienced users. Allow users to tailor the system to their frequent actions.
Content– make it relevant	Every extra unit of information in a dialogue competes with the relevant units of information and diminishes their relative visibility.
Cancel button– use sparingly	The Web is a navigation environment where users move between pages of information. Because hypertext navigation is the dominant user behavior, users have learned to rely on the *Back* button for getting out of unpleasant situations. Offer a *Cancel* button when users may fear that they have committed to something they want to avoid. Having an explicit way to *Cancel* provides an extra feeling of safety that is not afforded by simply leaving.

Implementation and Maintenance

Chapter 13

System Implementation

Chapter 14

Maintaining Information Systems

PART FIVE
Implementation and Maintenance

Implementation and maintenance are the last two phases of the systems development life cycle. The purpose of implementation is to build a properly working system, install it in the organization, replace old systems and work methods, finalize system and user documentation, train users, and prepare support systems to assist users. Implementation also involves closedown of the project, including evaluating personnel, reassigning staff, assessing the success of the project, and turning all resources over to those who will support and maintain the system. The purpose of maintenance is to fix and enhance the system to respond to problems and changing business conditions. Maintenance includes activities from all systems development phases. Maintenance also involves responding to requests to change the system, transforming requests into changes, designing the changes, and implementing them.

We address the variety of work done during system implementation in Chapter 13. For projects based on Agile Methodologies, coding and testing are done in concert with analysis and design, so systems resulting from such efforts will begin their implementation phases with coding and testing already completed. Projects based on traditional methodologies will begin implementation with detailed design specifications that are handed over to programming teams for coding and to quality assurance teams for testing. In Chapter 13, you will learn about testing systems and system components, and methods to ensure and measure software quality. Your role as a systems analyst may include developing a plan for testing, which includes developing all the test data required to exercise every part of the system. You start developing the test plan early in the project, usually during analysis, because testing requirements are highly related to system functional requirements. You also will learn how to document each test case and the results of each test. A testing plan usually follows a bottom-up approach, beginning with small modules, followed by extensive alpha testing by the programming group, beta testing with users, and final acceptance testing. Testing ensures the quality of the software by using measures and methods such as structured walk-throughs.

Installing a new system involves more than making technical changes to computer systems. Managing installation includes managing organizational changes as much as it does technical changes. We review several approaches to installation and several frameworks you can use to anticipate and control human and organizational resistance to change.

Documentation is extensive for any system. You have been developing most of the system documentation needed by system maintenance staff by keeping a thorough project workbook or CASE repository. You now need to finalize user documentation. In Chapter 13, we provide a generic outline for a user's guide as well as a wide range of guidelines you can use to develop high-quality user documentation. Remember, documentation must be tested for completeness, accuracy, and readability.

While documentation is being finalized, user support activities also need to be designed and implemented. Support includes training, whether through traditional, instructor-led classes; computer-based tutorials or e-learning; or vendor-provided training. Electronic performance support systems deliver on-demand training. Many types of training are available from various sources over the Internet or corporate intranets. Once trained, users will still encounter difficulties. Therefore, you, as an analyst, must consider ongoing support from help desks, newsletters, user groups, online bulletin boards, and other mechanisms, and these sources of support need to be tested and implemented. We conclude Chapter 13 with a brief review of project closedown activities because the end of implementation means the end of the project. We also provide an example of implementation for the Pine Valley Furniture WebStore.

After implementation, however, work on the system is just beginning. Today, as much as 80 percent of the life cycle cost of a system occurs after implementation. Maintenance handles updates to correct flaws and to accommodate new technologies as well as to meet new business conditions, regulations, and other requirements. In Chapter 14, you will learn about your role in systems maintenance.

There are four kinds of maintenance: corrective, adaptive, perfective, and preventive. You can help control the potentially monumental cost of a system by making systems maintainable. You can affect maintainability by reducing the number of defects, improving the skill of users, preparing high-quality documentation, and developing a sound system structure.

You may also be involved in establishing a maintenance group for a system. You will learn about different organizational structures for maintenance personnel and the reasons for each, and you will learn how to measure maintenance effectiveness. Configuration management and deciding how to handle change requests are important. You will learn how a systems librarian keeps track of baseline software modules, checks these out to maintenance staff, and then rebuilds systems. You will also learn about special issues for maintaining websites and read about an example of a maintenance situation for the Pine Valley Furniture WebStore.

Chapter 13 includes the final installment of the Petrie Electronics project case. This final case segment helps you to understand implementation issues in an organizational context.

CHAPTER THIRTEEN

System Implementation

Learning Objectives

After studying this chapter, you should be able to:

- Describe the process of coding, testing, and converting an organizational information system and outline the deliverables and outcomes of the process.

- Prepare a test plan for an information system.

- Apply four installation strategies: direct, parallel, single-location, and phased installation.

- List the deliverables for documenting the system and for training and supporting users.

- Compare the many modes available for organizational information system training, including self-training and electronic performance support systems.

- Discuss the issues of providing support for end users.

- Explain why system implementation sometimes fails.

- Describe the threats to system security and remedies that can be applied.

- Show how traditional implementation issues apply to electronic commerce applications.

Introduction

After maintenance, the implementation phase of the systems development life cycle (SDLC) is the most expensive and time-consuming phase of the entire life cycle. Implementation is expensive because so many people are involved in the process; it is time consuming because of all the work that has to be completed. In a traditional plan-driven systems development project, physical design specifications must be turned into working computer code, and the code must be tested until most of the errors have been detected and corrected. In a systems development project governed by Agile Methodologies, design, coding, and testing are done in concert, as you learned in previous chapters. Regardless of methodology used, once coding and testing are complete and the system is ready to "go live," it must be installed (or put into production),

481

user sites must be prepared for the new system, and users rely on the new system rather than the existing one to get their work done.

Implementing a new information system into an organizational context is not a mechanical process. The organizational context has been shaped and reshaped by the people who work in the organization. The work habits, beliefs, interrelationships, and personal goals of an organization's members all affect the implementation process. Although factors important to successful implementation have been identified, there are no sure recipes you can follow. During implementation, you must be attuned to key aspects of the organizational context, such as history, politics, and environmental demands—aspects that can contribute to implementation failure if ignored.

In this chapter, you will learn about the many activities that the implementation phase comprises. We will discuss coding, testing, installation, documentation, user training, support for a system after it is installed, and implementation success. Our intent is not to teach you how to program and test systems—most of you have already learned about writing and testing programs in the courses you took before this one. Rather, this chapter shows you where coding and testing fit in the overall scheme of implementation, especially in a traditional, plan-driven context. The chapter stresses the view of implementation as an organizational change process that is not always successful.

In addition, you will learn about providing documentation about the new system for the information systems personnel who will maintain the system and for the system's users. These same users must be trained to use what you have developed and installed in their workplace. Once training has ended and the system has become institutionalized, users will have questions about the system's implementation and how to use it effectively. You must provide a means for users to get answers to these questions and to identify needs for further training.

As a member of the system development team that developed and implemented the new system, your job is winding down now that installation and conversion are complete. The end of implementation marks the time for you to begin the process of project closedown. You read about project closedown in Chapter 3 when you learned about project management. At the end of this chapter, we will return to the topic of formally ending the systems development project.

After a brief overview of the coding, testing, and installation processes and the deliverables and outcomes from these processes, we will talk about software application testing. We then present the four types of installation: direct, parallel, single-location, and phased. You then will read about the process of documenting systems and training and supporting users as well as the deliverables from these processes. We then discuss the various types of documentation and numerous methods available for delivering training and support services. You will read about implementation as an organizational change process, with many organizational and people issues involved in the implementation effort. You will also read about the threats to security that organizations face and some of the things that can be done to make systems more secure. Finally, you will see how the implementation of an electronic commerce application is similar to the implementation of more traditional systems.

SYSTEM IMPLEMENTATION

System implementation is made up of many activities. The six major activities we are concerned with in this chapter are coding, testing, installation, documentation, training, and support (see Figure 13-1). The purpose of these steps is to convert the physical system specifications into working and reliable software and hardware, document the work that has been done, and provide help for current and future users and caretakers of the system. Coding and testing may have already been completed by this point if Agile Methodologies have been followed. Using a plan-driven methodology, coding and testing are often done by other project team members besides

FIGURE 13-1
Systems development life cycle with the
implementation phase highlighted

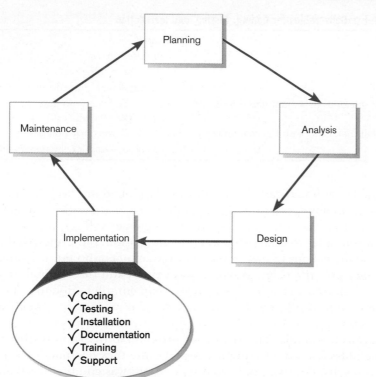

analysts, although analysts may do some programming. In any case, analysts are responsible for ensuring that all of these various activities are properly planned and executed. Next, we will briefly discuss these activities in two groups: (1) coding, testing, and installation and (2) documenting the system and training and supporting users.

Coding, Testing, and Installation Processes

Coding, as we mentioned before, is the process whereby the physical design specifications created by the analysis team are turned into working computer code by the programming team. Depending on the size and complexity of the system, coding can be an involved, intensive activity. Regardless of the development methodology followed, once coding has begun, the testing process can begin and proceed in parallel. As each program module is produced, it can be tested individually, then as part of a larger program, and then as part of a larger system. You will learn about the different strategies for testing later in this chapter. We should emphasize that, although testing is done during implementation, you must begin planning for testing earlier in the project. Planning involves determining what needs to be tested and collecting test data. This is often done during the analysis phase because testing requirements are related to system requirements.

Installation is the process during which the current system is replaced by the new system. This includes conversion of existing data, software, documentation, and work procedures to those consistent with the new system. Users must give up the old ways of doing their jobs, whether manual or automated, and adjust to accomplishing the same tasks with the new system. Users will sometimes resist these changes, and you must help them adjust. However, you cannot control all the dynamics of user–system interaction involved in the installation process.

Deliverables and Outcomes from Coding, Testing, and Installation

Table 13-1 shows the deliverables from the coding, testing, and installation processes. Some object-oriented languages, such as Eiffel, provide for documentation

TABLE 13-1 **Deliverables for Coding, Testing, and Installation**

1. Coding	3. Installation
a. Code	a. User guides
b. Program documentation	b. User training plan
2. Testing	c. Installation and conversion plan
a. Test scenarios (test plan) and test data	i. Software and hardware installation schedule
b. Results of program and system testing	ii. Data conversion plan
	iii. Site and facility remodeling plan

to be extracted automatically from software developed in Eiffel. Other languages, such as Java, employ specially designed utilities, such as JavaDocs, to generate documentation from the source code. Other languages will require more effort on the part of the coder to establish good documentation. But even well-documented code can be mysterious to maintenance programmers who must maintain the system for years after the original system was written and the original programmers have moved on to other jobs. Therefore, clear, complete documentation for all individual modules and programs is crucial to the system's continued smooth operation. Increasingly, CASE tools are used to maintain the documentation needed by systems professionals. The results of program and system testing are important deliverables from the testing process because they document the tests as well as the test results. For example, what type of test was conducted? What test data were used? How did the system handle the test? The answers to these questions can provide important information for system maintenance because changes will require retesting, and similar testing procedures will be used during the maintenance process.

The next two deliverables, user guides and the user training plan, result from the installation process. User guides provide information on how to use the new system, and the training plan is a strategy for training users so that they can quickly learn the new system. The development of the training plan probably began earlier in the project, and some training, on the concepts behind the new system, may have already taken place. During the early stages of implementation, the training plans are finalized and training on the use of the system begins. Similarly, the installation plan lays out a strategy for moving from the old system to the new, from the beginning to the end of the process. Installation includes installing the system (hardware and software) at central and user sites. The installation plan answers such questions as when the new system will be installed, which installation strategies will be used, who will be involved, what resources are required, which data will be converted and cleansed, and how long the installation process will take. It is not enough that the system is installed; users must actually use it.

As an analyst, your job is to ensure that all of these deliverables are produced and are done well. You may produce some of the deliverables, such as test data, user guides, and an installation plan; for other deliverables, such as code, you may only supervise or simply monitor their production or accomplishment. The extent of your implementation responsibilities will vary according to the size and standards of the organization you work for, but your ultimate role includes ensuring that all the implementation work leads to a system that meets the specifications developed in earlier project phases.

The Processes of Documenting the System, Training Users, and Supporting Users Although the process of documentation proceeds throughout the life cycle, it receives formal attention during the implementation phase because the end of implementation largely marks the end of the analysis team's involvement in systems development. As the team is getting ready to move on to new projects, you and the other analysts need to prepare documents that reveal all of the important information

TABLE 13-2 Deliverables for Documenting the System, Training, and Supporting Users

1. Documentation
   a. System documentation
   b. User documentation
2. User Training Plan
   a. Classes
   b. Tutorials
3. User Training Modules
   a. Training materials
   b. Computer-based training aids
4. User Support Plan
   a. Help desk
   b. Online help
   c. Bulletin boards and other support mechanisms

you have accumulated about this system during its development and implementation. There are two audiences for this final documentation: (1) the information systems personnel who will maintain the system throughout its productive life, and (2) the people who will use the system as part of their daily lives. The analysis team in a large organization can get help in preparing documentation from specialized staff in the information systems department.

Larger organizations also tend to provide training and support to computer users throughout the organization. Some of the training and support is very specific to particular application systems, whereas the rest is general to particular operating systems or off-the-shelf software packages. For example, it is common to find courses on Microsoft Windows® in organization-wide training facilities. Analysts are mostly uninvolved with general training and support, but they do work with corporate trainers to provide training and support tailored to particular computer applications they have helped to develop. Centralized information system training facilities tend to have specialized staff who can help with training and support issues. In smaller organizations that cannot afford to have well-staffed centralized training and support facilities, fellow users are the best source of training and support that users have, whether the software is customized or off the shelf.

Deliverables and Outcomes from Documenting the System, Training Users, and Supporting Users

Table 13-2 shows the deliverables from documenting the system, training users, and supporting users. At the very least, the development team must prepare user documentation. For most modern information systems, documentation includes any online help designed as part of the system interface. The development team should think through the user training process: Who should be trained? How much training is adequate for each training audience? What do different types of users need to learn during training? The training plan should be supplemented by actual training modules, or at least outlines of such modules, that at a minimum address the three questions stated previously. Finally, the development team should also deliver a user support plan that addresses issues such as how users will be able to find help once the information system has become integrated into the organization. The development team should consider a multitude of support mechanisms and modes of delivery. Each deliverable is addressed in more detail later in this chapter.

SOFTWARE APPLICATION TESTING

As we mentioned previously, in traditional plan-driven systems development projects, analysts prepare system specifications that are passed on to programmers for coding. Although coding takes considerable effort and skill, the practices and processes of writing code do not belong in this text. However, because software application testing is an activity that analysts plan (beginning in the analysis phase) and sometimes supervise, depending on organizational standards, you need to understand the essentials of the testing process. Although this section of the text focuses on testing from the perspective of traditional development practices, many of the same types of tests can be used during the analyze–design–code–test cycle common to the Agile Methodologies. Coding and testing in eXtreme Programming will be discussed briefly toward the end of this section on testing.

Software testing begins early in the SDLC, even though many of the actual testing activities are carried out during implementation. During analysis, you develop a master test plan. During design, you develop a unit test plan, an integration test plan, and a system test plan. During implementation, these various plans are put into effect and the actual testing is performed.

The purpose of these written test plans is to improve communication among all the people involved in testing the application software. The plan specifies what each person's role will be during testing. The test plans also serve as checklists you can use to determine whether the master test plan has been completed. The master test plan is not just a single document, but a collection of documents. Each of the component documents represents a complete test plan for one part of the system or for a particular type of test. Presenting a complete master test plan is far beyond the scope of this book. Refer to Mosley's (1993) *Handbook of MIS Software Application Testing* for a complete test plan, which comprises a 101-page appendix. To give you an idea of what a master test plan involves, we present an abbreviated table of contents of one in Table 13-3.

A master test plan is a project within the overall systems development project. Because at least some of the system testing will be done by people who have not been involved in the system development so far, the Introduction provides general information about the system and the need for testing. The Overall Plan and Testing Requirements sections are like a Baseline Project Plan for testing, with a schedule of events, resource requirements, and standards of practice outlined. Procedure Control explains how the testing is conducted, including how changes to fix errors will be documented. The fifth and final section explains each specific test necessary to validate that the system performs as expected.

Some organizations have specially trained personnel who supervise and support testing. Testing managers are responsible for developing test plans, establishing testing standards, integrating testing and development activities in the life cycle, and ensuring that test plans are completed. Testing specialists help develop test plans, create test cases and scenarios, execute the actual tests, and analyze and report test results.

TABLE 13-3   Table of Contents of a Master Test Plan

1. Introduction	4. Procedure Control
a. Description of system to be tested	a. Test initiation
b. Objectives of the test plan	b. Test execution
c. Method of testing	c. Test failure
d. Supporting documents	d. Access/change control
2. Overall Plan	e. Document control
a. Milestones, schedules, and locations	5. Test-Specific or Component-Specific Test Plans
b. Test materials	a. Objectives
i. Test plans	b. Software description
ii. Test cases	c. Method
iii. Test scenarios	d. Milestones, schedule, progression, and locations
iv. Test log	e. Requirements
c. Criteria for passing tests	f. Criteria for passing tests
3. Testing Requirements	g. Resulting test materials
a. Hardware	h. Execution control
b. Software	i. Attachments
c. Personnel	

(*Source:* Adapted from Mosley, 1993.)

TABLE 13-4   A Categorization of Test Types

	Manual	Automated
Static	Inspections	Syntax checking
Dynamic	Walk-throughs	Unit test
	Desk checking	Integration test
		System test

(*Source*: Adapted from Mosley, 1993.)

Seven Different Types of Tests

Software application testing is an umbrella term that covers several types of tests. Mosley (1993) organizes the types of tests according to whether they employ static or dynamic techniques and whether the test is automated or manual. Static testing means that the code being tested is not executed. The results of running the code are not an issue for that particular test. Dynamic testing, on the other hand, involves execution of the code. Automated testing means the computer conducts the test, whereas manual testing means that people complete the test. Using this framework, we can categorize the different types of tests, as shown in Table 13-4.

Let's examine each type of test in turn. **Inspections** are formal group activities where participants manually examine code for occurrences of well-known errors. Syntax, grammar, and some other routine errors can be checked by automated inspection software, so manual inspection checks are used for more subtle errors. Each programming language lends itself to certain types of errors that programmers make when coding, and these common errors are well-known and documented. Code inspection participants compare the code they are examining with a checklist of well-known errors for that particular language. Exactly what the code does is not investigated in an inspection. It has been estimated that code inspections detect from 60 to 90 percent of all software defects as well as provide programmers with feedback that enables them to avoid making the same types of errors in future work (Fagan, 1986). The inspection process can also be used for tasks such as design specifications.

Inspections
A testing technique in which participants examine program code for predictable language-specific errors.

Unlike inspections, what the code does is an important question in a walk-through. The use of structured walk-throughs is a very effective method of detecting errors in code. As you saw in Chapter 5, structured walk-throughs can be used to review many systems development deliverables, including logical and physical design specifications as well as code. Whereas specification walk-throughs tend to be formal reviews, code walk-throughs tend to be informal. Informality tends to make programmers less apprehensive about walk-throughs and helps increase their frequency. According to Yourdon (1989), code walk-throughs should be done frequently when the pieces of work reviewed are relatively small and before the work is formally tested. If walk-throughs are not held until the entire program is tested, the programmer will have already spent too much time looking for errors that the programming team could have found much more quickly. The programmer's time will have been wasted, and the other members of the team may become frustrated because they will not find as many errors as they would have if the walk-through had been conducted earlier. Further, the longer a program goes without being subjected to a walk-through, the more defensive the programmer becomes when the code is reviewed. Although each organization that uses walk-throughs conducts them differently, there is a basic structure that you can follow that works well (see Figure 13-2).

It should be stressed that the purpose of a walk-through is to detect errors, not to correct them. It is the programmer's job to correct the errors uncovered in a walk-through. Sometimes it can be difficult for the reviewers to refrain from suggesting ways to fix the problems they find in the code, but increased experience with the process can help change a reviewer's behavior.

FIGURE 13-2
Steps in a typical walk-through
(*Source:* Based on Yourdon, 1989.)

GUIDELINES FOR CONDUCTING A CODE WALK-THROUGH
1. Have the review meeting chaired by the project manager or chief programmer, who is also responsible for scheduling the meeting, reserving a room, setting the agenda, inviting participants, and so on.
2. The programmer presents his or her work to the reviewers. Discussion should be general during the presentation.
3. Following the general discussion, the programmer walks through the code in detail, focusing on the logic of the code rather than on specific test cases.
4. Reviewers ask to walk through specific test cases.
5. The chair resolves disagreements if the review team members cannot reach agreement among themselves and assigns duties, usually to the programmer, for making specific changes.
6. A second walk-through is then scheduled if needed.

Desk checking
A testing technique in which the program code is sequentially executed manually by the reviewer.

Unit testing
Each module is tested alone in an attempt to discover any errors in its code.

Integration testing
The process of bringing together all of the modules that a program comprises for testing purposes. Modules are typically integrated in a top-down, incremental fashion.

System testing
The bringing together of all of the programs that a system comprises for testing purposes. Programs are typically integrated in a top-down, incremental fashion.

Stub testing
A technique used in testing modules, especially where modules are written and tested in a top-down fashion, where a few lines of code are used to substitute for subordinate modules.

What the code does is important in **desk checking**, an informal process in which the programmer or someone else who understands the logic of the program works through the code with a paper and pencil. The programmer executes each instruction, using test cases that may or may not be written down. In one sense, the reviewer acts as the computer, mentally checking each step and its results for the entire set of computer instructions.

Among the list of automated testing techniques in Table 13-4, only one technique is static—syntax checking. Syntax checking is typically done by a compiler. Errors in syntax are uncovered but the code is not executed. For the other three automated techniques, the code is executed.

Unit testing, sometimes called *module testing*, is an automated technique whereby each module is tested alone in an attempt to discover any errors that may exist in the module's code. But because modules coexist and work with other modules in programs and the system, they must also be tested together in larger groups. Combining modules and testing them is called **integration testing**. Integration testing is gradual. First you test the coordinating module and only one of its subordinate modules. After the first test, you add one or two other subordinate modules from the same level. Once the program has been tested with the coordinating module and all of its immediately subordinate modules, you add modules from the next level and then test the program. You continue this procedure until the entire program has been tested as a unit. **System testing** is a similar process, but instead of integrating modules into programs for testing, you integrate programs into systems. System testing follows the same incremental logic that integration testing does. Under both integration and system testing, not only do individual modules and programs get tested many times, so do the interfaces between modules and programs.

Current practice calls for a top-down approach to writing and testing modules. Under a top-down approach, the coordinating module is written first. Then the modules at the next level in the structure chart are written, followed by the modules at the next level, and so on, until all of the modules in the system are done. Each module is tested as it is written. Because top-level modules contain many calls to subordinate modules, you may wonder how they can be tested if the lower-level modules haven't been written yet. The answer is **stub testing**. Stubs are two or three lines of code written by a programmer to stand in for the missing modules. During testing, the coordinating module calls the stub instead of the subordinate module. The stub accepts control and then returns it to the coordinating module.

System testing is more than simply expanded integration testing where you are testing the interfaces between programs in a system rather than testing the interfaces between modules in a program. System testing is also intended to demonstrate whether a system meets its objectives. This is not the same as testing a system to determine whether it meets requirements—that is the focus of acceptance testing, which will be discussed later. To verify that a system meets its objectives, system testing involves using nonlive test data in a nonlive testing environment. Nonlive means that the data and situation are artificial, developed specifically for testing purposes, although both the data and the environment are similar to what users would encounter in everyday system use. The system test is typically conducted by information systems personnel and led by the project team leader, although it can also be conducted by users under MIS guidance. The scenarios that form the basis for system tests are prepared as part of the master test plan.

The Testing Process

Up to this point, we have talked about the master test plan and seven different types of tests for software applications. We haven't said very much about the process of testing itself. There are two important things to remember about testing information systems:

1. The purpose of testing is to confirm that the system satisfies requirements.
2. Testing must be planned.

These two points have several implications for the testing process, regardless of the type of test being conducted. First, testing is not haphazard. You must pay attention to many different aspects of a system, such as response time, response to boundary data, response to no input, response to heavy volumes of input, and so on. You must test anything (within resource constraints) that could go wrong or be wrong with a system. At a minimum, you should test the most frequently used parts of the system and as many other paths throughout the system as time permits. Planning gives analysts and programmers an opportunity to think through all the potential problem areas, list these areas, and develop ways to test for problems. As indicated previously, one part of the master test plan is creating a set of test cases, each of which must be carefully documented (see Figure 13-3 for an outline of a test case description).

Pine Valley Furniture Company
Test Case Description

Test Case Number:
Date:
Test Case Description:

Program Name:
Testing State:
Test Case Prepared By:

Test Administrator:

Description of Test Data:

Expected Results:

Actual Results:

FIGURE 13-3
Test case description form
(*Source:* Adapted from Mosley, 1993.)

A test case is a specific scenario of transactions, queries, or navigation paths that represent a typical, critical, or abnormal use of the system. A test case should be repeatable so that it can be rerun as new versions of the software are tested. This is important for all code, whether written in-house, developed by a contractor, or purchased. Test cases need to determine that new software works with other existing software with which it must share data. Even though analysts often do not do the testing, systems analysts, because of their intimate knowledge of applications, often make up or find test data. The people who create the test cases should not be the same people as those who coded and tested the system. In addition to a description of each test case, there must also be a description of the test results, with an emphasis on how the actual results differed from the expected results (see Figure 13-4). This description will indicate why the results were different and what, if anything, should be done to change the software. This description will then suggest the need for retesting, possibly introducing new tests to discover the source of the differences.

One important reason to keep such a thorough description of test cases and results is so that testing can be repeated for each revision of an application. Although new versions of a system may necessitate new test data to validate new features of the application, previous test data usually can and should be reused. Results from the use of the test data with prior versions are compared to new versions to show that changes have not introduced new errors and that the behavior of the system, including response time, is no worse. A second implication for the testing process is that test cases must include illegal and out-of-range data. The system should be able to handle any possibility, no matter how unlikely; the only way to find out is to test.

Pine Valley Furniture Company
Test Case Results

Test Case Number:
Date:

Program Name:
Module Under Test:

Explanation of difference between actual and expected output:

Suggestions for next steps:

FIGURE 13-4
Test case results form
(*Source:* Adapted from Mosley, 1993.)

Testing often requires a great deal of labor. Manual code reviews can be very time consuming and tedious work; and, most importantly, are not always the best solution. As such, special purpose testing software, called a **testing harness**, is being developed for a variety of environments to help designers automatically review the quality of their code. In many situations, a testing harness will greatly enhance the testing process because they can automatically expand the scope of the tests beyond the current development platform as well as be run every time there is a new version of the software. For instance, with the testing harness called NUnit (see Figure 13-5), an open-source unit testing framework for .NET, a developer can answer questions such as how stable is the code? Does the code follow standard rules? Will the code work across multiple platforms? When deploying large-scale, multi-platform projects, automatic code review systems have become a necessity.

Testing harness
An automated testing environment used to review code for errors, standards violations, and other design flaws.

Combining Coding and Testing

Although coding and testing are in many ways part of the same process, it is not uncommon in large and complicated systems development environments to find the two practices separated from each other. Big companies and big projects often have dedicated testing staffs that develop test plans and then use the plans to test software after it has been written. You have already seen how many different types of testing there are, and you can deduce from that how elaborate and extensive testing can be. As you recall, with eXtreme Programming (XP) (Beck and Andres, 2004) and other Agile Methodologies, coding and testing are intimately related parts of the same process, and the programmers who write the code also write the tests. The general idea is that code is tested soon after it is written.

After testing, all of the code that works may be integrated at the end of each working day, and working versions of the system will be released frequently, as often as once per week in some cases. XP developers design and build working systems in very little time (relative to traditionally organized methods).

One particular technique used in XP to continually improve system quality is **refactoring**. Refactoring is nothing more than simplifying a system, typically after a new feature or set of features has been added. As more features are added to a system, it becomes more complex, and this complexity will be reflected in the code. After a time of increasing complexity, XP developers stop and redesign the system. The system must still pass the test cases written for it after it has been simplified, so rework continues until the tests can be passed. Different forms of refactoring include simplifying complex statements, abstracting solutions from reusable code, and removing duplicate code. Refactoring and the continuing simplification it implies reflect the iterative nature of XP and the other Agile Methodologies. As development progresses and the system gets closer to being ready for production, the iterations and the evolution of the system slow, a process Beck (2000) calls "productionizing."

Refactoring
Making a program simpler after adding a new feature.

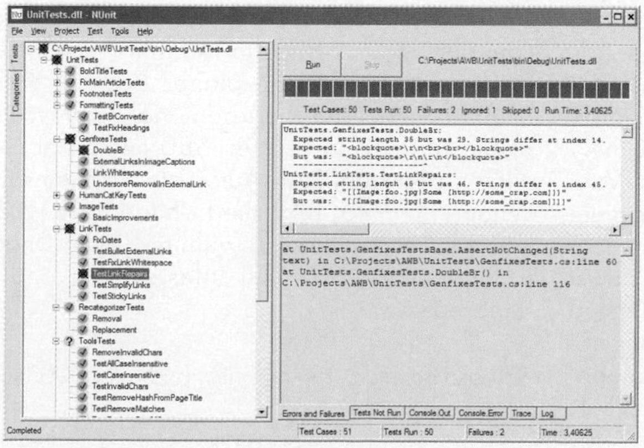

FIGURE 13-5
NUnit, a unit testing framework for .NET

A system ready to go into production is ready to be released to users, either customers ready to buy the software or internal users.

Acceptance Testing by Users

Once the system tests have been satisfactorily completed, the system is ready for **acceptance testing**, which is testing the system in the environment where it will eventually be used. Acceptance refers to the fact that users typically sign off on the system and "accept" it once they are satisfied with it. The purpose of acceptance testing is for users to determine whether the system meets their requirements. The extent of acceptance testing will vary with the organization and with the system in question. The most complete acceptance testing will include **alpha testing**, in which simulated but typical data are used for system testing; **beta testing**, in which live data are used in the users' real working environment; and a system audit conducted by the organization's internal auditors or by members of the quality assurance group.

During alpha testing, the entire system is implemented in a test environment to discover whether the system is overtly destructive to itself or to the rest of the environment. The types of tests performed during alpha testing include the following:

- Recovery testing—forces the software (or environment) to fail in order to verify that recovery is properly performed.
- Security testing—verifies that protection mechanisms built into the system will protect it from improper penetration.
- Stress testing—tries to break the system (e.g., what happens when a record is written to the database with incomplete information or what happens under extreme online transaction loads or with a large number of concurrent users).
- Performance testing—determines how the system performs in the range of possible environments in which it may be used (e.g., different hardware configurations, networks, operating systems, and so on); often the goal is to have the system perform with similar response time and other performance measures in each environment.

In beta testing, a subset of the intended users runs the system in the users' own environments using their own data. The intent of the beta test is to determine whether the software, documentation, technical support, and training activities work

---

Acceptance testing
The process whereby actual users test a completed information system, the end result of which is the users' acceptance of it.

Alpha testing
User testing of a completed information system using simulated data.

Beta testing
User testing of a completed information system using real data in the real user environment.

Bugs in the Baggage

Testing a complex software system can be a long and frustrating proposition. A case in point was the software used to control 4000 baggage cars at the Denver International Airport. Errors in the software put the airport's opening on hold for months, costing taxpayers $500,000 a day and turning airport bonds into junk. The airport was supposed to have opened in March 1994, but it did not open until February 1995 because of problems with the baggage-handling system. The system routinely damaged luggage and routed bags to the wrong flights. Various causes of the delay were identified, including last-minute design change requests from airport officials, and mechanical problems. The bottom-line lesson is that system designers must build in plenty of test and debugging time when scaling up proven technology into a much more complicated environment.

United Airlines, the major carrier at Denver International Airport, took over as systems integrator in October 1994.

At the same time, the City of Denver commissioned a traditional conveyor-belt baggage-handling system in the airport for an additional $50 million. When the airport opened in 1995 only United used the automated baggage system, and then only to ferry bags to its flights. The system was not able to ferry bags from the planes back to the airport. All the other carriers used the traditional conveyor system. The City of Denver tried to recover $80 million of the system's $193 million cost from BAE Automated Systems Inc., the system's vendor. In 1996, BAE sued United for $17.5 million and the city for $4.1 million in withheld fees. United countersued. In September 1997, all sides settled, and no details were released.

Sources: Bozman, 1994; Griffin and Leib, 2003; Schneier, 1993.

as intended. In essence, beta testing can be viewed as a rehearsal of the installation phase. Problems uncovered in alpha and beta testing in any of these areas must be corrected before users can accept the system. Systems analysts can tell many stories about long delays in final user acceptance due to system bugs (see the box "Bugs in the Baggage" for one famous incident).

INSTALLATION

The process of moving from the current information system to the new one is called **installation**. All employees who use a system, whether they were consulted during the development process or not, must give up their reliance on the current system and begin to rely on the new system. Four different approaches to installation have emerged over the years: direct, parallel, single-location, and phased (Figure 13-6). The approach an organization decides to use will depend on the scope and complexity of the change associated with the new system and the organization's risk aversion.

Direct Installation

The direct, or abrupt, approach to installation (also called "cold turkey") is as sudden as the name indicates: The old system is turned off and the new system is turned on (Figure 13-6a). Under **direct installation**, users are at the mercy of the new system. Any errors resulting from the new system will have a direct impact on the users and how they do their jobs and, in some cases—depending on the centrality of the system to the organization—on how the organization performs its business. If the new system fails, considerable delay may occur until the old system can again be made operational and business transactions are reentered to make the database up to date. For these reasons, direct installation can be very risky. Further, direct installation requires a complete installation of the whole system. For a large system, this may mean a long time until the new system can be installed, thus delaying system benefits or even missing the opportunities that motivated the system request. On the other hand, it is the least expensive installation method, and it creates considerable interest in making the installation a success. Sometimes, a direct installation is the only possible strategy if there is no way for the current and new systems to coexist, which they must do in some way in each of the other installation approaches.

Parallel Installation

Parallel installation is as riskless as direct installation is risky. Under parallel installation, the old system continues to run alongside the new system until users and management are satisfied that the new system is effectively performing its duties and the old system can be turned off (Figure 13-6b). All of the work done by the old system is concurrently performed by the new system. Outputs are compared (to the greatest extent possible) to help determine whether the new system is performing as well as the old. Errors discovered in the new system do not cost the organization much, if anything, because errors can be isolated and the business can be supported with the old system. Because all work is essentially done twice, a parallel installation can be very expensive; running two systems implies employing (and paying) two staffs to not only operate both systems, but also to maintain them. A parallel approach can also be confusing to users because they must deal with both systems. As with direct installation, there can be a considerable delay until the new system is completely ready for installation. A parallel approach may not be feasible, especially if the users of the system (such as customers) cannot tolerate redundant effort or if the size of the system (number of users or extent of features) is large.

Installation
The organizational process of changing over from the current information system to a new one.

Direct installation
Changing over from the old information system to a new one by turning off the old system when the new one is turned on.

Parallel installation
Running the old information system and the new one at the same time until management decides the old system can be turned off.

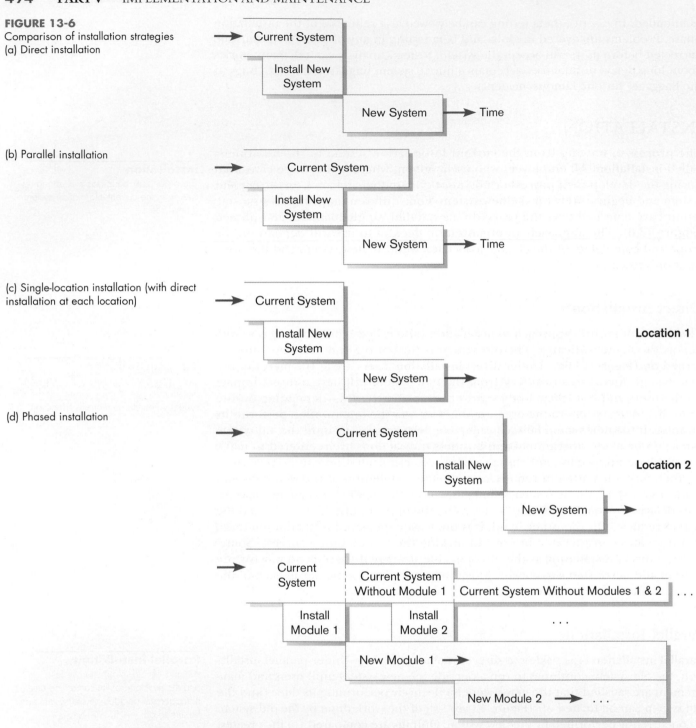

FIGURE 13-6
Comparison of installation strategies
(a) Direct installation

(b) Parallel installation

(c) Single-location installation (with direct installation at each location)

(d) Phased installation

Single-Location Installation

Single-location installation, also known as *location* or *pilot installation*, is a middle-of-the-road approach compared with direct and parallel installation. Rather than convert all of the organization at once, single-location installation involves changing from the current to the new system in only one place or in a series of separate sites over time. (Figure 13-6c depicts this approach for a simple situation of two locations.) The single location may be a branch office, a single factory, or one department, and the actual approach used for installation in that location may be any of the other approaches. The key advantage to single-location installation is that it limits potential damage and potential cost by limiting the effects to a single site. Once management has determined that installation has been successful at one location, the new system may be deployed in the rest of the organization, possibly continuing with installation at one location at a time. Success at the pilot site can be used to convince reluctant personnel at other sites that the system can be worthwhile for them as well. Problems with the system (the actual software as well as documentation, training, and support) can be resolved before deployment to other sites. Even though the single-location approach may be simpler for users, it still places a large burden on information systems (IS) staff to support two versions of the system. On the other hand, because problems are isolated at one site at a time, IS staff members can devote all of their efforts to success at the pilot site. Also, if different locations require sharing of data, extra programs will need to be written to synchronize the current and new systems; although this will happen transparently to users, it is extra work for IS staff. As with each of the other approaches (except phased installation), the whole system is installed; however, some parts of the organization will not get the benefits of the new system until the pilot installation has been completely tested.

Phased Installation

Phased installation, also called *staged installation*, is an incremental approach. With phased installation, the new system is brought online in functional components; different parts of the old and new systems are used in cooperation until the whole new system is installed. (Figure 13-6d shows the phase-in of the first two modules of a new system.) Phased installation, like single-location installation, is an attempt to limit the organization's exposure to risk, whether in terms of cost or disruption of the business. By converting gradually, the organization's risk is spread out over time and place. Also, a phased installation allows for some benefits from the new system before the whole system is ready. For example, new data-capture methods can be used before all reporting modules are ready. For a phased installation, the new and replaced systems must be able to coexist and probably share data. Thus, bridge programs connecting old and new databases and programs often must be built. Sometimes, the new and old systems are so incompatible (built using totally different structures) that pieces of the old system cannot be incrementally replaced, so this strategy is not feasible. A phased installation is akin to bringing out a sequence of releases of the system. Thus, a phased approach requires careful version control, repeated conversions at each phase, and a long period of change, which may be frustrating and confusing to users. On the other hand, each phase of change is smaller and more manageable for all involved.

Phased installation
Changing from the old information system to the new one incrementally, starting with one or a few functional components and then gradually extending the installation to cover the whole new system.

Planning Installation

Each installation strategy involves converting not only software, but also data and (potentially) hardware, documentation, work methods, job descriptions, offices and other facilities, training materials, business forms, and other aspects of the system. For example, it is necessary to recall or replace all the current system documentation and business forms, which suggests that the IS department must keep track of who has these items so that they can be notified and receive replacement items. In practice, you will rarely choose a single strategy to the exclusion of all others; most installations will rely on a combination of two or more approaches. For example, if you choose a single-location strategy, you have to decide how installation will proceed there and at subsequent sites. Will it be direct, parallel, or phased?

Of special interest in the installation process is the conversion of data. Because existing systems usually contain data required by the new system, current data must be made error free, unloaded from current files, combined with new data, and loaded into new files. Data may need to be reformatted to be consistent with more advanced data types supported by newer technology used to build the new system. New data fields may have to be entered in large quantities so that every record copied from the current system has all the new fields populated. Manual tasks, such as taking a physical inventory, may need to be done in order to validate data before they are transferred to the new files. The total data conversion process can be tedious. Furthermore, this process may require that current systems be shut off while the data are extracted so that updates to old data, which would contaminate the extract process, cannot occur.

Any decision that requires the current system to be shut down, in whole or in part, before the replacement system is in place must be done with care. Typically, off-hours are used for installations that require a lapse in system support. Whether a lapse in service is required or not, the installation schedule should be announced to users well in advance to let them plan their work schedules around outages in service and periods when their system support might be erratic. Successful installation steps should also be announced, and special procedures put in place so that users can easily inform you of problems they encounter during installation periods. You should also plan for emergency staff to be available in case of system failure so that business operations can be recovered and made operational as quickly as possible. Another consideration is the business cycle of the organization. Most organizations face heavy workloads at particular times of year and relatively light loads at other times. A well-known example is the retail industry, where the busiest time of year is the fall, right before the year's major gift-giving holidays. You wouldn't want to schedule installation of a new point-of-sale system to begin December 1 for a department store. Make sure you understand the cyclical nature of the business you are working with before you schedule installation.

Planning for installation may begin as early as the analysis of the organization supported by the system. Some installation activities, such as buying new hardware, remodeling facilities, validating data to be transferred to the new system, and collecting new data to be loaded into the new system, must be done before the software installation can occur. Often the project team leader is responsible for anticipating all installation tasks and assigns responsibility for each to different analysts.

Each installation process involves getting workers to change the way they work. As such, installation should be looked at not as simply installing a new computer system, but as an organizational change process. More than just a computer system is involved—you are also changing how people do their jobs and how the organization operates.

DOCUMENTING THE SYSTEM

In one sense, every systems development project is unique and will generate its own unique documentation. The approach taken by the development team, whether more traditional and plan oriented or more Agile, will also determine the amount and type of documentation that is generated. System development projects do have many similarities, however, which dictate that certain activities be undertaken and which of those activities must be documented. Bell and Evans (1989) illustrate how a generic SDLC maps onto a generic list of when specific systems development documentation elements are finalized (Table 13-5). As you compare the generic life cycle in Table 13-5 with the life cycle presented in this book, you will see that there are differences, but the general structure of both life cycles is the same because both include the basic phases of analysis, design, implementation, and project planning. Specific documentation will vary depending on the life cycle you are following, and the format and content of the documentation may be mandated by the organization for which you work. However, a basic outline of documentation can be adapted for specific needs, as shown in Table 13-5. Note that this table indicates when documentation is typically finalized; you should start developing documentation elements early, as the information needed is captured.

TABLE 13-5 SDLC and Generic Documentation Corresponding to Each Phase

Generic Life-Cycle Phase	Generic Document
Requirements Specification	System Requirements Specification
	Resource Requirements Specification
Project Control Structuring	Management Plan
	Engineering Change Proposal
System Development	
Architectural design	Architecture Design Document
Prototype design	Prototype Design Document
Detailed design and implementation	Detailed Design Document
Test specification	Test Specifications
Test implementation	Test Reports
System Delivery	User's Guide
	Release Description
	System Administrator's Guide
	Reference Guide
	Acceptance Sign-Off

(*Source:* Adapted from Bell and Evans, 1989.)

We can simplify the situation even more by dividing documentation into two basic types, **system documentation** and **user documentation**. System documentation records detailed information about a system's design specifications, its internal workings, and its functionality. In Table 13-5, all of the documentation listed (except for System Delivery) would qualify as system documentation. System documentation can be further divided into internal and external documentation (Martin and McClure, 1985). **Internal documentation** is part of the program source code or is generated at compile time. **External documentation** includes the outcome of all of the structured diagramming techniques you have studied in this book, such as data flow and entity-relationship diagrams. Although not part of the code itself, external documentation can provide useful information to the primary users of system documentation—maintenance programmers. In the past, external documentation was typically discarded after implementation, primarily because it was considered too costly to keep up to date, but today's environment makes it possible to maintain and update external documentation as long as desired.

Whereas system documentation is intended primarily for maintenance programmers (see Chapter 14), user documentation is intended primarily for users. An organization may have definitive standards on system documentation, often consistent with CASE tools and the systems development process. These standards may include the outline for the project dictionary and specific pieces of documentation within it. Standards for user documentation are not as explicit.

System documentation
Detailed information about a system's design specifications, its internal workings, and its functionality.

User documentation
Written or other visual information about an application system, how it works, and how to use it.

Internal documentation
System documentation that is part of the program source code or is generated at compile time.

External documentation
System documentation that includes the outcome of structured diagramming techniques such as data flow and entity-relationship diagrams.

User Documentation

User documentation consists of written or other visual information about an application system, how it works, and how to use it. An excerpt of online user documentation for Microsoft Office appears in Figure 13-7. Notice how the documentation is organized by topic. On the left, there are links that lead to documentation for the various Office tools. On the right are topics specific to Word. In between is a link to a tutorial video for Word users. This particular page is devoted to providing assistance, and the menu at the top also provides links to other relevant pages, like support, downloads, and templates. Such presentation methods have become standard for help files in online PC documentation.

Figure 13-7 shows the content of a user's guide, which is just one type of user documentation. Other types of user documentation include reference guides, quick

FIGURE 13-7
Example of online user documentation
(*Source:* Microsoft Corporation.)

reference guides, release descriptions, system administrator's guides, and acceptance sign-offs (Table 13-5). A reference guide consists of an exhaustive list of a system's functions and commands, usually in alphabetic order. Most online reference guides allow you to search by topic area or by typing in the first few letters of a keyword. Reference guides are very good for locating specific information; they are not as good for learning the broader picture of how to perform all of the steps required for a given task. A quick-reference guide provides essential information about operating a system in a short, concise format. When computer resources are shared and many users perform similar tasks on the same machines (as with airline reservation or mail-order catalog clerks), quick-reference guides are often printed on index cards or as small books and mounted on or near the computer terminal. An outline for a generic user's guide (from Bell and Evans, 1989) is shown in Table 13-6. The purpose of such a guide is to provide information on how users can use a computer system to perform specific tasks. The information in a user's guide is typically ordered by how often tasks are performed and by their complexity.

In Table 13-6, sections with an "n" and a title in square brackets mean that there are many such sections, each for a different topic. For example, for an accounting application, sections 4 and beyond might address topics such as entering a transaction in the ledger, closing the month, and printing reports. The items in parentheses are optional, included as necessary. An index becomes more important for larger user's guides. Figure 13-8 shows a reference guide for help with Microsoft Excel, this time featuring a list of help contents. This particular reference guide is intended for people who have never used Excel before. The organization of user's guides differs from one software product to the next. User guides also differ depending on the intended audience, whether novice or expert. You may want to compare the guide in Figure 13-8 with ones for other packages to identify differences.

A release description contains information about a new system release, including a list of documentation for the new release, features and enhancements, known problems and how they have been dealt with in the new release, and information about installation. A system administrator's guide is intended primarily for those who will install and administer a new system. It contains information about the network on which the system will run, software interfaces for peripherals such as printers, troubleshooting, and setting up user accounts. Finally, an acceptance sign-off allows users to test for proper system installation and then signify their acceptance of the new system with their signatures.

TABLE 13-6   Outline of a Generic User's Guide

Preface
1. Introduction
   1.1. Configurations
   1.2 Function flow
2. User interface
   2.1 Display screens
   2.2 Command types
3. Getting started
   3.1 Login
   3.2 Logout
   3.3 Save
   3.4 Error recovery
   3.n [Basic procedure name]
n. [Task name]
Appendix A—Error Messages
   ([Appendix])
Glossary
   Terms
   Acronyms
Index

(*Source:* Adapted from Bell and Evans, 1989.)

FIGURE 13-8
Contents of the help facility in
Microsoft Excel
(*Source:* Microsoft Corporation.)

TRAINING AND SUPPORTING USERS

Training and **support** are critical for the success of an information system. As the person whom the user holds responsible for the new system, you and other analysts on the project team must ensure that high-quality training and support are available. Although training and support can be talked about as if they are two separate things, in organizational practice the distinction between the two is not all that clear because the two sometimes overlap. After all, both deal with learning about computing.

It is clear that support mechanisms are also a good way to provide training, especially for intermittent users of a system (Eason, 1988). Intermittent or occasional system users are not interested in, nor would they profit from, typical user training methods. Intermittent users must be provided with "point-of-need support"—specific answers to specific questions at the time the answers are needed. A variety of mechanisms, such as the system interface itself and online help facilities, can be designed to provide both training and support at the same time.

Training Information Systems Users

Computer use requires skills, and training people to use computer applications can be expensive for organizations. Training of all types is a major activity in American corporations, but information systems training is often neglected. Many organizations tend to underinvest in computing skills training. It is true that some organizations institutionalize high levels of information system training, but many others offer no systematic training at all.

The type of training needed will vary by system type and user expertise. The list of potential topics from which you will determine if training will be useful include the following:

- Use of the system (e.g., how to enter a class registration request)
- General computer concepts (e.g., computer files and how to copy them)
- Information system concepts (e.g., batch processing)
- Organizational concepts (e.g., FIFO inventory accounting)
- System management (e.g., how to request changes to a system)
- System installation (e.g., how to reconcile current and new systems during phased installation)

Support
Providing ongoing educational and problem-solving assistance to information system users. For in-house developed systems, support materials and jobs will have to be prepared or designed as part of the implementation process.

As you can see from this partial list, many potential topics go beyond simply how to use the new system. It may be necessary for you to develop training for users in other areas so that users will be ready, conceptually and psychologically, to use the new system. Some training, such as concept training, should begin early in the project because this training can assist in the "unfreezing" (helping users let go of long-established work procedures) element of the organizational change process.

Each element of training can be delivered in a variety of ways. Table 13-7 lists the most common training methods used by information system departments. The most common delivery method for corporate training remains traditional instructor-led classroom training (U.S. GAO, 2003). Many times, users turn to the resident expert and to fellow users for training. Users are more likely to turn to local experts for help than to the organization's technical support staff because the local expert understands the users' primary work and the computer systems they use. Given their dependence on fellow users for training, it should not be surprising that end users describe their most common mode of computer training as self-training.

One conclusion from the experience with user training methods is that an effective strategy for training on a new system is to first train a few key users and then organize training programs and support mechanisms that involve these users to provide further training, both formal and on demand. Often, training is most effective if you customize it to particular user groups, and the lead trainers from these groups are in the best position to provide this training to their colleagues.

Increasingly, corporations are turning to e-learning as a key delivery mode for training. Although the term *e-learning* is not precisely defined, it generally means the same thing as distance learning; that is, a formalized learning system designed to be carried out remotely, using computer-based electronic communication. You may have taken a distance-learning course at your school, or you may have experience in on-campus classes with some of the dominant software packages used in e-learning, such as WebCT, Blackboard, or Desire2Learn. E-learning courses can be delivered over the Internet or over company intranets. Such courses can be purchased from vendors or prepared by the corporation's in-house training staff. E-learning is relatively inexpensive compared to traditional classroom training, and it has the additional advantage of being available anytime from just about anywhere. Students can also learn at their own pace. E-learning systems can make available several different elements that enhance the learning experience, including simulations, online access to mentors and experts, e-books, net meetings, and video on demand. Another trend in corporate training is blended learning, the combining of e-learning with instructor-led classroom training. A recent survey reported that over 80 percent of respondents were using e-learning or blended learning to train their employees (Kim et al., 2008). Half of the respondents in the study believed that e-learning would become the dominant training delivery method in their organizations.

Another training method listed in Table 13-7 is software help components. One common type of software help component is called an **electronic performance support system (EPSS)**. EPSSs are online help systems that go beyond simply providing help—they embed training directly into a software package (Cole et al., 1997). An EPSS may take on one or more forms: It can be an online tutorial, provide hypertext-based access to context-sensitive reference material, or consist of an expert system shell that acts as a coach. The main idea behind the development of an EPSS is that the user never has to leave the application to get the benefits of training. Users learn a new system or unfamiliar features at their own pace and on their own machines, without having to lose work time to remote group training sessions. Furthermore, this learning is on demand; a user completes the EPSS when he or she is most motivated to learn. EPSS is sometimes referred to as "just-in-time knowledge." Figure 13-9 shows the beginning of a tutorial for new users of Microsoft's Excel 2010. The tutorial is designed for users of past versions of Excel who are switching to Excel 2010. Users can go through the tutorial at their own pace, whenever they want, stopping and starting it as necessary.

TABLE 13-7 **Types of Training Methods**

Resident expert

Traditional instructor-led classroom training

E-learning/distance learning

Blended learning (combination of instructor-led and e-learning)

Software help components

External sources, such as vendors

Electronic performance support system (EPSS)
Component of a software package or an application in which training and educational information is embedded. An EPSS can take several forms, including a tutorial, an expert system shell, and hypertext jumps to reference materials.

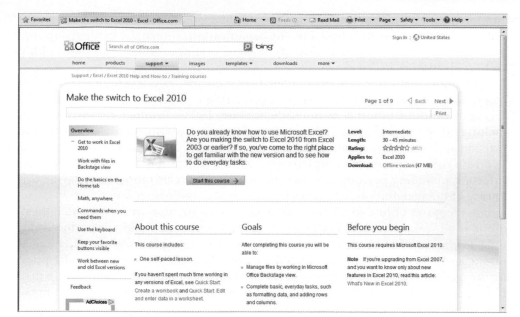

FIGURE 13-9
Beginning of a tutorial on Microsoft
Powerpoint

(*Source:* Microsoft Corporation.)

As both training and support for computing are increasingly able to be delivered online in modules, with some embedded in software packages and applications (as is the case for EPSS), the already blurred distinction between training and support blurs even more. Some of the issues most particular to computer user support are examined in the next section.

Supporting Information Systems Users

Historically, computing support for users has been provided in one of a few forms: on paper, through online versions of paper-based support, by third-party vendors, or by other people who work for the same organization. As we stated earlier, support, whatever its form, has often been inadequate for users' needs. Yet users consider support to be extremely important.

As computing spread throughout organizations, especially with the advent of personal computers, the need for support increased as more and more employees came to rely on computing to do their jobs. As organizations moved to client/server architectures, their need for support increased even more, and organizations began to rely more and more on vendor support (Crowley, 1993). This increased need for support came in part from the lack of standards governing client/server products and the resulting need to make equipment and software from different vendors compatible. Vendors are able to provide the necessary support, but as they have shifted their offerings from primarily expensive mainframe packages to inexpensive off-the-shelf software, they find they can no longer bear the cost of providing the support for free. Most vendors now charge for support, and many have instituted 900 numbers or sell customers unlimited support for a given monthly or annual charge.

Automating Support   In an attempt to cut the costs of providing support and to catch up with the demand for additional support services, vendors have automated many of their support offerings. Online support forums provide users access to information on new releases, bugs, and tips for more effective usage. Forums are offered over the Internet or over company intranets. On-demand fax allows users to order support information through an 800 number and receive that information instantly over their fax machines. Finally, voice-response systems allow users to navigate option menus that lead to prerecorded messages about usage, problems, and workarounds. Organizations have established similar support mechanisms for systems developed or purchased by the organization. Internal e-mail, group support systems, and office automation can be used to support such capabilities within an organization.

Vendors may offer support that enables users to access a vendor's knowledge bases, including electronic support services, a single point of contact, and priority access to vendor support personnel (Schneider, 1993). Product knowledge bases include all of the technical and support information about vendor products and provide additional information for on-site personnel to use in solving problems. Vendors routinely supply complete user and technical documentation over the Internet, including periodic updates, so that a user organization can provide this library of documentation, bug reports, workaround notices, and notes on undocumented features online to all internal users. Electronic support services include all of the vendor support services discussed earlier, but they are tailored specifically for the corporation. The single point of contact is a system engineer who is often based on-site and serves as a liaison between the corporation and the vendor. Finally, priority access means that corporate workers can always get help via telephone or e-mail from a person at the vendor company, usually within a prespecified response time of four hours or less.

Such vendor-enhanced support is especially appropriate in organizations where a wide variety of a particular vendor's products is in use, or where most in-house application development either uses the vendor's products as components of the larger system or where the vendor's products are themselves used as the basis for applications. An example of the former would be the case where an organization has set up a client/server architecture based on a particular vendor's SQL server and APIs. Which applications are developed in-house to run under the client/server architecture depends heavily on the server and APIs, and direct vendor support dealing with problems in these components would be very helpful to the enterprise information systems staff. An example of the second would include order entry and inventory control application systems developed using Microsoft's Access or Excel. In this case, the system developers and users, who are sometimes the same people for such package-based applications, can benefit considerably from directly questioning vendor representatives about their products.

Providing Support through a Help Desk   Whether assisted by vendors or going it alone, the center of support activities for a specific information system in many organizations is the help desk. A **help desk** is an information systems department function and is staffed by IS personnel. The help desk is the first place users should call when they need assistance with an information system. The help desk staff members either deal with the users' questions or refer the users to the most appropriate person.

For many years, a help desk was the dumping ground for people IS managers did not know what else to do with. Turnover rates were high because the help desk was sometimes little more than a complaints department, the pay was low, and burnout rates were high. In today's information-systems-dependent enterprises, however, this situation has changed. Help desks are gaining new respect as management comes to appreciate the special combination of technical skills and people skills needed to make good help desk staffers. The two most valued skills for help desk personnel are related to communication and customer service (Crowley, 1993).

Help desk personnel need to be good at communicating with users, listening to their problems, and intelligently communicating potential solutions. These personnel also need to understand the technology they are helping users with. It is crucial, however, that help desk personnel know when new systems and releases are being implemented and when users are being trained for new systems. Help desk personnel should be well trained on new systems. One sure recipe for disaster is to train users on new systems but not train the help desk personnel these same users will turn to for their support needs.

Help desk
A single point of contact for all user inquiries and problems about a particular information system or for all users in a particular department.

Support Issues for the Analyst to Consider

Support is more than just answering user questions about how to use a system to perform a particular task or about the system's functionality. Support also consists of tasks such as providing for recovery and backup, disaster recovery, and PC maintenance; writing newsletters and offering other types of proactive information sharing; and setting up user groups. It is the responsibility of analysts for a new system to be sure that all forms of support are in place before the system is installed.

For medium to large organizations with active information system functions, many of these issues are dealt with centrally. For example, users may be provided with backup software by the central information systems unit and a schedule for routine backup. Policies may also be in place for initiating recovery procedures in case of system failure. Similarly, disaster recovery plans are almost always established by the central IS unit. Information systems personnel in medium-to-large organizations are also routinely responsible for PC maintenance because the PCs belong to the enterprise. IS unit specialists might also be in charge of composing and transmitting newsletters or overseeing automated bulletin boards and organizing user groups.

When all of these (and more) services are provided by central IS, you must follow the proper procedures to include any new system and its users in the lists of those to whom support is provided. You must design training for the support staff on the new system and make sure that system documentation will be available to it. You must make the support staff aware of the installation schedule and keep these people informed as the system evolves. Similarly, any new hardware and off-the-shelf software has to be registered with the central IS authorities.

When there is no official IS support function to provide support services, you must devise a creative plan to provide as many services as possible. You may have to write backup and recovery procedures and schedules, and the users' departments may have to purchase and be responsible for the maintenance of their hardware. In some cases, software and hardware maintenance may have to be outsourced to vendors or other capable professionals. In such situations, user interaction and information dissemination may have to be more informal than formal: Informal user groups may meet over lunch or over a coffeepot rather than in officially formed and sanctioned forums.

ORGANIZATIONAL ISSUES IN SYSTEMS IMPLEMENTATION

Despite the best efforts of the systems development team to design and build a quality system and to manage the change process in the organization, the implementation effort sometimes fails. Sometimes employees will not use the new system that has been developed for them or, if they do use it, their level of satisfaction with it is very low. Why do systems implementation efforts fail? This question has been the subject of information systems research for over 50 years. In the first part of this section, we will try to provide some answers, looking at the factors that research has identified as important to implementation success. In the second part of this section, you will read about another important organizational issue for information systems, security. You will read about the various threats to the security of organizational systems and some of the remedies that can be applied to help deal with the problem.

Why Implementation Sometimes Fails

The conventional wisdom that has emerged over the years is that there are at least two conditions necessary for a successful implementation effort: management support of the system under development and the involvement of users in the development process (Ginzberg, 1981b). Conventional wisdom holds that if both of these conditions are met, you should have a successful implementation. But despite the

System Implementation Failures

Hershey, famous all over the world for its chocolate, faced a crisis in October 1999. October is one of the key months of the year for candy makers because of the Halloween holiday. In 1999, Hershey was having problems trying to get its candy delivered to warehouses and from there to retailers in time for Halloween. Hershey was having trouble getting orders into its new system and getting the order details to the warehouses for fulfillment. Its new $112 million order-fulfillment system, containing components from SAP, Seibel, and Manugistics, was not working correctly. The system was supposed to have been installed in April of that year, but conversion was delayed until July due to incomplete development and testing. The remaining problems with the system were not found until the next high-volume ordering event of the candy maker's year occurred, Halloween.

Another case of implementation failure involves SAP and the city of Richmond, California. Richmond began installing SAP in 2000. By mid-2004, the city had spent $4.5 million, and the implementation was still not complete. Instead of presenting the city with the functionality it wanted in SAP's R/3, some of the city's department heads said that the system had actually created more work for them. The finance director reported that using the system to prepare the budget had actually required hundreds of hours of extra work on the part of his staff. The planning director reported that the system fell far short of his needs for billing and revenue-tracking. While the city attorney contemplated a lawsuit against SAP and the Denver-based consulting company hired to help with implementation, the city's information technology director maintained that the system

implementation was not a failure at all. At the time, she said that the problems cited by staff were just the usual complaints from people not yet used to new technology. By the end of 2008, the city had decided to switch from SAP R/3 to a system called MUNIS, a system designed specifically for municipalities. The date for going live with MUNIS was set for January 1, 2009.

Avis Europe provides another example. In 2004, Avis Europe incurred a £45m charge as a result of shutting down its credit hire business and due to problems with information technology. In 2003, the company had announced it planned to implement PeopleSoft. A year later, it terminated the project due to delays and additional costs blamed on problems with the system's design and implementation. The cancellation took place before the system had been rolled out to any aspect of Avis's business, minimizing disruption to operations.

Sources: Harper, W. 2004. "System failure." *East Bay Express.* Available at *www.eastbayexpress.com/news/system_failure/Content? oid=287358.* Accessed March 17, 2009; "City Manager's Weekly Report for the week ending October 17th, 2008." Available at *www.ci.richmond.ca.us/Archive.asp?ADID=1931.* Accessed March 17, 2009; Best, J. 2004. "Avis bins PeopleSoft after £45m IT failure." *ZDNet Australia.* Available at *www.zdnet.com.au/news/business/soa/Avis-bins-PeopleSoft-system-after-45m-IT-failure/0,139023166,139164049,00.htm.* Accessed March 17, 2009.

support and active participation of management and users, information systems implementation sometimes fails (see the box "System Implementation Failures" for examples).

Management support and user involvement are important to implementation success, but they may be overrated compared to other factors that are also important. Research has shown that the link between user involvement and implementation success is sometimes weak (Ives and Olson, 1984). User involvement can help reduce the risk of failure when the system is complex, but user participation in the development process only makes failure more likely when there are financial and time constraints in the development process (Tait and Vessey, 1988). Information systems implementation failures are too common, and the implementation process is too complicated, for the conventional wisdom to be completely correct.

Over the years, other studies have found evidence of additional factors that are important to a successful implementation process. Three such factors are: commitment to the project, commitment to change, and the extent of project definition and planning (Ginzberg, 1981b). Commitment to the project involves managing the systems development project so that the problem being solved is well understood and the system being developed to deal with the problem actually solves it. Commitment to change involves being willing to change behaviors, procedures, and other aspects of the organization. The extent of project definition and planning is a measure of how well the project was planned. The more extensive the planning effort is, the less likely implementation failure is. Still another important factor related to implementation success is user expectations (Ginzberg, 1981a). The more realistic a user's early expectations about a new system and its capabilities are, the more likely it is that the user will be satisfied with the new system and actually use it.

Although there are many ways to determine if an implementation has been successful, the two most common and trusted are the extent to which the system is used, and the users' satisfaction with the system (Lucas, 1997). Lucas, who has studied information systems implementation in depth, identified six factors that influence the extent to which a system is used (1997):

1. *User's personal stake.* How important the domain of the system is for the user; in other words, how relevant the system is to the work the user performs. The user's personal stake in the system is itself influenced by the level of support management provides for implementation and by the urgency to the user of the problem addressed by the system. The higher the level of management support and the more urgent the problem, the higher the user's personal stake in the system.
2. *System characteristics.* Includes aspects of the system's design such as ease of use, reliability, and relevance to the task the system supports.
3. *User demographics.* Characteristics of the user, such as age and degree of computer experience.
4. *Organizational support.* These are the same issues of support you read about earlier in this chapter. The better the system support, the more likely an individual will be to use the system.
5. *Performance.* What individuals can do with a system to support their work will have an impact on extent of system use. The more users can do with a system and the more creative ways they can develop to benefit from the system, the more they will use it. The relationship between performance and use goes both ways. The higher the levels of performance, the more use. The more use, the greater the performance.
6. *Satisfaction.* Use and satisfaction also represent a two-way relationship. The more satisfied the users are with the system, the more they will use it. The more they use it, the more satisfied they will be.

The factors identified by Lucas and the relationships they have to each other are shown in the model in Figure 13-10. In the model, it is easier to see the relationships among the various factors, such as how management support and problem urgency affect the user's personal stake in the system. Notice also that the arrows that show the relationships between use and performance and satisfaction have two heads, illustrating the two-way relationships between these factors.

It should be clear that, as an analyst and as someone responsible for the successful implementation of an information system, you have more control over some factors than others. For example, you will have considerable influence over the

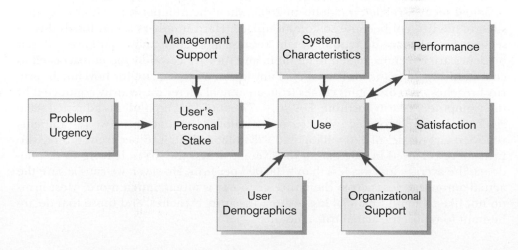

FIGURE 13-10
Implementation success

(*Source:* Adapted from Lucas, H. C. 1997. *Information Technology for Management.* New York: McGraw-Hill, with the permission of the McGraw-Hill Companies. All rights reserved.)

system's characteristics, and you may have some influence over the levels of support that will be provided for users of the system. You have no direct control over a user's demographics, personal stake in the system, management support, or the urgency of the problem to the user. This doesn't mean you can ignore factors that you can't change. On the contrary, you need to understand these factors very well because you will have to balance them with the factors you can change in your system design and in your implementation strategy. You may not be able to change a user's demographics or personal stake in a system, but you can help design the system and your implementation strategy with these factors in mind.

The factors mentioned so far are straightforward. For example, a lack of computer experience can make a user hesitant, inefficient, and ineffective with a system, leading to a system that is not providing its full potential benefit. If top management does not seem to care about the system, why should subordinates care? However, additional factors can be categorized as political and might be more hidden, difficult to affect, and even unrelated to the system that is being implemented, yet instrumental to the system's success.

The basis for political factors is that individuals who work in an organization have their own self-interested goals, which they pursue in addition to the goals of their departments and of their organizations. For example, people might act to increase their own power relative to that of their co-workers; at other times, people will act to prevent co-workers with more power (such as bosses) from using that power or from gaining more. Because information is power, information systems often are seen as instruments of one's ability to influence and exert power. It is helpful to understand the history and politics around an information system, and to deal with negative political factors as well as the more objective and operational ones. Sometimes political interpretations provide better explanations for the implementation process and why events took the course they did.

Once an information system has been successfully implemented, the importance of documentation grows. A successfully implemented system becomes part of the daily work lives of an organization's employees. Many of those employees will use the system, but others will maintain it and keep it running.

Security Issues

The security of information systems has become an increasingly important issue for organizations and their management. According to CERT/CC (Computer Emergency Readiness Team/Coordination Center) at Carnegie Mellon University, the number of unique system vulnerabilities cataloged in 2007 was 7236. That number is seven times greater than the 1090 vulnerabilities reported in 2000. A vulnerability is a weakness in a system that can be readily exploited by someone who knows about it and knows how to take advantage of it. CERT/CC stopped reporting the number of actual security-related incidents in 2003, when the number hit 137,539, because such incidents had become so commonplace. Hard numbers about losses due to security breaches are difficult to obtain because most companies that have suffered breaches are too embarrassed to admit it, and they are certainly too embarrassed to communicate the actual dollar value of any losses. One estimate for how much security breaches cost companies comes from an annual survey on security conducted by the Computer Security Institute. For 2010, 77 out of 149 respondents admitted security breaches and provided estimates of their dollar loss amounts. Two respondents reported large losses of $20 million and $25 million (USD), respectively. If these two firms were removed in calculating the average across the firms, leaving 75 respondents, the average loss was less than $100,000 per firm. However, we can be sure the actual amount of loss across the entire economy is much, much more. Most firms do not like to admit financial losses due to security breaches, and those that do are hesitant to report the actual true amounts.

If organizations are victims of security breaches, what are the sources of these threats? Table 13-8 provides some of the answers. As you might expect, a large proportion of the firms that responded to this survey admitted to having been attacked by the external threats of computer viruses. This would also include Trojan horses, worms, and other kinds of what is called *malware* (**mal**icious soft**ware**). Other external sources of security threats mentioned in the table include laptop theft, system penetration, and denial of service. Denial of service is a popular tactic used to prevent access to a website, orchestrated through sending the website server more messages than it can handle in a short period of time. Note that although external threats are common, internal threats are common as well. The fifth and seventh most common threats involved insider abuse. Employee abuse includes such seemingly innocent activities as sending personal e-mail on company systems or surfing the Internet for personal use during work hours. Although these activities may not damage company systems, they do use company resources that may be needed for work. Downloading large music or video files during work hours on company equipment could actually impede work because downloading large files can consume bandwidth and slow work processes. Unauthorized access to information or privilege escalation by insiders is more devious, as these activities are committed with the intent to harm the firm.

Companies can act, and most do, to deal with information security issues. Although most spend a small proportion of their information systems budgets on security, some companies spend more than 10 percent of their IS budgets (Table 13-8). Smaller companies spend more than larger companies because smaller companies do not benefit from economies of scale in their security spending. You probably spend $30 to $50 per year yourself to protect your desktop or laptop computer by licensing security software and keeping it up to date. When companies and individuals start to think about systems security, they first think about technological solutions to the problem (Schneier, 2000). Common solutions, in addition to anti-malware software, include firewalls, anti-spyware software, and intruder detection software.

Firewalls, used by 95 percent of firms, are built to keep intruders out. A firewall is a set of related programs that protects the resources of a network from users from other networks. Basically, a firewall works closely with a router program to examine each network packet to determine whether to forward it toward its destination. A firewall is often installed in a specially designated computer separate from the rest of the network so that no incoming request can get directly at private network resources. Intrusion detection software (IDS) is used to detect intruders when they make it past the firewall. Intrusion detection systems are used by 62 percent of companies responding to the CSI survey (Table 13-8). An IDS monitors an organization's network traffic looking for unauthorized activity. Such detection systems typically use a signature-based approach or an anomaly-based approach. In a signature-based approach, the IDS searches for known forms of attack much like a virus detection system searches for known viruses.

Yet the weakest link in any computer defense is the people who use the computer system. For example, many system users fail to use good passwords: they may tell other people (including strangers) their passwords, or write their passwords on sticky notes they post on their computer monitors. The best defensive technology in the world cannot overcome human laziness and negligence. Experts argue that the human aspect of computer security can be dealt with through the implementation of procedures and policies regarding user behaviors (Denning, 1999; Mitnick and Simon, 2002). Such policies involve system users not giving out passwords, changing passwords regularly, keeping operating system and virus detection software updated, and so on. Sound systems security practice demands the effective use of appropriate information technologies as well as the diligent involvement of employees and decision makers in the defense of organization information technology assets.

TABLE 13-8 Selected Statistics on IT Security

Nature of Breaches	
Malware infection	67%
Being fraudulently represented as sender of phishing messages	39%
Laptop/mobile theft	34%
Bots/zombies within the organization	29%
Insider abuse of internet access or e-mail	25%
Denial of service	17%
Unauthorized access or privilege escalation by insider	13%
Proportion of IT Budget Devoted to Security	
Proportion spending more than 10%	19%
Proportion spending between 2% and 10%	40%
Proportion spending between 1% and 2%	16%
Proportion spending less than 1%	10%
Security Technologies Used	
Antivirus software	97%
Firewalls	95%
Anti-spyware software	85%
Virtual private network	79%
Vulnerability/Patch management	68%
Encryption of data in transit	66%
Intrusion detection	62%

(*Source:* Data from Computer Security Institute [Richardson], 2011.)

ELECTRONIC COMMERCE APPLICATION: SYSTEM IMPLEMENTATION AND OPERATION FOR PINE VALLEY FURNITURE'S WEBSTORE

Like many other analysis and design activities, system implementation and operation of an Internet-based electronic commerce application is no different than the processes followed for other types of applications. Previously, you read how Jim Woo and the Pine Valley Furniture (PVF) development team transformed the conceptual data model for the WebStore into a set of normalized relations. Here we will examine how the WebStore system was tested before it was installed and brought online.

The programming of all WebStore software modules has been completed. The programmers have extensively tested each unique module, and it is now time to perform a systemwide test of the WebStore. In this section, we will examine how test cases were developed, how bugs were recorded and fixed, and how alpha and beta testing were conducted.

Developing Test Cases for the WebStore

To begin the systemwide testing process, Jim and the PVF development team developed test cases to examine every aspect of the system. Jim knew that system testing, like all other aspects of the SDLC, needed to be a very structured and planned process. Before opening the WebStore to the general public, every module and component of the system needed to be tested within a controlled environment. Based on his experience in implementing other systems, Jim felt that they would need to develop approximately 150 to 200 separate test cases to fully examine the WebStore. To help focus the development of test cases and to assign primary responsibility to members of his team to specific areas of the system, Jim developed the following list of testing categories:

- Simple functionality: Add to cart, list section, calculate tax, change personal data
- Multiple functionality: Add item to cart and change quantity, create user account, change address
- Function chains: Add item to cart, check out, create user account, purchase
- Elective functions: Returned items, lost shipments, item out of stock
- Emergency/crisis: Missing orders, hardware failure, security attacks

The development group broke into five separate teams, each working to develop an extensive set of cases for each of the testing categories. Each team had one day to develop their test cases. Once developed, each team would lead a walkthrough so that everyone would know the totality of the testing process and to facilitate extensive feedback to each team so that the testing process would be as comprehensive as possible. To make this point, Jim stated, "What happens when a customer repeatedly enters the same product into the shopping cart? Can we handle that? What happens when the customer repeatedly enters and then removes a single product? Can we handle that? Although some of these things are unlikely to ever occur, we need to be confident that the system is robust to any type of customer interaction. We must develop every test case necessary to give us confidence that the system will operate as intended, 24-7-365!"

A big part of successful system testing is to make sure that no information is lost and that all tests are described in a consistent way. To achieve this, Jim provided all teams with standard forms for documenting each case and for recording the results of each test. This form had the following sections:

- Test Case ID
- Category/Objective of Test
- Description
- System Version

- Completion Date
- Participant(s)
- Machine Characteristics (processor, operating system, memory, browser, etc.)
- Test Result
- Comments

The teams also developed standard codes for each general type of test, and these were used to create the Test Case ID. For example, all tests related to "Simple Functionality" were given an ID with SF as a prefix and a number as the suffix (e.g., SF001). The teams also developed standards for categorizing tests, listing objectives, and writing other test form contents. Establishing these standards ensured that the testing process would be consistently documented.

Bug Tracking and System Evolution   An outcome of the testing process is the identification of system bugs. Consequently, in addition to setting a standard method for writing and documenting test cases, Jim and the teams established several other rules to ensure a smooth testing process. Experienced developers have long known that an accurate bug-tracking process is essential for rapid troubleshooting and repair during the testing process. You can think of bug tracking as creating a "paper trail" that makes it much easier for programmers to find and repair the bug. To make sure that all bugs were documented in a similar way, the team developed a bug-tracking form that had the following categories:

- Bug Number (simple incremental number)
- Test Case ID That Generated the Bug
- Is the Bug Replicable?
- Effects
- Description
- Resolution
- Resolution Date
- Comments

The PVF development team agreed that bug fixes would be made in batches because all test cases would have to be redone every time the software was changed. The redoing of all the test cases each time the software is changed is done to ensure that in the process of fixing the bug, no other bugs are introduced into the system. As the system moves along in the testing process—as batches of bugs are fixed—the version number of the software is incremented. During the development and testing phases, the version is typically below the "1.0" first release version.

Alpha and Beta Testing the WebStore

After completing all system test cases and resolving all known bugs, Jim moved the WebStore into the alpha-testing phase, in which the entire PVF development team as well as personnel around the company would put the WebStore through its paces. To motivate employees throughout the company to actively participate in testing the WebStore, several creative promotions and giveaways were held. All employees were given a T-shirt that said, "I shop at the WebStore, do you?" Additionally, all employees were given $100 to shop at the WebStore and were offered a free lunch for their entire department if they found a system bug while shopping on the system. Also during alpha testing, the development team conducted extensive recovery, security, stress, and performance testing. Table 13-9 provides a sample of the types of tests performed.

After completing alpha testing, PVF recruited several of their established customers to help in beta testing the WebStore. As real-world customers used the system, Jim was able to monitor the system and fine-tune the servers for optimal system performance. As the system moved through the testing process, fewer and fewer bugs were found. After several days of "clean" usage, Jim felt confident that it was time to open the WebStore for business.

TABLE 13-9   Sample of Tests Conducted on the WebStore during Alpha Testing

Test Type	Tests Performed
Recovery	• Unplug main server to test power backup system • Switch off main server to test the automatic switching to backup server
Security	• Try to purchase without being a customer • Try to examine server directory files both within the PVF domain and when connecting from an outside Internet service provider
Stress	• Have multiple users simultaneously establish accounts, process purchases, add to shopping cart, remove from shopping cart, and so on
Performance	• Examine response time using different connection speeds, processors, memory, browsers, and other system configurations • Examine response time when backing up server data

WebStore Installation

Throughout the testing process, Jim kept PVF management aware of each success and failure. Fortunately, because Jim and the development team followed a structured and disciplined development process, there were far more successes than failures. In fact, he was confident that the WebStore was ready to go online and would recommend to PVF's top management that it was time to "flip the switch" and let the world enter the WebStore.

PROJECT CLOSEDOWN

In Chapter 3, you learned about the various phases of project management, from project initiation to closing down the project. If you are the project manager and you have successfully guided your project through all of the phases of the SDLC presented so far in this book, you are now ready to close down your project. Although the maintenance phase is just about to begin, the development project itself is over. As you will see in the next chapter, maintenance can be thought of as a series of smaller development projects, each with its own series of project management phases.

As you recall from Chapter 3, your first task in closing down the project involves many different activities, from dealing with project personnel to planning a celebration of the project's ending. You will likely have to evaluate your team members, reassign most to other projects, and perhaps terminate others. As project manager, you will also have to notify all of the affected parties that the development project is ending and that you are now switching to maintenance mode.

Your second task is to conduct post-project reviews with both your management and your customers. In some organizations, these post-project reviews will follow formal procedures and may involve internal or EDP (electronic data processing) auditors. The point of a project review is to critique the project, its methods, its deliverables, and its management. You can learn many lessons to improve future projects from a thorough post-project review.

The third major task in project closedown is closing out the customer contract. Any contract that has been in effect between you and your customers during the project (or as the basis for the project) must be completed, typically through the consent of all contractually involved parties. This may involve a formal "signing off" by the clients stating that your work is complete and acceptable. The maintenance phase will typically be covered under new contractual agreements. If your customer is outside your organization, you will also likely negotiate a separate support agreement.

As an analyst member of the development team, your job on this particular project ends during project closedown. You will likely be reassigned to another project dealing with an organizational problem. Maintenance on your new system will begin and continue without you. To complete our consideration of the SDLC, however, we will cover the maintenance phase and its component tasks in Chapter 14.

SUMMARY

This chapter presented an overview of the various aspects of the systems implementation process. You studied seven different types of testing: (1) code inspections, in which the code is examined for well-known errors; (2) walkthroughs, when a group manually examines what the code is supposed to do; (3) desk checking, when an individual mentally executes the computer instructions; (4) syntax checking, typically done by a compiler; (5) unit or module testing; (6) integration testing, in which modules are combined and tested together until the entire program has been tested as a whole; and (7) system testing, in which programs are combined to be tested as a system and where the system's meeting of its objectives is examined. You also learned about acceptance testing, in which users test the system for its ability to meet their requirements, using live data in a live environment.

You read about four types of installation: (1) direct, when the old system is shut off just as the new one is turned on; (2) parallel, when both old and new systems are run together until it is clear the new system is ready to be used exclusively; (3) single-location, when one site is selected to test the new system; and (4) phased, when the system is installed bit by bit.

You learned about four types of documentation: (1) system documentation, which describes in detail the design of a system and its specifications; (2) internal documentation, that part of system documentation that is included in the code itself or that emerges at compile time; (3) external documentation, the part of system documentation that includes the output of diagramming techniques such as data flow and entity-relationship diagramming; and (4) user documentation, which describes a system and how to use it for the system's users.

Computer training has typically been provided in classes and tutorials. Although there is some evidence that lectures have their place in teaching people about computing and information systems, the current emphasis in training is on automated delivery methods, such as online reference facilities, multimedia training, and EPSSs. The latter embed training in the applications themselves in an attempt to make training a seamless part of using an application for daily operations. The emphasis in support is also on providing online delivery, including online support forums. As organizations move toward client/server architectures, they rely more on vendors for support. Vendors provide many online support services, and they work with customers to bring many aspects of online support in-house. A help desk provides aid to users in a particular department or for a particular system.

You saw how information systems researchers have been trying to explain what constitutes a successful implementation. If there is a single main point in this chapter, it is that implementation is a complicated process, from managing programmer teams, to the politics that influence what happens to a system after it has been successfully implemented, to planning and implementing useful training and support mechanisms. Analysts have many factors to identify and manage for a successful system implementation. Successful implementation rarely happens by accident or occurs in a totally predictable manner. The first step in a successful implementation effort may be realizing just that fact. Once systems are implemented, organizations have to deal with threats from both inside and outside the organization to the systems' security. Although technology such as virus-detection software and firewalls can be employed to help secure systems, good security also requires policies and procedures that guide employees in proper system usage.

In many ways, the implementation of an Internet-based system is no different. Just as much careful attention, if not more, has to be paid to the details of an Internet implementation as to a traditional system. Successful implementation for an Internet-based system is not an accident either.

KEY TERMS

1. Acceptance testing
2. Alpha testing
3. Beta testing
4. Desk checking
5. Direct installation
6. Electronic performance support system (EPSS)
7. External documentation
8. Help desk
9. Inspections
10. Installation
11. Integration testing
12. Internal documentation
13. Parallel installation
14. Phased installation
15. Refactoring
16. Single-location installation
17. Stub testing
18. Support
19. System documentation
20. System testing
21. Testing harness
22. Unit testing
23. User documentation

Match each of the key terms above with the definition that best fits it.

_____ A testing technique in which participants examine program code for predictable language-specific errors.

_____ A testing technique in which the program code is sequentially executed manually by the reviewer.

_____ Component of a software package or application in which training and educational information are embedded. It may include a tutorial, expert system, and hypertext jumps to reference materials.

_____ Written or other visual information about an application system, how it works, and how to use it.

_____ Changing over from the old information system to a new one by turning off the old system when the new one is turned on.

_____ Each module is tested alone in an attempt to discover any errors in its code; also called module testing.

_____ The organizational process of changing over from the current information system to a new one.

_____ System documentation that includes the outcome of structured diagramming techniques, such as data flow and entity-relationship diagrams.

_____ The process whereby actual users test a completed information system, the end result of which is the users' acceptance of it.

_____ Detailed information about a system's design specifications, its internal workings, and its functionality.

_____ Running the old information system and the new one at the same time until management decides the old system can be turned off.

_____ The process of bringing together all of the modules that a program comprises for testing purposes. Modules are typically integrated in a top-down, incremental fashion.

_____ A technique used in testing modules, especially modules that are written and tested in a top-down fashion, where a few lines of code are used to substitute for subordinate modules.

_____ Changing from the old information system to the new one incrementally, starting with one or a few functional components and then gradually extending the installation to cover the whole new system.

_____ Bringing together all of the programs that a system comprises for testing purposes. Programs are typically integrated in a top-down, incremental fashion.

_____ System documentation that is part of the program source code or is generated at compile time.

_____ Providing ongoing educational and problem-solving assistance to information system users. Support material and jobs must be designed along with the associated information system.

_____ User testing of a completed information system using real data in the real user environment.

_____ Trying out a new information system at one site and using the experience to decide if and how the new system should be deployed throughout the organization.

_____ User testing of a completed information system using simulated data.

_____ A single point of contact for all user inquiries and problems about a particular information system or for all users in a particular department.

_____ Making a program simple after adding a new feature.

_____ An automated testing environment used to review code for errors, standards violations, and other design flaws.

REVIEW QUESTIONS

1. What are the deliverables from coding, testing, and installation?

2. Explain the code-testing process.

3. What are structured walk-throughs for code? What is their purpose? How are they conducted? How are they different from code inspections?

4. What are the four approaches to installation? Which is the most expensive? Which is the most risky? How does an organization decide which approach to use?

5. What is the conventional wisdom about implementation success?

6. List and define the factors that are important to successful implementation efforts.

7. Explain Lucas's model of implementation success.

8. What is the difference between system documentation and user documentation?

9. What are the common methods of computer training?

10. What is self-training?

11. What is e-learning?

12. What proof do you have that individual differences matter in computer training?

13. Why do corporations rely so heavily on vendor support?

14. Describe the delivery methods many vendors employ for providing support.

15. Describe the various roles typically found in a help desk function.

16. What are the common security threats to systems? How can they be addressed?

PROBLEMS AND EXERCISES

1. Prepare a testing strategy or plan for PVF's Purchasing Fulfillment System.

2. Which installation strategy would you recommend for PVF's Purchasing Fulfillment System? Which would you recommend for Hoosier Burger's inventory control system? If you recommended different approaches, please explain why. How is PVF's case different from Hoosier Burger's?

3. Develop a table that compares the four installation strategies, showing the pros and cons of each. Try to make a direct comparison when a pro of one is a con of another.

4. One of the most difficult aspects of using the single-location approach to installation is choosing an appropriate location. What factors should be considered in picking a pilot site?

5. You have been a user of many information systems including, possibly, a class registration system at your school, a bank account system, a word-processing system, and an airline reservation system. Pick a system you have used and assume you were involved in the beta testing of that system. What criteria would you have used to judge whether this system was ready for general distribution?

6. Why is it important to keep a history of test cases and the results of those test cases even after a system has been revised several times?

7. How much documentation is enough?

8. What is the purpose of electronic performance support systems? How would you design one to support a word-processing package? A database package?

9. Discuss the role of a centralized training and support facility in a modern organization. Given advances in technology and the prevalence of self-training and consulting among computing end users, how can such a centralized facility continue to justify its existence?

10. Is it good or bad for corporations to rely on vendors for computing support? List arguments both for and against reliance on vendors as part of your answer.

11. Suppose you were responsible for establishing a training program for users of Hoosier Burger's inventory control system (described in previous chapters). Which forms of training would you use? Why?

12. Suppose you were responsible for establishing a help desk for users of Hoosier Burger's inventory control system (described in previous chapters). Which support system elements would you create to help users be effective? Why?

13. Your university or school probably has some form of micro-computer center or help desk for students. What functions does this center perform? How do these functions compare to those outlined in this chapter?

14. Suppose you were responsible for organizing the user documentation for Hoosier Burger's inventory control system (described in previous chapters). Write an outline that shows the documentation you would suggest be created, and generate the table of contents or outline for each element of this documentation.

15. What types of security policies and procedures does your university have in place for campus information systems?

FIELD EXERCISES

1. Interview someone you know or have access to who works for a medium to large organization. Ask for details on a specific instance of organizational change: What changed? How did it happen? Was it well planned or ad hoc? How were people in the organization affected? How easy was it for employees to move from the old situation to the new one?

2. Reexamine the data you collected in the interview in Field Exercise 1. This time, analyze the data from a political perspective. How well does the political model explain how the organization dealt with change? Explain why the political model does or does not fit.

3. Ask a systems analyst you know or have access to about implementation. Ask what the analyst believes is necessary for a successful implementation.

4. Prepare a research report on successful and unsuccessful information system implementations. After you have found information on two or three examples of both successful and unsuccessful system implementations, try to find similarities and differences among the examples of each type of implementation. Do you detect any patterns? Can you add

any factors important to success that were not mentioned in this chapter?

5. Talk with people you know who use computers in their work. Ask them to get copies of the user documentation they rely on for the systems they use at work. Analyze the documentation. Would you consider it good or bad? Support your answer. Whether good or bad, how might you improve it?

6. Volunteer to work for a shift at a help desk at your school's computer center. Keep a journal of your experiences. What kind of users did you have to deal with? What kinds of questions did you get? Do you think help desk work is easy or hard? What skills are needed by someone in this position?

7. Let's say your professor has asked you to help him or her train a new secretary on how to prepare class notes for electronic distribution to class members. Your professor uses word-processing software and an e-mail package to prepare and distribute the notes. Assume the secretary knows nothing about either package. Prepare a user task guide that shows the secretary how to complete this task.

REFERENCES

Beck, K. and C. Andres. 2004. *eXtreme Programming eXplained.* Upper Saddle River, NJ: Addison-Wesley.

Bell, P., and C. Evans. 1989. *Mastering Documentation.* New York: John Wiley & Sons.

Bozman, J. S. 1994. "United to Simplify Denver's Troubled Baggage Project." *ComputerWorld* 10(10): 76.

CERT/CC. *www.cert.org/.* Accessed March 3, 2009.

Cole, K., O. Fischer, and P. Saltzman. 1997. "Just-in-Time Knowledge Delivery." *Communications of the ACM* 40(7): 49–53.

Crowley, A. 1993. "The Help Desk Gains Respect." *PC Week* 10 (November 15): 138.

Denning, D.E. 1999. *Information Warfare and Security.* Boston: Addison-Wesley.

Downes, S. 2005. "E-learning 2.0." *E-learn Magazine.* Available at *elearning.org.* Accessed February 18, 2009.

Eason, K. 1988. *Information Technology and Organisational Change.* London: Taylor & Francis.

Fagan, M. E. 1986. "Advances in Software Inspections." *IEEE Transactions on Software Engineering* 12(7): 744–51.

Ginzberg, M. J. 1981a. "Early Diagnosis of MIS Implementation Failure: Promising Results and Unanswered Questions." *Management Science* 27(4): 459–78.

Ginzberg, M. J. 1981b. "Key Recurrent Issues in the MIS Implementation Process." *MIS Quarterly* 5(2): 47–59.

Griffin, G., and J. Leib. 2003. "Webb Blasts United." DenverPost.com. *www.denverpost.com/Stories/0,1413,36 %257E26385%257E1514432,00.html.* July 16. Accessed February 18, 2009.

Ives, B., and M. H. Olson. 1984. "User Involvement and MIS Success: A Review of Research." *Management Science* 30(5): 586–603.

Kim, K-J., C. J. Bonk, and Oh, E. 2008. "The Present and Future State of Blended Learning in Workplace Learning Setting in the United States." *Performance Improvement* 47(8): 5-16.

Lucas, H. C. 1997. *Information Technology for Management.* New York: McGraw-Hill.

Martin, J., and C. McClure. 1985. *Structured Techniques for Computing.* Upper Saddle River, NJ: Prentice Hall.

Mitnick, K. D., and W. L. Simon. 2002. *The Art of Deception.* New York: John Wiley & Sons.

Mosley, D. J. 1993. *The Handbook of MIS Application Software Testing.* Upper Saddle River, NJ: Prentice Hall/Yourdon Press.

Richardson, R. 2011. *2010–2011 CSI Computer Crime and Security Survey.* Computer Security Institute. Available at *www.GoSCI .com.*

Schneider, J. 1993. "Shouldering the Burden of Support." *PC Week* 10 (November 15): 123, 129.

Schneier, B. 2000. *Secrets and Lies: Digital Security in a Networked World.* New York: John Wiley & Sons.

Schurr, A. 1993. "Support is No. 1." *PC Week* (November 15): 126.

SearchSecurity.com. *www.searchsecurity.com.* Accessed February 18, 2009.

Tait, P., and I. Vessey. 1988. "The Effect of User Involvement on System Success: A Contingency Approach." *MIS Quarterly* 12(1): 91–108.

Torkzadeh, G., and W. J. Doll. 1993. "The Place and Value of Documentation in End-User Computing." *Information & Management* 24(3): 147–58.

United States General Accounting Office (U.S. GAO). 2003. "Information Technology Training: Practices of Leading Private-Sector Companies." Available at *www.gao.gov/ getrpt?GAO-03-390.* Accessed February 18, 2009.

Yourdon, E. 1989. *Managing the Structured Techniques,* 4th ed. Upper Saddle River, NJ: Prentice Hall.

PETRIE ELECTRONICS

Chapter 13: System Implementation

Jim Watanabe was in his new car, driving down I-5, on his way to work. He dreaded the phone call he knew he was going to have to make.

The original go-live date for a pilot implementation of Petrie Electronics' new customer relationship management (CRM) system was July 31. That was only six weeks away, and Jim knew there was no way they were going to be ready. The XRA CRM they were licensing turned out to be a lot more complex than they had thought. They were behind schedule in implementing it. Sanjay Agarwal, who was a member of Jim's team and who was in charge of systems integration for Petrie, wanted Jim to hire some consultants with XRA experience to help with implementation. So far, Jim had been able to stay under budget, but missing his deadlines and hiring some consultants would push him over his budget limit.

It didn't help that John Smith, head of marketing, kept submitting requests for changes to the original specifications for the customer loyalty program. As specified in the project charter, the new system was supposed to track customer purchases, assign points for cumulative purchases, and allow points to be redeemed for "rewards" at local stores. The team had determined that those rewards would take the form of dollars-off coupons. Customers who enrolled in the program would be given accounts which they could access from Petrie's website. When they signed on, they could check their account activity to see how many points they had accumulated. If they had earned enough points, they were rewarded with a coupon. If they wanted to use the coupon, they would have to print it out on their home printers and bring it in to a store to use on a purchase. The team had decided long ago that keeping everything electronic saved Petrie the considerable costs of printing and mailing coupons to customers.

But now marketing had put in a change request that would give customers a choice of having coupons mailed to them automatically or printing them from the website at home. This option, although nice for customers, added complexity to the XRA system implementation, and it added to the costs of operation. Jim had also learned yesterday from the marketing representative on his team, Sally Fukuyama, that now Smith wanted another change. Now he wanted customers to be able to use the coupons for online purchases from Petrie's website. This change added a whole new layer of complexity, affecting Petrie's existing systems for ordering online, in addition to altering yet again the implementation of the XRA CRM.

As if that wasn't enough, Juanita Lopez was now telling Jim that she would not be ready to let the team pilot the system in her Irvine store. Juanita was saying her store would not be ready by the end of July. Maybe that wouldn't matter, since they were going to miss the go-live date for the pilot. But Juanita was hinting she would not be ready for months after that. It seemed as if she didn't want her store to be used for the pilot at all. Jim didn't understand it. But maybe he should try to find another store to use as the pilot site.

Jim was almost at his exit. Soon he would be at the office, and he would have to call Ella Whinston and tell her the status of the project. He would have to tell her that they would miss the go-live date, but in a way it didn't matter since he didn't have a pilot location to go live at. In addition to going over schedule, he was going to have to go over budget, too. He didn't see any way they would be ready for the pilot anywhere close to when they had scheduled, unless he hired the consultants Sanjay wanted. And he would have to stop the latest change request filed by marketing. Even more important, he would have to keep the rumored change request, about using coupons for online purchases, from being submitted in the first place.

Maybe, just maybe, if he could hire the consultants, fight off the change requests, and get Juanita to cooperate, they might be ready to go live with a pilot in Irvine on October 15. That gave him four months to complete the project. He and the team were going to have to work hard to make that happen.

Jim realized he had missed his exit. Great, he thought, I hope it gets better from here.

Case Questions

1. Why don't information systems projects work out as planned? What causes the differences between the plan and reality?
2. Why is it important to document change requests? What happens if a development team doesn't?
3. When a project is late, do you think that adding more people to do the work helps or not? Justify your answer.
4. What is the role of a pilot project in information systems analysis? Why do you think the Petrie's team decided to do a pilot project before rolling out the customer loyalty system for everyone?
5. Information systems development projects are said to fail if they are late, go over budget, or do not contain all of the functionality they were designed to have. Is the customer loyalty program a failure? Justify your answer. If not, how can failure be prevented? Is it important to avert failure? Why or why not?

CHAPTER FOURTEEN

Maintaining Information Systems

Learning Objectives

After studying this chapter, you should be able to:

- Explain and contrast four types of maintenance.

- Describe several factors that influence the cost of maintaining an information system and apply these factors to the design of maintainable systems.

- Describe maintenance management issues, including alternative organizational structures, quality measurement, processes for handling change requests, and configuration management.

- Explain the role of CASE tools in maintaining information systems.

Introduction

In this chapter, we discuss systems maintenance, the largest systems development expenditure for many organizations. In fact, more programmers today work on maintenance activities than work on new development. Your first job after graduation may very well be as a maintenance programmer/analyst. This disproportionate distribution of maintenance programmers is interesting because software does not wear out in a physical manner as do buildings and machines.

There is no single reason why software is maintained; however, most reasons relate to a desire to evolve system functionality in order to overcome internal processing errors or to better support changing business needs. Thus, maintenance is a fact of life for most systems. This means that maintenance can begin soon after the system is installed. As with the initial design of a system, maintenance activities are not limited only to software changes, but include changes to hardware and business procedures. A question many people have about maintenance relates to how long organizations should maintain a system. Five years? Ten years? Longer? There is no simple answer to this question, but it is most often an issue of economics. In other words, at what point does it make financial sense to discontinue evolving an older system and build or purchase a new one? The focus of a great deal of upper IS management attention is devoted to assessing the trade-offs between maintenance and new development. In this chapter, we will provide you with a better understanding of the maintenance process and describe the types of issues that must be considered when maintaining systems.

In this chapter, we also briefly describe the systems maintenance process and the deliverables and outcomes from this process. This is followed by a detailed discussion contrasting the types of maintenance, an overview of critical management issues, and a description of the role of Computer-Aided Software Engineering (CASE) and automated development tools in the maintenance process. Finally, we describe the process of maintaining Web-based applications, including an example for the Pine Valley Furniture's (PVF's) WebStore application.

MAINTAINING INFORMATION SYSTEMS

Once an information system is installed, the system is essentially in the maintenance phase of the systems development life cycle (SDLC). When a system is in the maintenance phase, some person within the systems development group is responsible for collecting maintenance requests from system users and other interested parties, such as system auditors, data center and network management staff, and data analysts. Once collected, each request is analyzed to better understand how it will alter the system and what business benefits and necessities will result from such a change. If the change request is approved, a system change is designed and then implemented. As with the initial development of the system, implemented changes are formally reviewed and tested before installation into operational systems.

The Process of Maintaining Information Systems

As we can see in Figure 14-1, the maintenance phase is the last phase of the SDLC. It is here that the SDLC becomes a cycle, with the last activity leading back to the first. This means that the process of maintaining an information system is the process of returning to the beginning of the SDLC and repeating development steps until the change is implemented.

Also shown in Figure 14-1, four major activities occur within maintenance:

1. Obtaining maintenance requests
2. Transforming requests into changes
3. Designing changes
4. Implementing changes

FIGURE 14-1
Systems development life cycle

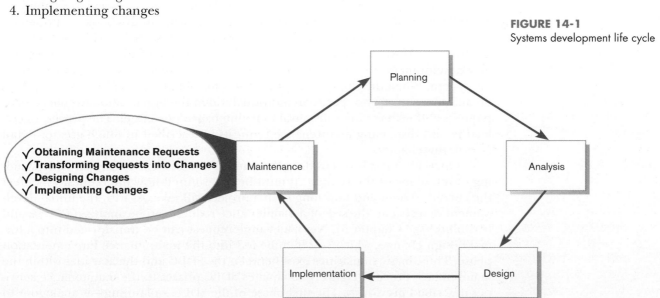

FIGURE 14-2
System Service Request for purchasing
fulfillment system (Pine Valley Furniture)

Pine Valley Furniture
System Service Request

REQUESTED BY   Juanita Lopez   DATE   November 5, 2014

DEPARTMENT   Purchasing, Manufacturing Support

LOCATION   Headquarters, 1-322

CONTACT   Tel: 4-3267   FAX: 4-3270   e-mail: jlopez

TYPE OF REQUEST   URGENCY

[X]   New System   [   ]   Immediate—Operations are impaired or
                                                    opportunity lost
[   ]   System Enhancement   [   ]   Problems exist, but can be worked around
[   ]   System Error Correction   [X]   Business losses can be tolerated until new
                                                    system is installed

PROBLEM STATEMENT

Sales growth at PVF has caused greater volume of work for the manufacturing support unit within Purchasing. Further, more concentration on customer service has reduced manufacturing lead times, which puts more pressure on purchasing activities. In addition, cost-cutting measures force Purchasing to be more aggressive in negotiating terms with vendors, improving delivery times, and lowering our investments in inventory. The current modest systems support for manufacturing purchasing is not responsive to these new business conditions. Data are not available, information cannot be summarized, supplier orders cannot be adequately tracked, and commodity buying is not well supported. PVF is spending too much on raw materials and not being responsive to manufacturing needs.

SERVICE REQUEST

I request a thorough analysis of our current operations with the intent to design and build a completely new information system. This system should handle all purchasing transactions, support display and reporting of critical purchasing data, and assist purchasing agents in commodity buying.

IS LIAISON   Chris Martin (Tel: 4-6204   FAX: 4-6200   e-mail: cmartin)

SPONSOR   Sal Divario, Director, Purchasing

-------------------------- TO BE COMPLETED BY SYSTEMS PRIORITY BOARD --------------------------
[   ]   Request approved   Assigned to _____
                               Start date _____
[   ]   Recommend revision
[   ]   Suggest user development
[   ]   Reject for reason _____

Obtaining maintenance requests requires that a formal process be established whereby users can submit system change requests. Earlier in this book, we presented a user request document called a System Service Request (SSR), which is shown in Figure 14-2. Most companies have some sort of document like an SSR to request new development, to report problems, or to request new features within an existing system. When developing the procedures for obtaining maintenance requests, organizations must also specify an individual within the organization to collect these requests and manage their dispersal to maintenance personnel. The process of collecting and dispersing maintenance requests is described in much greater detail later in this chapter.

Once a request is received, analysis must be conducted to gain an understanding of the scope of the request. It must be determined how the request will affect the current system and how long such a project will take. As with the initial development of a system, the size of a maintenance request can be analyzed for risk and feasibility (see Chapter 5). Next, a change request can be transformed into a formal design change, which can then be fed into the maintenance implementation phase. Thus, many similarities exist between the SDLC and the activities within the maintenance process. Figure 14-3 equates SDLC phases to the maintenance activities described previously. The first phase of the SDLC—planning—is analogous to

FIGURE 14-3
Maintenance activities parallel those of the SDLC

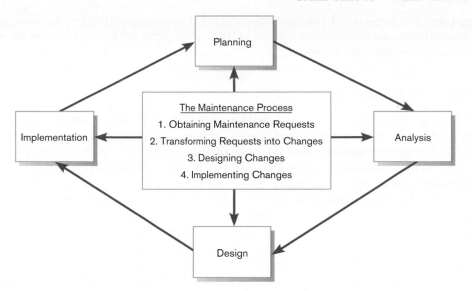

the maintenance process of obtaining a maintenance request (step 1). The SDLC analysis phase is analogous to the maintenance process of transforming requests into a specific system change (step 2). The SDLC design phase, of course, equates to the designing changes process (step 3). Finally, the SDLC phase implementation equates to step 4, implementing changes. This similarity between the maintenance process and the SDLC is no accident. The concepts and techniques used to initially develop a system are also used to maintain it.

Deliverables and Outcomes

Because the maintenance phase of the SDLC is basically a subset of the activities of the entire development process, the deliverables and outcomes from the process are the development of a new version of the software and new versions of all design documents developed or modified during the maintenance process. This means that all documents created or modified during the maintenance effort, including the system itself, represent the deliverables and outcomes of the process. Those programs and documents that did not change may also be part of the new system. Because most organizations archive prior versions of systems, all prior programs and documents must be kept to ensure the proper versioning of the system. This enables prior versions of the system to be re-created if needed. A more detailed discussion of configuration management and change control is presented later in this chapter.

Because of the similarities among the steps, deliverables, and outcomes of new development and maintenance, you may be wondering how to distinguish between these two processes. One difference is that maintenance reuses most existing system modules in producing the new system version. Other distinctions are that a new system is developed when there is a change in the hardware or software platform or when fundamental assumptions and properties of the data, logic, or process models change.

CONDUCTING SYSTEMS MAINTENANCE

A significant portion of the expenditures for information systems within organizations does not go to the development of new systems but to the maintenance of existing systems. We will describe various types of maintenance, factors influencing the complexity and cost of maintenance, alternatives for managing maintenance, and the role of CASE tools in maintenance. Given that maintenance activities

TABLE 14-1 **Types of Maintenance**

Type	Description
Corrective	Repair design and programming errors
Adaptive	Modify system to environmental changes
Perfective	Evolve system to solve new problems or take advantage of new opportunities
Preventive	Safeguard system from future problems

Maintenance
Changes made to a system to fix or enhance its functionality.

Corrective maintenance
Changes made to a system to repair flaws in its design, coding, or implementation.

Adaptive maintenance
Changes made to a system to evolve its functionality to changing business needs or technologies.

Perfective maintenance
Changes made to a system to add new features or to improve performance.

Preventive maintenance
Changes made to a system to avoid possible future problems.

consume the majority of information-systems-related expenditures, gaining an understanding of these topics will yield numerous benefits to your career as an information systems professional.

Types of Maintenance

You can perform several types of **maintenance** on an information system (see Table 14-1). By maintenance, we mean the fixing or enhancing of an information system. **Corrective maintenance** refers to changes made to repair defects in the design, coding, or implementation of the system. For example, if you had recently purchased a new home, corrective maintenance would involve repairs made to things that had never worked as designed, such as a faulty electrical outlet or a misaligned door. Most corrective maintenance problems surface soon after installation. When corrective maintenance problems surface, they are typically urgent and need to be resolved to curtail possible interruptions in normal business activities. Of all types of maintenance, corrective accounts for as much as 75 percent of all maintenance activity (Andrews and Leventhal, 1993; Pressman, 2005). This is unfortunate because corrective maintenance adds little or no value to the organization; it simply focuses on removing defects from an existing system without adding new functionality (see Figure 14-4).

Adaptive maintenance involves making changes to an information system to evolve its functionality to changing business needs or to migrate it to a different operating environment. Within a home, adaptive maintenance might be adding storm windows to improve the cooling performance of an air conditioner. Adaptive maintenance is usually less urgent than corrective maintenance because business and technical changes typically occur over some period of time. Contrary to corrective maintenance, adaptive maintenance is generally a small part of an organization's maintenance effort, but it adds value to the organization.

Perfective maintenance involves making enhancements to improve processing performance or interface usability or to add desired, but not necessarily required, system features ("bells and whistles"). In our home example, perfective maintenance would be adding a new room. Many systems professionals feel that perfective maintenance is not really maintenance but rather new development.

Preventive maintenance involves changes made to a system to reduce the chance of future system failure. An example of preventive maintenance might be to increase the number of records that a system can process far beyond what is currently needed or to generalize how a system sends report information to a printer so that the system can easily adapt to changes in printer technology. In our home example, preventive maintenance could be painting the exterior to better protect the home from severe weather conditions. As with adaptive maintenance, both perfective and preventive maintenance are typically a much lower priority than corrective maintenance. Over the life of a system, corrective maintenance is most likely to occur after initial system installation or after major system changes. This means

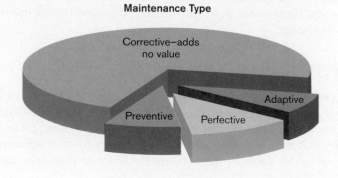

FIGURE 14-4
Value and non-value adding of different types of maintenance
(*Sources:* Based on Andrews and Leventhal, 1993; Pressman, 2005.)

that adaptive, perfective, and preventive maintenance activities can lead to corrective maintenance activities if not carefully designed and implemented.

The Cost of Maintenance

Information systems maintenance costs are a significant expenditure. For some organizations, as much as 60 to 80 percent of their information systems budget is allocated to maintenance activities (Kaplan, 2002). These huge maintenance costs are due to the fact that many organizations have accumulated more and more older, so-called legacy systems that require more and more maintenance (see Figure 14-5). More maintenance means more maintenance work for programmers. For systems developed in-house, on average, 52 percent of a company's programmers are assigned to maintain existing software (Lytton, 2001). In situations where a company has not developed its systems in-house but instead has licensed software, as in the case of ERP systems, maintenance costs remain high. The standard cost of maintenance for most ERP vendors is 22 percent annually (Nash, 2010). In addition, about one-third of the costs of establishing and keeping a presence on the Web go to programming maintenance (Legard, 2000). These high costs associated with maintenance mean that you must understand the factors influencing the maintainability of systems. Maintainability is the ease with which software can be understood, corrected, adapted, and enhanced. Systems with low maintainability result in uncontrollable maintenance expenses.

Numerous factors influence the **maintainability** of a system. These factors, or cost elements, determine the extent to which a system has high or low maintainability. Of these factors, three are most significant: the number of latent defects, the number of customers, and documentation quality. The others—personnel, tools, and software structure—have noticeable, but less, influence.

- *Latent defects.* This is the number of unknown errors existing in the system after it is installed. Because corrective maintenance accounts for most maintenance activity, the number of latent defects in a system influences most of the costs associated with maintaining a system.
- *Number of customers for a given system.* In general, the greater the number of customers, the greater the maintenance costs. For example, if a system has only one customer, problem and change requests will come from only one source. Also, training, error reporting, and support will be simpler. Maintenance requests are less likely to be contradictory or incompatible.

Maintainability
The ease with which software can be understood, corrected, adapted, and enhanced.

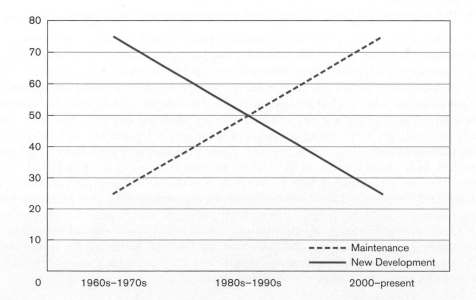

FIGURE 14-5
New development versus maintenance as a percentage of the software budget over the years

(*Source:* Based on Pressman, 2005.)

FIGURE 14-6

Quality documentation eases
maintenance

(*Source:* Based on Hanna, M. 1992. "Using Documentation as a Life-Cycle Tool." *Software Magazine* [December]: 41–46.)

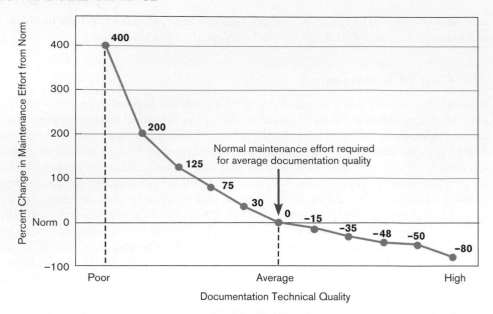

- *Quality of system documentation.* Without quality documentation, maintenance efforts can increase exponentially. For example, Figure 14-6 shows that the system maintenance effort takes 400 percent longer with poor quality documentation. High-quality documentation leads to an 80 percent reduction in the system maintenance effort when compared with average-quality documentation. In other words, quality documentation makes it easier to find code that needs to be changed and to understand how the code needs to be changed. Good documentation also explains why a system does what it does and why alternatives were not feasible, which saves wasted maintenance efforts.
- *Maintenance personnel.* In some organizations, the best programmers are assigned to maintenance. Highly skilled programmers are needed because the maintenance programmer is typically not the original programmer and must quickly understand and carefully change the software.
- *Tools.* Tools that can automatically produce system documentation where none exists can also lower maintenance costs. Also, tools that can automatically generate new code based on system specification changes can dramatically reduce maintenance time and costs.
- *Well-structured programs.* Well-designed programs are easier to understand and fix.

Since the mid-1990s, many organizations have taken a new approach to managing maintenance costs. Rather than develop custom systems internally or through contractors, they have chosen to buy packaged application software. Although vendors of packaged software charge an annual maintenance fee for updates, these charges are more predictable and lower than for custom-developed systems (Worthen, 2003). However, internal maintenance work may still be necessary when using packages. One major maintenance task is to make the packaged software compatible with other packages and internally developed systems with which it must cooperate. When new releases of the purchased packages appear, maintenance may be needed to make all the packages continue to share and exchange data. Some companies are minimizing this effort by buying comprehensive packages, such as ERP packages, which provide information services for a wide range of organizational functions (from human resources to accounting, manufacturing, and sales and marketing). Although the initial costs to install such ERP packages can be significant, they promise great potential for drastically reducing system maintenance costs.

TABLE 14-2   Advantages and Disadvantages of Different Maintenance Organizational Structures

Type	Advantages	Disadvantages
Separate	Formal transfer of systems between groups improves the system and documentation quality	All things cannot be documented, so the maintenance group may not know critical information about the system
Combined	Maintenance group knows or has access to all assumptions and decisions behind the system's original design	Documentation and testing thoroughness may suffer due to a lack of a formal transfer of responsibility
Functional	Personnel have a vested interest in effectively maintaining the system and have a better understanding of functional requirements	Personnel may have limited job mobility and lack access to adequate human and technical resources

Managing Maintenance

As maintenance activities consume more and more of the systems development budget, maintenance management has become increasingly important. Today, far more programmers worldwide are working on maintenance than on new development. In other words, *maintenance* is the largest segment of programming personnel, and this implies the need for careful management. We will address this concern by discussing several topics related to the effective management of systems maintenance.

Managing Maintenance Personnel   One concern with managing maintenance relates to personnel management. Historically, many organizations had a "maintenance group" that was separate from the "development group." With the increased number of maintenance personnel, the development of formal methodologies and tools, changing organizational forms, end-user computing, and the widespread use of very high-level languages for the development of some systems, organizations have rethought the organization of maintenance and development personnel. In other words, should the maintenance group be separated from the development group? Or should the same people who build the system also maintain it? A third option is to let the primary end users of the system in the functional units of the business have their own maintenance personnel. The advantages and disadvantages to each of these organizational structures are summarized in Table 14-2.

In addition to the advantages and disadvantages listed in Table 14-2, there are numerous other reasons why organizations should be concerned with how they manage and assign maintenance personnel. One key issue is that many systems professionals don't want to perform maintenance because they feel that it is more exciting to build something new than change an existing system (Martin et al., 2008). In other words, maintenance work is often viewed as "cleaning up someone else's mess." Also, organizations have historically provided greater rewards and job opportunities to those performing new development, thus making people shy away from maintenance-type careers. As a result, no matter how an organization chooses to manage its maintenance group—separate, combined, or functional—it is now common to rotate individuals in and out of maintenance activities. This rotation is believed to lessen the negative feelings about maintenance work and to give personnel a greater understanding of the difficulties of and relationships between new development and maintenance.

Measuring Maintenance Effectiveness   A second management issue is the measurement of maintenance effectiveness. As with the effective management of personnel, the measurement of maintenance activities is fundamental to understanding the

FIGURE 14-7
How the mean time between failures
should change over time

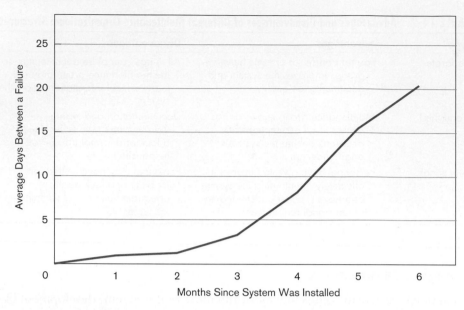

quality of the development and maintenance efforts. To measure effectiveness, you must measure the following factors:

- Number of failures
- Time between each failure
- Type of failure

Measuring the number of and time between failures will provide you with the basis to calculate a widely used measure of system quality. This metric is referred to as the **mean time between failures (MTBF)**. As its name implies, the MTBF metric shows the average length of time between the identification of one system failure and the next. Over time, you should expect the MTBF value to rapidly increase after a few months of use (and corrective maintenance) of the system (see Figure 14-7 for an example of the relationship between MTBF and age of a system). If the MTBF does not rapidly increase over time, it will be a signal to management that major problems exist within the system that are not being adequately resolved through the maintenance process.

A more revealing method of measurement is to examine the failures that are occurring. Over time, logging the types of failures will provide a very clear picture of where, when, and how failures occur. For example, knowing that a system repeatedly fails to log new account information to the database when a particular customer is using the system can provide invaluable information to the maintenance personnel. Were the users adequately trained? Is there something unique about this user? Is there something unique about an installation that is causing the failure? What activities were being performed when the system failed?

Tracking the types of failures also provides important management information for future projects. For example, if a higher frequency of errors occurs when a particular development environment is used, such information can help guide personnel assignments; training courses; or the avoidance of a particular package, language, or environment during future development. The primary lesson here is that without measuring and tracking maintenance activities, you cannot gain the knowledge to improve or know how well you are doing relative to the past. To effectively manage and to continuously improve, you must measure and assess performance over time.

Controlling Maintenance Requests Another maintenance activity is managing maintenance requests. There are various types of maintenance requests—some correct minor or severe defects in the systems, whereas others improve or extend system

Mean time between failures (MTBF)

A measurement of error occurrences that can be tracked over time to indicate the quality of a system.

FIGURE 14-8
How to prioritize maintenance requests

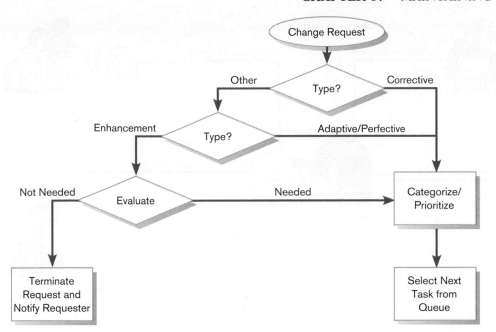

functionality. From a management perspective, a key issue is deciding which requests to perform and which to ignore. Because some requests will be more critical than others, some method of prioritizing requests must be determined.

Figure 14-8 shows a flowchart that suggests one possible method you could apply for dealing with maintenance change requests. First, you must determine the type of request. If, for example, the request is an error—that is, a corrective maintenance request—then the flowchart shows that the request is placed in the queue of tasks waiting to be performed on the system. For an error of high severity, repairs to remove it must be made as soon as possible. If, however, the error is considered "non-severe," then the change request can be categorized and prioritized based upon its type and relative importance.

If the change request is not an error, then you must determine whether the request is to adapt the system to technology changes and/or business requirements, perfect its operation in some way, or enhance the system so that it will provide new business functionality. Enhancement-type requests must first be evaluated to see whether they are aligned with future business and information systems' plans. If not, the request will be rejected and the requester will be informed. If the enhancement appears to be aligned with business and information systems plans, it can then be prioritized and placed into the queue of future tasks. Part of the prioritization process includes estimating the scope and feasibility of the change. Techniques used for assessing the scope and feasibility of entire projects should be used when assessing maintenance requests (see Chapter 5).

Managing the queue of pending tasks is an important activity. The queue of maintenance tasks for a given system is dynamic—growing and shrinking based upon business changes and errors. In fact, some lower-priority change requests may never be accomplished because only a limited number of changes can be accomplished at a given time. In other words, changes in business needs between the time the request was made and when the task finally rises to the top of the queue may result in the request being deemed unnecessary or no longer important given current business directions. Thus, managing the queue of pending tasks is an important activity.

To better understand the flow of a change request, see Figure 14-9. Initially, an organizational group that uses the system will make a request to change the system. This request flows to the project manager of the system (labeled 1). The project manager evaluates the request in relation to the existing system and pending changes and forwards the results of this evaluation to the system priority board

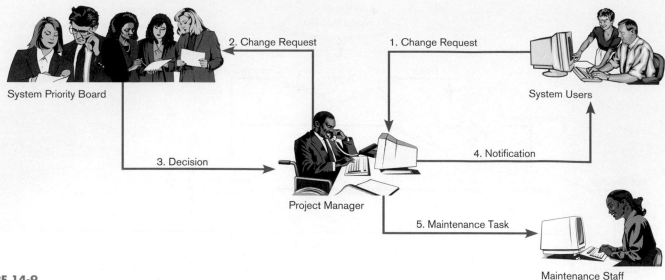

FIGURE 14-9
How a maintenance request moves through an organization

(labeled 2). This evaluation will also include a feasibility analysis that includes estimates of project scope, resource requirements, risks, and other relevant factors. The board evaluates, categorizes, and prioritizes the request in relation to both the strategic and information systems plans of the organization (labeled 3). If the board decides to kill the request, the project manager informs the requester and explains the rationale for the decision (labeled 4). If the request is accepted, it is placed in the queue of pending tasks. The project manager then assigns tasks to maintenance personnel based upon their availability and task priority (labeled 5). On a periodic basis, the project manager prepares a report of all pending tasks in the change request queue. This report is forwarded to the priority board where they reevaluate the requests in light of the current business conditions. This process may result in removing some requests or reprioritizing others.

Although each change request goes through the approval process depicted in Figure 14-9, changes are usually implemented in batches, forming a new release of the software. It is too difficult to manage a lot of small changes. Further, batching changes can reduce maintenance work when several change requests affect the same or highly related modules. Frequent releases of new system versions may also confuse users if the appearance of displays, reports, or data entry screens changes.

Configuration Management   A final aspect of managing maintenance is **configuration management**, which is the process of ensuring that only authorized changes are made to a system. Once a system has been implemented and installed, the programming code used to construct the system represents the **baseline modules** of the system. The baseline modules are the software modules for the most recent version of a system whereby each module has passed the organization's quality assurance process and documentation standards. A **system librarian** controls the checking out and checking in of the baseline source code modules. If maintenance personnel are assigned to make changes to a system, they must first check out a copy of the baseline system modules—no one is allowed to directly modify the baseline modules. Only those modules that have been tested and have gone through a formal check-in process can reside in the library. Before any code can be checked back in to the librarian, the code must pass the quality control procedures, testing, and documentation standards established by the organization.

When various maintenance personnel working on different maintenance tasks complete each task, the librarian notifies those still working that updates have been

Configuration management
The process of ensuring that only authorized changes are made to a system.

Baseline modules
Software modules that have been tested, documented, and approved to be included in the most recently created version of a system.

System librarian
A person responsible for controlling the checking out and checking in of baseline modules for a system when a system is being developed or maintained.

Configuration Management Tools

There are two general kinds of configuration management tools: revision control and source code control. With revision control tools, each system module file is "frozen" (unchangeable) to a specific version level or is designated as "floating" (changeable), meaning a programmer may check out, lock, and modify a specific system module. Only the most recent version of a module—and the specific changes made to it—are stored in the library; all previous versions of a module can be reconstructed, if needed, by applying any changes to the module in reverse order. Source code control tools extend revision control to not only a single module, but to any interrelated files to the module being changed. Configuration management tools have become invaluable to the system maintenance process by facilitating the rebuilding of any historic version of a system by recompiling the proper source code modules. Configuration management tools allow you to trace an executable code module back to its original source code version, greatly speeding the identification of programming errors.

made to the baseline modules. This means that all tasks being worked on must now incorporate the latest baseline modules before being approved for check-in. Following a formal process of checking modules out and in, a system librarian helps to ensure that only tested and approved modules become part of the baseline system. It is also the responsibility of the librarian to keep copies of all prior versions of all system modules, including the **build routines** needed to construct *any version* of the system that has *ever* existed. It may be important to reconstruct old versions of the system if new ones fail or to support users that cannot run newer versions on their computer system.

Build routines
Guidelines that list the instructions to construct an executable system from the baseline source code.

Special software systems have been created to manage system configuration and version control activities (see the box "Configuration Management Tools"). This software is becoming increasingly necessary as the change control process becomes ever more complicated in organizations deploying several different networks, operating systems, languages, and database management systems in which there may be many concurrent versions of an application, each for a different platform. One function of this software is to control access to modules in the system library. Each time a module is checked out or in, this activity is recorded, after being authorized by the librarian. This software helps the librarian to track that all necessary steps have been followed before a new module is released to production, including all integration tests, documentation updates, and approvals.

Role of CASE and Automated Development Tools in Maintenance

In traditional systems development, much of the time is spent on coding and testing. When software changes are approved, code is first changed and then tested. Once the functionality of the code is assured, the documentation and specification documents are updated to reflect system changes. Over time, the process of keeping all system documentation "current" can be a very boring and time-consuming activity that is often neglected. This neglect makes future maintenance by the same or *different* programmers difficult at best.

A primary objective of using CASE and other automated tools for systems development and maintenance is to radically change the way in which code and documentation are modified and updated. When using an integrated development environment, analysts maintain design documents such as data flow diagrams and screen designs, not source code. In other words, design documents are modified and then code generators automatically create a new version of the system from these updated designs. Also, because the changes are made at the design specification level, most documentation changes, such as an updated data flow diagram, will have already been completed during the maintenance process itself. Thus, one of the biggest advantages to using a CASE tool, for example, is its usefulness in system maintenance.

In addition to using general automated tools for maintenance, two special-purpose tools, reverse engineering and reengineering tools, are used to maintain older systems that have incomplete documentation or that were developed prior to CASE

tool use. These tools are often referred to as *design recovery* tools because their primary benefit is to create high-level design documents of a program by reading and analyzing its source code.

Reverse engineering tools are those that can create a representation of a system or program module at a design level of abstraction. For example, reverse engineering tools read program source code as input; perform an analysis; and extract information such as program control structures, data structures, and data flow. Once a program is represented at a design level using both graphical and textual representations, the design can be more effectively restructured to current business needs or programming practices by an analyst. For example, Microsoft's Visual Studio .NET can be used to reverse engineer applications into UML or other development diagrams (see Figure 14-10).

Similarly, **reengineering** tools extend reverse engineering tools by automatically (or interactively with a systems analyst) altering an existing system in an effort to improve its quality or performance. As reverse and reengineering capabilities are included in more popular development environments, the ability to extend the life and evolve the capabilities of existing systems will be enhanced (Orr, 2005).

Reverse engineering
Automated tools that read program source code as input and create graphical and textual representations of design-level information such as program control structures, data structures, logical flow, and data flow.

Reengineering
Automated tools that read program source code as input; perform an analysis of the program's data and logic; and then automatically, or interactively with a systems analyst, alter an existing system in an effort to improve its quality or performance.

WEBSITE MAINTENANCE

All of the discussion on maintenance in this chapter applies to any type of information system, no matter what platform it runs on. However, some special issues and procedures apply to websites, based on their special nature and operational status. These issues and procedures include the following:

- *24/7/365.* Most websites are never purposely unavailable. In fact, an e-commerce website has the advantage of continuous operation. Thus, maintenance of pages and the overall site usually must be done without taking the site offline. However, it may be necessary to lock out use of pages in a portion of a website while changes are made to those pages. This can be done by inserting a "Temporarily Out of Service" notice on the main page of the section being maintained and disabling all links within that segment. Alternatively, references to the main page of the section can be temporarily rerouted to an alternative location where the current pages are kept while maintenance is performed to create new versions of those pages. The really tricky part is keeping the site consistent for a user during a session; that is, it can be confusing to a user to see two different versions of a page during the same online session. Browser caching functions may bring up an old version of a page even when that page changes during the session. One precaution against this confusion is locking, as explained previously. Another approach is to not lock a page being changed, but to include a date and time stamp of the most recent change. This gives the page visitor an indication of the change, which may reduce confusion.

- *Check for broken links.* Arguably the most common maintenance issue for any website (besides changing the content of the site) is validating that links from site pages (especially for links that go outside the source site) are still accurate. Periodic checks need to be performed to make sure active pages are found from all links—this can be done via software such as LinkAlarm (*www.linkalarm.com*) and Doctor HTML (*www.fixingyourwebsite.com*). Note the irony of any potentially changing URL in a published paper or textbook! In addition, periodic human checks need to be performed to make sure that the content found at a still-existing referenced page is still the intended content.
- *HTML validation.* Before modified or new pages are published, these pages should be processed by a code validation routine to ensure that all the code, including applets, works. If you are using HTML, XML, script, or another editor, such a feature is likely built into the software.
- *Reregistration.* It may be necessary to reregister a website with search engines when the content of your site changes significantly. Reregistration may be necessary in order for visitors to find your site based on the new or changed content.
- *Future editions.* One of the most important issues to address to ensure effective website use is to avoid confusing visitors. In particular, frequent visitors can become confused if the site is constantly changing. To avoid confusion, you can post indications of future enhancements to the site and, as with all information systems, you can batch changes to reduce the frequency of site changes.

In addition, various website design decisions can greatly influence a site's maintainability; refer to Chapter 12 to review guidelines for effective website design.

ELECTRONIC COMMERCE APPLICATION: MAINTAINING AN INFORMATION SYSTEM FOR PINE VALLEY FURNITURE'S WEBSTORE

In this chapter, we examined various aspects of conducting system maintenance. Maintenance is a natural part of the life of any system. In this section, we conclude our discussion of PVF's WebStore by examining a maintenance activity for this system.

Maintaining Pine Valley Furniture's WebStore

Early on a Saturday evening, Jackie Judson, vice president of marketing at PVF, was reviewing new-product content that was recently posted on the company's electronic commerce website, the WebStore. She was working on Saturday evening because she was leaving the next day for a long-overdue, two-week vacation to the Black Hills of South Dakota. Before she could leave, however, she wanted to review the appearance and layout of the pages.

Midway through the review process, pages from the WebStore began to load very slowly. Finally, after requesting detailed information on a particular product, the WebStore simply stopped working. The title bar on her Web browser reported the error:

Cannot Find Server

Given that her plane for Rapid City left in less than 12 hours, Jackie wanted to review the content and needed to figure out some way to overcome this catastrophic system error. Her first thought was to walk over to the offices of the information systems development group within the same building. When she did, she found no one there. Her next idea was to contact Jim Woo, senior systems analyst and the project manager for the WebStore system. She placed a call to Jim's home and found that he was at the grocery store but would be home soon. Jackie left a message for Jim to call her ASAP at the office.

Within 30 minutes, Jim returned the call and was on his way into the office to help Jackie. Although not a common occurrence, this was not the first time that Jim had gone into the office to assist users when systems failed during off hours. Before leaving for the office, he connected to the Internet and also found the WebStore to be unavailable. Because PVF outsourced the hosting of the WebStore to an outside vendor, Jim immediately notified the vendor that the WebStore was down. The vendor was a local Internet service provider (ISP) that had a long-term relationship with PVF to provide Internet access, but it had limited experience with hosting commercial websites. Jim was informed that a system "glitch" was responsible for the outage and that the WebStore would be online within the next hour or so. Unfortunately, this was not the first time the WebStore had failed, and Jim felt powerless. More than ever before, he believed that PVF needed to find a better way to learn about system failures and, more important, it needed an improved confidence that the system would operate reliably.

On Monday morning, Jim requested a meeting with several senior PVF managers. At this meeting, he posed the following questions:

- How much is our website worth?
- How much does it cost our company when our website goes down?
- How reliable does our website need to be?

These questions encouraged a spirited discussion. Everyone agreed that the WebStore was "critical" to PVF's future and unanimously agreed that it was "unacceptable" for the site to be down. One manager summarized the feeling of the group by stating, "I cannot think of a single valid excuse for the system to crash . . . our customers have incredibly high expectations of us . . . one major mishap could prove disastrous for PVF."

Jim outlined to the group what he felt the next steps needed to be. "I believe that the root of our problem is with our contract with our Web hosting company. Specifically, we need to renegotiate our contract, or find a different vendor, so that it includes wording to reflect our expectations of service. Our current agreement does not address how emergencies are responded to or what remedies we have for continued system failures. The question we must also address is the cost differences between having a site that operates 99 percent of the time versus one that operates 99.999 percent of the time. I believe," he continued, "that it could increase our costs tens of thousands of dollars per month to guarantee extremely high levels of system reliability."

At the conclusion of the meeting, the senior managers unanimously agreed that Jim should immediately develop a plan for addressing the WebStore's service level problems. To begin this process, Jim prepared a detailed list of specific vendor services they desired. He felt a very specific list was needed so that the relative costs for different services and varying levels of service (e.g., response times for system failures and penalties for noncompliance) could be discussed.

When asked by a colleague what type of maintenance was being performed on the WebStore to improve system reliability, Jim had to pause and think. "Well, it is clearly *adaptive* in that we plan to migrate the system to a more reliable environment.

It is also *perfective* and *preventive.* . . . It is perfective in that we want to make some operational changes that will improve system performance, and it is also preventive given that one of our objectives is to reduce the likelihood of system failures." Through this discussion it became clear to Jim that the system was much larger than simply the HTML used to construct the WebStore; it also included the hardware, system software, procedures, and response team in place to deal with unforeseen events. Although he had heard it said many times before, Jim now understood that a successful system reflected all of its various aspects.

SUMMARY

Maintenance is the final phase in the SDLC. During maintenance, systems are changed to rectify internal processing errors or to extend the functionality of the system. Maintenance is where a majority of the financial investment in a system occurs and can span more than 20 years. More and more information systems professionals have devoted their careers to systems maintenance and, as more systems move from initial development into operational use, it is likely that even more professionals will in the future.

It is during maintenance that the SDLC becomes a life cycle because requests to change a system must first be approved, planned, analyzed, designed, and then implemented. In some special cases, when business operations are impaired due to an internal system error, quick fixes can be made. This, of course, circumvents the normal maintenance process. After quick fixes are made, maintenance personnel must back up and perform a thorough analysis of the problem to make sure that the correction conforms to normal systems development standards for design, programming, testing, and documentation. Maintenance requests can be one of four types: corrective, adaptive, perfective, and preventive.

How a system is designed and implemented can greatly impact the cost of performing maintenance. The number of unknown errors in a system when it is installed is a primary factor in determining the cost of maintenance. Other factors, such as the number of separate customers and the quality of documentation, significantly influence maintenance costs.

Another maintenance management issue relates to understanding how to measure the quality of the maintenance effort. Most organizations track the frequency, time, and type of each failure and compare performance over time. Because limited resources preclude organizations from performing all maintenance requests, some formal process for reviewing requests must be established to make sure that only those requests deemed consistent with organizational and information systems plans are performed. A central source, usually a project manager, is used to collect maintenance requests. When requests are submitted, this person forwards each request to a committee charged with assessing its merit. Once assessed, the project manager assigns higher-priority activities to maintenance personnel.

Maintenance personnel must be prevented from making unapproved changes to a system. To do this, most organizations assign one member of the maintenance staff, typically a senior programmer or analyst, to serve as the system librarian. The librarian controls the checking out and checking in of system modules to ensure that appropriate procedures for performing maintenance, such as adequate testing and documentation, are adhered to.

CASE tools are actively employed during maintenance to enable maintenance to be performed on design documents rather than on low-level source code. Reverse engineering and reengineering CASE tools are used to recover design specifications of older systems that were not constructed using CASE tools or for systems with inadequate design specifications. Once recovered, these older systems can then be changed at the design level rather than the source code level, yielding a significant improvement in maintenance personnel productivity.

Website maintenance involves some special attention, including: 24/7/365 operation, checking for broken external links, validating code changes before publishing new or revised pages, reregistration of the website for search engines, and avoiding visitor confusion by previewing future changes.

KEY TERMS

1. Adaptive maintenance
2. Baseline modules
3. Build routines
4. Configuration management
5. Corrective maintenance
6. Maintainability
7. Maintenance
8. Mean time between failures (MTBF)
9. Perfective maintenance
10. Preventive maintenance
11. Reengineering
12. Reverse engineering
13. System librarian

Match each of the key terms above with the definition that best fits it.

_____ Changes made to a system to fix or enhance its functionality.

_____ Changes made to a system to repair flaws in its design, coding, or implementation.

_____ Changes made to a system to evolve its functionality to changing business needs or technologies.

_____ Changes made to a system to add new features or to improve performance.

_____ Changes made to a system to avoid possible future problems.

_____ The ease with which software can be understood, corrected, adapted, and enhanced.

_____ A measurement of error occurrences that can be tracked over time to indicate the quality of a system.

_____ The process of ensuring that only authorized changes are made to a system.

_____ Software modules that have been tested, documented, and approved to be included in the most recently created version of a system.

_____ A person responsible for controlling the checking out and checking in of baseline modules for a system when a system is being developed or maintained.

_____ Guidelines that list the instructions to construct an executable system from the baseline source code.

_____ Automated tools that read program source code as input and create graphical and textual representations of design-level information such as program control structures, data structures, logical flow, and data flow.

_____ Automated tools that read program source code as input; perform an analysis of the program's data and logic; and then automatically, or interactively with a systems analyst, alter an existing system in an effort to improve its quality or performance.

REVIEW QUESTIONS

1. Contrast the following terms:
   a. Adaptive maintenance, corrective maintenance, perfective maintenance, preventive maintenance
   b. Baseline modules, build routines, system librarian
   c. Maintenance, maintainability

2. List the steps in the maintenance process and contrast them with the phases of the SDLC.

3. What are the different types of maintenance and how do they differ?

4. Describe the factors that influence the cost of maintenance. Are any factors more important than others? Why?

5. Describe three ways for organizing maintenance personnel and contrast the advantages and disadvantages of each approach.

6. What types of measurements must be taken to gain an understanding of the effectiveness of maintenance? Why is tracking mean time between failures an important measurement?

7. What managerial issues can be better understood by measuring maintenance effectiveness?

8. Describe the process for controlling maintenance requests. Should all requests be handled in the same way or are there situations when you should be able to circumvent the process? If so, when and why?

9. What is meant by configuration management? Why do you think organizations have adopted the approach of using a systems librarian?

10. How are CASE tools used in the maintenance of information systems?

11. What is the difference between reverse engineering and re-engineering CASE tools?

12. What are some special maintenance issues and procedures that are especially relevant for websites?

PROBLEMS AND EXERCISES

1. Maintenance has been presented as both the final stage of the SDLC (see Figure 14-1) and as a process similar to the SDLC (see Figure 14-3). Why does it make sense to talk about maintenance in both of these ways? Do you see a conflict in looking at maintenance in both ways?

2. In what ways is a request to change an information system handled differently from a request for a new information system?

3. According to Figure 14-4, corrective maintenance is by far the most frequent form of maintenance. What can you do as a systems analyst to reduce this form of maintenance?

4. What other or additional information should be collected on a System Service Request (see Figure 14-2) for maintenance?

5. Briefly discuss how a systems analyst can manage each of the six cost elements of maintenance described in this chapter.

6. Suppose you were a system librarian. Using entity-relationship diagramming notation, describe the database you would need to keep track of the information necessary in your job. Consider operational, managerial, and planning aspects of the job.

7. Suppose an information system were developed following a RAD approach (see Chapter 1). How might maintenance be different if the system had been developed following the traditional life cycle? Why?

8. Software configuration management is similar to configuration management in any engineering environment. For example, the product design engineers for a refrigerator need to coordinate dynamic changes in compressors, power supplies, electronic controls, interior features, and exterior designs as innovations to each occur. How do such product design engineers manage the configuration of their products? What similar practices do systems analysts and librarians have to follow?

9. In the section on PVF's WebStore, it is mentioned that Jim Woo will prepare a list of ISP Web-hosting services. Prepare such a list for website maintenance. Contrast the responsibilities of the ISP with those of PVF.

FIELD EXERCISES

1. Study an information systems department with which you are familiar or to which you have access. How does this department organize for maintenance? Has this department adopted one of the three approaches outlined in Table 14-2 or does it use some other approach? Talk with a senior manager in this department to discover how well this maintenance organization structure works.

2. Study an information systems department with which you are familiar or to which you have access. How does this department measure the effectiveness of systems maintenance? What specific metrics are used and how are these metrics used to effect changes in maintenance practices? If there is a history of measurements over several years, how can changes in the measurements be explained?

3. With the help of other students or your instructor, contact a system librarian in an organization. What is this person's job description? What tools does this person use to help him or her in the job? To whom does this person report? What previous jobs did this person hold, and to what job does this person expect to be promoted in the near future?

4. Interview the Webmaster at a company where you work or where you have a contact. Investigate the procedures followed to maintain this website. Document these procedures. What differences or enhancements did you find, compared with the special website maintenance issues and procedures listed in this chapter?

REFERENCES

Andrews, D. C., and N. S. Leventhal. 1993. *Fusion: Integrating IE, CASE, JAD: A Handbook for Reengineering the Systems Organization.* Upper Saddle River, NJ: Prentice Hall.

Hanna, M. 1992. "Using Documentation as a Life-Cycle Tool." *Software Magazine* 6(December): 41–46.

Kaplan, S. 2002. "Now Is the Time to Pull the Plug on Your Legacy Apps." *CIO Magazine,* March 15. Available at (*www.cio.com/archive/031502/infrastructure.html.* Accessed February 14, 2009.

Legard, D. 2000. "Study: Online Maintenance Costs Just Keep Growing." *InfoWorld,* November 8. Available at (*archive.infoworld.com/articles/hn/xml/00/11/08/001108hnmaintain.xml.* Accessed February 14, 2009.

Lytton, N. 2001. "Maintenance Dollars at Work." *ComputerWorld,* July 16. Available at (*www.computerworld.com/managementtopics/roi/story/0,10801,62240,00.html.* Accessed February 14, 2009.

Martin, E. W., C. V. Brown, D. W. DeHayes, J. A. Hoffer, and W. C. Perkins. 2008. *Managing Information Technology: What Managers Need to Know,* 6th ed. Upper Saddle River, NJ: Prentice Hall.

Nash, K. 2010. "ERP: How and Why You Need to Manage It Differently." Available at (*http://www.cio.com/article/526213/ERP_How_and_Why_You_Need_to_Manage_It_Differently?page=1&taxonomyId=3000.* Accessed April 26, 2012.

Orr, S. 2005. "Design and Generate Code with Visio." Dotnetjunkies, March 7. Available at (*www.dotnetjunkies.com/Article/74851895-C4D4-4F11-956D-A27D849E4A62.dcik.* Accessed February 14, 2009.

Pressman, R. S. 2005. *Software Engineering,* 6th ed. New York: McGraw-Hill.

Worthen, B. 2003. "No Tolerance for High Maintenance." *CIO Magazine,* July 1. Available at (*www.cio.com/archive/060103/vendor.html.* Accessed February 14, 2009.

GLOSSARY OF TERMS

Abstract class   A class that has no direct instance, but whose descendents may have direct instances. (8)

Abstract operation   Defines the form or protocol of the operation, but not its implementation. (8)

Acceptance testing   The process whereby actual users test a completed information system, the end result of which is the users' acceptance of it. (13)

Action stubs   That part of a decision table that lists the actions that result for a given set of conditions. (7)

Activation   The time period during which an object performs an operation. (7)

Activity   In business process modeling, an action that must take place for a process to be completed. (7)

Activity diagram   Shows the conditional logic for the sequence of system activities needed to accomplish a business process. (7)

Actor   An external entity that interacts with a system. (7)

Adaptive maintenance   Changes made to a system to evolve its functionality to changing business needs or technologies. (14)

Affinity clustering   The process of arranging planning matrix information so that clusters of information with a predetermined level or type of affinity are placed next to each other on a matrix report. (4)

Aggregation   A part-of relationship between a component object and an aggregate object. (8)

Alpha testing   User testing of a completed information system using simulated data. (13)

Analysis   The second phase of the SDLC in which system requirements are studied and structured. (1)

Application program interface (API)   Software building blocks that are used to ensure that common system capabilities, such as user interfaces and printing, as well as modules are standardized to facilitate data exchange between clients and servers. (12)

Application server   A computing server where data analysis functions primarily reside. (12)

Application software   Computer software designed to support organizational functions or processes. (1)

Association   A named relationship between or among object classes. (8)

Association role   The end of an association where it connects to a class. (8)

Associative class   An association that has attributes or operations of its own or that participates in relationships with other classes. (8)

Associative entity   An entity type that associates the instances of one or more entity types and contains attributes that are peculiar to the relationship between those entity instances; also called a *gerund*. (8)

Asynchronous message   A message in which the sender does not have to wait for the recipient to handle the message. (7)

Attribute   A named property or characteristic of an entity that is of interest to the organization. (8)

Balancing   The conservation of inputs and outputs to a DFD process when that process is decomposed to a lower level. (7)

Baseline modules   Software modules that have been tested, documented, and approved to be included in the most recently created version of a system. (14)

Baseline Project Plan (BPP)   A major outcome and deliverable from the project initiation and planning phase that contains the best estimate of a project's scope, benefits, costs, risks, and resource requirements. (5)

Behavior   Represents how an object acts and reacts. (8)

Beta testing   User testing of a completed information system using real data in the real user environment. (13)

Binary relationship   A relationship between instances of two entity types. This is the most common type of relationship encountered in data modeling. (8)

Bottom-up planning   A generic information systems planning methodology that identifies and defines IS development projects based upon solving operational business problems or taking advantage of some business opportunities. (4)

Break-even analysis   A type of cost-benefit analysis to identify at what point (if ever) benefits equal costs. (5)

Build routines   Guidelines that list the instructions to construct an executable system from the baseline source code. (14)

Business case   The justification for an information system, presented in terms of the tangible and intangible economic benefits and costs and the technical and organizational feasibility of the proposed system. (5)

Business process reengineering (BPR) The search for, and implementation of, radical change in business processes to achieve breakthrough improvements in products and services. (6)

Business-to-business (B2B) Electronic commerce between business partners, such as suppliers and intermediaries (4)

Business-to-consumer (B2C) Electronic commerce between businesses and consumers. (4)

Business-to-employee (B2E) Electronic commerce between businesses and their employees. (4)

Calculated field A field that can be derived from other database fields. Also known as a computed field or a derived field. (9)

Candidate key An attribute (or combination of attributes) that uniquely identifies each instance of an entity type. (8)

Cardinality The number of instances of entity B that can (or must) be associated with each instance of entity A. (8)

Cascading Style Sheets (CSSs) A set of style rules that tells a Web browser how to present a document. (12)

Class diagram Shows the static structure of an object-oriented model the object classes, their internal structure, and the relationships in which they participate. (8)

Class-scope attribute An attribute of a class that specifies a value common to an entire class, rather than a specific value for an instance. (8)

Class-scope operation An operation that applies to a class rather than an object instance. (8)

Client The (front-end) portion of the client/server database system that provides the user interface and data manipulation functions. (12)

Client/server architecture A LAN-based computing environment in which a central database server or engine performs all database commands sent to it from client workstations, and application programs on each client concentrate on user interface functions. (12)

Closed-ended questions Questions in interviews that ask those responding to choose from among a set of specified responses. (6)

Cloud computing The provision of computing resources, including applications, over the Internet, so customers do not have to invest in the computing infrastructure needed to run and maintain the resources. (2)

Command language interaction A human–computer interaction method whereby users enter explicit statements into a system to invoke operations. (11)

Competitive strategy The method by which an organization attempts to achieve its mission and objectives. (4)

Composite attribute An attribute that has meaningful component parts. (8)

Composition A part-of relationship in which parts belong to only one whole object, and the parts live and die with the whole object. (8)

Computer-aided software engineering (CASE) tools Software tools that provide automated support for some portion of the systems development process. (1)

Conceptual data model A detailed model that captures the overall structure of organizational data that is independent of any database management system or other implementation considerations. (8)

Concrete class A class that can have direct instances. (8)

Condition stubs That part of a decision table that lists the conditions relevant to the decision. (7)

Configuration management The process of ensuring that only authorized changes are made to a system. (14)

Constructor operation An operation that creates a new instance of a class. (8)

Content management system (CMS) A special type of software application for collecting, organizing, and publishing website content. (12)

Context development A method that helps analysts to better understand how a system fits within existing business activities and data. (12)

Context diagram An overview of an organizational system that shows the system boundaries, external entities that interact with the system, and the major information flows between the entities and the system. (7)

Cookie crumbs The technique of placing "tabs" on a Web page that show a user where he or she is on a site and where he or she has been. (11)

Corporate strategic planning An ongoing process that defines the mission, objectives, and strategies of an organization. (4)

Corrective maintenance Changes made to a system to repair flaws in its design, coding, or implementation. (14)

Critical path The shortest time in which a project can be completed. (3)

Critical path scheduling A scheduling technique whose order and duration of a sequence of task activities directly affect the completion date of a project. (3)

Customization Internet sites that allow users to customize the content and look of the site based on their personal preferences. (12)

Data flow diagram (DFD) A picture of the movement of data between external entities and the processes and data stores within a system. (7)

Data mart A data warehouse that is limited in scope; its data are obtained by selecting and (where appropriate) summarizing data from the enterprise data warehouse. (12)

Data store Data at rest, which may take the form of many different physical representations. (7)

Data type   A coding scheme recognized by system software for representing organizational data. (9)

Data warehouse   A subject-oriented, integrated, time-variant, nonvolatile collection of data used in support of management decision making. (12)

Database engine   The (back-end) portion of the client/server database system running on the server that provides database processing and shared access functions. (12)

Decision table   A matrix representation of the logic of a decision, which specifies the possible conditions for the decision and the resulting actions. (7)

Default value   A value a field will assume unless an explicit value is entered for that field. (9)

Degree   The number of entity types that participate in a relationship. (8)

Deliverable   An end product of an SDLC phase. (3)

Denormalization   The process of splitting or combining normalized relations into physical tables based on affinity of use of rows and fields. (9)

Derived attribute   An attribute whose value can be computed from related attribute values. (8)

Design   The third phase of the SDLC in which the description of the recommended solution is converted into logical and then physical system specifications. (1)

Desk checking   A testing technique in which the program code is sequentially executed manually by the reviewer. (13)

DFD completeness   The extent to which all necessary components of a DFD have been included and fully described. (7)

DFD consistency   The extent to which information contained on one level of a set of nested DFDs is also included on other levels. (7)

Dialogue   The sequence of interaction between a user and a system. (11)

Dialogue diagramming   A formal method for designing and representing human–computer dialogues using box and line diagrams. (11)

Direct installation   Changing over from the old information system to a new one by turning off the old system when the new one is turned on. (13)

Discount rate   The rate of return used to compute the present value of future cash flows. (5)

Disjoint rule   Specifies that if an entity instance of the supertype is a member of one subtype, it cannot simultaneously be a member of any other subtype. (8)

Disruptive technologies   Technologies that enable the breaking of long-held business rules that inhibit organizations from making radical business changes. (6)

Domain   The set of all data types and values that an attribute can assume. (8)

Domain naming system (BIND)   A method for translating Internet domain names into Internet Protocol (IP) addresses. BIND stands for Berkeley Internet Name Domain. (12)

Drop-down menu   A menu-positioning method that places the access point of the menu near the top line of the display; when accessed, menus open by dropping down onto the display. (11)

Economic feasibility   A process of identifying the financial benefits and costs associated with a development project. (5)

Electronic commerce (EC)   Internet-based communication to support day-to-day business activities. (4)

Electronic data interchange (EDI)   The use of telecommunications technologies to directly transfer business documents between organizations. (4)

Electronic performance support system (EPSS)   Component of a software package or an application in which training and educational information is embedded. An EPSS can take several forms, including a tutorial, an expert system shell, and hypertext jumps to reference materials. (13)

Encapsulation   The technique of hiding the internal implementation details of an object from its external view. (8)

Enterprise data warehouse (EDW)   A centralized, integrated data warehouse that is the control point and single source of all data made available to end users for decision support applications throughout the entire organization. (12)

Enterprise resource planning (ERP) system   A system that integrates individual traditional business functions into a series of modules so that a single transaction occurs seamlessly within a single information system rather than several separate systems. (2)

Entity instance   A single occurrence of an entity type. Also known as an instance. (8)

Entity type   A collection of entities that share common properties or characteristics. (8)

Entity-relationship data model (E-R model)   A detailed, logical representation of the entities, associations, and data elements for an organization or business area. (8)

Entity-relationship diagram (E-R diagram)   A graphical representation of an E-R model. (8)

Event   In business process modeling, a trigger that initiates the start of a process. (7)

Extend relationship   An association between two use cases where one adds new behaviors or actions to the other. (7)

eXtensible Markup Language (XML)   An Internet-authoring language that allows designers to create customized tags, enabling the definition, transmission, validation, and interpretation of data between applications. (12)

eXtensible Style Language (XSL) A specification for separating style from content when generating HTML documents. (12)

Extension The set of behaviors or functions in a use case that follow exceptions to the main success scenario. (7)

External documentation System documentation that includes the outcome of structured diagramming techniques such as data flow and entity-relationship diagrams. (13)

Feasibility study A study that determines if the proposed information system makes sense for the organization from an economic and operational standpoint. (3)

Field The smallest unit of named application data recognized by system software. (9)

File organization A technique for physically arranging the records of a file. (9)

File server A device that manages file operations and is shared by each client PC attached to a LAN. (12)

Flow In business process modeling, it shows the sequence of action in a process. (7)

Foreign key An attribute that appears as a nonprimary key attribute in one relation and as a primary key attribute (or part of a primary key) in another relation. (9)

Form A business document that contains some predefined data and may include some areas where additional data are to be filled in. An instance of a form is typically based on one database record. (10)

Form interaction A highly intuitive human–computer interaction method whereby data fields are formatted in a manner similar to paper-based forms. (11)

Formal system The official way a system works as described in organizational documentation. (6)

Functional decomposition An iterative process of breaking the description of a system down into finer and finer detail, which creates a set of charts in which one process on a given chart is explained in greater detail on another chart. (7)

Functional dependency A constraint between two attributes in which the value of one attribute is determined by the value of another attribute. (9)

Gantt chart A graphical representation of a project that shows each task as a horizontal bar whose length is proportional to its time for completion. (3)

Gap analysis The process of discovering discrepancies between two or more sets of DFDs or discrepancies within a single DFD. (7)

Gateway In business process modeling, a decision point. (7)

Hashed file organization A file organization in which the address of each row is determined using an algorithm. (9)

Help desk A single point of contact for all user inquiries and problems about a particular information system or for all users in a particular department. (13)

Homonym A single attribute name that is used for two or more different attributes. (9)

Hypertext Markup Language (HTML) The standard language for representing content on the Web through the use of hundreds of command tags. (12)

Hypertext Transfer Protocol (HTTP) A communication protocol for exchanging information on the Internet. (12)

Icon Graphical picture that represents specific functions within a system. (11)

Identifier A candidate key that has been selected as the unique, identifying characteristic for an entity type. (8)

Implementation The fourth phase of the SDLC in which the information system is coded, tested, installed, and supported in the organization. (1)

Include relationship An association between two use cases where one use case uses the functionality contained in the other. (7)

Incremental commitment A strategy in systems analysis and design in which the project is reviewed after each phase and continuation of the project is rejustified. (4)

Index A table used to determine the location of rows in a file that satisfy some condition. (9)

Indexed file organization A file organization in which rows are stored either sequentially or nonsequentially, and an index is created that allows software to locate individual rows. (9)

Indifferent condition In a decision table, a condition whose value does not affect which actions are taken for two or more rules. (7)

Informal system The way a system actually works. (6)

Information systems analysis and design The complex organizational process whereby computer-based information systems are developed and maintained. (1)

Informational systems Systems designed to support decision making based on stable point-in-time or historical data. (12)

Information systems planning (ISP) An orderly means of assessing the information needs of an organization and defining the systems, databases, and technologies that will best satisfy those needs. (4)

Inheritance The property that occurs when entity types or object classes are arranged in a hierarchy and each entity type or object class assumes the attributes and methods of its ancestors, that is, those higher up in the hierarchy. Inheritance allows new but related classes to be derived from existing classes. (1)

Inspections A testing technique in which participants examine program code for predictable language-specific errors. (13)

Installation The organizational process of changing over from the current information system to a new one. (13)

Intangible benefit A benefit derived from the creation of an information system that cannot be easily measured in dollars or with certainty. (5)

Intangible cost A cost associated with an information system that cannot be easily measured in terms of dollars or with certainty. (5)

Integration depth A measurement of how far into the existing technology infrastructure a system penetrates. (12)

Integration testing The process of bringing together all of the modules that a program comprises for testing purposes. Modules are typically integrated in a top-down, incremental fashion. (13)

Interface A method by which users interact with an information system. (11)

Internal documentation System documentation that is part of the program source code or is generated at compile time. (13)

Internet A large, worldwide network of networks that use a common protocol to communicate with each other. (4)

JAD session leader The trained individual who plans and leads Joint Application Design sessions. (6)

Joint Application Design (JAD) A structured process in which users, managers, and analysts work together for several days in a series of intensive meetings to specify or review system requirements. (6)

Key business processes The structured, measured set of activities designed to produce a specific output for a particular customer or market. (6)

Legal and contractual feasibility The process of assessing potential legal and contractual ramifications due to the construction of a system. (5)

Level Perspective from which a use case description is written, typically ranging from high level to extremely detailed. (7)

Level-0 diagram A DFD that represents a system's major processes, data flows, and data stores at a high level of detail. (7)

Level-n diagram A DFD that is the result of *n* nested decompositions from a process on a level-0 diagram. (7)

Lightweight graphics The use of small, simple images to allow a Web page to be displayed more quickly. (10)

Local area network(LAN) The cabling, hardware, and software used to connect workstations, computers, and file servers located in a confined geographical area (typically within one building or campus). (12)

Logical design The part of the design phase of the SDLC in which all functional features of the system chosen for development in analysis are described independently of any computer platform. (1)

Maintainability The ease with which software can be understood, corrected, adapted, and enhanced. (14)

Maintenance The final phase of the SDLC in which an information system is systematically repaired and improved. (1) (14)

Mean time between failures (MTBF) A measurement of error occurrences that can be tracked over time to indicate the quality of a system. (14)

Menu interaction A human–computer interaction method in which a list of system options is provided and a specific command is invoked by user selection of a menu option. (11)

Method The implementation of an operation. (8)

Middleware A combination of hardware, software, and communication technologies that brings data management, presentation, and analysis together into a three-tiered client/server environment. (12)

Minimal guarantee The least amount promised to the stakeholder by a use case. (7)

Mission statement A statement that makes it clear what business a company is in. (4)

Multiplicity A specification that indicates how many objects participate in a given relationship. (8)

Multivalued attribute An attribute that may take on more than one value for each entity instance. (8)

Natural language interaction A human–computer interaction method whereby inputs to and outputs from a computer-based application are in a conventional spoken language such as English. (11)

Network diagram A diagram that depicts project tasks and their interrelationships. (3)

Nominal Group Technique(NGT) A facilitated process that supports idea generation by groups. At the beginning of the process, group members work alone to generate ideas, which are then pooled under the guidance of a trained facilitator. (6)

Normalization The process of converting complex data structures into simple, stable data structures. (9)

Null value A special field value, distinct from zero, blank, or any other value, that indicates that the value for the field is missing or otherwise unknown. (9)

Object A structure that encapsulates (or packages) attributes and methods that operate on those attributes. An object is an abstraction of a real-world thing in which data and processes are placed together to model the structure and behavior of the real-world object. (1) (8)

Object class A logical grouping of objects that have the same (or similar) attributes, relationships, and behaviors; also called *class*. (1) (8)

Object-based interaction A human–computer interaction method in which symbols are used to represent commands or functions. (11)

Objective statements A series of statements that express an organization's qualitative and quantitative goals for reaching a desired future position. (4)

Object-oriented analysis and design (OOAD) Systems development methodologies and techniques based on objects rather than data or processes. (1)

One-time cost A cost associated with project start-up and development or system start-up. (5)

Online analytical processing (OLAP) The use of graphical software tools that provide complex analysis of data stored in a database. (12)

Online transaction processing (OLTP) The immediate automated responses to the requests of users. (12)

Open-ended questions Questions in interviews that have no prespecified answers. (6)

Operation A function or a service that is provided by all the instances of a class. (8)

Operational feasibility The process of assessing the degree to which a proposed system solves business problems or takes advantage of business opportunities. (5)

Operational systems Systems that are used to interact with customers and run a business in real time. (12)

Optional attribute An attribute that may not have a value for every entity instance. (8)

Organizational breadth A measurement that tracks the core business functions affected by a system. (12)

Outsourcing The practice of turning over responsibility for some to all of an organization's information systems applications and operations to an outside firm. (2)

Overlap rule Specifies that an entity instance can simultaneously be a member of two (or more) subtypes. (8)

Parallel installation Running the old information system and the new one at the same time until management decides the old system can be turned off. (13)

Partial specialization rule Specifies that an entity instance of the supertype does not have to belong to any subtype. (8)

Perfective maintenance Changes made to a system to add new features or to improve performance. (14)

Personalization Providing Internet content to a user based upon knowledge of that customer. (12)

PERT (Program Evaluation Review Technique) A technique that uses optimistic, pessimistic, and realistic time estimates to calculate the expected time for a particular task. (3)

Phased installation Changing from the old information system to the new one incrementally, starting with one or a few functional components and then gradually extending the installation to cover the whole new system. (13)

Physical design The part of the design phase of the SDLC in which the logical specifications of the system from logical design are transformed into technology-specific details from which all programming and system construction can be accomplished. (1)

Physical file A named set of table rows stored in a contiguous section of secondary memory. (9)

Physical table A named set of rows and columns that specifies the fields in each row of the table. (9)

Planning The first phase of the SDLC in which an organization's total information system needs are identified, analyzed, prioritized, and arranged. (1)

Pointer A field of data that can be used to locate a related field or row of data. (9)

Political feasibility The process of evaluating how key stakeholders within the organization view the proposed system. (5)

Polymorphism The same operation may apply to two or more classes in different ways. (8)

Pool In business process modeling, a way to encapsulate a process that has two or more participants. (7)

Pop-up menu A menu-positioning method that places a menu near the current cursor position. (11)

Preconditions Things that must be true before a use case can start. (7)

Present value The current value of a future cash flow. (5)

Preventive maintenance Changes made to a system to avoid possible future problems. (14)

Primary key An attribute (or combination of attributes) whose value is unique across all occurrences of a relation. (9)

Primitive DFD The lowest level of decomposition for a DFD. (7)

Process The work or actions performed on data so that they are transformed, stored, or distributed. (7)

Project A planned undertaking of related activities to reach an objective that has a beginning and an end. (3)

Project charter A short document prepared for the customer during project initiation that describes what the project will deliver and outlines generally at a high level all work required to complete the project. (3)

Project closedown The final phase of the project management process that focuses on bringing a project to an end. (3)

Project execution The third phase of the project management process in which the plans created in the prior phases (project initiation and planning) are put into action. (3)

Project initiation The first phase of the project management process in which activities are performed to assess the size, scope, and complexity of the project and to establish procedures to support later project activities. (3)

Project management A controlled process of initiating, planning, executing, and closing down a project. (3)

Project manager A systems analyst, with a diverse set of skills—management, leadership, technical, conflict management, and customer relationship—who is responsible for initiating, planning, executing, and closing down a project. (3)

Project planning The second phase of the project management process that focuses on defining clear, discrete activities and the work needed to complete each activity within a single project. (3)

Project Scope Statement (PSS) A document prepared for the customer that describes what the project will deliver and outlines generally at a high level all work required to complete the project. (5)

Project workbook An online or hard-copy repository for all project correspondence, inputs, outputs, deliverables, procedures, and standards that is used for performing project audits, orienting new team members, communicating with management and customers, identifying future projects, and performing post-project reviews. (3)

Prototyping An iterative process of systems development in which requirements are converted to a working system that is continually revised through close collaboration between an analyst and users. (6)

Query operation An operation that accesses the state of an object but does not alter the state. (8)

Rapid Application Development (RAD) Systems development methodology created to radically decrease the time needed to design and implement information systems. RAD relies on extensive user involvement, prototyping, integrated CASE tools, and code generators. (1)

Rational Unified Process (RUP) An object-oriented systems development methodology. RUP establishes four phases of development: inception, elaboration, construction, and transition. Each phase is organized into a number of separate iterations. (1)

Recurring cost A cost resulting from the ongoing evolution and use of a system. (5)

Recursive foreign key A foreign key in a relation that references the primary key values of that same relation. (9)

Reengineering Automated tools that read program source code as input; perform an analysis of the program's data and logic; and then automatically, or interactively with a systems analyst, alter an existing system in an effort to improve its quality or performance. (14)

Refactoring Making a program simpler after adding a new feature. (13)

Referential integrity A rule that states that either each foreign key value must match a primary key value in another relation or the foreign key value must be null (i.e., have no value). (9)

Relation A named, two-dimensional table of data. Each relation consists of a set of named columns and an arbitrary number of unnamed rows. (9)

Relational database model Data represented as a set of related tables or relations. (9)

Relationship An association between the instance of one or more entity types that is of interest to the organization. (8)

Repeating group A set of two or more multivalued attributes that are logically related. (8)

Report A business document that contains only predefined data; it is a passive document used solely for reading or viewing. A report typically contains data from many unrelated records or transactions. (10)

Request for proposal (RFP) A document provided to vendors that asks them to propose hardware and system software that will meet the requirements of a new system. (2)

Required attribute An attribute that must have a value for every entity instance. (8)

Resources Any person, group of people, piece of equipment, or material used in accomplishing an activity. (3)

Reuse The use of previously written software resources, especially objects and components, in new applications. (2)

Reverse engineering Automated tools that read program source code as input and create graphical and textual representations of design-level information such as program control structures, data structures, logical flow, and data flow. (14)

Rules That part of a decision table that specifies which actions are to be followed for a given set of conditions. (7)

Schedule feasibility The process of assessing the degree to which the potential time frame and completion dates for all major activities within a project meet organizational deadlines and constraints for affecting change. (5)

Scribe The person who makes detailed notes of the happenings at a Joint Application Design session. (6)

Second normal form (2NF) A relation is in second normal form if every nonprimary key attribute is functionally dependent on the whole primary key. (9)

Secondary key One or a combination of fields for which more than one row may have the same combination of values. (9)

Sequence diagram Depicts the interactions among objects during a certain period of time. (7)

Sequential file organization A file organization in which rows in a file are stored in sequence according to a primary key value. (9)

Simple message A message that transfers control from the sender to the recipient without describing the details of the communication. (7)

Single-location installation Trying out a new information system at one site and using the experience to decide if and how the new system should be deployed throughout the organization. (13)

Slack time The amount of time that an activity can be delayed without delaying the project. (3)

Source/sink The origin and/or destination of data; sometimes referred to as *external entities*. (7)

Stakeholder People who have a vested interest in the system being developed. (7)

State Encompasses an object's properties (attributes and relationships) and the values of those properties. (8)

Stub testing A technique used in testing modules, especially where modules are written and tested in a top-down fashion, where a few lines of code are used to substitute for subordinate modules. (13)

Subtype A subgrouping of the entities in an entity type that is meaningful to the organization and that shares common attributes or relationships distinct from other subgroupings. (8)

Success guarantee What a use case must do effectively in order to satisfy stakeholders. (7)

Supertype A generic entity type that has a relationship with one or more subtypes. (8)

Support Providing ongoing educational and problem-solving assistance to information system users. For in-house developed systems, support materials and jobs will have to be prepared or designed as part of the implementation process. (13)

Swimlane In business process modeling, a way to visually encapsulate a process. (7)

Synchronous message A type of message in which the caller has to wait for the receiving object to finish executing the called operation before it can resume execution itself. (7)

Synonym Two different names that are used for the same attribute. (9)

System documentation Detailed information about a system's design specifications, its internal workings, and its functionality. (13)

System librarian A person responsible for controlling the checking out and checking in of baseline modules for a system when a system is being developed or maintained. (14)

System testing Bringing together of all of the programs that a system comprises for testing purposes. Programs are typically integrated in a top-down, incremental fashion. (13)

Systems analyst The organizational role most responsible for the analysis and design of information systems. (1)

Systems development life cycle (SDLC) The traditional methodology used to develop, maintain, and replace information systems. (1)

Systems development methodology A standard process followed in an organization to conduct all the steps necessary to analyze, design, implement, and maintain information systems. (1)

Tangible benefit A benefit derived from the creation of an information system that can be measured in dollars and with certainty. (5)

Tangible cost A cost associated with an information system that can be measured in dollars and with certainty. (5)

Technical feasibility A process of assessing the development organization's ability to construct a proposed system. (5)

Template-based HTML Templates to display and process common attributes of higher-level, more abstract items. (10)

Ternary relationship A simultaneous relationship among instance of three entity types. (8)

Testing harness An automated testing environment used to review code for errors, standards violations, and other design flaws. (13)

Thin client A client device designed so that most processing and data storage occur on the server. (12)

Third normal form (3NF) A relation is in second normal form and has no functional (transitive) dependencies between two (or more) nonprimary key attributes. (9)

Three-tiered client/server Advanced client/server architectures in which there are three logical and distinct applications—data management, presentation, and analysis—that are combined to create a single information system. (12)

Time value of money (TVM) The concept that money available today is worth more than the same amount tomorrow. (5)

Top-down planning A generic information systems planning methodology that attempts to gain a broad understanding of the information systems needs of the entire organization. (4)

Total specialization rule Specifies that each entity instance of the supertype must be a member of some subtype of the relationship. (8)

Trigger Event that initiates a use case. (7)

Triggering operation (trigger) An assertion or rule that governs the validity of data manipulation operations such as insert, update, and delete; also called a *trigger*. (8)

Unary relationship A relationship between instances of one entity type; also called *recursive relationship*. (8)

Unit testing Each module is tested alone in an attempt to discover any errors in its code. (13)

Update operation An operation that alters the state of an object. (8)

Usability An overall evaluation of how a system performs in supporting a particular user for a particular task. (10)

Use case A depiction of a system's behavior or functionality under various conditions as the system responds to requests from users. (7)

Use case diagram A picture showing system behavior, along with the key actors that interact with the system. (7)

User documentation Written or other visual information about an application system, how it works, and how to use it. (13)

Value chain analysis Analyzing an organization's activities to determine where value is added to products and/or services and the costs incurred for doing so; usually also includes a comparison with the activities, added value, and costs of other organizations for the purpose of making improvements in the organization's operations and performance. (4)

Walk-through A peer group review of any product created during the systems development process; also called a *structured walk-through*. (5)

Well-structured relation A relation that contains a minimum amount of redundancy and that allows users to insert, modify, and delete the rows without error or inconsistencies; also known as a *table*. (9)

Work breakdown structure The process of dividing the project into manageable tasks and logically ordering them to ensure a smooth evolution between tasks. (3)

GLOSSARY OF ACRONYMS

2NF	Second Normal Form		**IE**	Information Engineering
3NF	Third Normal Form		**I/O**	Input/Output
API	Application Program Interface		**IP**	Internet Protocol
ASP	Application Service Provider		**IS**	Information Systems
B2B	Business-to-Business		**ISP**	Information Systems Planning
B2C	Business-to-Consumer		**ISP**	Internet Service Provider
B2E	Business-to-Employee		**IT**	Information Technology
BEA	Break-Even Analysis		**JAD**	Joint Application Design
BEC	Broadway Entertainment Company, Inc.		**LAN**	Local Area Network
BIND	Berkeley Internet Name Domain		**LDM**	Logical Data Model
BPMN	Business Process Modeling Notation		**LF**	Late Finish
BPP	Baseline Project Plan		**MIS**	Management Information Systems
BPR	Business Process Reengineering		**MTBF**	Mean Time Between Failures
BSP	Business Systems Planning		**NGT**	Nominal Group Technique
CASE	Computer-Aided Software Engineering		**NPV**	Net Present Value
CERT/CC	Computer Emergency Readiness Team/ Coordination Center		**ODBC**	Open Database Connectivity
			OLAP	Online Analytical Processing
CIO	Chief Information Officer		**OLTP**	Online Transaction Processing
CMS	Content Management System		**OOAD**	Object-Oriented Analysis and Design
COCOMO	Constructive Cost Model		**PDA**	Personal Digital Assistant
CSS	Cascading Style Sheet		**PE**	Petrie Electronics
CTS	Customer Tracking System		**PERT**	Program Evaluation Review Technique
DBMS	Database Management System		**PIP**	Project Initiation and Planning
DFD	Data Flow Diagram		**POS**	Point-of-Sale
DOS	Disk Operating System		**PSS**	Project Scope Statement
EC	Electronic Commerce		**PVF**	Pine Valley Furniture
EDI	Electronic Data Interchange		**RAD**	Rapid Application Development
EDW	Enterprise Data Warehouse		**RAM**	Random Access Memory
EER	Extended Entity Relationship		**RFP**	Request for Proposal
EF	Early Finish		**RFQ**	Request for Quote
EPSS	Electronic Performance Support System		**ROI**	Return on Investment
E-R	Entity Relationship		**RUP**	Rational Unified Process
ERP	Enterprise Resource Planning		**SAP**	Systems, Applications, and Products
ET	Estimated Time		**SDLC**	Systems Development Life Cycle
GUI	Graphical User Interface		**SNA**	System Network Architecture
HTML	Hypertext Markup Language		**SPTS**	Sales Promotion Tracking System
HTTP	Hypertext Transfer Protocol		**SQL**	Structured Query Language
IDS	Intrusion Detection Software		**SSR**	System Service Request

SysML	Systems Modeling Language	**WBS**	Work Breakdown Structure
TE	Earliest Expected Completion Time	**WML**	Wireless Markup Language
TL	Latest Expected Completion Time	**XML**	eXtensible Markup Language
TVM	Time Value of Money	**XSL**	eXtensible Style Language
UML	Unified Modeling Language		

INDEX

A

Abstract class, 324
Abstract operation, 325
Acceptance testing, 492–493
Access, Microsoft, 344–345
Account, 355
Action stubs, 233
Activation, 268
Activity, business process modeling, 276
Activity diagrams, 262–264
Actor, 248
Adaptive maintenance, 520
Ad hoc reuse, 69
Advertising, 469–472
Affinity clustering, 132
Aggregation, 326
Agile methodologies
    activities in, 40
    analysis-design-code-test cycle, 200
    continual user involvement, 199–200
    development of, 45–48
    requirements determination using, 199–203
    traditional methods v., 48
Agile Usage-Centered Design, 200–201
Alpha testing, 492, 509
Amazon.com, 460, 468
Analysis. *See also* Object-oriented analysis and design
    break-even, 152
    cost-benefit, 88–89, 147
    DFDs in, 228–233
    gap, 231
    introduction to, 177–178
    packaged conceptual data models facilitating, 306–308
    paralysis, 180
    phase, 36, 38
    of procedures and documents, for requirements determination, 187–191
    value chain, 121
    weighted multicriteria, 122–123
Analysis-design-code-test loop, 40, 51, 200
Analytical processing, 464–465
Application program interface (API), 449
Application server, 450–451
Application service provider (ASP), 33
Application software, 30
ArgoUML, 42–43
Association, 319–321
Association role, 320
Associative classes, 321–323, 330
Associative entities, 299–301, 312
Asynchronous message, 268
Attributes, 291
    class-scope, 324
    in E-R modeling, 291–294
    stereotypes represented for, 323

B

Back-end functions, 448
Balancing, 223–224
Baseline modules, 526
Baseline Project Plan (BPP)
    building and reviewing, 157–167
    definition, 144

feasibility assessment section of, 145–157, 158, 161
introduction section of, 159
management issues section of, 158, 161–162
outline of, 158
in planning, 89–90, 144, 158–167
reviewing, 162–167
System Description section of, 159–161
Beck, Kent, 201, 491
Behavior, 318
Benefits
    in cost-benefit analysis, 88–89, 147
    intangible, 148
    project, 147–148, 168
    tangible, 147
Berkeley Internet Name Domain. *See* Domain naming system
Beta testing, 492, 509
Bill-of-materials structure, 295–297, 353
Binary relationship, 296, 321, 351–353
BIND. *See* Domain naming system
Boeing, 200
Bookmarks, 461
Bottom-up approach, 129, 288
Bottom-up source, 120, 123, 124
Boundary, system, 250
BPR. *See* Business Process Reengineering (BPR)
Break-even analysis (BEA), 152–153
Broadway Entertainment Company, Inc. (BEC), 72, 140
Budget, preliminary, 88–89
Bug tracking, 509
Build routines, 527
Business By Design, 61
Business case, 144
Business process modeling, 275–278
    activity, 276
    event, 276
    example of, 278
    flow, 276
    gateway, 276
    introduction, 275
    notation, 276–277
    pool, 277
    swimlane, 277
Business process modeling notation (BPMN), 276–277
    recruiting process with, 278
Business Process Reengineering (BPR)
    definition of, 197
    DFDs in, 231–233
    in requirements determination, 197–199
Business rules, 303–306
Business Systems Planning (BSP), 128
Business-to-business (B2B), 134
Business-to-consumer (B2C), 134
Business-to-employee (B2E), 134

C

Calculated fields, 360
Candidate keys, 292
Cardinality, 297–298, 301
Careers, 30
Cascading Style Sheet (CSS), 456–457
CASE. *See* Computer-aided software engineering

Charter, project, 81
Chief information officer (CIO), 118
China, 57
Class diagramming, 283
    associative classes in, 321–323
    definition of, 318
    for Hoosier Burger, 328–329, 357–358
    objects and classes in, 328–329
    OOAD and, 328–329
    operation types in, 319
Classifying, 120–121
Class-scope attribute, 324
Class-scope operation, 319
Class/subclass relationships, 356
Clickstream, 467
Client, 448
Client/server architecture, 444
    designing systems for, 448–453
    file server v., 450
    three-tiered, 451
    types of, 452–453
Closed-ended questions, 181, 183
Cloud computing, 61–63, 64
CMS. *See* Content management system
Code walkthrough, 488
Coding, 36–37, 40, 48, 50
    in analysis-design-code-test, 200
    in analysis-design-code-test loop, 40, 51
    in implementation, 483–485
    sheets, 383, 397
    techniques, 360–361
    testing and, 48, 491–492
ColdFusion, 33
Color, 390–391
Command language interaction, 409–410
Communication, 87, 91, 161–162
Competitive strategy, 126
Complete keyword, 324
Completeness, DFD, 228–229
Component-based development, 67
Composite attribute, 293
Composite partitioning, 364
Composition, 326
Computer-aided software engineering (CASE), 32–33, 40, 42–45
    DFD completeness and, 228–229
    in JAD, 194–195
    in maintenance, 527–528
    repository of, 231, 284, 306
    in SDLC, 42–44
Conceptual data modeling
    for BEC, 341
    business rules and, 303–306
    concrete class, 324
    data requirements structuring and, 284–288, 295–302, 306–308
    definition of, 284
    deliverables, outcomes and, 285–286
    electronic commerce application, 308–311
    E-R modeling and, 295–302
    for Hoosier Burger, 326–329
    information gathered for, 286–288
    packaged, 306–308
    process of, 284–285
    by PVF, 308–311

requirements determination questions for, 287
SDLC and, 285
Concrete class, 324
Condition stubs, 233
Configuration management, 526–537
Connections, 250
Consistency, DFD, 229
Constantine, Larry, 200
Constraints, 131–132
COnstructive COst MOdel (COCOMO), 85
Constructor operation, 319
Content management system (CMS), 469
Context development, 462
Context diagram, 217–219
Contract, closing of, 92
Contractual feasibility, 157
Cookie crumbs, 436–438
Coordinator, 163
Corporate planning
ISP and, 124–134
strategic, 125–127
Corporate restructuring, 443
Corporate strategic planning, 125–127
Corrective maintenance, 520
Cost-benefit analysis, 88–89, 147
Costs, 65
maintenance, 521–523
project, 148–150, 168
tangible, 148
Critical path, 97–98
Critical path scheduling, 94
Critical success factors, 126
Customer bookmarks, 461
Customer loyalty, 458–460
Customer relationship management (CRM), 515
Customization, 460

D

Data. *See also* Conceptual data modeling; Data flow
diagram; Data requirements structuring;
Online data management
entry forms, 384, 422–425
errors, 423–424, 426
flow, 215
input, 384, 423–425
integrity, 361–362, 401
mart, 466
redundancy, 125
store, 215
type, 342, 359–361
warehouse, 465–468
Database
distributed, 452–453
engine, 448
management of, 32, 424
ODBC and, 451
patterns, 306–308
in relational database model, 345–346, 378–378
Database design
deliverables and outcomes in, 342–345
electronic commerce application, 371–373
E-R diagrams transformed into relations in,
350–354
fields designed in, 359–362
for Hoosier Burger, 356–359, 370–371
introduction to, 339–340
merging relations in, 354–366
normalization and, 338, 347–349
for Petrie Electronics, 378–379
physical, 370–371, 378–379
physical file and, 359, 364
physical tables designed in, 362–369
process of, 341–342
purposes of, 339–340
for PVF 381–383, 342–345

relational database model in, 388–389, 345–346
steps of, 340
Database management systems (DBMS), 32, 424
Data Entity-to-Information System, 131
Data entry forms, 384
functional capabilities for, 422
in interface and dialogue design, 422–425
structuring of, 435–436, 422–423
Data flow diagram (DFD)
in analysis process, 228–233
application, 308
balancing, 223–224
in BPR, 231–233
completeness, 228–229
consistency, 229
decomposition of, 220–223
definition of, 213, 215–217
drawing guidelines for, 228–230
electronic commerce application, 236–238,
308–309
flexibility of, 338
form and report design and, 381
for Hoosier Burger, 225–228, 327
iterative development of, 229–230
level-0, 218–219, 222–224, 227–229, 237
level-*n*, 222, 224
mechanics, 214–225
OLTP and, 463
primitive, 222, 230
process modeling using, 236–238, 309
process requirements structuring and, 214–225
rules of, 219–220, 225
structure and, 282
symbols in, 215–217
timing, 229
unbalanced, 224
Data requirements structuring. *See also* Entity-
relationship modeling
conceptual data modeling and, 284–288,
295–302, 306–308
E-R modeling and, 283, 288–302
introduction to, 282–283
DBMS. *See* Database management systems
Deadlines, 42
Debugger, 490
Decision tables
definition of, 233
Hoosier Burger, 235–236, 326–327
logic modeling with, 233–236
in requirements structuring, 233–236
rules, 233
Decomposition, functional, 220
Default value, 361
Degree, of relationships, 295–297
Deliverables, 76
conceptual data modeling and, 285–286
in database design, 342–345
design, 445
E-R diagram, 285–286
in form and report design, 383–386
in identification and selection of projects, 123–124
in implementation, 483–485
in interface and dialogue design, 408–409
maintenance, 519
in PIP, 144–145
for process modeling, 214
for requirements determination, 179–180
Denormalization, 362–365
Department of Justice, SDLC of, 34
Derived attribute, 294
Design, 383. *See also* Database design; Distributed and
Internet systems design; Form and report
design; Interface and dialogue design
in analysis-design-code-test, 200
in analysis-design-code-test loop, 40, 51

logical, 36–37
phase, 36, 38, 338, 381, 444
physical, 36–37
specifications, 384–386, 409
Designed reuse, 69
Design recovery tools, 528
Desk checking, 487–488
Development group, 119–120, 154–155
DFD. *See* Data flow diagram
Dialogue box. *See* Pop-up menu
Dialogue diagramming, 430, 432
Dialogues. *See* Interface and dialogue design
Direct installation, 493
Direct observation, 185–187
Discount rate, 151
Disjoint keyword, 324
Disjoint rule, 302–303
Disruptive technologies, 198–199
organizational rules, eliminated by, 199
Distributed advertisement server, 469–472
Distributed and Internet systems design
alternative designs, 449–453
client/server architecture in, 448–453
consistency in, 456–458
deliverables and outcomes in, 445
distributed systems in, 446–453
electronic commerce application, 469–472
Internet systems in, 453–469
introduction to, 443–444
LAN and, 446–448
online data management in, 461–469
for Petrie Electronics, 476–477
process of, 444–445
for PVF, 447, 461–464, 469–472
quality, 459
single location systems *v.*, 445
summary of, 472
Web site management and, 458–461
Distributed database, 452–453
Distributed function, 452–453
Distributed presentation, 452–453
Distributed processing, 452–453
Documentation, 65
analysis of, 187–191
finalized, 480
process of, 484–485
quality, 522
system, 496–499
user, 497–499
Domain, 304–305
Domain naming system (BIND), 454
Downsizing, 443
Drop-down menu, 413

E

EC. *See* Electronic commerce
Economic feasibility, 145–154
EDI. *See* Electronic data interchange
EDW. *See* Enterprise data warehouse
e-learning, 500
Electronic commerce (EC)
conceptual data modeling application, 295–302
database design application, 371–373
design application of, 469–472
DFD application, 236–238, 308
form and report design application, 399–401
implementation application, 508–510
maintenance application, 529–531
in PIP, 167–169
process modeling application, 236–238,
256–259, 308–309
project identification and selection application,
134–135
requirements determination application,
203–205

Electronic data interchange (EDI), 134–135
Electronic performance support system (EPSS), 500
Electronic reports and forms, 396–397, 419–420
Encapsulation, 319
Enterprise data warehouse (EDW), 466
Enterprise resource planning (ERP) systems, 60–61, 66, 117–118
Entities, 289–291, 299–301, 312, 350
Entity class. *See* Entity type
Entity instance, 289–290
Entity-relationship (E-R) diagramming, 283, 288
Entity-relationship (E-R) modeling
    attributes in, 291–294
    business rules and, 303–304
    candidate keys and identifiers in, 292
    conceptual data modeling and, 295–302
    data requirements structuring and, 283, 288–302
    definition of, 288
    deliverables from, 285–286
    entities in, 289–291, 299–301, 350
    flexibility of, 338
    generalization and, 323–326
    introduction to, 288–294
    normalization and, 350–354
    for Petrie Electronics, 334
    PVF, 310–311
    relationships in, 294
    supertypes and subtypes represented in, 302–303
    transformed into relations, 350–354
Entity type, 289
EPSS. *See* Electronic performance support system
E-R modeling. *See* Entity-relationship modeling
ERP systems. *See* Enterprise resource planning systems
Errors
    data, 423–424, 426
    Web site, 400, 436
Event, business process modeling, 276
Evolutionary prototyping, 196, 205
Excel, 498, 500
Execution, project, 89–91
Expected time durations, 94, 95
Extend relationship, 250
eXtensible Markup Language (XML), 455, 457
eXtensible Style Language (XSL), 456
Extensions, 255
External documentation, 497
eXtreme Programming (XP), 40, 47–48, 201–203, 491

F

Facilitated reuse, 69
Feasibility
    assessment, 145–157, 158, 161, 168
    economic, 145–154
    legal and contractual, 157
    operational, 156
    in PIP, 145–154
    political, 157
    schedule, 156–157
    study, 77
    technical, 154–155
Fields
    calculated, 360
    design of, 359–362
File
    controls, 368–369
    organization, 342, 364, 366–369
    physical, 359, 364
    server, 446–448, 450
Firefox, 411, 429
Firewall, 507
Flexibility, 65

Flow, business process modeling, 276
Foreign key, 349, 353
Form, 382
Formal system, 189
Form and report design
    characteristics for consideration in, 399
    deliverables and outcomes in, 383–386
    DFDs and, 381
    electronic commerce application, 399–401
    formatting in, 386–397
    introduction to, 380–382
    paper *v.* electronic, 396–397, 419–420
    process of, 382–383
    prototype in, 383
    for PVF, 385–386, 388–390, 394–397, 399–401
    usability and, 397–399
    users and, 382–383, 398–399
Formatting
    color in, 390–391
    in form and report design, 386–397
    guidelines, 387–388, 391, 393, 395
    highlighting information in, 388–390
    of lists, 392–396
    of tables, 392–396
    text displayed in, 391–392
    usability and, 398
Forms. *See also* Data entry forms
    data integrity rules and, 401
    definition of, 380, 382
    electronic, 396–397, 419–420
    interaction and, 415–416
    paper, 396–397, 419–420
    review, 162–164
Free slack, 99
Front-end applications, 448
Functional decomposition, 130, 220
Functional dependence, 347–348
Functionality, 64, 65
Function-to-Objective matrix, 130
Function-to-Process matrix, 130

G

Gantt chart, 84, 86, 90, 93–97, 101
Gap analysis, 231
Gateway, business process modeling, 276
General Electric (GE), 56
Generalization, 323–326
Generic strategies, 126
Google Advanced Search Engine, 416
Google Apps, 62
Graphical user interface (GUI) environments, 433–435
Graphics tablet, 417–418
Graphs, 395–397
Group interview, 184–185
GUI. *See* Graphical user interface environments

H

Hard problems, completion of, 109
Hardware devices, for interaction, 417–418
Hashed file organization, 367, 368–369
Hash partitioning, 364
Help, 426–428
    desk, 502
    template-based, 401
    validation, 529
Highlighting, 388–390
Homonyms, 355–356
Hoosier Burger
    class diagram for, 328–329, 357–358
    conceptual data modeling for, 326–329
    database design for, 356–359, 370–371
    decision table of, 235–236, 326–327
    DFD example of, 225–228, 327
    use case for, 251, 292

HTML. *See* Hypertext Markup Language
HTTP. *See* Hypertext Transfer Protocol
Human Resources (HR), 278
Hypertext Markup Language (HTML), 427–429
    CSS and, 456–457
    definition of, 454–455
    in interface and dialogue design, 426–429
    types of, 428
    usable, 427
    validation, 529
Hypertext Transfer Protocol (HTTP), 454

I

IBM, 57, 59, 62, 65, 192, 198
IBM Credit Corporation, 231–233
Icons, 415
Identification and selection, of projects
    classifying and ranking in, 120–121
    corporate planning and, 124–134
    deliverables and outcomes in, 123–124
    development projects, 118–124
    electronic commerce application, 134–135
    introduction to, 116–118
    methods of, 119–120
    need for, 125
    potential development projects, 119–120
    process of, 119–123
    from SSR, 116
Identifiers, 292
Impartiality, 179
Impertinence, 178
Implementation
    coding in, 483–485
    deliverables and outcomes in, 483–485
    electronic commerce application, 508–510
    failure of, 503–506
    installation in, 483–485, 493–496, 510
    introduction to, 481–482
    organizational issues in, 503–507
    overview of, 480
    for Petrie Electronics, 515
    phase, 36–38, 480–483
    project closedown and, 510
    for PVF, 489–490, 508–510
    security and, 506–507
    success of, 505–506
    summary of, 511
    system documentation in, 496–499
    testing in, 482–493, 508–509
    user training and supporting in, 484–485, 499–503
Inception phase, 111
Include relationship, 250–252
Incomplete keyword, 324
Incremental commitment, 123, 165, 214
Index, 366, 374
Indexed file organization, 366–369
India, 57–58
Indifferent condition, 234
Industry-specific data models, 307
Informal system, 189
Informational systems, 464
Information Engineering (IE), 128
Information systems analysis and design, 30. *See also* Systems development
Information systems planning (ISP), 127–134
    corporate planning and, 124–134
    definition of, 127
    process of, 128
Information System-to-Objective, 131
Information technology (IT)
    careers in, 30
    outsourcing and, 56–58
    services firms, 58–59, 64
Inheritance, 49, 323

In-house development, 63–64
Inspections, 487
Installation, 37, 480
    definition of, 493
    in implementation, 483–485, 493–496, 510
    strategies, 493–496
Intangible benefits, 148
Integration depth, 462
Integration problems, 355–356
Integration testing, 488
Interaction methods and devices
    command language interaction, 409–415
    form interaction, 415
    hardware options, 417–418
    in interface and dialogue design, 409–418
    menu interaction, 410–415
    natural language interaction, 415–416
    object-based interaction, 415
Interface and dialogue design
    data entry and, 422–425
    deliverables and outcomes in, 408–409
    dialogues in, 407, 429–438, 441–442
    electronic commerce application, 435–438
    feedback in, 425–426
    guidelines for, 430
    in GUI environments, 433–435
    help and, 426–429
    interaction methods and devices in, 409–418
    interfaces in, 409, 418–429, 433–438, 441–442
    introduction to, 407–408
    layouts in, 419–422
    for Petrie Electronics, 441–442
    process of, 408
    prototyping in, 433
    for PVF, 418–420, 429–432, 435–438
    summary of, 438
    usability of, 427, 433
Internal documentation, 497
Internet. See also Distributed and Internet systems
            design
    basics, 134–135
    cloud computing and, 61–63, 64
    definition of, 134
    incorporation of, 29–30, 32
Internet systems design
    content and display separated in, 454–455
    future evolution of, 455
    standards driving, 454
Interviewing
    groups, 184–185
    guidelines, 181, 183–184
    outline, 181–182
    questions chosen in, 181–183
    for requirements determination, 180–185
Intranet, 134
Intrusion detection software (IDS), 507
Inventory Information, 204–205
Inventory information, for WebStore, 204–205
Invoice, 189
ISP. See Information systems planning
IS staff, 193
IS Steering Committee, 140
Iteration Planning Game, 202–203
Iterations, project, 110
Iterative development, 47, 49
Iterative development, of DFD, 229–230
IT services firms, 58–59, 64

J
J. Lyons & Sons, 56
JAD. See Joint Application Design
Joint Application Design (JAD), 192–195, 286
    CASE tools during, 194–195
    for PVF, 202–203
    for requirements determination, 177

session leader, 192
    taking part in, 194
Joystick, 417–418

K
Keyboards, 417–418
Key business processes, 198
Kia Motors, 127

L
Labeling, 394
LAN. See Local area network
LAN-based DBMS, 444, 447
Layout
    characteristics, 203, 204
    in interface and dialogue design, 419–422
    Web page, 400
LDM. See Logical data model
Legal and contractual feasibility, 157
Level, in written use cases, 253–254, 256
Level-0 diagram, 218–219, 222–224,
        227–229, 237
Level-n diagram, 222, 224
Light pen, 345–418
Lightweight graphics, 401
Link titles, 458, 459
Linux, 63
Listening, 180–184, 183
Lists, formatting of, 392–396
Local area network (LAN), 134, 446–448. See also
            Client/server architecture
Location-to-Function matrix, 130
Location-to-Unit matrix, 130
Logical database design. See Database design
Logical data model (LDM), 307
Logical design, 36–37
Logic modeling, with decision tables, 233–236
Long-term planning, 82–83, 133
Low-cost producer, 127
Loyalty, 458–460

M
Maintainability, 521
Maintenance, 520, 523
    CASE in, 527–528
    cost of, 521–523
    deliverables and outcomes of, 519
    effectiveness, 523–524
    electronic commerce application of, 529–531
    introduction to, 516–517
    management of, 523–527
    oracle, 163
    overview of, 480
    personnel, 522, 523
    phase, 38, 480, 517–519
    process of, 517–519
    PVF, 518, 529–531
    requests, 524–526
    summary of, 531
    types of, 520–521
    Website, 528–529
Malware, 522
Managed reuse, 69
Management issues section, of BPP, 158,
        161–162
Management procedures, 80
Management reporting component, 471–472
Managers, 193
Mandatory cardinality, 298
Many-to-many relationship, 364, 365
Master test plan, 486
Matrices
    communication Matrix for, 162
    planning, 131–132
    Project Communication, 162

risk, 155–156
    task responsibility, 162
Matsutoya Corporation, 72
Maximum cardinality, 298
Mean time between failures (MTBF), 524
Menus
    drop-down, 413
    guidelines for, 413, 476
    interaction, 410–415
    navigation driven by, 436–438
    pop-up, 412–413
    ribbon, 411
    single-level, 412
    Visual Basic .NET for building, 414–415
Merging transaction, 464–465
Method, 325
Methodologies, 30. See also Agile methodologies
Microsoft
    Access, 344–345
    Excel, 498, 500
    HTML Help, 427–429
    Office, 410–411, 413
    PowerPoint, 167, 411, 501
    Project, 59–60, 94, 99–102
    Security Development Lifecycle (SDL), 39
    training in, SDL, 39
    Visual Basic .NET, 414–415
    Visual Studio .NET, 433, 528
    Word, 410, 413, 435
Middleware, 451
Minimal guarantee, 254
Minimum cardinality, 298
Mission statement, 125–126
Modality, 435
Module testing. See Unit testing
Mosley, D. J., 486
Mouse, 417–418
MTBF. See Mean time between failures
Multiplicity, 320
Multivalued attribute, 292–293

N
Natural language interaction, 415–416
Navigation
    characteristics, 203, 204
    flow, 421
    menu-driven, 436–438
Nearshoring, 57–58
Net present value (NPV), 151, 153
Network diagram, 86–87, 90, 93–99, 101
NGT. See Nominal Group Technique
Niche, 127
Nielsen, Jakob, 399, 476
Nominal Group Technique (NGT), 185
Nonintegrated systems, 117
Nonintelligent primary key, 379
Nonkeys, dependencies between, 356
Normalization, 339
    database design and, 338, 347–349
    definition of, 338, 347
    denormalization and, 362–365
    E-R diagrams and, 350–354
    rules of, 347
    second normal form and, 347–348
    third normal form and, 347–349
Normalized relations, 342
Normalized tables, 339
Null value, 361–362
NUnit, 491

O
Object, 48–49, 317–319
Object-based interaction, 415
Object classes, 49, 318
Objective statements, 126

Object Management Group (OMG), 275
Object modeling
aggregation represented in, 326
associations in, 319–321
associative classes in, 321–323
generalization represented in, 323–326
objects and classes in, 317–319
OOAD and, 282–329
operation types in, 319
stereotypes represented for attributes in, 323
Object-oriented analysis and design (OOAD),
48–50. *See also* Use case
activity diagrams and, 262–264
class diagrams and, 317–329
object modeling and, 317–329
project management and, 107–113
sequence diagrams and, 266–273
system components in, 107–109
Observation, 185–187, 190
ODBC. *See* Open Database Connectivity
Office, Microsoft, 410–411, 413
Off-the-shelf software, 60, 64–66
OLAP. *See* Online analytical processing
OLTP. *See* Online transaction processing
One-time costs, 148–150
One-to-one relationship, 364
Online analytical processing (OLAP), 464
Online data management
CMS for, 468–469
context development in, 462
data warehousing in, 465–468
in distributed and Internet systems design,
461–469
OLAP in, 464
OLTP in, 462–464
processing in, 464–465
Online transaction processing (OLTP), 462–464
OOAD. *See* Object-oriented analysis and design
Open Database Connectivity (ODBC), 451
Open-ended questions, 181, 183
Open source software, 63
Operation, 318
Operational feasibility, 156
Operational systems, 464
Optional attribute, 293
Optional cardinality, 298
Oracle, 32, 33, 59, 61
Organizational breadth, 462
Outcomes
conceptual data modeling and, 285–286
in database design, 342–345
design, 445
in form and report design, 383–386
in implementation, 483–485
in interface and dialogue design, 408–409
maintenance, 519
in PIP, 144–145
for process modeling, 214
in project identification and selection, 123–124
for requirements determination, 179–180
Outsourcing, 56–58, 160
Overlapping keyword, 324
Overlap rule, 303

P
Packaged conceptual data models, 306–308
Packaged software producers, 59–60, 64
Page titles, 457–458
Palm Pilots, 33
Paper reports and forms, 396–397, 420
Parallel installation, 493–494
Partial specialization rule, 302
Passwords, 461, 507
Patton, Jeff, 201

PeopleSoft, 61
Perfective maintenance, 520
Performance testing, 492
Personalization, 460
PERT. *See* Program Evaluation Review Technique
Petrie Electronics
alternatives for, 210
constraints for, 210
customer loyalty project, 210
database design for, 378–379
data requirements for, 333
design for, 476–477
determining systems requirements, 210–211
E-R diagram for, 333–334
form and report design for, 405–406
identifying and selecting system development
projects, 140
implementation for, 515
interface and dialogue design for, 441–442
initiating and planning systems development
projects, 173
introduction, 72
managing the information systems project, 114
matrix for consumers, 334
preliminary design, 441
requirements for, 210
scope statement for, 174
structuring system process requirements,
280–281
XRA system, 333, 441
Phased installation, 494, 495
Physical database design, 370–371, 378–379
Physical design, 36–37
Physical file, 359, 364
Physical tables
definition of, 362
designing of, 362–369
rows in, 364–368
Pilot installation. *See* Single-location installation
Pine Valley Furniture (PVF)
background of, 74–75
coding and compression techniques and,
360–361
conceptual data modeling and, 308–311
current situation of, 129–130
database design for, 342–345, 371–373
design for, 447, 461–464, 469–472
E-R diagram for, 310–311
feasibility assessment by, 145–153, 155, 168
file server and, 447
form and report design for, 385–386, 388–390,
394–397, 399–401
implementation for, 489–490, 508–510
interface and dialogue design for, 418–420,
429–432, 435–438
JAD session of, 203–205
maintenance for, 518, 529–531
mission statement of, 125–126
objective statement of, 126
process modeling using DFDs and, 236–238
project management for, 74–75, 76, 79–81, 82,
88–89, 95–99
prototype for, 469–472
SSR, 145, 518
walkthrough of, 162–165
WebStore of, 135, 167–169, 203–205, 237–238,
257–259, 308–311, 399–401, 435–438,
461–464, 469–472, 508–510, 529–531
PIP. *See* Project initiation and planning
Planning. *See also* Baseline Project Plan; Project
initiation and planning
bottom-up, 129
BSP, 128
corporate, 124–134

corporate strategic, 125–127
ERP, 60–61, 66, 117–118
information systems planning (ISP), 124–134
long-term, 82–83, 133
outline for, 132–133
overview of, 116
phase, 35–36, 38
presentation, 166
project, 81–89
in project management, 81–89
for resources, 60–61, 85–86, 118
short-term, 83, 133
top-down, 129
Planning Game, 201–203
Pocket PCs, 33
Pointer, 365
Political feasibility, 157
Polymorphism, 325
Pool, business process modeling, 277
Pop-up menu, 412–413
Porter, Michael, 126
Postproject reviews, 92
PowerBuilder, 32
PowerPoint, 167, 411
Preconditions, 254
Preliminary investigation, 35–36
Preliminary schedule, 86–87
Presentations, 167
Presenter, 163
Present value, 151
Preventive maintenance, 520–521
Primary key, 342, 347–348, 371
Primitive DFD, 222, 230
Procedures
analysis of, 187–191
communication, 91
example of, 188
management, 80
project, 88
Process, 215–217, 219–220
Process modeling
deliverables and outcomes for, 214
with DFDs, 236–238, 309
electronic commerce application, 236–238,
256–259, 309
process requirements structuring and, 236–238
for PVF, 236–238
with use cases, 256–259
Process requirements structuring
decision tables in, 233–236
DFDs and, 214–225
introduction to, 212–213
process modeling and, 213–214, 236–238
Process-to-Data Entity, 131
Process-to-Information System, 131
Procter & Gamble (P&G), 199
Product differentiation, 126, 127
Product focus, 126, 127
Program Evaluation Review Technique
(PERT), 95
Programming languages, 32, 49
Project, Microsoft, 59–60, 94, 99–102
Project dictionary, 43
Project initiation and planning (PIP), 79–81
building and reviewing BPP in, 157–167
challenge of, 142–143
deliverables and outcomes in, 144–145
electronic commerce and, 167–169
elements of, 144
end of, 142
feasibility assessed in, 145–157
installation, 494, 495–496
introduction to, 141–142
master test plan in, 486

process of, 143–145
scope, 159–162
Project management
closedown in, 91–92
definition of, 78
execution in, 89–91
initiation in, 79–81
OOAD and, 107–113
planning in, 81–89
project manager's role in, 73–74
PVF, 79–81, 83, 88, 90, 95–99, 125
software, 99–102
Project manager
activities and skills of, 79
definition of, 75
as juggler, 78
role of, 73–74
Projects. *See also* Identification and selection, of
    projects
benefits of, 147–148, 168
charter, 81–82
closedown of, 510
costs of, 148–150, 168
definition of, 76
flow, 133
initiation, 79–81
iterations, 110
planning, 81–89
report, 101–102
scope, 83, 89
size of, 154
standards and procedures, 88
starting date, 100
status, 91
structure of, 154
tasks, 84–85, 100–101
workbook, 80, 91
Project Scope Statement (PSS), 89, 144–145,
    159–161
for the customer tracking systems, 160
defined, 145
Prompting cues, 426
Prototyping
defined, 195
in dialogue design, 433
evolutionary, 196, 205
in form and report design, 383
McConnell's evolutionary model, 196
methodology, 195
process, 44–45
PVF, 469–472
in requirements determination, 195–197
throwaway, 196–197
PSS. *See* Project Scope Statement
PVF. *See* Pine Valley Furniture
PVF WebStore, 135, 167–169, 203–205, 237–238,
    257–259
for workbook, 80

Q

Query operation, 319
Questions, interview, 181–183

R

RAM. *See* Random access memory
Random access memory (RAM), 447
Range control, 361
Range partitioning, 363
Ranking, 120–121
Rapid Application Development (RAD), 40, 42,
    44–45
life cycle, 45
Rational Unified Process (RUP), 49
Recovery testing, 492

Recurring cost, 149
Recursive foreign key, 353
Recursive relationship, 295–296, 353
Redundancy, 125, 346
Reengineering, 528
Refactoring, 491
Referential integrity, 349, 361
Reframing, 179
Relation
definition of, 345
E-R diagrams transformed into, 350–354
merging of, 354–356
well-structured, 346
Relational database model, 345–346, 378–379
Relationships, 294
binary, 296, 321, 351–353
cardinalities in, 297–298, 301
class/subclass, 356
degree of, 295–297, 301
in E-R modeling, 294
extend, 250
include, 250–252
many-to-many, 364, 365
naming and defining, 299
one-to-one, 364
representation of, 351–354
ternary, 296–297, 301, 322
unary, 295–297, 353–354
Remote data management, 452–453
Remote presentation, 452–453
Repeating group, 293
Reports, 190–191. *See also* Form and report design
definition of, 380, 382
electronic, 396–397
paper, 396–397
project, 101–102
types of, 382
Repository, 43, 231, 284, 306
Request for proposal (RFP), 66
Request for quote (RFQ), 66
Required attribute, 293
Requirements determination
with agile methodologies, 199–203
BPR in, 197–199
conceptual data modeling and, 287
contemporary methods for, 192–197
deliverables and outcomes for, 179–180
direct observation for, 185–187
electronic commerce application, 203–205
JAD for, 177, 192–195
procedures and documents analysis for,
    187–191
process of, 178–179
prototyping during, 195–197
radical methods for, 197–199
traditional methods for, 180
Requirements structuring. *See* Data requirements
    structuring; Process requirements
    structuring
Reregistration, 529
Resources
assigning and billing, 102
availability of, 121
definition of, 94
planning for, 60–61, 85–86, 118
Return on Investment (ROI), 151, 153
Reuse, 67–69
Reverse engineering, 528
Review, of BPP, 157–167
Ribbon menu, 411
Risk
assessment, 88, 154–155
factors, 154–155
identification, 88

matrix, 155–156
technical difficulty and, 121
Rolls-Royce, 126–127
Rows, table, 364–368
Royce, W. W., 41

S

Sales Promotion Tracking System (SPTS), 95–99
SAP AG, 32, 59, 61
Scheduling
charts, 90
critical path, 94
feasibility of, 156–157
methods, 101–102
preliminary, 86–87
of project plan, 92–99
Schuster, Alex, 74
Scribe, 193
SDLC. *See* Systems development life cycle
Search engine referrals, 461
Secondary key, 366, 371
Second normal form (2NF), 347–348
Secretary, 163
Security
implementation and, 506–507
system, 461
testing, 492
Security Development Lifecycle (SDL), 39. *See also*
    Systems development life cycle (SDLC)
Microsoft's, lifecycle, 39
products of, 38
Self-adaptive software development, 46–47
Self-delegation, 269
Sequence, dialogue, 429–431
Sequence diagrams
definition of, 267
dynamic modeling with, 267–268
for Hoosier Burger, 271–273
introduction to, 266
OOAD and, 266–273
use case designed with, 267–273
Sequential file organization, 366, 367, 369
Short-term planning, 83, 133
Simple message, 268
Single-level menu, 412
Single-location installation, 494–495
Slack time, 98–99
Software. *See also* Computer-aided software
    engineering
application, 30
application testing, 485–491
engineering, 42
leading firms, specialization, 59
malware, 507
off-the-shelf, 60, 64–66
open source, 63
outsourcing and, 56–58
packaged, 59–60, 64
project management, 99–102
reuse, 67–69
self-adaptive development of, 46–47
sources, 28, 55–56, 58–64
validating information about, 66–67
Software as a service (SaaS), 62
Source/sink, 215–217
Sponsor, 193
Staged installation. *See* Phased installation
Stakeholders, 254
Standards bearer, 163
State, 318
Status information, 425–426
Steering committees, 119–120
Stereotypes, 323
Story Cards, 202

Strategic alignment, 121
Stress testing, 492
Structured walkthroughs, 162
Stub testing, 488
Subtypes, 302–303
Success guarantee, 254
Superclass, 305
Supertypes, 302–303
Support, 499
   automated, 501–502
   user, 484–485, 499–503
   vendor, 65, 501–502
Swimlane, 262
   business process modeling, 277
Symbolic debugger, 490
Synchronous message, 268
Synonyms, 355
System boundary, 250
System components, 107–109
System Description section, of BPP, 159–161
System documentation, 497
System evolution, 509
System librarian, 526
System prototype evolution, 205
Systems analyst
   characteristics of, 178–179
   definition of, 31
   life as, 28
   primary responsibility of, 30
   support issues and, 503
Systems analysts, 193
Systems development
   definition of, 29–30
   evolution of, 28, 32–33
   foundations for, 28
   heart of, 40–42
   improvement of, 42–45
   methodology, 33
   modern approach to, 28, 32–33
   organizational approach to, 31
   possible costs of, 149
   purposes of, 77
   reasons for, 35–36
   requests for, 119
   speed in, 28
Systems development life cycle (SDLC). *See also*
     Analysis; Design; Implementation; Mainte-
     nance; Planning
   approach to, 50–51
   book guide based on, 35
   CASE in, 42–44
   circular process of, 33–34
   conceptual data modeling and, 284–285, 341
   definition of, 33
   of Department of Justice's systems, 34
   design specifications and, 384–386
   nature of, 28
   project manager in, 73
   specialized, 39–40
   traditional waterfall, 41–42, 199
System security, 461
Systems engineering, 42
System Service Request (SSR), 76–77, 80, 81, 83, 92
   project identification from, 116
   PVF, 145–146, 518
Systems integration, 32
System testing, 488

T

Tables
   decision, 233–236, 326–327
   formatting of, 392–396
   graphs v., 395–396

   physical, 362–369
   rows of, 364–368
Tangible benefits, 147
Tangible costs, 148
Target situation, 131–132
Task Cards, 202
Task responsibility matrix, 162
Technical difficulty, 121
Technical feasibility, 154–155
Techniques, 31
Template-based HTML, 401
Ternary relationship, 296–297, 301, 322
Test case, 489–490, 508–509
Testing
   acceptance, 492–493
   alpha, 492, 509
   in analysis-design-code-test loop, 40, 51, 200
   approach to, 37
   beta, 492, 509
   coding and, 48, 491–492
   in implementation, 482–493, 508–509
   master test plan in, 486
   for Petrie Electronics, 515
   process, 489–491
   security, 492
   software application, 485–491
   testing harness, 491
   types of, 487–489
   validation, 424–425
Testing harness, 491
Text display, 391–392
"The Agile Manifesto," 46
Thin clients, 455–456
Thin-client technologies, 455
Third normal form (3NF), 347–349
3NF. *See* Third normal form
Three-tiered client/server, 451
Throwaway prototyping, 196–197
Time value of money (TVM), 150–154
Timing, DFD, 229
Tools, 31
Top-down approach, 287
Top-down planning, 129
Top-down source, 120, 123, 124
Total specialization rule, 302
Touch screen, 417–418
Trackball, 417–418
Traditional waterfall SDLC, 41–42, 199
Training, user, 484–485, 499–503
Transition strategy, 132
Transitive dependency, 348
Trends, 131–132
Triggering operation (trigger), 255, 305–306
Trustworthiness, 458–460
Two-level architecture, 465–466
2NF. *See* Second normal form

U

UML. *See* Unified modeling language
Unary relationship, 295–297, 353–354
Unbalanced DFD, 224
Unified modeling language (UML), 285
Unit testing, 488
Unit-to-Function matrix, 130
Universal data models, 306–307
Update operation, 319
Upsizing, 443
Usability
   definition of, 397
   in form and report design, 397–399
   format and, 398
   help, 427
   in interface and dialogue design, 427, 433

Use case
   definitions, 248–252
   diagram, 248–252, 255, 257
   goals, 253
   for Hoosier Burger, 251, 252
   process modeling using, 256–259
   sequence diagrams and, 267–273
   symbols, 249–252
   template, 252–256
   written, 252–256, 258
User(s), 193
   acceptance testing by, 492–493
   continual involvement of, 199–200
   direct observation of, 185–187
   documentation, 497–498
   form and report design and, 382–383, 398–399
   friendliness, 398
   groups of, 154–155
   personal stake of, 505
   training and supporting, 484–485, 499–503

V

Validation tests and techniques, 424–425
Value chain analysis, 121
Vendor support, 65, 501–502
View integration, 373
Visual Basic, 32
Visual Basic .NET, 414–415
Visual Studio .NET, 433, 528
Voice device, 417–418

W

Walkthroughs
   action list for, 165
   code, 488
   defined, 162
   for electronic commerce system, 168
   PVF, 162–163, 166
   review form for, 162–165
   structured, 162
Warehouse, data, 465–468
Warning messages, 426
Waterfall SDLC, 41–42, 199
Weak entity, 350
Web sites
   Amazon.com, 460, 468
   CMS for, 469
   consistency of, 456–458
   errors in, 400, 436
   guidelines for, 399–400, 476–477
   layout of, 400
   living forever, 460–461
   maintenance of, 528–529
   management of, 458–461
   Nielsen, Jakob, 476
   PVF WebStore, 308–311, 399–401, 435–438,
     461–464, 469–472, 508–510, 529–531
   security of, 461
Weighted multicriteria analysis, 122–123
Well-structured relation, 346
Wireless system components, 33
Word, Microsoft, 410, 413, 435
Workbook, 43
   project, 80, 89
Work breakdown structure, 84, 85, 86, 95
Work results, 91
Written use cases, 252–256, 258

X

XML. *See* eXtensible Markup Language
XRA system, 515
XSL. *See* eXtensible Style Language